AF528886

Introduction to Nano Science and Technologies

Introduction to
Nano Science and Technologies

Prof. Anjaneyulu Yerramilli
Ph.D., FRSC (London)
Professor & Director, Center for Energy Studies,
K L University, Green Fields, Vaddeswaram,
Guntur Dist., India.

A unit of **BSP Books Pvt. Ltd.**
4-4-309/316, Giriraj Lane, Sultan Bazar,
Hyderabad - 500 095.
Phone : 040 - 23445605, 23445688

Introduction to Nano Science and Technologies ***by Anjaneyulu Yerramilli***

Published by:

A unit of **BSP Books Pvt., Ltd.**

4-4-309/316, Giriraj Lane, Sultan Bazar,
Hyderabad - 500 095
Phone : 040 - 23445605, 23445688
e-mail : info@bspbooks.net

ISBN: 978-93-5230-134-8 (HB)

Dedication,

This book is dedicated to my wife Mrs.Y.S.V. Ratnakumari, daughters Dr. Sudha Yerramilli, Ph.D, Jackson State University, Jackson, MS, USA, and Dr. Madhavi Chada, M.D; Physician, Newbridge Spine & Pain Center LLC, Maryland, USA for their constant support and encouragement without which completing this book would not have been possible.

सी.एस.आई.आर. - राष्ट्रीय पर्यावरण अभियांत्रिकी अनुसंधान संस्थान

CSIR - National Environmental Engineering Research Institute

(वैज्ञानिक तथा औद्योगिक अनुसंधान परिषद् / Council of Scientific & Industrial Research)

डॉ. सतीश आर. वटे

निदेशक

Dr. Satish R. Wate

Director

Foreword

Nanotechnology is often described as 21" Century emerging technology as it not only holds promise for society, but also is capable of revolutionizing our approach to common problems. It offers new innovative ways for production of nano structures by controlling and manipulating the fundamental building blocks of matter at nanoscale that is, at the atomic/molecular level. This will greatly help in developing unique nano composite materials which are stronger, lighter. more durable, highly reactive, more sieve-like, better electrical conductors and so on with potential for various special technological applications. Products, processes and applications of nanotechnology are also expected to contribute significantly to environmental and climate protection by saving raw materials, energy and water as well as by reducing greenhouse gases and hazardous wastes. While there is clear evidence that nanotechnology can contribute to human welfare, it is also recognized that it can also cause some adverse effects on environment and human health and not all of the effects of nanomaterials in the environment are currently known. It is possible that nanoparticles could accumulate in different organisms, become incorporated in living tissue and move up the food chain in a process known as bioaccumulation.

A major challenge that nanotechnology now faces is the education and training of a new generation of skilled workers. In-view of intense inter disciplinary nature of nanotechnologies, extensive education and research programs that emphasize the unique properties of nanomaterials linking with biological, physical, engineering and computer sciences are required to be developed by every country. Investments are to be made for training nanotechnology workforce (undergraduates, graduate students, post doctoral fellows, and budding scientists) to achieve competitive scientific and technological excellence. India is also focusing on application of nanotechnology for addressing challenges in Energy and Environment sector besides IT related sectors.

I am sure this book on "Introduction to Nano Science and Technologies" which presents an overview of basic fundamental concepts governing nano science and technologies along with state-of-art nanotechnologies which find wide applications in many advanced scientific areas, is the need of the hour and will be very useful in providing latest technological information for various education and training programs to be designed/pursued in nanotechnologies.

Dr. Anjaneyulu Yeramilli, with his rich experience as a teacher, researcher and author of a number of books on Energy and Environment, presented in this book very illustratively the unusual and unique properties of nano materials/composites/structures/devices, their production processes, underlying interdisciplinary science concepts, their emerging unique tailor made applications in various fields like energy, environment, medicine, pharmaceuticals, information technology, electronic, communications, defense etc. This book also covers a number of Laboratory Exercises explaining nano Science/Technology concepts, which will be very useful for both teachers and students at under-graduate/graduate level and this, in my opinion is a special feature of the book that must be highlighted.

In this context I congratulate Professor Anjaneyulu Yerramilli for his excellent efforts in bringing out this very informative book on "Introduction to Nano Science and Technologies" and I strongly feel that the book can serve as an excellent Resource material/Text Book for young researchers, undergraduate/graduate students working in Engineering and Science disciplines.

Dr. S.R. Wate
Director, NEERI

November 2015

K L E F University

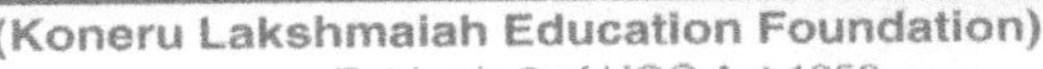

(Koneru Lakshmaiah Education Foundation)

Estd. u/s 3 of UGC Act 1956

D. No.: 29-36-38, Museum Road,
Governorpet, Vijayawada - 520 002
Phone No. : 0866 - 2577715
Campus : 0863 - 2399999

Er. Koneru Satyanarayana
President

Prologue

Rapid advances in Nano Science & Technology are making us to realize potential benefits in many aspects of our daily lives like energy savings, alternative energy supplies, efficient use of raw materials, environmental protection, agriculture applications, medical breakthroughs etc, leading to tremendous expansion of career opportunities. However a major challenge for nanotechnology development is the need for corresponding science, engineering and medical education efforts for generation of well trained skilled workers. By implementing nanotechnology into the undergraduate education program at the early stage an engaging environment will be created in which students broaden their horizons for science, math, engineering and technology, while preparing them for their places in the future Nano-driven employment. Students need an interdisciplinary education in the basic sciences, the engineering sciences, and the information sciences, as well as an understanding of the relationships of these fields to nanotechnology for which proper resource materials are in great need.

Currently there is a dearth of textbooks in Nano sciences and technology particularly tailored to the needs of undergraduate/graduate students. In this context I am extremely happy to note that one of our distinguished Professor Dr. Anjaneyulu Yerramilli Director of our Center for Energy Studies authored this book on "Introduction to Nano Science and Technologies" which provide the student the necessary information to have a firm grasp of fundamental aspects in this emerging discipline.

Dr. Anjaneyulu Yerramilli, with his rich experience as a teacher researcher and author of a number books on Energy and Environment, made good efforts in bringing out this book illustrating how at Nanoscale various physico chemical and biological systems exhibit interestingly different properties that are being exploited for various emerging Nano technologies like energy, environment, medicine, pharmaceuticals, information technology, electronic, communications, defense etc. Further a special Chapter on demonstrative Laboratory Exercises in Nano-technologies may be attractive for both students and Teachers dealing these courses. I am sure this book may inspire many students to choose their career in this emerging Technology of 21st Century.

I congratulate Professor Anjaneyulu Yerramilli for his sincere efforts in bringing out this very informative book on Nano Science and Technology and wish many more to come from this experienced author.

Er. Koneru Satyanarayana

Preface

Nano Science and Technologies are highly interdisciplinary in nature, a convergence of a number of disciplines such as physics, chemistry, biology, biomedicine, information technology, photonics, electronics, robotics, materials science etc. into a single engineering paradigm where all the phenomenon are re-configured at molecular level. As Nanotechnology impact is growing in magnitude in every activity of our growth its scope is expanding tremendously worldwide and it is one of the hottest career options available to Engineering and Science graduates globally and in India also.

The enormous challenges presented by Nano Science and Nanotechnology education at Undergraduate and Graduate levels requires information on state-of-art nano-technologies explaining the fundamental concepts which reveal the unusual properties of nano materials. As a teacher involved for the last two decades in research and teaching in nano technology, an attempt is made to address the above issues in this book on "Introduction to Nano Science and Technologies". This book covers (in 15 Chapters) all the basic concepts of nano Science and advances in nano Technologies, providing students with the broad knowledge base required for an introduction of the unique properties of nano structured materials.

Chapter 1 deals with basic concepts of nanoscience and nanotechnologies, history and classification of nano materials, natural/non engineered nano materials, their unique mechanical, physico-chemical, electrical, optical and magnetic properties. This chapter also gives an overview of their current industrial applications, improving product performance, process engineering, design and fabrication techniques for engineered nano materials, top-down and bottom-up techniques, resource management, environmental issues, self assembly, future nano technologies, molecular manufacturing, safety and health risks of nano materials.

State of Art technologies of various potential areas in Nano Science and Technologies like Carbon Nano Tubes (CNT) their special properties, production processes and applications (**Chapter 2**), Nano – Micro Fabrication techniques **(Chapter 3)** like top down lithography and bottom up deposition - thin films, Chemical Vapor deposition (CVD), Micro Electro Mechanical Systems (MEMS) and Nano-Electro Mechanical Systems (NEMS) (**Chapter 4**), their fabrication techniques, their special applications, industrial process control challenges and future developments are presented in Chapters **2-4** explaining the fundamental concepts with illustrative examples.

For scaling up nano technologies for industrial processes the biggest challenge is quality control and characterization using advanced instruments. Various quality control processes adopted by different countries, Standards development, details of various important instruments (Metrology) like Scanning Electron Microscope (SEM), Atomic Force Microscope (AFM), Scanning Tunneling Microscope (STM), Transmission electron microscopy (TEM) which are used for characterization of nano materials/structures are presented in (**Chapter 5**), Emerging nano technologies using polymeric/organic materials, liquid crystals, their nano composites and nano ferrofluids which find special applications in defense, electronics, communications, sensors, bio-medical areas etc., are discussed in (**Chapter 6).**

This book also covers important technical information on the role of surface and colloid chemistry in nano technology **(Chapter 7)**, molecular manufacturing, its benefits, medical uses, environmental applications the salient aspects of Drexler-Smalley debate - realistic projections on Molecular nano Technology, future Projections on Molecular manufacturing, **(Chapter 8)** nano Size (Quantum) effects on Semiconductors their Optical and electronic properties, semiconductor nano crystals (Quantum Dots : QDs), and impact of Nano Fabrication Techniques on Moore's Law (**Chapter 9**) the fundamental principles of quantum computing techniques (**Chapter 10**), emerging technologies using Quantum Dots and Nano-photonics (**Chapter 11**). Thin films applications, their deposition Processes. (**Chapter12**), various Convergent nano technologies , their overall potential , their future visions for unique applications for improvement of Human Performance and Quality of Life **(Chapter 13**). This information will be very useful for undergraduate and graduate students for getting comprehensive understanding on emerging trends in the application of nano technologies.

Chapter 14 presents opportunities and issues/concerns in the application of nano technologies. Challenges to be addressed and strategies to be adopted by developed and developing countries in promotion of nano technologies are discussed in detail which will be useful for researchers/administrators for proper planning for realizing the full benefits of newly emerging nano technologies for sustainable growth. Information on Current Career opportunities in Nano Technologies, *US Scenario and* Indian Scenario is also provided in this chapter which will be useful for students to be better informed for choosing their career in this important emerging technological field

As much information is not available on laboratory exercises in this field **Chapter 15** is devoted for describing a number of Laboratory Exercises giving complete details on experimental procedures, expected results and how to do data analysis explaining various nano Science/Technology concepts. This information will be greatly useful for both teachers and students to plan and execute laboratory exercises in this important field even in a moderately equipped labs at Undergraduate/graduate level.

All the information given in various Chapters in this book are presented lucidly describing extraordinary properties of nano materials, composites, and structures/devices, underlying basic science concepts, their emerging unique tailor made applications in advanced fields like communications, defense, biomedical areas with illustrative examples and data. Each chapter is provided with concerned references, bibliography for further reading and typical Questions.

I hope this book can serve as a Text Book/Resource material in Nano Science & Technologies for under graduate/graduate students and researchers /scientists working in related Engineering and Science disciplines.

-Author

Acknowledgements

The permission provided by several publishers and authors to refer their content as sources shown at appropriate places in the text and according permission to use some generic figures, flow charts and data is gratefully acknowledged.

I am greatly indebted to our Honble President KL University, Er. Koneru. Satyanarayana garu, for his keen interest in Nano Technologies and for his continuous support and encouragement in writing this book.

My special thanks are due to Dr. L. S. S. Reddy, Ph. D, Vice Chancellor K. L. University for his great administrative support in all my efforts.

The support given by Dr. Srinivas. Desamsetty, Ph.D, Project Scientist C, NCMRWF, Ministry of Earth Sciences, New Delhi, in collection of literature and in formatting the manuscript is greatly appreciated.

-Author

Contents

Chapter 10: Quantum Computing

Chapter 11: Quantum Dots, Nano-Photonics and their Applications

Chapter 12: Thin Films

Chapter 13: Nano-Bio-Info-Cogno (NBIC) Convergent Technologies

Chapter 14: Opportunities and Challenges in Newly Emerging Nano-Technologies and Strategies for their Utilization for Sustainable Development

1

Nano Science and Technologies

1.1 Concepts of Nano Science and Nanotechnologies

1.1.1 Introduction

Nano science deals with the study of materials at atomic, molecular and structures level with at least one dimension roughly between 1 and 100 nanometers where their properties differ significantly from those at a macro scale.

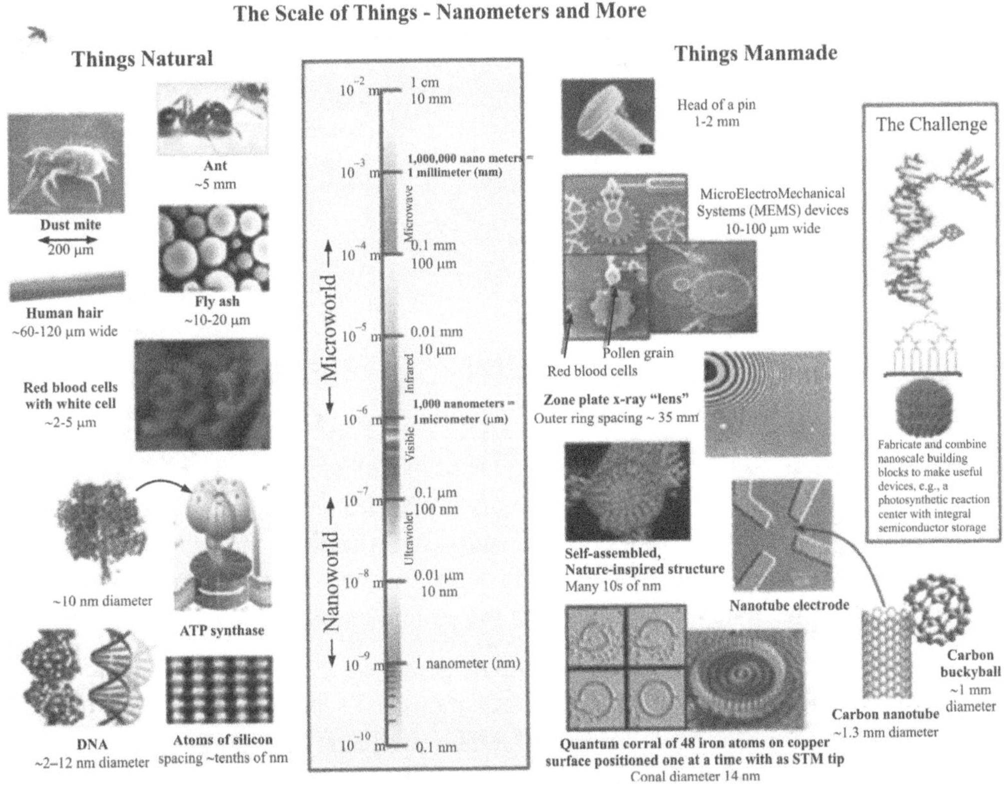

Figure 1.1 Scale of things in micro and nano world (1)

The prefix "nano" means one billionth. One nanometer (abbreviated as 1 nm) is 1/1,000,000,000 of a meter, which is close to 1/1,000,000,000 of a yard. To get a sense of the nano scale, a human hair measures 50,000 nanometers across and a bacterial cell measures a few hundred nanometers across. The smallest things visualized with the unaided human eye are 10,000 nanometers across. Just ten hydrogen atoms in a line make up one nanometer. To provide better idea the scale of various natural and man-made things which fall in micro and nano scale are presented in Fig. 1.1.

1.1.2 What is Nano Science?

Nano Science being an interdisciplinary science became one of the leading scientific fields today since it combines knowledge from the fields of Physics, Chemistry, Biology, Medicine, Informatics, and Engineering. Nano science expands the borders of material science and engineering by adding biology, biochemistry, information science to physics and chemistry disciplines. Nano science is thus a "horizontal-integrating interdisciplinary science that cuts across all vertical sciences and engineering disciplines" (Figure 1.2).

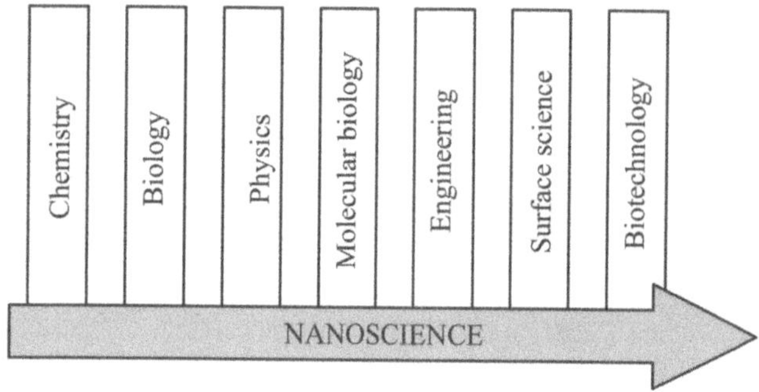

Figure 1.2 Nano science -horizontally integrating multi disciplinary science (*Source:* L. Filipponi, iNANO, Aarhus University, Creative Commons Share Alike 3.0)

1.1.3 What are Nano Technologies?

"Nanotechnologies are the design, characterization, production and application of structures, devices and systems by controlling shape and size at nanometre scale."

The fundamental objective of nanotechnology is to model, simulate, design and manufacture nanostructures and nano-devices with extraordinary properties and assemble them economically into a working system with revolutionary functional abilities. Nanotechnology offers new innovative ways of material development by controlling and manipulating the fundamental building blocks of matter at nanoscale, that is, at the atomic/molecular level.

Nanotechnologies are also horizontal-enabling convergent technologies (Figure 1.3). They are "horizontal" because they cut across numerous industrial sectors; they are "enabling" since they provide the platform, the tools to realise certain products; and are "convergent" because they bring together sectors of science that were previously separated.

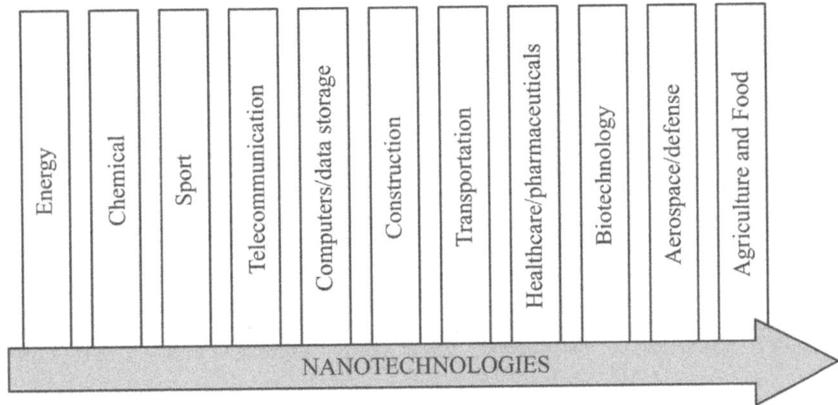

Figure 1.3 Nano Technologies-horizontal converging technologies
(*Source:* L. Filipponi, iNANO, Aarhus University, Creative Commons Share Alike 3.0)

1.1.3.1 Scope and Potential of Nano Technologies

Nanotechnologies are newly emerging technologies with great potential to lead and provide innovative breakthroughs that can be applied to real life. Novel nano and biomaterials, and nano devices can be fabricated and controlled by nanotechnology tools and techniques, which investigate and tune the properties, responses, and functions of living and non-living matter, at sizes below 100 nm. The application and use of nanomaterials in electronic and mechanical devices, in optical and magnetic components, quantum computing, tissue engineering, and other biotechnologies, with smallest features, widths well below 100 nm, are the economically most important parts of the nanotechnology nowadays and presumably in the near future. The number of nano products is rapidly growing since more and more nano engineered materials are reaching the global market.

1.1.4 Definition of Nanotechnology

US National Science and Technology Council (2) defines nanotechnology as : "*The essence of nanotechnology is the ability to work at the molecular level, atom by atom, to create large structures with fundamentally new molecular organization. The aim is to exploit these properties by gaining control of structures and devices at atomic, molecular, and supramolecular levels and to learn to efficiently manufacture and use these devices*".

In short, nanotechnology is the ability to build micro and macro materials and products with atomic precision.

United States National Science Foundation (3) defines nanoscience / nanotechnology as studies that deal with materials and systems having the following key properties.

1. **Dimension:** At least one dimension from 1 to 100 nanometers (*nm*).
2. **Process:** Designed with methodologies that shows fundamental control over the physical and chemical attributes of molecular-scale structures.
3. **Building block property:** They can be combined to form larger structures.

Nano science, in a general sense, is quite natural in microbiological sciences considering that the sizes of many bioparticles dealt with (like enzymes, viruses, etc) fall within the nanometer range.

Our ability to control and manipulate nanostructures will make it possible to exploit new physical, biological and chemical properties of systems that are intermediate in size, between single atoms, molecules and bulk materials.

As per British Standards Institute (BSI) the following definitions are also used.

Nano material: Material with one or more external dimensions, or an internal structure, which could exhibit novel characteristics compared to the same material without nanoscale features.

Nano particle: Particle with one or more dimensions at the nanoscale.

Nano composite: Composite in which at least one of the phases has at least one dimension on the nanoscale.

Nano structured material: Having a structure at the nanoscale,

1.1.5 How Laws of Physics differ at the Nano Scale-Quantum Size Effects

Nano materials are closer in size to single atoms and molecules than to bulk materials. Due to the very small size of nano materials, their mass will also be extremely small resulting gravitational forces to be negligible while electromagnetic forces become dominant in determining their behavior which can be explained by Quantum mechanics theory. Quantum mechanics is a scientific model that was developed for describing the motion and energy of atoms and electrons.

1.1.5.1 Wave-Corpuscle Duality of Matter

Wave-corpuscle duality of matter for objects of very small mass, such as the electron, will have a more pronounced effect. Thus, electrons exhibit wave behaviour and their position is represented by a wave (probability) function.

According to Heisenberg's uncertainty principle (quantum mechanical approach) the position and momentum of an object cannot simultaneously and precisely be determined [4].

According to Werner Heisenberg "The more precisely the **position** is determined, the less precisely the **momentum** is known". Heisenberg's Uncertainty Principle applies only to the subatomic particles like electron, positron, photon, etc. It does not forbid the possibility of nanotechnology, which has to do with the position and momentum of such large particles like atoms and molecules. This is because the mass of the atoms and molecules are quite large and the quantum mechanical calculation by the Heisenberg Uncertainty Principle (4) places no limit on how well atoms and molecules can be held in place [Figure 1.4].

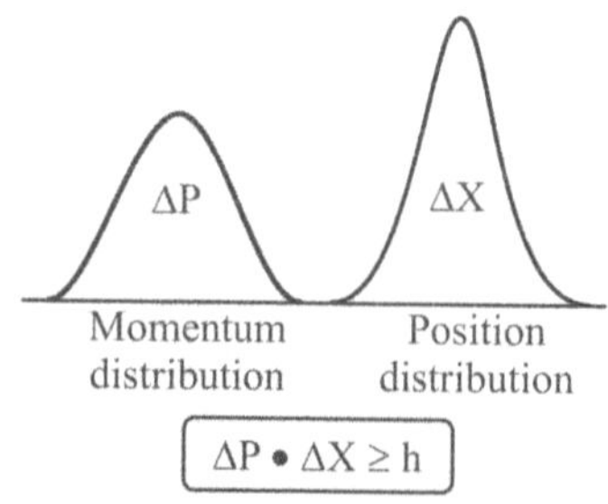

Figure 1.4 Heisenberg uncertainty principle

1.1.5.2 Tunneling Effect

Tunneling is the penetration of an electron into an energy region that is classically forbidden (5). Electron (or quantum) tunneling is attained when a particle (an electron) with lower kinetic energy is able to exist on the other side of an energy barrier with higher potential energy, thus defying a fundamental law of classic mechanics.

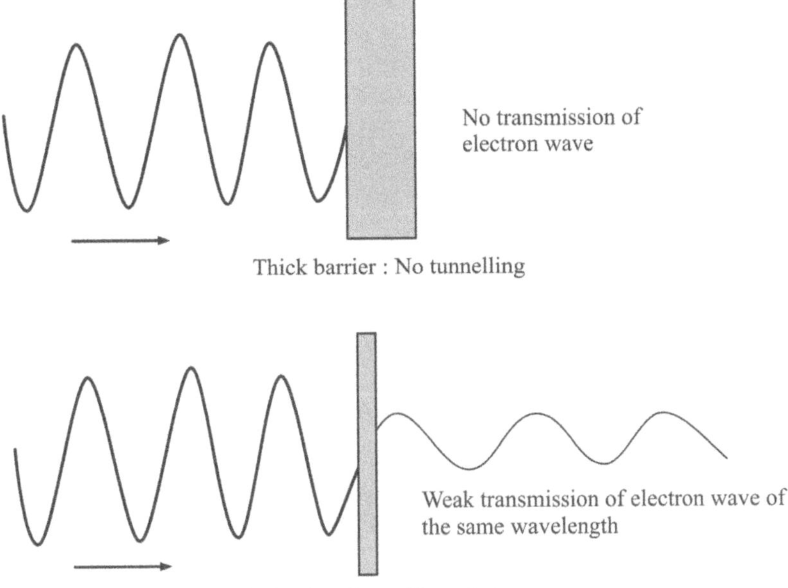

Figure 1.5 Quantum tunneling

Classic physics states that a body can pass a barrier (potential barrier) only if it has enough energy to 'jump' over it. Therefore, if the object has lower energy than that needed to jump over the energy barrier (the 'obstacle'), in classic physics, the probability of finding the object on the other side of the barrier is zero. In quantum physics, a particle with energy less than that required to jump the barrier has a finite probability of being found on the other side of the barrier. As shown in Figure 1.5 one can imagine that the particle passes into a 'virtual tunnel' through the barrier. However it should be noted that in order to have a tunnel effect, the 'thickness' of the barrier (i.e. energy potential) must be comparable to the wavelength of the particle and, therefore, this effect is observed only at nano metre level. The intensity of the current generated by the tunneling electron gives information about the thickness of the boundary. This principle is used in 'Scanning Tunneling Microscopy' (STM). Conductive surfaces with nano scale texture can be imaged by measuring the tunneling current between a non-contact surface and the substrate. Hence the STM has become an important research tool in nanotechnology for characterization of surfaces with nano scale resolution.

1.1.5.3 Quantum Confinement

In Nano particles/crystals, the electronic energy levels are not continuous as in the bulk but are discrete (finite density of states), because of the confinement of the electronic wave function to the physical dimensions of the particles. This phenomenon is called Quantum confinement and therefore Nano Crystals are also referred to as quantum dots (QDs).

In a nano material, such as a metal, electrons are confined in space rather than free to move in the bulk of the material. The quantum mechanical (wavelike) properties of electrons inside matter are influenced by variations on the nanoscale. By nanoscale design of materials it is possible to vary their micro and macroscopic properties, such as charge capacity, magnetization and melting temperature, without changing their chemical composition. For example, the ability of nano gold as well as of other noble metals and semiconductors relies on quantum confinement which is a very successful model for describing the size dependent electronic structure of nanometer sized materials. According to this theory electrons are confined in all three dimensions causing matter to behave completely different in terms of its optical and electronic properties. When the dimension of a material approaches the electron wavelength in one or more dimensions, quantum mechanical characteristics of the electrons that are not manifest in the bulk material can start to contribute to or even dominate the physical properties of the material.

1.1.5.4 Reasons for Quantization of Energy in Nano Materials

In nano particles size quantisation arises because the size of a nanoparticle is comparable to the de Broglie wavelength of its charge carriers (*i.e.,* electrons and holes). (Quantum dots are nanomaterials that display the effect of quantization of energy). Due to the spatial confinement of the charge carriers, the edge of the valance and conduction bands split into discrete, quantized, electronic levels. These electronic levels are similar to those in atoms and molecules. The spacing of the electronic levels and the bandgap increases with decreasing particle size. (Figure 1.6). This is because the electron hole pairs are now much closer together and the Coulombic interaction between them can no longer be neglected giving an overall higher kinetic energy.

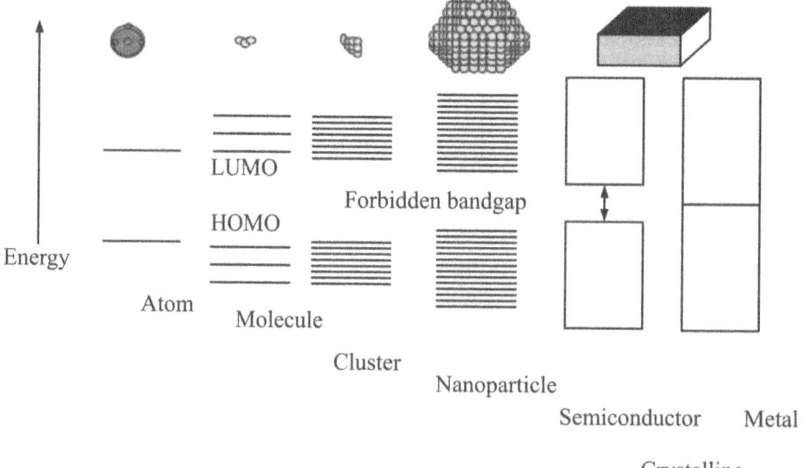

Figure 1.6 Size dependent quantization of energy levels HOMO (highest occupied molecular orbital) and LUMO (lowest unoccupied molecular orbital)

1.1.5.5 Random Molecular Motion

Molecules move due to their kinetic energy (assuming the sample is above absolute zero). This is called random molecular motion and is always present. At the macro scale, this motion is very small compared to the sizes of the objects and thus is not influential on how the object moves. At the nanoscale, however, these motions can be of the same scale as the size of the particles and thus have an important influence on how they behave. One example of a random kinetic motion is Brownian motion.

1.1.6 Quantum Size Effects on Physical Properties

Thus when the dimension of a material approaches the electron wavelength in one or more dimensions, quantum mechanical characteristics of the electrons that are not manifest in the bulk material can start to contribute to or even dominate the physical properties of the material.

Both physical and chemical properties are derived from atomic and molecular origin in a complex way. For example the electronic and optical properties and the chemical reactivity of small clusters are completely different from the better known property of each component in the bulk or at extended surfaces. Nano particles with a diameter between one and several tens of nanometers possess an electronic structure that is intermediate of the discrete electronic level of an atom or molecule and the band structure of a bulk material. The resulting size dependent role of physical properties is called the quantum size effect. Depending on their chemical composition the particles in the nano range display physical, chemical and biological properties which can be manipulated for the desired objectives. It is predicted that the behavior of the particles at the macro and nano range may be very different and the same material in bulk.

The origin of the quantum size effects strongly depends on the type of bonding in the crystal also. Complex quantum mechanical models are required to predict the evolution of such properties with particle size, and typically very well defined conditions are needed to compare experiments and theoretical predictions.

Nano materials have a relatively larger surface area when compared to the same mass of bulk material produced in a larger form. Nano particles can make materials more chemically reactive and affect their strength or electrical properties. The following materials based on the size and shape, fall under Nano materials which have great industrial importance.

(a) Nano particles
(b) Nano capsules
(c) Nano fibers
(d) Nano wires
(e) Fullerenes (carbon 60)
(f) Nano tubes
(g) Nano springs
(h) Nano belts
(i) Quantum dots
(j) Nano fluids etc

1.1.6.1 Different Categories of Quantum Confined Nano Materials

Under the influence of quantization of energy levels nanoscale materials will have special properties of great importance and can be grouped into the following categories based on the degree freedom of electrons in these nano structures.

1. **Three Dimensional Systems:** Nano particles which have all three dimensions-length, breadth and height.

   Nano particles are of interest because of the new properties (such as chemical reactivity and optical behaviour) that they exhibit compared with larger particles of the same materials. For example, titanium dioxide and zinc oxide become transparent at the nanoscale and have found application in sunscreens.

   Nano particles have a range of potential applications: In the short-term application such as in cosmetics, textiles and paints. In the longer term applications such as drug delivery where they could be to used deliver drugs to a specific site in the body.

   Nano particles can also be arranged into layers on surfaces, providing a large surface area and hence enhanced activity, relevant to a range of potential applications such as catalysts

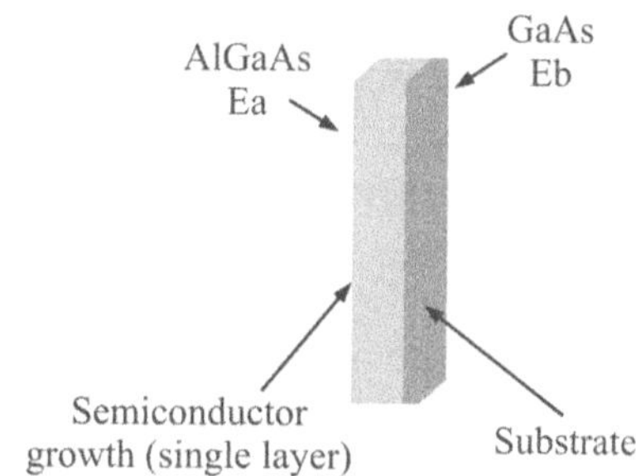

Figure 1.7 Single quantum well : AlGaAs on a GaAs

2. **Two Dimensional Quantum Well Systems:** They will have only length and breadth for example, nano wires and nano tubes: Quantum well is a two dimensional system where the electron can move in two directions and restricted in one direction. They are called Quantum because; the electronic property is quantized as the spatial distance is very very small.

   In Quantum Well one dimension is reduced to nano range while the other two dimensions remain large. Quantum well are formed in semiconductor by having a material like gallium arsenide sandwiched between two layers of a material with a wider band gap like aluminium arsenide. These structures can be grown molecular beam epitaxy or chemical vapor deposition with control of layer thickness down to monolayers.

   ***Single Quantum Well (SQW)*:** A single quantum well structure is formed by hetro epitaxial growth by an ultra thin layer of binary, tertiary, quaternary, or quintanary, semi conductor alloys on a suitable substrate. Examples are InP on a GaAs (InP/GaAS), AlGaAs on a GaAs (Figure 1.7) and Si on Ge(Si/Ge) systems.

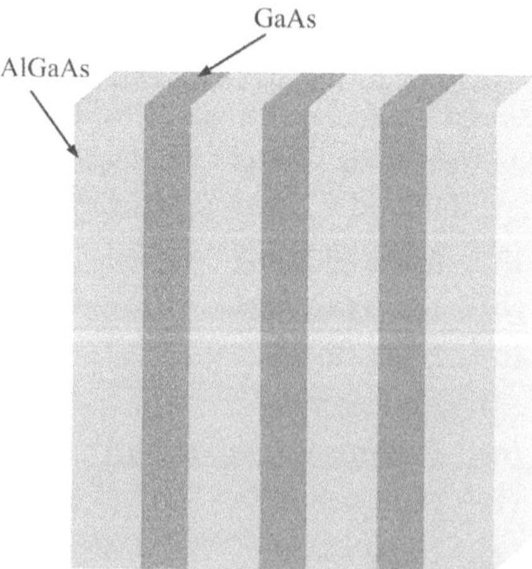

Figure 1.8 AlGaAs/GaAs/AlGaAs double quantum well

***Double Quantum Well (DQW) Systems*:** A double quantum structure is a hetero structure in which a small band gap ultra thin layer is sandwiched between wide band gap layers. Often the wide band gap materials are of the same semiconductor. An example is: AlGaAs/GaAs/AlGaAs double quantum well (Figure 1.8) (system(DQW).

***Multiple Quantum Well System (MQW)*:** A multiple quantum system is a hetero structure in which alternating layers of small band gap and wide band gap layers are grown on each other (Figure 1.9). Often the wide band gap materials form a barrier to prevent communication between the electrons in various quantum wells.

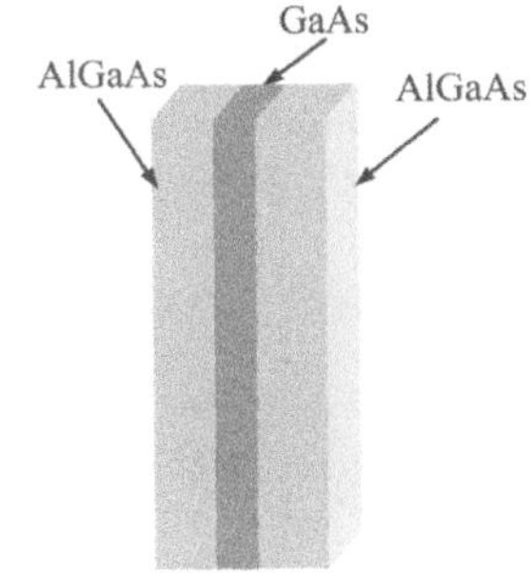

Figure 1.9 AlGaAs/GaAs – multiple quantum well system

3. **One Dimension:** It has only one parameter either length (or) breadth (or) height (example: very thin surface coatings.

   Quantum Wire is a also one-dimensional system where the electron can move in one direction and restricted in two directions.

   Quantum Wires: Quantum wires are ultra fine wires or linear arrays of Nano dots, formed by self-assembly. They can be made from a wide range of materials such as Semiconductor Nanowires made of silicon, gallium nitride and indium phosphide. Nanowires have potential applications in high-density data storage, either as magnetic read heads or as patterned storage media, in electronic and opto-electronic Nano devices, for metallic, interconnects of quantum devices and Nano devices. Nano wires can be prepared by growth techniques such as: Chemical Vapour deposition (CVD) Electroplating.

   In quantum wires two dimension are reduced and on dimension remains large.

   Therefore, the electrical resistivity of quantum wire can not be calculated using conventional formula (where ρ is the resistivity, l is the length, and A is the cross-sectional area of the wire).

$$R = \rho \frac{l}{A}$$

   This equation is not valid for quantum wires.

For calculating electrical resistivity of a conducting material we need two dimensions which are not present in a quantum wire and so it will not obey conventional formula applicable for bulk materials.

General properties of Nano wire: Diameter – tens of nanometers, Single crystal formation common crystallographic orientation along the nanowire axis, Minimal defects within wire, Minimal irregularities within nanowire arrays. Some examples of nano wires are shown in Figure 1.10

Magnetic Nano Wires: Example: Cobalt, gold, copper and cobalt-copper nanowire arrays (Figure 1.10) important for storage device applications.

Figure 1.10 Cobalt nanowires on Si substrate

4. **Zero Dimension: Quantum Dots:** In a quantum dot all three dimensions are reduced to zero. These belong to the extreme case of size reduction in which all three dimensions reach the low nano meter range. A quantum dot is a semiconductor nano structure that confines the motion of conduction band electrons, valence band holes or exitons (bound pair of conduction band electrons and valence band holes) in all three spatial directions (Figure 1.11).

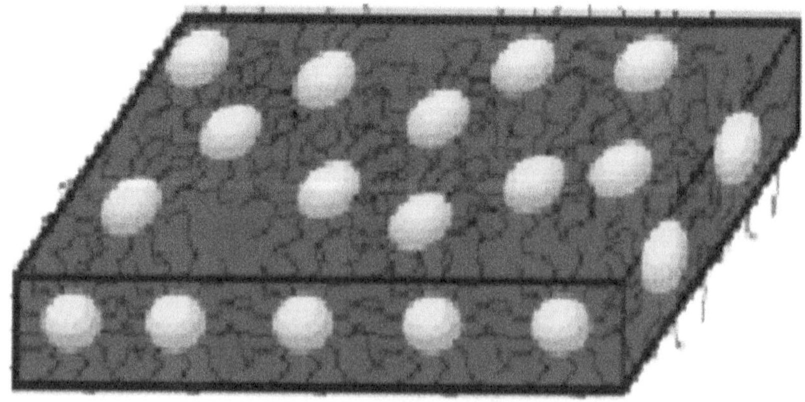

Figure 1.11 Quantum dots

The quantum confinement effects with variation of sizes are summarized in Figure 1.12.

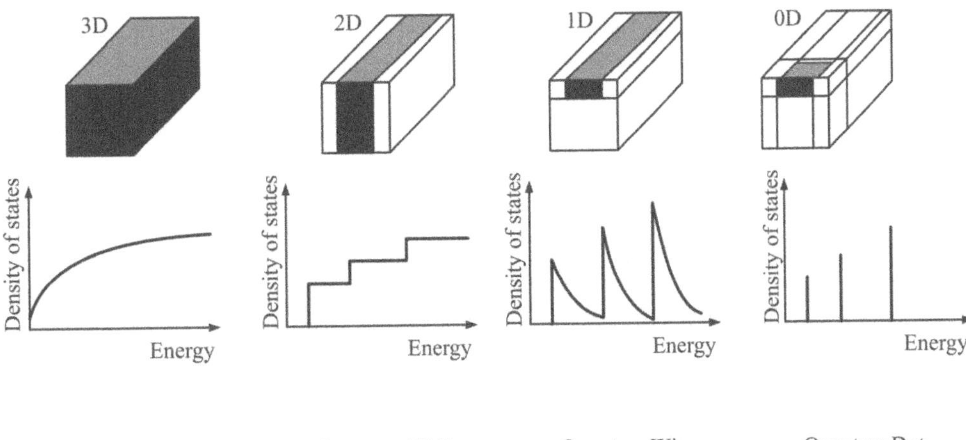

Bulk Quantum Well Quantum Wire Quantum Dot

- If a bulk metal is made thinner and thinner, until the electrons can move only in two dimensions (instead of 3), then it is "2D quantum confinement."
- Next level is 'quantum wire'
- Ultimately 'quantum dot'

Figure 1.12 Summary of quantum confinement in with variation of size of nano particle

1.1.7 What is Nano Chemistry?

Nanomaterials often arise from a number of molecules held together or large molecules that assume specific three-dimensional structures through intermolecular bonding (macromolecules). Therefore, nanoscience also deals with supramolecular chemistry (i.e., the chemistry that deals with interactions among molecules), which is just a sub-area of the general field called 'chemistry'. In these macromolecules, intermolecular bonding and surfaces often plays a crucial role and perform numerous functions as given below

(i) Intermolecular bondings, such as hydrogen bonding and van der Waals bonding are weak interactions but in a large number they can have a total energy that can be quite significant. Consider, for instance, the structure of DNA (which has a cross-section of 2 nm): the two helixes are held together by numerous hydrogen bonds. This point becomes particularly relevant in nanoscience, where materials can have very large surface areas and, consequently, small forces can be applied to very large areas .

(ii) In nanoscience, macromolecules are often considered as 'devices' that, for instance, can trap or release a specific ion under certain environmental conditions (pH etc.). A biological example of such a macro- molecule is ferritin.

Therefore, in nano science and nanotechnologies, where molecules can themselves be devices, bonds may also be device components. One area of study is for example, the use of molecules as molecular switches, actuators and electronic wires; physical and chemical properties depend on many of its surface properties.

(iii) Surfaces perform numerous functions: they keep things in or out; they allow the flow of a material or energy across an interface; they can initiate or terminate a chemical reaction as in the case of catalysts. Nano Surface chemistry deals with chemical, physical and biological interactions on nano surfaces. In this context, the term interface, rather than surface, is often used, to emphasise the fact that it is a boundary between two phases: the material and the surrounding environment (liquid, solid or gas).

(iv) If a bulk material is subdivided into an ensemble of individual nanomaterials, the total volume remains the same, but the collective surface area is greatly increased.

(v) In nano surface chemistry the chemical groups at the material interface determine its properties. Properties like catalytic reactivity, electrical resistivity, adhesion, gas storage and chemical reactivity depend on the nature of the interface.

(vi) This has a profound effect on reactions that occur at the surface such as such as catalysis reactions, detection reactions, and reactions that, to be initiated, require the physical adsorption of certain species at the material's surface.

(vii) Atoms and molecules that exist at the surface or at an interface are different from the same atoms or molecules that exist in the interior of a material. This is true for any material. Atoms and molecules at the interface have enhanced reactivity and a greater tendency to agglomerate: surface atoms and molecules are unstable, they have high surface energy.

(viii) Nano materials are inherently unstable, therefore there are various methods that nanomaterials adopt to minimise their inherent high surface energy. One of the ways of reducing the surface energy in nanoparticles is agglomeration. Surface energy is an additive quantity. The surface of 10 identical nanoparticles is equal to the sum of the surface energy of each individual nanoparticle. If these were to agglomerate, and become one large particle, the over- all surface energy would be reduced.

(ix) Nano particles have a strong intrinsic tendency to agglomerate. To avoid this, surfactants can be used. This also explains why when nanoparticles are used in research and industry they are often immobilised on a solid support or mixed within a matrix. Even in commercial products that claim to contain nanoparticles (such as sunscreens) microscope images show that they are actually present in the form of agglomerates of > 100 nm dimensions.

1.2 Classification of Nano Materials

Nano materials can be classified based on the basis of their origin as given below

(i) Natural/non engineered nano sized materials available in the environment like proteins, viruses,

(ii) Engineered nano materials which are produced by fabrication process by manipulating/controlling/modifying nano characteristics of materials like nano composites, nano fibers, nano tubes devices, quantum dots etc.

1.2.1 Natural/Non Engineered Nano Materials

Natural nano materials are those available in nature (Plant, animal and mineral), without human modification or processing and which possess un-usual properties because of their inherent nano structure. The arrangement of tens to hundreds of molecules into shapes and forms in the nanoscale range provide nano structure to biological molecules. The interaction of light water and other materials with these nanostructures gives the natural materials some remarkable properties. In nature, we encounter some outstanding solutions to complex problems in the form of fine nanostructures with which precise functions are associated.

1.2.1.1 Nature's Examples for Non Engineered Nano Materials

The physical origins of the remarkable properties of many biological materials are due to complex, often hierarchical structures. They are characterized by a surprising level of adaptability and multifunctionality. These materials can provide a model for designing radically improved artificial materials for many applications, such as solar cells, fuel cells, textiles, drug delivery systems, etc.

Further in the formation and functioning of these natural nano structure characteristic follows very simple laws given below which are highly inspiring to follow

(i) Nature runs on sunlight and uses only the energy it needs. Natural nanomaterials are extremely energy efficient.

(ii) Nature organize materials to function and recycles everything — waste products are minimized in nature.

(iii) Nature rewards cooperation although it encourages diversity and local expertise.

1.2.1.1.1 Examples of Natural Nano Materials

The following are some examples of naturally available nano materials

(i) ***Nano particles from natural erosion and volcanic activity*:** Minerals, such as clays, are nanostructured: clays are a type of layered silicate characterized by a fine 2D crystal structure. Among clays, mica has been the most studied.

Mica is made up of large sheets of silicate held together by relatively strong bonds. Smectic clays, such as montmorillonite, have relatively weak bonds between layers. Each layer consists of two sheets of silica held together by cations such as Li^+, Na^+, K^+ and Ca^{2+}. The presence of the cations is necessary to compensate for the overall negative charge of the single layers. The layers are 20-200 nm in diameter laterally and form into aggregates called tactoids, which can be about 1 nm or more thick.

(ii) ***Naturally occurring clays include montmorillonite (MMT) and hecrite*:** The fine nanostructure of clays determines their properties. When water is added, the clay swells, but the volume change is rather unusual. It is several times the original volume due to the 'opening' up of the layered structure by the water molecules that replace the cations. Clay swelling is a significant factor in soil stability and must be taken into account when building roads etc.

(iii) ***Natural colloids, such as milk and blood (liquid colloids), fog (aerosol type), gelatin (gel type)*:** In these materials when nano particles are dispersed in the medium (liquid or gas) but do not form a solution, rather a colloid. All these materials have the characteristic of scattering light and often their colour (as in the case of blood and milk) is due to the scattering of light by the nano particles that make them up.

(iv) ***Mineralised natural materials, such as shells, corals and bones*:** Many of these materials are formed by calcium carbonate crystals that self-assemble together with other natural materials, such as polymers, to form fascinating three-dimensional architectures. For instance, a shell is grown by a layer of cells that first lays down a coating of protein supported by a polysaccharide polymer like chitin. The proteins act like a nano-assembly mechanism to control the growth of carbonate crystals. Around each crystal remains a honey comb –like matrix of protein and Chitin. This relatively 'flexible envelope' is fundamental for the mechanical properties of the shell and mitigates cracking. The size of each crystal is around 100 nm. The result is that the nacre of mollusc shells has extraordinary physical properties (strength, resistance to compression, etc.).

(v) ***Materials like skin, claws, beaks, feathers, horns, hair*:** These materials are made largely of very flexible proteins like keratin, elastin and collagen. Keratins have a large glycine and alanine content.

(vi) ***Insect wings and opals*:** The colours seen in opals and butterflies are directly related to their fine structure, which reveals packed nanostructures that act like a diffraction grid and induce iridescence. In the case of opals, this is due to packed silica spheres in the nano metre range, uniform in size and arranged in layers.

Spider silk: Silk is the material with the greatest known strength — about five times that of steel of the same weight. The extraordinary properties of spider silk are due to the proteins that make up the silk (mainly fibroin) and its supramolecular organization which is at the nano scale level.

(vii) ***Lotus leaves and similar (nasturtium)*:** The nano structure of these leaves is responsible for their extraordinary surface properties and their ability to 'self-clean'. The leaves of the lotus plant have the outstanding characteristic of totally repelling water because they are super hydrophobic (Figure 1.13).

Self-cleaning properties of the lotus plant are produced by a combination of the micro structure of leaves and epidermal cells on their rough surfaces which are covered with wax crystals(6). These crystals provide a water-repellent layer, which is enhanced by the roughness of the surface, making it a super hydrophobic surface, with a contact angle of about 150°. Figure 1.13 presents the progressive magnification of lotus leaf where on the right, nanocrystals a few tens of nanometres in size are shown

The consequence is that water droplets roll off the leaf surface and, in doing so, drag dirt away from it, as shown in Figure 1.13 This effect, 'self-cleaning', renders the lotus leaf clean and resistant to dirt. The same effect is found in other leaves such as those of nasturtium (Tropaeolum) and some Cannas, and in some animals such as the water strider.

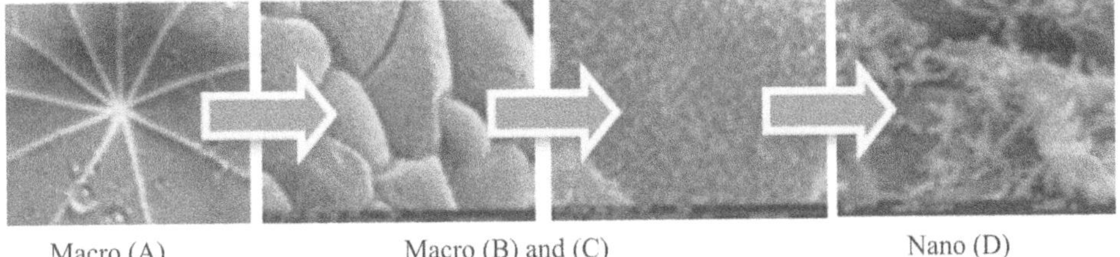

Figure 1.13 Progressive magnification of lotus leaf
(Image credit (A): A.Snyder, Exploratorium; (B, C): A.Marshall, Stanford University, (D): A. Otten and S. Herminghaus, Göttingen, Germany)

(viii) ***Nano Adhesives of Lizard (Geckos') Feet*:** The structure of the Lizard foot is on of the best example for the relationship between function and nanostructure. The ability of Lizards to walk upside down, against gravity, even on wet or dirty surfaces is mainly due to the nanostructure of their feet. (Figure 1.14)

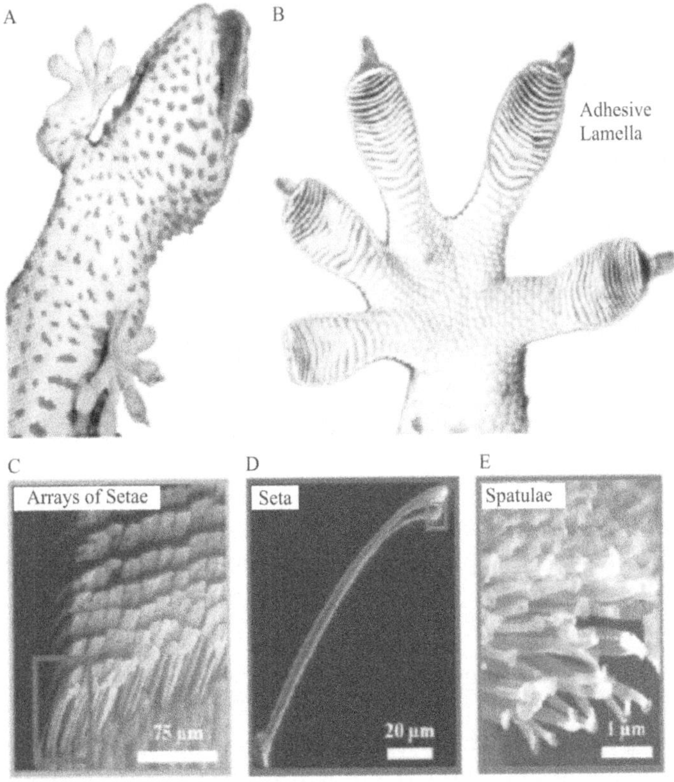

Figure 1.14 Lizards amazing feet with 10 N adhesive force (7)

The adhesion Characters of Lizard Feet and its mechanism are summarized in Table 1.1.

Table 1.1 Adhesive characters of lizard feet and mechanism of adhesion (8)

A lizard from southeast Asia which...	
Can generate – 10 B of adhesive force	
Can run up to – 1 m/s	
Can generate sheer stress of – 0.1 N mm^{-2} (–1 atm)	
Can walk on Any surfaces (hydrophobic/hydrophilic/rough/smooth/charged/uncharged...)	
What is the mechanism for such an amazing adhesion?	
– micro-suction?	No. adhesion works in vacuum.
– friction?	No. measured friction constant too low.
– micro-interlocking?	No. it walks on very smooth surface.
– capillary force?	No. it walks in ionized air.
– adhesion by glue?	No. there are no skin glands on their feet.

A lizard can cling to virtually any surface at any orientation; walk on smooth or rough surfaces, even upside down on a glass surface; and walk on a dirty or wet surface maintaining full contact and adhesion to it. As it walks, a lizard does not secrete any sticky substance, and its feet do not have any suction-like features (even at microscopic sizes). The reason for the lizards amazing properties lies in the nanostructures that are present on its feet. The lizard foot has a series of small ridges called scansors which contain numerous projections called setae. Each seta is about 100 μm long and has a diameter of about 5 μm. There are about half a million of these setae on the foot of a lizard. Each seta is further subdivided into about a thousand 200 nm-wide projections called spatulae (Figure 1.14). As a result, the total surface area of the lizards feet is enormous. The lizard spatulae are very flexible, so they essentially mould themselves into the molecular structure of any surface. The result is a strong adhesion which is entirely due to van der Waals forces. A single seta can resist a force of 200 μN, or approximately 10 atmospheres of stress. This is a very good example of the effect of large surface area on small forces. Another very interesting property of lizards is that their feet don't get dirty as they walk, even if they walk on a surface covered with sand, dirt, water, etc. Their feet stay clean even on dirty surfaces and full adhesion is maintained. The phenomenon has been investigated and it was found that the feet remain clean because it is more energetically favourable for particles to be deposited on the surface than to remain adhering to the lizard spatulae. If a lizard walks over a dirty surface, it takes only few steps for its feet to be totally clean again, and adhesion is not compromised.

Nano Photonic Crystals of the wings of butterflies: Butterflies often display extraordinary colours and iridescence due to nano photonic crystals. The shift in colour of butterfly when observed at different angles which is a consequence of the wing surface and its interaction with light. The wings of Morpho rhetenor are formed by rows of scales arranged like tiles on a roof (Figure 1.15). Each scale is about 70 × 200 μm and has a smaller structure on its surface, a very intricate and highly ordered nano metre organisation of ridges. Each ridge is about 800 μm wide. The spaces between them form a natural photonic crystal that can generate constructive and destructive Interference. The SEM (Figure 1.15) analysis of the cross-section of the ridges on the wings shows and an even more intricate structure that looks like fir trees.

Figure 1.15 SEM analysis of the cross-section of the ridges on the wings of butterfly (Images credit: (far left): Wiki Commons, Creative Commons Attribution Share Alike 3.0; (all other images): S. Yoshioka, Osaka University,)

Learning nanotechnologies from nature: Natural nano materials are of interest not only to understand (and appreciate) the amazing properties of biological materials but also to gather inspiration for the design and engineering of new materials with advanced properties. Some examples of nano technologies developed by mimicking natural systems are presented in Table 1.2.

Table 1.2 Biomimetics by inspiration from natural materials

Biomimetric Material	Inspired from
Polymers	Substructure of nacre
Structural elements	Wood, ligaments and bone
Electrical conduction	Eels and nervous system
Photoemission	Deep-sea fish and glow-worms
Photonic crystals	Butterfly and bird wings
Hydrophobic surfaces	Lotus leaves and human skin
Adhesives	Geckos' feet
High tensile strength fibre	Spider silk
Artificial intelligence and computing	Human brain

1.2.2 Engineered Nano Structured Materials

1.2.2.1 Classification of Engineered Nano Materials/Devices

Nano structures which are produced by controlled manipulation of their microstructure at atomic level are termed as Engineered Nano Materials. The quantum confined nano materials discussed earlier also fall under this category.

These are classified by Gleiter into three categories (9)

1. Nano materials /structures/devices with reduced dimensions or dimensionality in the form of isolated substrate supported, or embedded nano sized particles or thin wires or thin films come under first category. The fabrication techniques normally used to attain this kind of nano structure are Chemical vapor deposition, (CVD), Physical Vapor Deposition (PVD), various aerosol techniques precipitation from the vapor, super saturation liquids

or solids. The examples of technological application of this type of nano materials are catalysts, semiconductor devices utilizing multi layer quantum well structures

2. Nano materials/devices in which nano sized microstructure is limited to a thin surface region of a bulk material fall under second category. CVD. PVD, ion implantation and laser beam treatments are widely used procedures to modify chemical composition and atomic structure of solid surfaces at nano scale level. Surfaces with enhanced corrosion resistance, hardness, ware resistance or protective coatings are examples widely used present day technologies. In these technologies the properties of a thin surface layer is improved by creating a nano meter sized micro structure in a thin surface region. Such process or nano devices/systems find wide application in next generation electronic devices such as highly integrated circuits, terabit memories, single electron transistors, quantum computers etc.
3. The third category consists of bulk solids with a nano meter scale micro structure. Those are the solids in which chemical composition, the atomic arrangement and or the size of building blocks (Crystalites or atomic or molecular groups) forming the solids varies on a length scale of few nano meters through the bulk.

Another classification of engineered nano structure materials/systems/as defined by Richard W. Siegel depends on the number of dimensions which lie within nano meter range

(i) 3D systems confined in all three dimensions e.g., Structures typically composed of consolidated equiaxed crystallites

(ii) 2D systems confined in two dimensions ex: filamentary structures where length is substantially greater than the cross sectional dimensions

(iii) 1D systems confined in one dimension ex: layered or laminate structures

(iv) zero-D –zero structures such as pores, nano particles quantum dots etc., which are discussed earlier pages.

Thus three dimensional structures or bulk materials with a nano meter sized micro structure are assembled of nano meter sized building blocks or grains that are mostly crystallites. The schematic model of various nano structured material is shown in Figure 1.16.

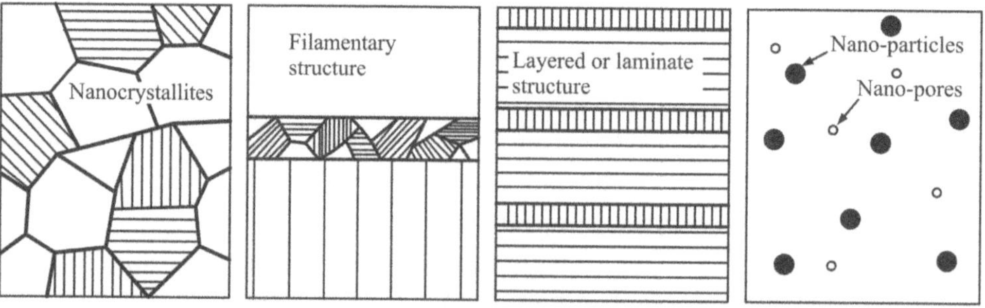

Figure 1.16 Schematic of nano structured materials with 3D, 2D, 1D and zero D confinement of electrons

These building blocks differ in their atomic structure, their crystallographic orientation and/or their chemical composition. Incoherent or coherent interfaces may be formed between them depending on an atomic structure, the crystallographic orientation and/or chemical composition or adjacent crystallites. In other words materials assembled of nano meter sized building blocks are micro structurally heterogeneous consisting of building blocks (crystallites) and the regions between adjacent blocks (e.g., grain boundaries)

The latter three categories can be further grouped into four families as shown in Figure 1.17.

In the most simple case all grains and interfacial regions have the same chemical composition. e.g., Semicrystalline polymers (consisting of stacked lamellae separated by non-crystalline region), multilayers of thin film crystallites separated by an amorphous layer (a-Si:N:H/nc-Si) etc.

In the second case, materials are classified with different chemical composition of grains. Possibly quantum well structures are the best example of this family.

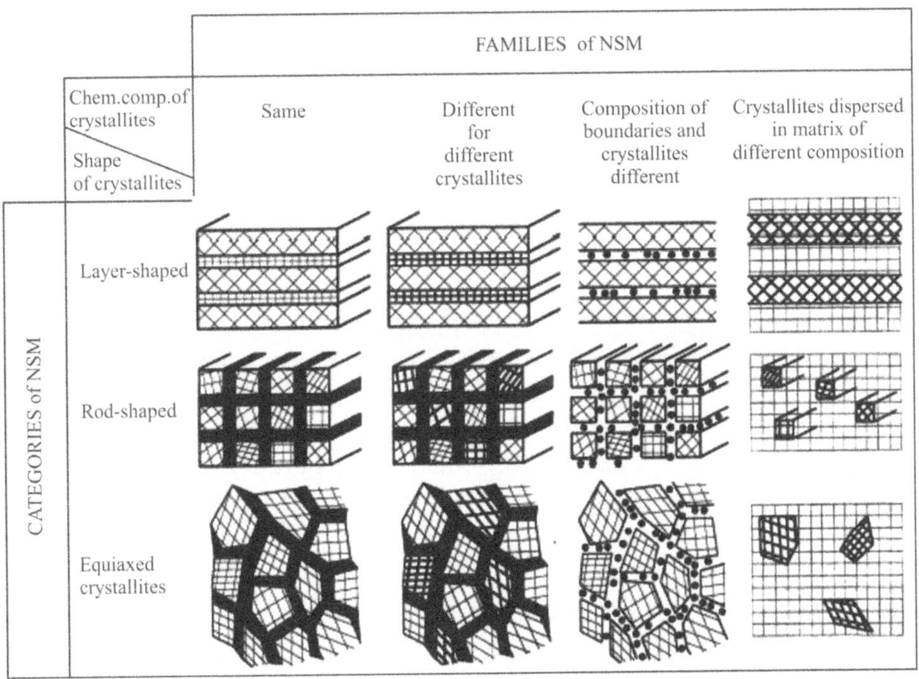

Figure 1.17 Relation between categories & families of nano structured materials (NSM) (Gleiter H., Acta Mater., 2000, vol. 48).

In the third family includes all materials that have a different chemical composition of its forming matter (including different interfaces) e.g., ceramic of alumina with Ga in its interface

The fourth family includes all nanomaterials formed by nanometer sized grains (layers, rods or equiaxed crystallites) dispersed in a matrix of different chemical composition. Precipitation hardened alloys typically belong to this family. E.g., Nanometer sized Ni_3Al precipitates dispersed in a nickel matrix-generated by annealing a supersaturated Ni-Al solid solution- are an example of such alloys. Most high-temperature materials used in modern jet engines are based on precipitation-hardened Ni_3Al/Ni alloys.

1.3 History of Nano Technology

Nanotechnology is often regarded as an outcome of research in the latter part of the twentieth century, a product of the drive towards miniaturization led by the semiconductor industry. However, in a broader sense, nanotechnology has been in practice unrealized as such, for a long time. Two thousand years ago the ancient Greeks used a permanent hair-dying recipe that worked by depositing 5 nm lead sulfide crystals inside hair. High-quality steel made in India before the turn of the first millennium has been shown to contain — and owe its outstanding properties to carbide structures similar to modern carbon nanotubes. Medieval artists colored stained glass using metal nanoparticles.

The difference between these ancient examples of "nanotechnology" and the current situation is the ability to understand — or at least embark on a path towards understanding the fundamental principles underlying nano technological behavior, the ability to assess the current state of knowledge, and the ability to systematically plan for the future based on that knowledge.

The history of nanotechnology is generally understood to have begun in December 1959 when physicist Richard Feynman gave a speech, "There's Plenty of Room at the Bottom" (10), at an American Physical Society meeting at the California Institute of Technology in which he identified the potential of nanotechnology. Feynman said it should be possible to build machines small enough to manufacture objects with atomic precision, and that if information could be written on an atomic scale, "all of the information that man has carefully accumulated in all the books in the world can be written in a cube of material one two-hundredths of an inch wide—about the size of the smallest piece of dust visible to the human eye." He claimed that there were no physical laws preventing such achievements, while noting that physical properties would change in importance (e.g., gravity becoming less important), though surface phenomena would begin to dominate behavior. It is suggested to name the nanometer scale as the *Feynman* (φ nman) *scale* after Feynman's great contributions to nanotechnology (1 *Feynman* [φ] $\equiv 10^{-9}$ *meter* $=10^{-3}$ *Micron* [μ]=10 *Angstroms* [Å]).

In 1974, Norio Taniguchi first used the word "nanotechnology" (11), in regard to an ion sputter machine, to refer to "production technology to get the extra-high accuracy and ultra- fine dimensions, i.e., the preciseness and fineness of the order of one nanometer." In the 1980s, Eric Drexler authored the landmark book on nanotechnology, "Engines of Creation" (12) in which the concept of molecular manufacturing was introduced to the public at large. It is due to Drexler that much of the public's imagination has been captured by the potential of nanotechnology and nano manufacturing. In 1985, fullerenes, or "buckyballs," were discovered (13). By the 1990s, nanotechnology was advancing rapidly. In 1990, the first academic nanotechnology journal was published, in 1993 the first Feynman Prize was awarded, and by 2000 President Bill Clinton announced the U.S. National Nanotechnology Initiative (NNI). NNI and other nanotechnology proponents now anticipate the development of nano-enabled tools to help address many current challenges facing the United States and the international community, in the following areas

(a) Clean, secure, affordable energy;
(b) Stronger, lighter, more durable materials;
(c) Low-cost filters to provide clean drinking water;
(d) Medical devices and drugs to detect and treat diseases more effectively with fewer side effects;
(e) Lighting that uses a fraction of the energy associated with conventional systems;
(f) Sensors to detect and identify harmful chemical and biological agents; and

Techniques to clean up harmful chemicals in the environment

1.4 Unique Physico-Chemical, Mechanical, Electrical, Optical and Magnetic Properties of Nano Materials

1.4.1 Reasons for Nano Materials Exhibiting different Properties from Bulk

The unique (different) physical properties of nanomaterials in comparison with bulk are mainly occur due to the following reasons

1. Gravitational forces become negligible and electromagnetic forces dominate,
2. Quantum mechanics is the model used to describe motion and energy instead of the classical mechanics model,
3. Greater surface to volume ratios,
4. Large fraction of surface atoms,
5. Large surface energy,
6. Spatial confinement, and
7. Reduced imperfections.

The most fundamental properties of nano materials and devices depend on their size. For example, a nanoscale wire or circuit component does not necessarily obey Ohm's law (as discussed earlier), the foundation of modern electronics. Ohm's law relates current, voltage, and resistance, but it depends on the concept of electrons flowing down a wire like water down a river, which they cannot do if a wire is just one atom wide and the electrons need to traverse it one by one. This coupling of size with the most fundamental chemical, electrical, and physical properties of materials is key to all nano science. Anything smaller than a nanometer in size is just a loose atom or small molecule floating in space as a little dilute speck of vapor. So nanostructures aren't just smaller than anything we've made before, they are the smallest solid things it is possible to make. Additionally, the nanoscale is unique because it is the size scale (Figure 1.18) where the familiar day-to-day properties of materials like conductivity, hardness, or melting point meet the more exotic properties of the atomic and molecular world such as wave-particle duality and quantum effects. Similarly at the nano scale dimension, the principles of classical physics are no longer capable of describing their behavior (movement, energy, etc). At these dimensions, quantum mechanics principles apply. The same material (i.e., gold) at the nanoscale can have properties like optical, mechanical, electrical, etc., which are very different from (even opposite to) the properties the material has at the macro scale (bulk).

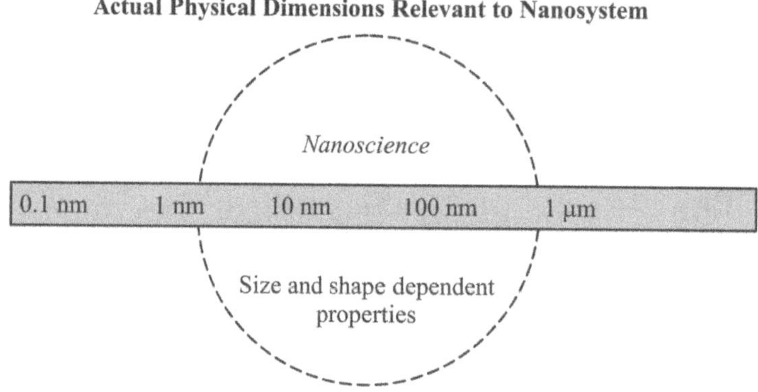

Figure 1.18 Length scale where materials exhibit size dependent properties

1.4.2 More % Surface Atoms in Nano Systems

The macroscopic physical properties of a substance (melting point, boiling point, conductivity etc.) are determined by studying a pure sample in quantities big enough to be measured under normal laboratory conditions. One mole of any material contains 6.022×10^{23} molecules (Avagadros number); Properties are usually measured by looking at large ($\sim 10^{23}$) aggregations of atoms or molecules. We assume that the result should be true for any size of group of water molecules. This is not correct for many materials: as the size of the material is reduced in the nano scale regime only hundreds to tens of thousand atoms will be present with large proportion of surface atoms (Example1). It is possible that the same material will display totally different properties (different melting point, conductivity etc). This is because matter at the nano scale no longer follows Newtonian physics but rather quantum mechanics.

Example 1: Typical Nano system of carbon may contain 100 to few thousand molecules at the surface which are less bounded by bulk atoms (Figure 1.19)

C Atom details

Bohr radius = 0.5292Å ≈ 0.05 nm

C atom (VdW radius) = 0.17 nm

In a 1nm line: 3C atoms

In a $1\text{nm} \times 1\text{nm}$ surface: 9C atoms

In a 1nm × 1nm × 1nm cube: 27 C atoms

In a 100 nm × 100 nm × 100 nm cube: 2.7×10^7 C atoms

In a 1m × 1m × 1m cube: 2.7 × 1028 C atoms

Example 2: Nano Systems % of Surface atoms

Example of Gold Nano particle

Sphere of radius 12.5 nm contains total approx. 480,000 atoms.

Surface contains approx. 48,000 atoms.

So, approx. 10% atoms are on the surface

Sphere of radius 5 nm contains total approx. 32,000 atom surface contains approx. 8000 atoms

So, approx. 25% atoms are on the surface.

Surface atoms will have un-used electrons which make them very reactive and excellent platform for catalysis

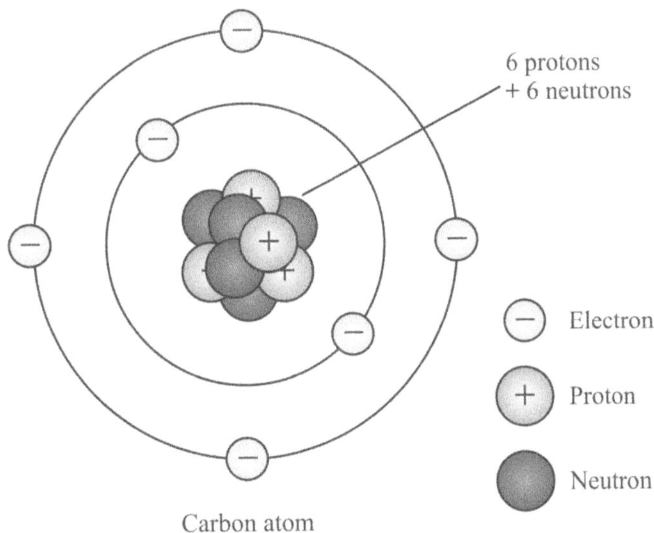

Figure 1.19 Electronic structure of carbon atom

1.4.3 Unique Physical Properties of Nano Materials

1.4.3.1 Lower Melting Point or Phase Transition Temperature with Reduced Lattice Constants

Melting Point is the temperature at which the atoms, ions, or molecules in a substance have enough energy to overcome the intermolecular forces that hold them in a "fixed" position in a solid. Surface atoms require less energy to move because they are in contact with fewer atoms of the substance (Figure 1.20).

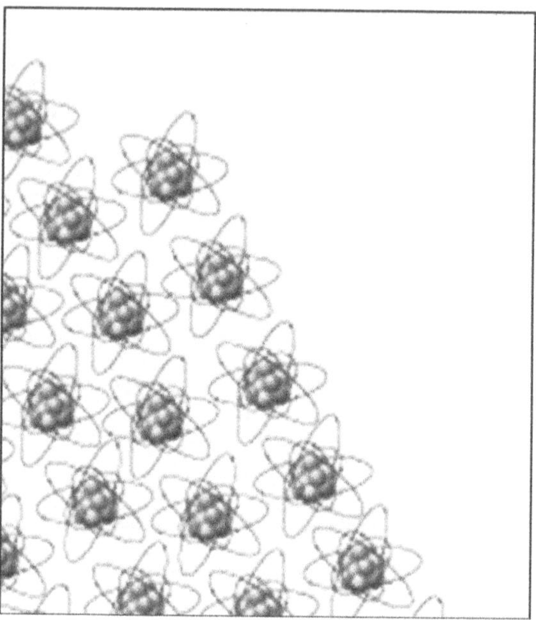

Figure 1.20 Nano structure showing that surface atoms are bound by lower number of atoms than in the bulk (14)

Low melting points for nano materials

Nano materials exhibit significantly lower melting point or phase transition temperature and appreciably reduced lattice constants. As can be seen in Figure 1.21 the melting points of many nano particles decrease dramatically when their size gets below 5 nm

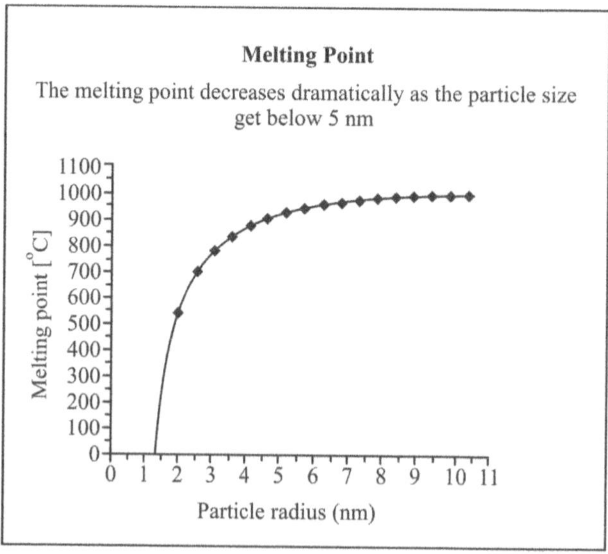

Figure 1.21 Dramatic decrease of melting point of nano particles below 5 nm (15)

This is due to a huge fraction of surface atoms out of total number of atoms

As an example the variation of melting points of gold as a function of particle diameter are presented in Figure 1.22.

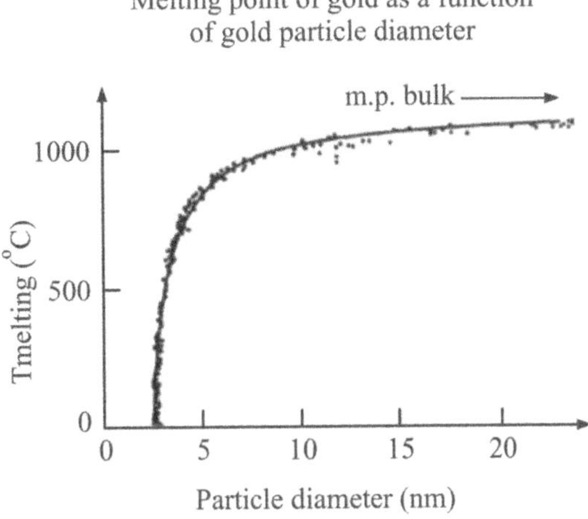

Figure 1.22 Decrease in melting point of gold nano particles as their size decreases (16)

1.4.3.2 Gravitational and Electromagnetic Forces at Nano Scale

Due to the smallness of nanomaterials, their mass is extremely small and gravitational forces become negligible. Instead electromagnetic forces will become dominant in determining the behaviour of atoms and molecules.

Gravitational force is a function of mass and distance and is weak between (low-mass) Nano sized particles (Figure 1.23).

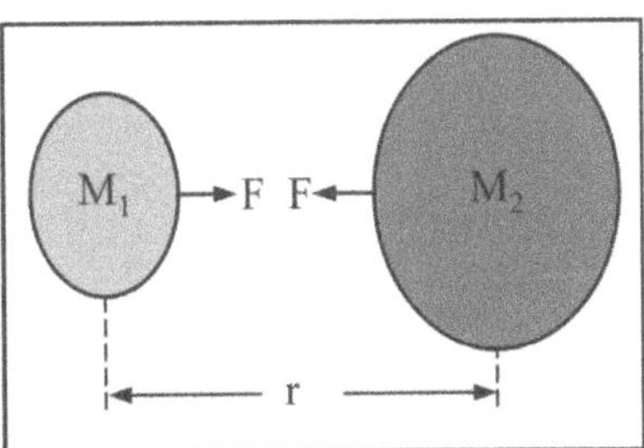

Figure 1.23 Gravitational force decreases with size (17)

At the nano scale, inertia and gravity would make no difference. The bacterium swimming through the water comes to a stop in a distance less than the diameter of the hydrogen atom. Attractive forces, such as van der Waals forces, and viscous forces between small objects are much stronger than the forces of gravity and inertia at that scale. As a result, all of these molecules and machines and cell parts are in constant motion, being pushed and pulled around in quick, random trajectories.

Electromagnetic force is a function of charge and distance is not affected by mass, so it can be very strong even when we have Nano sized particles (Figure 1.24).

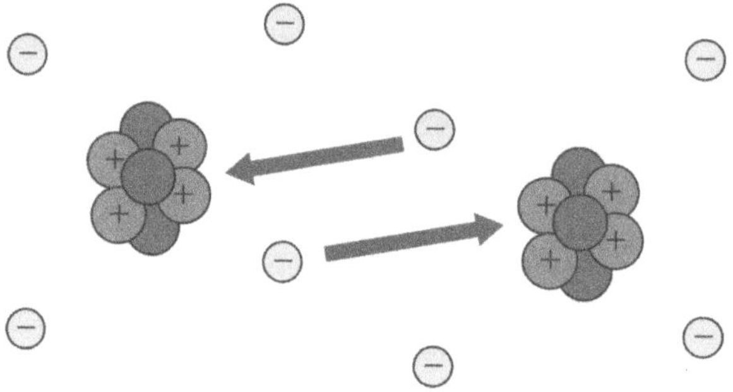

Figure 1.24 Electromagnetic forces become dominant as the size decreases in nano particles (18)

1.4.3.3 Enhanced Mechanical Strength

Nano materials exhibit enhancement in the mechanical strength (by one or two orders of magnitude) which may be simply due to the reduced probability of defects. While there will be more room for deformation in bulk material the nano material being tightly packed offer higher mechanical strength and tensile strength.

Some nano materials have inherent exceptional mechanical properties which are connected to their structure. One such material is carbon nanotubes: these are extremely small tubes having the same honeycomb structure of graphite, but with different properties compared to graphite. They can be single-walled or multi-walled, as illustrated in Figure 1.25.

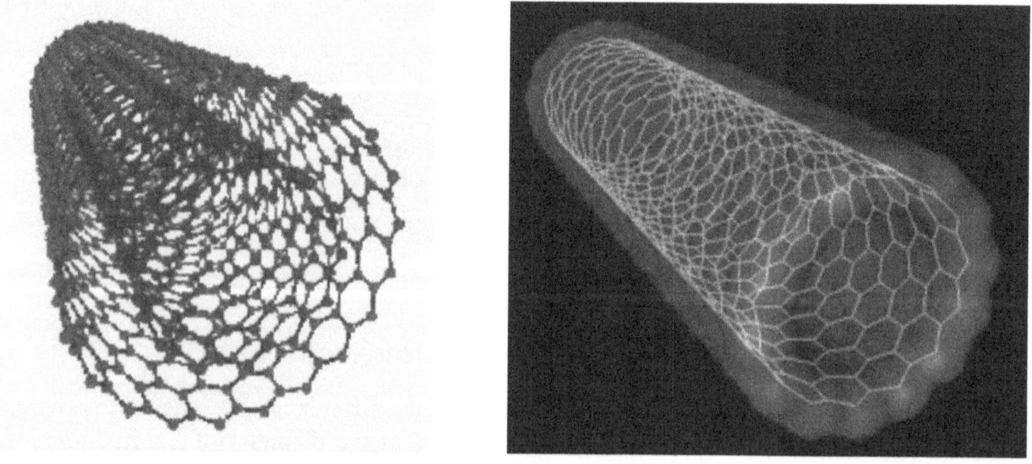

Figure 1.25 Single walled and multi walled carbon nano tubes

The chemical bonding of nanotubes is composed entirely of sp2 bonds (carbon double bonds) similar to those of graphite (whereas, in diamonds, all bonds are sp3). This bonding structure, which is stronger than the sp3 bonds found in diamonds, provides the molecules with their unique strength. Nanotubes naturally align themselves into 'ropes' held together by van der Waals forces.

Theoretical predictions say that carbon nanotubes are 100 times stronger than steel but only one sixth of its weight. Therefore, they are ideal in lightweight construction, for instance in the automotive and aviation industries. Carbon nanotubes are already used in some consumer products, such as tennis rackets to add strength (without compromising weight). The mechanical properties of carbon nanotubes are summarised in Table 1.3(19).

Table 1.3 Mechanical properties of carbon nano tubes

Material	Young's Modulus (GPa)	Tensile Strength (GPa)	Density (g/cm^3)
Single-wall nonotube	800	> 30	1.8
Multi-wall nanotube	800	> 30	2.6
Diamond	11400	> 20	3.52
Graphite	8	0.2	2.25
Steel	208	0.4	7.8
Wood	16	0.008	0.6

Young's modulus is a measure of how stiff, or elastic, a material is. The higher the value, the less a material deforms when a force is applied. Tensile strength describes the maximum force that can be applied per unit area before the material snaps or breaks. A third interesting measure of a material is its density, which gives an idea of how light it is. From Table 1.3, it can be seen that wood is very light but weak, whereas nanotubes are many times stronger than steel and yet much lighter

1.4.3.4 Improving Existing Materials

Nanomaterials can also be used to improve the mechanical properties of existing materials. In this case, nano composites are formed. One example is nano crystalline materials, which are polycrystalline (i.e. made of many crystals which are identical but connected without orientation) and defined as materials with grain sizes from a few nano metres up to 100 nm. In contrast, the grain size in industrial metallic materials is about 10 000 nm or greater. These materials generally show improved mechanical properties (toughness, hardness, etc.).

Within the crystalline structures, there can be defects (cracks or dislocations). If subjected to mechanical stress, the polycrystalline materials can fracture because these defects allow the crack to propagate. To impede the movement of cracks and dislocations tiny particles (nanoparticles) of another material can be added in the lattice.

Nano crystalline materials can have vastly improved mechanical, magnetic, electrical, and catalytic properties and greater corrosion resistance compared to conventional materials with large grains.

Nano crystalline materials are already used in the form of hard coatings to protect tools in industry, such as cutting tools, where they are making a great impact in improving tool performance and lifespan. Nanoparticles added to a ceramic, such as alumina also increases its mechanical properties and the same applies to the addition of nanomaterials (such as carbon nanotubes) to polymers. The possibility of using nanomaterials for improving the mechanical properties of materials involves both inorganic and organic (polymeric) composites.

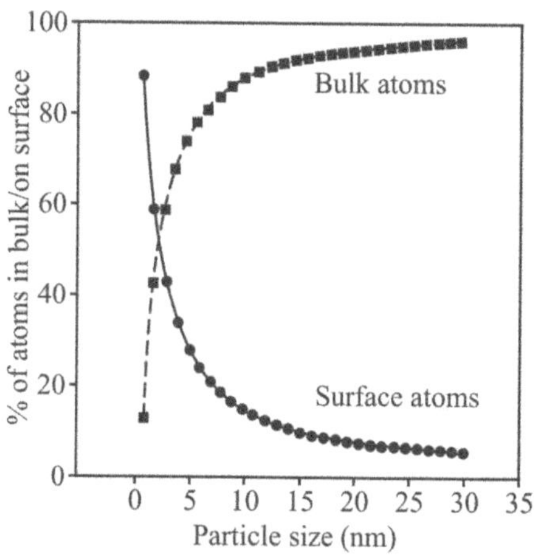

Figure 1.26 Increase in surface area and number of surface atoms with decreasing size of nano particles

1.4.3.5 Increased Surface to Volume Ratio and Enhanced Surface Energy

One of the distinguishing properties of nanomaterials is that they have an increased surface area resulting higher surface energy, large surface active molecules and greater reactivity than corresponding bulk materials.

A greater amount of a substance comes in contact with surrounding material. This results in better catalysts, since a greater proportion of the material is exposed for potential reaction.

The 'active surface' increases when the size of the increased active surface of a nano catalysts is decreased.

The smaller the catalyst particles, the greater the ratio of surface-to-volume (Figure 1.26).

The higher is the catalysts' active surface, the greater is the surface reactivity.

Total Surface area = height × width × number of sides × number of boxes

Total Volume = height × width × length × number of boxes

Surface to volume ratio = total surface area/total volume

Let us consider one Cubic Volume shown in Figure 1.27 its the Surface Area is 6 m^2.

When it is divided into eight pieces its Surface Area becomes 12 m^2, similarly. When the same volume is divided into 27 pieces its Surface Area becomes 18 m^2.

Thus we find that when the given volume is divided into smaller pieces the Surface Area increases. Hence as particle size decreases a greater proportion of atoms are found at the surface compared to those inside (Figure 1.27). Nano particles have a much greater surface area per given volume compared with larger particles. It makes materials more chemically reactive.

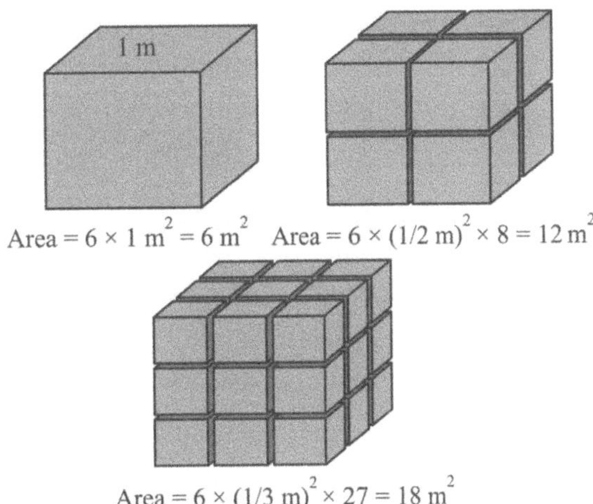

Figure 1.27 Surface area of cubes of 1m,1/2m and 1/3m lengths of each side

This increased reactivity for surface area to volume ratio is widely taken advantage in nature, one biological example being the body's digestive system. Within the small intestine, there are millions of folds and sub folds that increase the surface area of the inner lining of the digestive tract. These folds allow more nutrients and chemicals to be absorbed at the same time, greatly increasing our body's efficiency and the rate at which we digest food.

Thus nano particles, nanostructures and nano materials possess a large fraction of surface atoms per unit volume. The ratio of surface molecules to interior molecules changes dramatically (Table 1.4) if one successively divides a macroscopic object into smaller parts. The total surface energy increases with the overall surface area, which in turn is strongly dependent on the dimension of the material.

Table 1.4 Variation in cross section, total mass, no of molecules and fraction of molecules at the surface of a nano cluster with size (19)

Size (mm)	Cross section (10^{-18} m^2)	Mass (10^{-36} kg)	No. of molecules	Fraction of molecules of surface (%)
0.5	0.2	0.65	1	
1	0.8	5.2	8	100
2	3.2	42	64	90
5	20	650	1000	50
10	80	5200	8000	25
20	320	42000	64000	12

Surface Energy γ is the energy required (or produced) to create a unit area of a new surface, therefore,

$$\gamma = (\partial G/ \partial A)ni, T, P$$

G: Gibs free energy of the surface

A: Surface area of the particles

Due to the dangling bonds on the surface, surface atoms or molecules are under an inwardly directed force and the bond distance between molecules and the sub-surface molecules becomes smaller. An extra force or energy is required to pull the surface to its original positions.

This energy is defined as surface energy, γ

$$\gamma = N_b \varepsilon \rho_a / 2 \quad \text{.....(1.1)}$$

Nb: number of broken bonds

ε: bond strength

ρa = number of atoms or molecules per unit surface area = density

Table 1.5 presents the (typical data of nano $CaCO_3$) how the surface energy increases with decreasing size in nano systems.

Table 1.5 Surface area and surface energy variation with size of nano $CaCO_3$ (20)

Size (nm)	Surface area ($m^2\ mol^{-1}$)	Surface energy ($J\ mol^{-1}$)
1	1.11×10^9	2.55×10^4
2	5.07×10^8	1.17×10^4
5	2.21×10^8	5.09×10^3
10	1.11×10^8	2.55×10^3
20	5.07×10^7	1.17×10^3
10^2	1.11×10^7	2.55×10^2
10^3 (1 μm)	1.11×10^6	2.55×10

1.4.3.5.1 Nano Particle-Nano Particle Interactions

At the nano scale, particle-particle interactions are either dominated by weak Van der Waals forces, stronger polar and electrostatic interactions or covalent interactions.

Depending on the viscosity and polarisability of the fluid, particle aggregation is determined by the interparticle interaction.

By the modification of the surface layer, the tendency of a colloid to coagulate can be enhanced or hindered.

For nano particles suspended in air, charges can be accumulated by physical processes such as glow discharge or photoemission.

In liquids, particle charge can be stabilised by electrochemical processes at surfaces.

Thus nano particle-nano particle interaction forces and nano particle-fluid interactions are of key importance to describe physical and chemical processes, and the temporal evolution of free nanoparticles.

They remain difficult to characterise due to the small amount of molecules involved in the surface active layer.

Both surface energy, charge and solvation are relevant parameters to be considered. Due to the crucial role of the nano particle-nano particle interaction and the nanoparticle-fluid interaction, the term free nanoparticle can be easily misunderstood.

The interaction forces, either attractive or repulsive, crucially determine the fate of individual and collective nanoparticles. This interaction between nanoparticles resulting in aggregates and/or agglomerates may influence on their behaviour.

In gas suspensions, aggregation is crucially determined by the size and diffusion, and coagulation typically occurs faster than in the liquid phase as the sticking coefficient is closer to unity than in liquids

1.4.3.6 Special Optical Properties of Nano Materials

Some nanomaterials display very different optical properties, such as colour and transparency, compared to bulk materials (21).

1.4.3.6.1 Interaction of Light with Matter

The 'colour' of a material is a function of the interaction between the light and the object. In general light (I) incident on a material can be transmitted (T), absorbed (A) or reflected (R):

$$I = T + A + R$$

As the size of the materials is reduced, scattering (S) of light can also contribute to its colour (or transparency).

Reflection occurs when light strikes a smooth surface and the incident wave is directed back into the original medium. The reflected wave will have same geometrical structure as the incident wave.

Absorption (A) is a process that involves energy transformation. The energy levels of a substance determine the wavelengths of light that can be absorbed. It is a molecular phenomenon, dependent on the chemical identity and structure of the substance (not on the size of the molecules or clusters), and involves electronic transitions, vibrations and rotations. Chromophores and fluorephores are examples of organic materials that have specific electronic transitions.

Transmission (T) is the ability of light to pass through a material: it is complementary to absorption.

Transmission of light is what is left after reflection, scattering and absorption have occurred.

Scattering (S) is the phenomenon that occurs when radiation hits a structure with dimensions comparable to the incident wavelength. Therefore, it is a physical process that depends on cluster size, the refractive index of the cluster and the refractive index of the suspension medium. It is a physical interaction only, no energy transformations occur during scattering (as opposed to absorption), energy is simply redirected in many directions. The wavelength of incoming light and that of the outgoing light are the same. After the light hits the clusters in the colloid and is redirected once, it can encounter another cluster and be redirected again. This phenomenon is called multiple scattering. At the 'macro' level, the overall effect can be that light is sent back the way it came (back scattering) or moves forward in the same direction it was moving initially (front scattering). Maximum scattering occurs for wavelengths twice as large as the cluster size. Therefore, if the cluster is about 200 nm, maximum scattering is observed at 400 nm, which lies within the range of the visible spectrum.

The formula indicated above still holds if scattering occurs. Scattering simply contributes to the 'reflection' (back scattering) and 'transmission' (front scattering) parts of the equation. Light that has been absorbed cannot be scattered.

1.4.3.6.2 Colour Generation from Nano Particles and Nanostructures

Nano materials in general can have peculiar optical properties as a result of the way light interacts with their fine nanostructure. One of the distinguishing properties of metal nano particles in general is their optical properties, which are different from those of their bulk counterpart. This is due to an effect called localized surface plasmon resonance.

The free electrons in the metal (d electrons in silver and gold) are free to travel through the material. The mean free path in gold and silver is ~50 nm, therefore in particles smaller than this, no scattering is expected from the bulk. Thus, all interactions are expected to be with the surface. When the wavelength of light is much larger than the nanoparticle size it can set up standing resonance conditions as represented in Figure 1.28.

Light in resonance with the surface plasmon oscillation causes the free-electrons in the metal to oscillate. As the wave front of the light passes, the electron density in the particle is polarized to one surface and oscillates in resonance with the light's frequency causing a standing oscillation. The resonance condition is determined from absorption and scattering spectroscopy and is found to depend on the shape, size, and dielectric constants of both the metal and the surrounding material. This is referred to as the localized surface plasmon resonance(LSPR), since it is located at the surface. As the shape or size of the nanoparticle changes, the surface geometry changes causing a shift in the electric field density on the surface. This causes a change in the oscillation frequency of the electrons, generating different cross-sections for the optical properties including absorption and scattering (Figure 1.29).

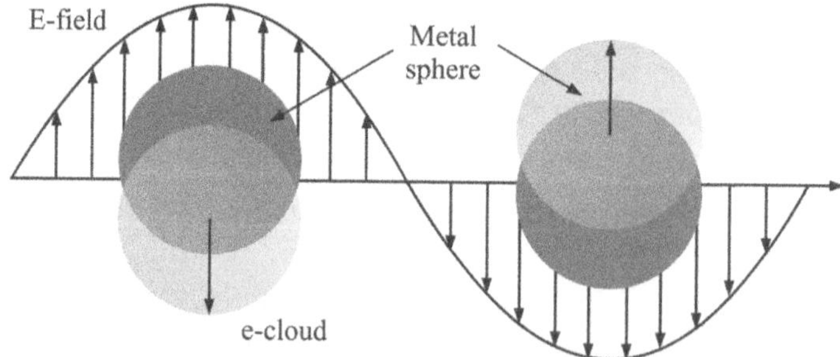

Figure 1.28 Schematic of plasmon oscillation for a sphere, showing the displacement of the conduction electron charge cloud relative to the nuclei (21).

Changing the dielectric constant of the surrounding material will have an effect on the oscillation frequency due to the varying ability of the surface to accommodate electron charge density from the nanoparticles.

Changing the solvent will change the dielectric constant, but the capping material is most important in determining the shift of the plasmon resonance due to the local nature of its effect on the surface of the nanoparticle. Chemically bonded molecules can be detected by the observed change they induce in the electron density on the surface, which results in a shift in the surface plasmon absorption maximum (Figure 1.29). This is the basis for the use of noble metal nano particles as sensitive sensors.

Thus LSPR energy is sensitive to the dielectric function of the material and the surroundings decrease in size and to the shape and size of the nanoparticle. This means that if a ligand, such a protein, attaches to the surface of the metal nanoparticle, its LSPR energy changes. Similarly, the LSPR effect is sensitive to other variations such as the distance between the nanoparticles, which can be changed by the presence of surfactants or ions. The LSPR effect has been observed not only on metal nanoparticles but also in nano rings, voids in metal films and other nanostructures.

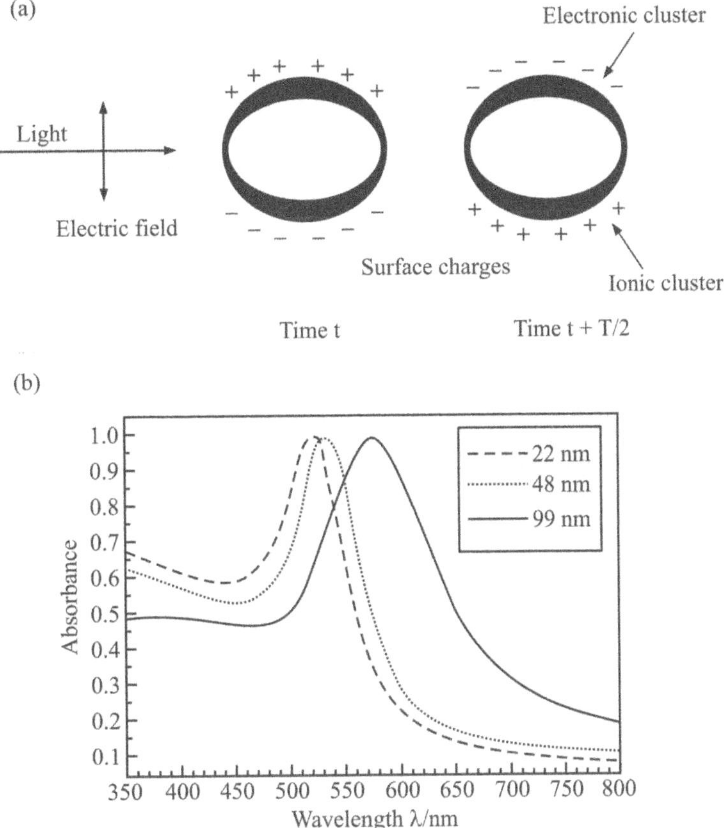

Figure 1.29 LPSR absorption spectra of spherical gold nano particles and their size dependence (21)

One of the consequences of the LSPR effect in metal nanoparticles is that they have very strong visible absorption due to the resonant coherent oscillation of the plasmons. As a result, colloids of metal nanoparticles such as gold can display different colours (Figure 1.30).

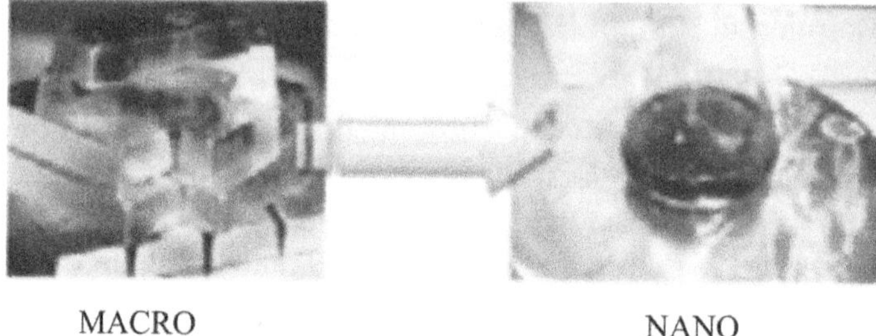

Figure 1.30 (macro (bulk) gold-yellow and nano Gold (1.5 nm) ruby red (22). (Image. LPillppooni iNano Arhus University Creative common Share Alike 3.0)

Figure 1.31 shows how colloids of gold and silver nano particles change their colour with size. The properties of metal nanoparticles make them useful in sensing.

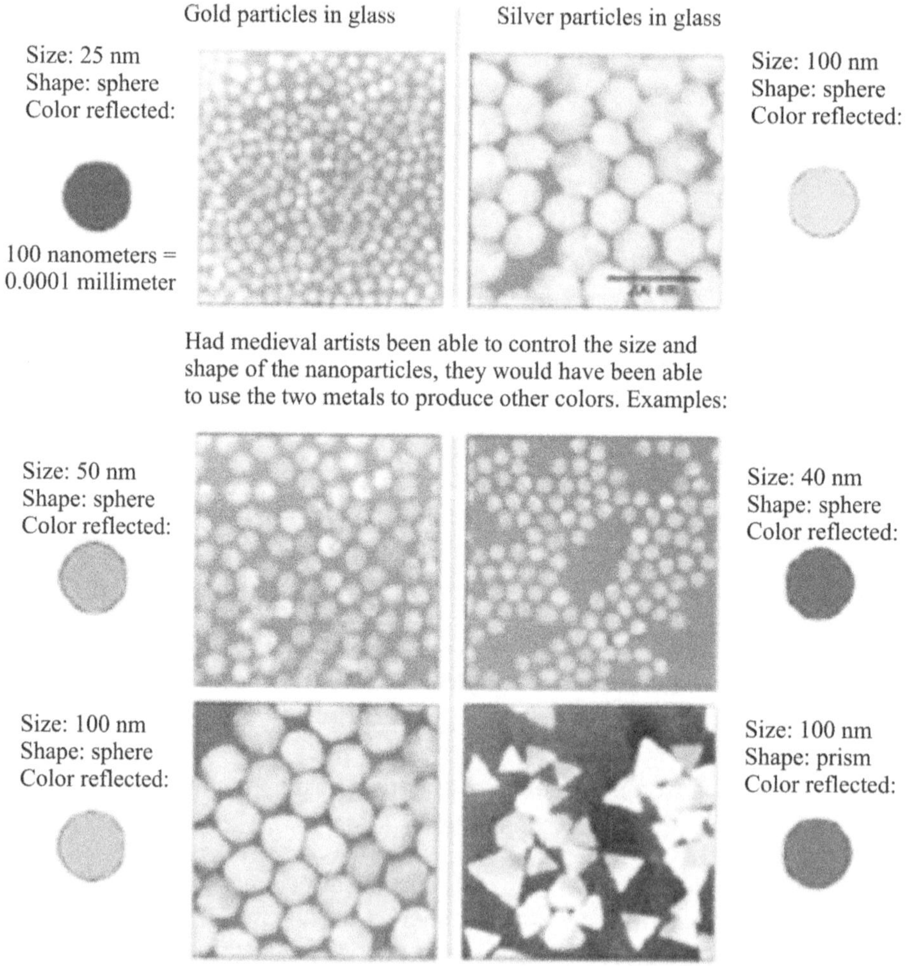

Gold particles in glass

Silver particles in glass

Size: 25 nm
Shape: sphere
Color reflected:

100 nanometers = 0.0001 millimeter

Size: 100 nm
Shape: sphere
Color reflected:

Had medieval artists been able to control the size and shape of the nanoparticles, they would have been able to use the two metals to produce other colors. Examples:

Size: 50 nm
Shape: sphere
Color reflected:

Size: 40 nm
Shape: sphere
Color reflected:

Size: 100 nm
Shape: sphere
Color reflected:

Size: 100 nm
Shape: prism
Color reflected:

Figure 1.31 Changes in colours of colloids of gold and silver with size

1.4.3.6.3 Quantization and Energy Level Spacing

As already mentioned, nano sized semiconductors have quantized energy states; therefore, the conduction and the valence bands split and become discrete. Charge transfer occurs between these discrete levels, so only wavelengths of a certain dimension are absorbed: as a consequence, emission is monochromatic. Quantum confinement causes the energy of the band gap to increase: therefore, more energy is needed in order to be absorbed by the band gap of the material. Higher energy means shorter wavelength (blue shift). The same applies to the wavelength of the fluorescent light emitted from the nano sized material, which will be higher, so the same blue shift will occur. Tuning the size of the semiconductor nanocrystal is a means of tuning the band gap and, therefore, the wavelength absorbed/emitted by the crystal. As a result, the same material (CdSe) emits different colours depending on its size (Figure 1.32). As the size is reduced, the electronic states may be viewed as superpositions of bulk states.

Hence, there is a shift to higher energy, the development of discrete features in the spectra, and concentration of the oscillator strength into just a few transitions. Qualitatively, all of these effects can be readily observed in the spectra of Figure 1.32, which show data for CdSe QDs are currently used as an alternative to conventional dyes in fluorescence microscopy and in other methods where dyes are used (e.g., dye-sensitised solar cells). QDs are also being studied as alternative light-emitting sources

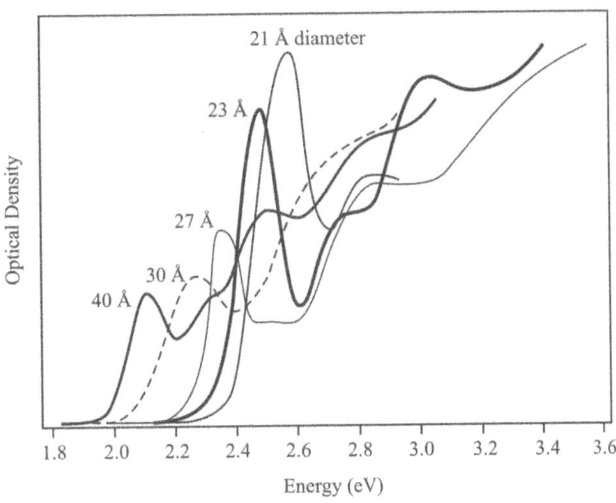

Figure 1.32 Optical absorption vs size for CdSe nano crystals shows the shift to higher energy in smaller sizes, as well as the development of discrete structure in the spectra and the concentration of oscillator strength into just a few transitions (23).

1.4.3.6.4 Electrical and Electronic Properties of Nano Materials

As discussed earlier electrical conductivity decreases with a reduced dimension due to increased surface scattering, while electrical conductivity of nanomaterial could be enhanced appreciably, due to the better ordering of microstructure. Nano structuring of bulk magnetic materials can be used to design the magnetization curve of the material, leading to soft or hard magnets with improved properties.

For metals, the electron mean free path (MFP) determines the thermal and electrical conductivity and affects the color of the metal. For most of the metals, MFP is of the order of 5–50 nm. Reducing further this threshold, the electrons begin to scatter off the crystal surface, and the resistivity of the particles increases. For very small metal particles, the conduction and valence bands begin to break down into discrete levels. For gold particles, this causes a change in color from red to orange at sizes around 1.5 nm.

Depending on their chemical composition optical and electronic properties have generated much attention in studies of properties of particles approaching molecular dimensions.

1.4.3.6.4.1 Electrical Properties of Metal Nano Particles

Electrical conductivity of bulk metals is based on their electronic band structures, and the mobility of electrons is related to their mean free path between two collisions with the lattice. The collective motion of electrons in a bulk metal obeys Ohm's law,

$V = RI$, where, V is the applied voltage, R is the resistance of the material and I is the current. As the electronic band structure changes into discrete energy levels, Ohm's law is no longer valid. If one electron is transferred to a small particle, the Coulomb energy of the latter increases by $EC = e^2/2C$, where C is the capacitance of the particle. If the temperature is low such that $kT < e^2/2C$, single electron tunneling processes are observed. Thermal motion of the atoms in the particle can initiate a change in the charge and the Coulomb energy so that further electrons may tunnel uncontrolled. Hence, the I-V characteristic of a quantum dot is not linear, but staircase-like. No current flows up to $VC = \pm e/2C$. If this value is reached, an electron can be transferred. Following this, an electron tunnelling process occurs if the Coulomb energy of the particle is compensated by an external voltage of $V = \pm ne/2C$. This behaviour is called Coulomb blockade. The charging energy increases with decreasing the size of the quantum dot.

Some nano materials exhibit electrical properties that are absolutely exceptional. Their electrical properties are related to their unique structure. Two of these are fullerenes and carbon nanotubes. For instance, carbon nanotubes can be conductors or semiconductors depending on their nanostructure. Carbon nanotubes are long, thin cylinders of carbon. Their electrical properties change with diameter, "twist", and number of walls. They can be either conducting or semi-conducting in their electrical behavior.

Applications of Special Electrical Properties of Nano Materials

The following are some examples where the special electrical conductivity of nano materials are exploited for micro electronics applications

- The use of carbon nanotubes in semiconductor chips;
- Research into the use of a variety of nanomaterials in lighting technologies (light emitting diodes or LEDs and organic light emitting diodes or OLEDs), with commercial use expected in the near future;

- Use of 'quantum dots' in lasers, along with ongoing research into application of other nano materials in laser technology;
- A variety of nano materials used in lithium-ion batteries, or which are being researched for this use;
- Potential use of carbon nanotubes and other nanomaterials in fuel cells and by the solar industry for use in photovoltaics.
- Research into use of nanomaterials to produce lead-free solder, as well as the development of solder-free assembly technology.

The tremendous efforts and progress have been made in molecular electronics. In molecular electronics single molecules are expected to be able to control electron transport which offers the promise of exploring the molecular functions of electronic devices and molecules can now be crafted into a working circuit as shown in Figure 1.33.

When the molecules are biologically active bioelectronics devices could be developed. In molecular electronics control over the electronic energy levels at the surface of conventional semiconductors and metals is achieved by assembling on the solid surfaces, poorly orgnized. partial monolayers of molecules instead of more commonly used ideal ones. Once those surfaces become interfaces these layers exert electrostatic rather than electro dynamic control over the resulting devices based on both on the electrical monopole and dipole effects of the molecules. Thus electronic transport devices incorporating organic molecules can be constructed without current flow through these molecules.

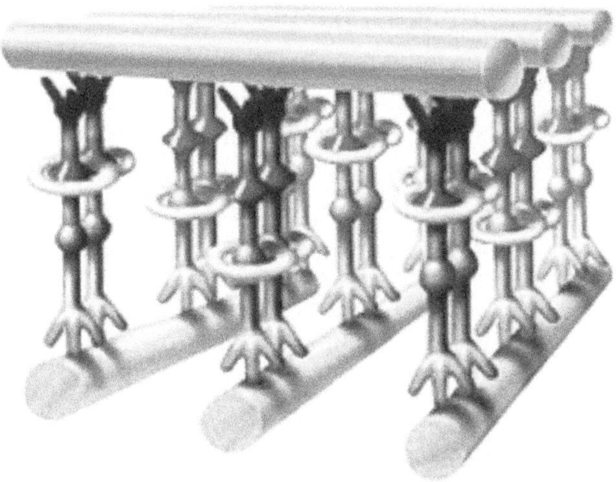

Figure 1.33 Schematic of molecules crafted into working circuit (24)

Nanoelectronic devices have the advantage of having higher integrating densities and also exhibit interesting properties in the radiofrequency region, thereby offering challenges and opportunities alike.

1.4.3.6.5 Magnetic Properties of Nano Particles

Magnetic materials are those that exist in a state of permanent magnetization without the need to apply a field. The strength of magnet is measured in terms of saturation magnetization and coercivity (surface area per unit volume) of the grains. There are three categories of magnetism diamagnetism, paramagnetism, and ferromagnetism. Diamagnetism is a fundamental property of all atoms and the magnetization is very small and opposed to the applied magnetic fields direction. However many materials exhibit para magnetism (In materials with un paired electron) where a magnetization develops parallel to the applied magnetic field as the field increased from zero.

Ferromagnetism is the property of those materials which are intrinsically magnetically ordered which develop spontaneously magnetization without the need to apply field. The ordering mechanism is the quantum mechanical exchange mechanism.

In general the magnetic behavior of a material depends on the structure of the material and on its temperature. In order to 'feel' a magnetic field, a material must have a non-zero net spin (unpaired electrons)(transition metals). The typical size of expected magnetic domains is around 1 μm. When the size of a magnet is reduced, the number of surface atoms becomes an important fraction of the total number of atoms, surface effects become important, and quantum effects start to prevail. When the size of these domains reaches the nanoscale, these materials show new properties due to quantum confinement, for example the giant magneto resistance effect (GMR). This is a fundamental nano effect which is now being used in modern data storage devices.

In magnetic nano particles, the energy of magnetic anisotropy may be that small that the vector of magnetization fluctuates thermally; this is called super paramagnetism. Such a material is free of reminisce, and coercivity. Touching superparamagnetic particles are loosing this special property by interaction, except the particles are kept at distance. Combining particles with high energy of anisotropy with super para magnetic ones leads to a new class of permanent magnetic materials.

Magnetic nano particles (MNPs) are of particular interest in fundamental importance, as well as a strong technological interest, ranging from fields as diverse as medicine applications to potential high density magnetic recording memories, and soil treatments. They exhibit very different macroscopic magnetic properties when compared with their bulk counterpart. Most of these differences are due to the effects of the small scale, leading to a significant increase in the fraction of atoms on the surface, as compared with those in the particle. In the case of MNPs, the effects of mutual magnetic interactions have to be added, in a complex way, to those of the reduced dimensions (25). Magnetic nano particles (MNPs) have potential for use in a number of biomedical applications, ranging from drug localization and magnetic hyperthermia to gene transfection and enhancement of medical images (26).

Magnetic particles have received a considerable attention due to their potential employment in the fields of high-density magnetic storage devices (due to their size and anisotropic behavior, which permits the use of smaller bit size that favors the attainable recording density), contrast enhancers for magnetic resonance imaging, magnetic vectors for cell targeting and drug delivery, wiring materials for audio and radio frequency transformers, high temperature space power systems etc.

Historically, the first magnetic nanoparticles used in magnetic fluids was magnetite (Fe_3O_4 in 1960 by NASA (27, 28). In Chapter VI more details on Ferro fluids are presented.

Different types of magnetic material include some alloys such has FePt, NiPt and NiPd (29,30), which can be used in the magnetic fluid and ferrites: MFe_2O_4 (M = Mn, Ni, Co, Zn, Fe) (31,32,33), γ-Fe_2O_3 (34).

Many magnets made up of nano crystalline yittrium samarium-cobalt grains possess very unusual magnetic properties due to their large surface area. Typical applications of these high power rare earth magnets include power generators in submarines, automobile alternators, Land based power generators and motors for ships, ultra sensitive instruments, MRIs etc.

1.4.3.6.6 Catalytic Properties of Nano Materials

A catalyst is a substance that increases a chemical reaction rate without being consumed or chemically altered. Nature's catalysis are called enzymes and are able to assemble specific end products, always finding pathways by which reactions take place with minimum energy consumption.

Man-made catalysts are not so energy efficient: they are often made of metal particles fixed on an oxide surface, working on a hot reactant stream (to reduce the phenomenon, 'catalyst poisoning', which occurs when species dispersed in the atmosphere, such as CO, occupy the active sites of the catalysts is used).

One of the most important properties of a catalyst is its active surface where the reaction takes place. The 'active surface' increases when the size of the catalysts is decreased: the smaller the catalyst particles, the greater the surface-to-volume ratio.

The higher the catalysts' active surface, the greater the surface reactivity. Research has shown that the spatial organisation of the active sites in a catalyst is also important. Both properties (nanoparticle size and molecular structure/ distribution) can be controlled using nanotechnology. Hence, this technology has great potential to expand catalyst design with benefits for the chemical, petroleum, automotive, pharmaceutical and food industries. The use of nanoparticles that have catalytic properties allows a drastic reduction in the amount of material used, with resulting economic and environmental benefits.

A good example of how nano science can impact the development of catalytic materials is that of gold. Bulk gold noble metal is a stable non toxic and resistant to oxidation and chemical attack.

On the other hand, nanoscale gold particles can catalyse chemical reactions. It has been found that finely dispersed gold nanoparticles on oxide supports are catalytically very active. In many cases the catalytcal activity and selectivity of dispersed gold nanoparticles exceed those of the commonly used transition metal catalysts such as platinum, rhodium palladium. This is an exciting result because metals like platinum and palladium (commonly used in catalysis such as car catalytic converters) are toxic and are also very rare metals, hence very expensive.

Due to their large surface, nanoparticles made of transition element oxides exhibit interesting catalytic properties. In special cases, catalysis may be enhanced and more specific by decorating these particles with gold or platinum clusters.

Due to their enhanced chemical activity, nano structural materials can be used as *catalysts* to react with such noxious and toxic gases as carbon monoxide and nitrogen oxide in automobile catalytic converters and power generation equipment to prevent environmental pollution arising from burning gasoline and coal.

Fuel cell technology is another important application of the noble metal nanoparticles relating the catalysis of the reactions. In the present, the fuel cell catalysts are based on platinum group metals (PGM). Pt and Pt-Ru alloys are some of the most frequently used catalysts from this group. In fact, the use of these metals is one major factor for cell costs, which has been one of the major drawbacks preventing it from growing into a more important technology. One possibility to produce economical catalysts is the use of bimetallic nano particles.

1.4.3.6.7 Nano Biotechnology Applications

Nano biotechnology converges the angstrom, nano and micro scale worlds with the exploitation of physical principles, chemical synthesis capabilities and functional properties of biological nanostructures. Nature has made highly precise and functional nanostructures: DNA, proteins, membranes, filaments and cellular components. These biological nanostructures typically consist of simple molecular building blocks of limited chemical diversity arranged into vast numbers of complex three dimensional architectures and dynamic interaction patterns

A key feature of biological entities is the systematic organization of matter on the nanoscale. Developments in nano science and nano technology would allow us to place man-made nanoscale things inside living cells.

Nano particles may be of the same dimensions as some biological molecules such as proteins and nucleic acids. Many of these biomolecules consist of long macromolecular chains which are folded and shaped by cooperative and weak interaction between side groups, H-bridges and salt bridges. Here, functionalized nanoparticles, such as colloidal gold (35), may intrude into the complex folded structures (36, 37).

Evidence for such interactions is seen from the experience with immune labelling (38) and related surface functionalization techniques to target nanoparticles to biomolecules as markers for high resolution Transmission Electron Microscopy and optical imaging systems. Other nanoparticle systems which are established for research purposes in cell systems include quantum dots (39) and magnetic nano particles (40).

It would also make it possible to make new materials using the self-assembly features of nature. Thus nano bitechnology will be a powerful combination of biology with materials science.

1.4.4 Summary of Importance of Properties of Nano Materials and their Applications

Thus the promise and essence of the nanoscale science and technology is based on the demonstrated fact that materials at the nanoscale have properties (i.e., chemical, electrical, magnetic, mechanical and optical) quite different from the bulk materials as discussed above Some of such properties are, somehow, intermediate between properties of the smallest elements (atoms and molecules) from which they can be composed of, and those of the macroscopic materials. Compared to bulk materials, it is demonstrated that nanoparticles possess enhanced performance properties when they are used in similar applications.

New phenomena are being observed and exploited for various applications and some examples are summarized below.

- Atomic diffusion through interfaces becomes an efficient mechanism of transport of matter at relatively low temperatures comparing to conventional matter. This effect can be used for example to considerably increase the sensitivity of a gas sensor, the kinetics of hydrogen diffusion in a hydrogen storage device, or lower the operation temperature of Solid Oxide Fuel Cells.
- If the crystal size is smaller than the electron mean free path, the electronic conductivity and the temperature coefficient is found to decrease because of grain boundary scattering.
- Phonon spectra are modified due to the effect of surfaces and the small size of particles – so called phonon confinement effects.
- Band gap changes in nano sized semiconductor particles leads to a blue shift of luminescence.
- Size induced control of luminescence relaxation in oxide nanoparticles leads to changes in the optical properties and therefore interesting optoelectronic device applications.
- Reduction of the size of quantum dots to the point where only a few electrons are present in each one is the basic technology for spintronics.
- Surface effects in magnetic materials control the magnetic properties of thin layers, leading to more efficient data storage devices and more sensitive magnetic sensors. Examples are Giant Magnetic Resistance, GMR.
- For metals with grain size less than 100 nm, mechanical properties change strongly because of the contribution of grain boundaries. This mechanism can lead to the development of materials with superior strength and ductility, and thus improved service performance.
- Thermodynamic phase equilibria are shifted or changed due to the contribution of interfaces or interface related strains, to the free energy of the system. This allows production of new non-equilibrium materials which exhibit properties not previously known.

- Tribological properties are changed tremendously when the interacting materials are in the nanoscale. These changes facilitate reduced friction and wear in MEMS applications, microsystems and microsurgical instruments.

Some special applications of nano material properties are presented in Table 1.6.

Table 1.6 Effects of nanomaterials and applications due to the reduced dimension

Effect of Nanoscale	Applications
Higher surface-to-volume ratio, enhanced reactivity	Catalysis, solar cells, batteries, gas sensors
Lower percolation threshold	Conductivity of materials, sensors
Increased hardness/wear resistance with decreasing grain size	Hard coatings, tools, protection layers
Narrower bandgap with decreasing grain size	Opto-electronics
Higher resistivity with decreasing grain size	Electronics, passive components, sensors
Improved atomic transport kinetics	Batteries, hydrogen storage
Lower melting and sintering temperature	Processing of materials, low temperature sintering materials
Improved reliability, fatigue	Electronic components, MEMS

1.5 Current Industrial Applications of Nano Technologies

Presently nanotechnologies became part of convergent new technologies, which offer synergies between nanotechnology, biotechnology, information technology, and cognitive sciences (such as psychology, neuroscience, biology or computer science). Each of these is currently progressing at a rapid rate, experiencing qualitative advancements, and interacting with more established fields such as mathematics, environmental technologies.

Figure 1.34 presents broad areas where nano technologies have great potential for industrial applications.

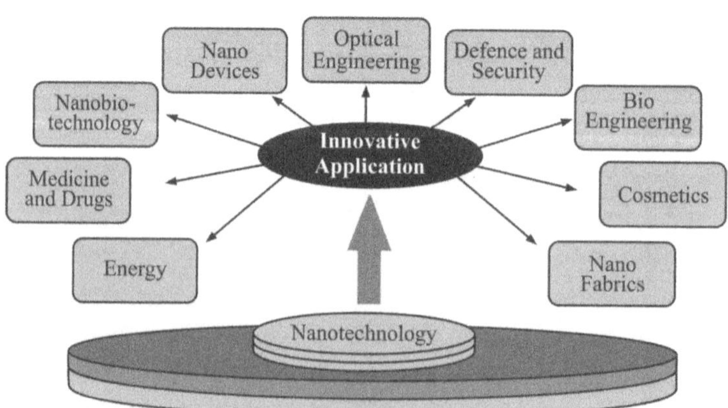

Figure 1.34 Broad areas of industrial applications of nano technologies

As can be seen from Figure 1.34 Nanotechnology is entering into all aspects of science and technology including, but not limited to, aerospace, agriculture, bioengineering, biology, energy, the environment, materials, manufacturing, medicine, military science and technology. It is truly an atomic and molecular manufacturing approach for building chemically and physically stable structures one atom or one molecule at a time. Presently some of the active nanotechnology research areas include nanolithography, nano devices, nano robotics, nano computers, nano powders, nano structured catalysts and nano porous materials, molecular manufacturing, diamond, carbon nanotube and fullerene products, nano layers, molecular nanotechnology, nano medicine, nano biology, organic nanostructures to name a few.

1.5.1 Improving Product Performance

Currently nano technologies are being used for improving the product performance in the following industrial process.

(i) Wear-resistant tires made by combining nanoscale particles of inorganic clays with polymers as well as other nanoparticle reinforced materials,

(ii) Greatly improved printing brought about by nanoscale particles that have the best properties of both dyes and pigments as well as advanced ink jet systems,

(iii) Vastly improved new generation of lasers, magnetic disk heads, nano layers with selective optical barriers and systems on a chip made by controlling layer thickness to better than a nanometer,

(iv) Production of advanced chemical and bio-detectors,

(v) Nano particles to be used in medicine with vastly advanced drug delivery and drug targeting capabilities,

(vi) Chemical-mechanical polishing with nano particle slurries, hard coatings and high hardness cutting tools

(vii) Nanotechnologies currently used by leading businesses and industrial process for a variety of technical and innovative applications. Examples include:

  (a) Exxon Mobil is using zeolites, minerals with pore sizes of less than 1 nm, as a more efficient catalyst to break down or crack large hydrocarbon molecules to form gasoline.

  (b) IBM has added nano scale layering to disk drives, thus exploiting the giant magneto resistive effect to attain highly dense data storage.

  (c) Gilead Sciences is using nanotechnology in the form of lipid spheres, also known as liposomes, which measure about 100 nm in diameter, to encase an anticancer drug to treat the AIDS-related Kaposi's sarcoma.

(d) Carbon Nanotechnologies, a company co-founded by buckyball discoverer Richard E. Smalley, is making carbon nanotubes more affordable by using a new and more efficient manufacturing process.

(e) Nano phase Technologies is utilizing nano crystalline particles, incorporated into other materials, to produce tough ceramics, transparent sun blocks, and catalysts for environmental uses, among other applications.

1.6 Engineering Principles and Initiatives of Nano Technology

1.6.1 Process Engineering, Design and Manufacturing

Nanotechnology is the engineered manipulation of atoms and molecules in a user defined and repeatable manner to build objects with certain desired properties. These nano size building blocks are intermediate systems in size lying between atoms and small molecules and microscopic and macroscopic system. These building blocks contain a limited and countable number of atoms. They can be synthesized and designed atom by atom. They constitute the means of our entry into new realms of nanoscience and nanotechnology. Molecular Building Blocks (MBBs) are distinguished for their unique properties (e.g., graphite, fullerene molecules made of a number of carbon atoms, e.g., C60, C70, C240, etc.

In the past, engineers and material scientists have been involved extensively with the characterization of given materials. Mechanics and materials engineering are really two sides of a coin, closely integrated and related. With the availability of advanced computing and new developments in material science, researchers can now characterize processes and design and manufacture materials with desirable performance and properties. One of the challenges is to model short-term nano/micro-scale material behavior, through meso-scale and macro-scale behavior into long-term structural systems performance. This will provide in better understanding and design of materials and structures across all physical scales, even though the seamless and realistic modeling of different scales from the nano-level to the system integration-level is not yet attainable.

1.6.2 Fabrication Techniques for Engineered Nano Materials

There are a wide variety of techniques that are capable of creating nanostructures with various degrees of quality, speed and cost. These manufacturing approaches fall under two categories: 'bottom-up', and 'top-down'. In recent years the limits of each approach, in terms of feature size and quality that can be achieved, have started to converge. A diagram illustrating some of the types of materials and products that these two approaches are used is shown in Figure 1.35.

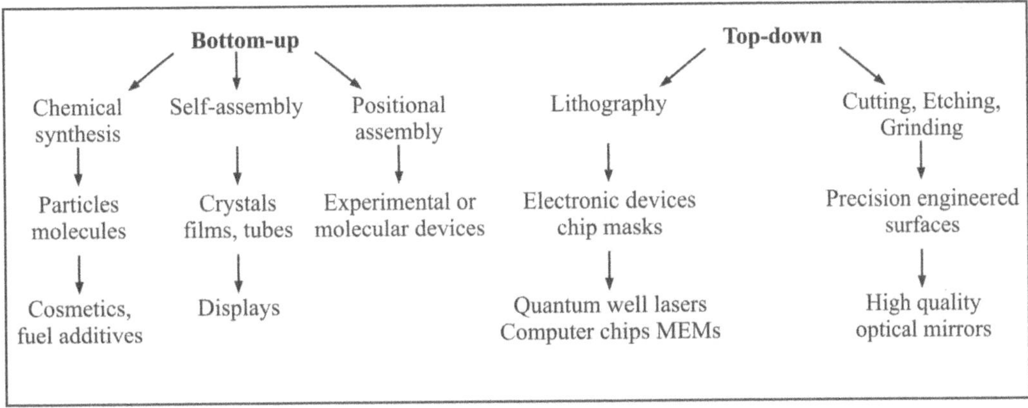

Figure 1.35 The use of bottom-up and top-down techniques in manufacturing

1.6.2.1 Bottom-Up Manufacturing

Bottom-up manufacturing involves the building of structures, atom-by- atom or molecule-by-molecule. The wide variety of approaches towards achieving this goal can be split into three categories: chemical synthesis, self-assembly, and positional assembly. As discussed below, positional assembly (with its many practical drawbacks as a manufacturing tool) is the only technique in which single atoms or molecules can be placed deliberately one-by-one. More typically, large numbers of atoms, molecules or particles are used or created by chemical synthesis, and then arranged through naturally occurring processes into a desired structure.

1.6.2.1.1 Chemical Synthesis

Chemical synthesis is a method of producing raw materials, such as molecules or particles, which can then be used either directly in products in their bulk disordered form, or as the building blocks of more advanced ordered materials, produced using the techniques.

The generic chemical process of nanomaterial manufacture, and applications involve precursors, primary/secondary building units, reactions in liquid/vapour phase, nano material production and characterization (Figure 1.36) in which all general production engineering principles apply.

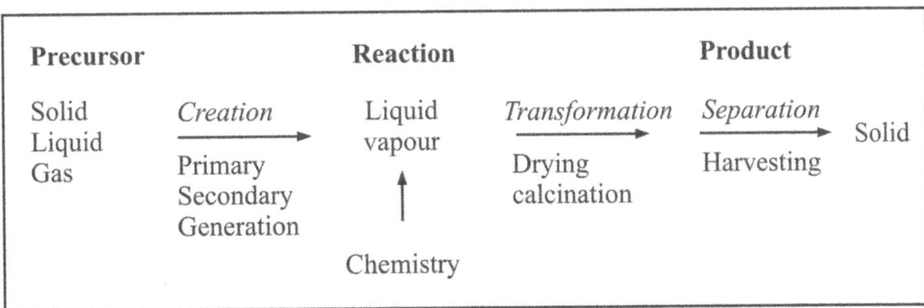

Figure 1.36 The generic processes that are involved in the production of nano particles

1.6.2.1.2 Unit Operations

The precursor phase is the starting point, and the material can be in any physical state (or multiphase) or spatial arrangement to other components. The first step is the creation of a new phase or state where the nanoparticles either form or can be formed by a chemical step. In other words, the phase change itself could bring about nanoparticle formation (rare but possible) although generally the circumstances are created whereby nanoparticles can be made, for example vaporisation of a precursor mixture. Once in a state where nanoparticles can be made, usually a chemical reaction of some description is performed to generate the desired material. A further phase transformation or even solid-state reaction may be necessary to produce the final product. Potential exposure of the workforce to nanoparticles is likely to be greatest when these materials are processed in a gaseous environment; in such cases worker exposure will need to be monitored closely. However, nanoparticles have a tendency to agglomerate, and are therefore often manufactured from a liquid phase as this enables surface energies to be better controlled, reducing agglomeration. This also reduces the potential exposure level of workers. Processing and handling ability is very important for nanomaterials: mixing nanoscale particles together before agglomerating and (for example) sintering can generate wholly new complex nano phase materials which could not be made by any other method. Most genuinely nanoscale and nanostructured materials, however, are still at the laboratory scale of synthesis (kilograms per day scale of operation or even less).

1.6.2.1.3 Resource Management and Environmental Issues

The potential benefits of nanotechnologies should be assessed in terms of life cycle assessment (LCA) (sometimes referred to as 'cradle-to-grave' analysis). LCA is the systematic analysis of the resource usages (for example, energy, water, raw materials) and emissions over the complete supply chain from the 'cradle' of primary resources to the 'grave' of recycling.

1.6.2.1.4 Self Assembly

Self assembly is a bottom-up production technique in which atoms or molecules arrange themselves into ordered nanoscale structures by physical or chemical interactions between the units. The formation of salt crystals and snowflakes, with their intricate structure, are examples of self-assembly processes. Although self assembly has occurred in nature for thousands of years, the use of self assembly in industry is relatively new. There is an economic and environmental interest in processes through which materials or product components essentially form themselves, creating less waste and using less energy. However, current understanding extends only to the creation of fairly rudimentary systems. Improved understanding of thermodynamic and kinetic processes at the nanoscale, enabled through advances in the characterisation techniques, and improved computer modelling, are expected to aid the development of more complex systems. One potential engineered processing technique involves the use of an external force or field (for example, electric or magnetic) to accelerate the often slow self-assembly process, which is attractive in an industrial context. This is known as directed self assembly.

1.6.2.1.5 Positional Assembly

The final bottom-up technique is positional assembly, whereby atoms, molecules or clusters are deliberately manipulated and positioned one-by-one. Techniques such as SPM for work on surfaces, or optical tweezers in free space, are used for this. Positional assembly is extremely laborious and is currently not suitable as an atomic-scale industrial process. The utility and strength of SPM in industry lie in their ability to characterise and measure surfaces with atomic-level precision, rather than as fabrication tools.

The fact that very rudimentary structures can be fabricated atom-by-atom has lead to speculation that tiny nanoscale machines could be made which could be used in parallel to manufacture materials atom-by-atom. The idea is to fabricate one or a few machines (or assemblers) that would first make copies of themselves, and then go on to make materials in parallel, in principle solving the problem of slow production speed.

1.6.2.2 Top-down Manufacturing

Top-down manufacturing involves starting with a larger piece of material and etching, milling or machining a nanostructure from it by removing material (as, for example, in circuits on microchips). This can be done by using techniques such as precision engineering and lithography, and has been developed and refined by the semiconductor industry over the past 30 years. These methods are discussed in detail in Chapter 4.

Top-down methods offer reliability and device complexity, although they are generally higher in energy usage, and produce more waste than bottom-up methods. The production of computer chips, for example, is not yet possible through bottom-up methods; however, techniques using bottom-up (or hybrid top-down/bottom-up) methods are under exploration.

1.6.3 Precision Engineering

In general, ultra-precision engineering and manufacture underpin much of the micro-electronics industry in everything from the production of the flat low-damage semiconductor wafers used as substrates for computer chips, to the mechanical stages used to position the wafers, to the manufacture of the precision optics used to print the patterns on the wafers. In addition, the techniques of ultra-precision engineering are used in a variety of consumer products such as computer hard disks, CD and DVD players.

Ultra-precision machine tools can now achieve very high performance in terms of both the accuracy with which form can be defined (up to 1 part in 10^7, or better than 100 nm over distances of tens of centimeters) and the surface finishes that can be achieved (0.5-1 nm root mean square surface roughness), although these are currently on simple shape surfaces and with low output levels. This capability, which is bringing benefits in several areas has been achieved through a combination of advances. These include: the use of advanced materials for cutting tools, based on diamond or cubic boron nitride; very stiff, precise machine tool structures; new linear and rotary bearing designs employing fluid films; and sensors for size control combined with numerical control and advanced servo-drive technologies. Very precise process and temperature control is needed to achieve this performance (the latter being of the order of $\pm 0.01°C$).

1.6.3.1 Lithography

Manufacturing in the Integrated Chip Technology (ICT) sector predominantly involves lithographic processes that pattern a semiconductor wafer in a sequence of fabrication steps.

Lithography involves the patterning of a surface through exposure to light, ions or electrons, and then subsequent etching and/or deposition of material on to that surface to produce the desired device. (See Chapter 4). The ability to pattern features in the nano metre range is fundamental to the success of the IT industry and the ITRS roadmap. The main lithographic tools can be conveniently separated into methods that use a focused beam of electrons or ions to write patterns, and those that rely on the projection of light through a mask to define a pattern over a complete semiconductor wafer. Electron- and ion-based methods are both capable of making sub-10 nm structures (with electron beam lithography having the greatest routine resolution), but they are too slow to be used directly in production. Optical lithography is used for production of semiconductor devices. Although it does not have the resolution of the beam-based techniques, it provides rapid throughput and cost-effective manufacture. Electron beam lithography is primarily used to fabricate the masks used for optical lithography, and ion beam techniques are mostly used to repair masks and for specialist device applications.

The requirement for ever-shrinking device structures has placed enormous technical demands on optical lithographic process, as the nanostructures have length scales similar to or less than the wavelength of the illuminating light (ultraviolet). Despite these difficulties, the ITRS roadmap implicitly expects optical lithography to keep track of future device dimensions until 2016 when the target critical device dimension reaches 22 nm.

Techniques developed in the microelectronics industry have also enabled the miniaturisation of small mechanical moving devices (MEMS), which in turn have lead to research into NEMS. MEMS technology seeks to exploit and extend the capabilities that have been provided by silicon integrated circuit manufacturing from one of making chips for electronic signal processing to the provision of on-chip sensing and/or actuation through the use of moving mechanical parts. Some MEMS technologies are starting to attain maturity in manufacture (for example, MEMS accelerometers are used widely in air-bag sensors). However, there are currently difficulties in the reproducible large-scale manufacture of more complex MEMS systems. Although not strictly a 'nanotechnology' as defined in this report, MEMS, NEMS and the technologies used to make them are used extensively in techniques that can access and exploit the nanoscale (such as SPMs or lab-on-a-chip and bio sensing). The reducing dimensional tolerances (less than 100 nm) being provided by modern lithographic patterning techniques are now enabling the production of structures of such small dimensions that they are becoming a legitimate part of nanotechnologies in their own right.

1.6.4 Convergence of Top-Down and Bottom-up Techniques

The relationship between top-down and bottom-up manufacturing is illustrated in Figure 1.37. The 'top-down' section is an updated version of the diagram produced by Norio Taniguchi. The 'bottom up' section illustrates how bottom-up processes have evolved to control ever-larger structures through advances in chemical processing. Now the dimensions that can be controlled by either approach are of a similar order, and this is leading to exciting new hybrid methods of manufacture.

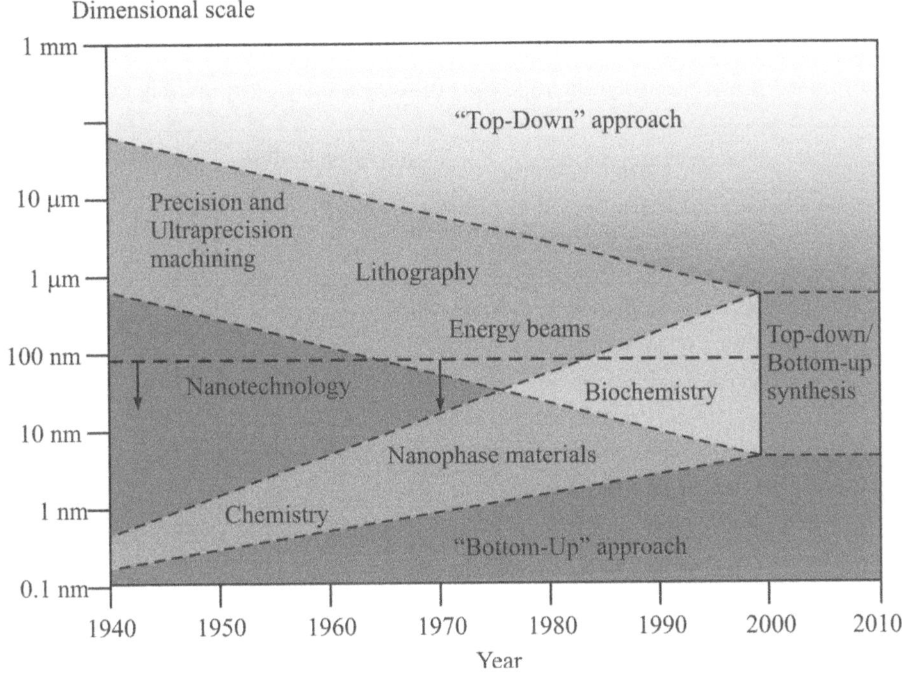

Figure 1.37 The convergence of top-down and bottom-up production techniques (41)

1.6.5 Characterisation

The characterization of materials like the determination of their shape, size, distribution, mechanical and chemical properties is an important part of the industrial process. It serves two broad purposes: as quality control, and as part of the research and development of new processes, materials and products. Nanotechnology 'breakthrough' had occurred in the tools used to observe and measure properties and processes at the nanoscale level. Sophisticated tools, such as the STM, AFM and TEM, enable surface and interfacial characterisation of materials at the nanoscale, allowing individual atoms to be observed and analysed. (These characterization techniques are discussed in Chapter 5). This will lead to greater understanding of the relationship between form and material properties, and enabling the control of processes at the nanoscale and the design materials with specific properties. However, the commercialisation of such advanced functional materials requires that they can be made in a predictable, reliable way, and in sufficient quantities. Until that is achieved production will be limited to academia and R&D departments within industry.

Table 1.7 gives estimates of current and future production of nanomaterials. Metal oxides, such as titanium dioxide, zinc oxide, silicon dioxide, aluminium oxide, zirconia and iron oxide, are currently the most commercially important nano particles.

Table 1.7 Estimated global production rates for various nanomaterials and devices based on international chemical journals and reviews and market research (BCC 2001). These rates are intended for guidance only, as validated numbers are commercially confidential.

Application	Material/device	Production 2005-2010	Projected (tones/annum) 2011-2020
Structural applications	Ceramics, catalysts, composites, coatings, thin films, powders, metals	10^3	10^4-10^5
Skincare products	Metal oxides (titanium dioxide, zinc oxide, iron oxide)	10^3	10^3 or more
ICT	Single wall nanotubes, nano electronics, opto-electro materials (titanium dioxide, zinc oxide, iron oxide), organic light-emitting diodes (OLEDs)	10^3	10^3 or more
Biotechnology	Nanoencapsulates, targeted drug delivery, bio-compatible, quantum dots, composites, biosensors	1	10
Instruments, sensors, characterization	MEMS, NEMS, SPM, dip-pen lithography direct write tools	10^2	10^2-10^3
Environmental	Nanofiltration, membranes	10^2	10^3-10^4

1.7 Future Nano Technologies

1.7.1 Precision Engineering

There are strong drivers to reduce tolerances in engineering, including miniaturisation, improved wear and reliability characteristics, automated assembly and greater interchangeability, reduced waste and requirement for re-work. As the trend towards miniaturisation continues, research and the industrial application of energy beam processing methods will increase, driven in particular by the electronics and computer industries. Techniques such as electron beam lithography (EBL), focused ion beam (FIB), reactive ion etching (RIE) and femtosecond pulsed laser ablation are becoming more accurate and cheaper to apply in a production context. Some examples of future applications of high-precision engineering are given below.

1.7.2 Integrated Chip Technologies (ICT)

The machines used to fabricate chips depend fundamentally upon the use of ultra-high precision techniques for their manufacture and nano metrology techniques for their operation. The manufacture of larger-diameter semiconductor wafers with improved flatness and reduced sub-surface damage should lead to improved device yields and reduced costs.

1.7.3 Optics

Innovative ductile-mode grinding processes, together with electrolytic in-process dressing (ELID), should result in the elimination of polishing in the production of high-quality optical devices. This is likely to be of particular importance in the production of the optics for extra-large astronomical telescopes such as the proposed 50 m and 100 m systems (Euro50 and OWL), which will consist of many individually figured segments (Shore et al 2003).

1.7.4 Transport

Precision-machined parts should be more reliable, because of reduced wear, requiring fewer replacement parts and less energy consumption. For example, the ability to produce surfaces with controlled textures through finishing to 10 nm average roughness followed by laser surface treatment is expected to lead to improved power transmission trains with losses through slip reduced by up to 50%. Precision manufacturing is predicted to lead to weight reductions in airframe wings and to improve the performance of internal combustion chambers.

1.7.5 Medical

It is hoped that the use of ultra-precision machining techniques to produce improved surface finishes on prosthetic implants should lead to lower wear and better reliability.

It is hoped that advances in precision engineering will enable the reduction of environmental impacts by, for example, reducing the use of lubricants.

1.7.6 The Chemical Industry

Long-term goal within the chemical industry is to use nanoscale 'building blocks' to assemble organised nano structures, that can in turn be manufactured into commercially useful products. From an understanding of the chemistry and physics of nanoscale materials, and top-down/bottom-up modelling and measurement, industry will concentrate on processes that use manufacturing at the nanoscale in a way that preserves the desired effect and function as nanoscale components are combined into macroscale materials and products. This will involve the development of technologies based on self-assembling materials, or more probably on directed-assembly methods, which allow for some form of massively parallel production, along with modelling and measuring tools. The vision is the manufacture of reproducible, accurate and designable nanomaterials.

The time-scale for the commercial exploitation of these types of highly organised structures or quantum materials is approximately 2020 and beyond, for use in the biotechnology and IT sectors. These materials will be extremely valuable, in excess of $1,000,000 per tonne, with the production rates of the order of 10–10,000 tonnes a year. The price is expected to remain relatively high because, though the effect of the nanomaterial will be to add value to consumer products, it will only form a tiny fraction of the final product as sold.

The desired functionality is created through exploiting structure – property relationships. Measurement, modelling and simulation are essential for the characterisation and subsequent control of property and functional performance and therefore the production of desirable materials. The development of measurement tools for use at the nanoscale will move from laboratory-based characterisation to in-line and on-line methods of monitoring and controlling accuracy at the 6-sigma level (99.9997% accurate) in terms of reproducible structure, texture and surface properties. The use of computer simulation based on advanced structure – property – process predictor codes will become the key technology for manufacture-by-design, where the characteristics of the material are effectively 'dialled up' through morphology, texture, structure and reactivity based on the interaction of materials across molecular- and nano-length scales. The structure or form of the material then dictates the processing options for economic, reliable and reproducible operation. The combination of measurement, modelling and manufacturing technologies will be the basis for intelligent material systems. It is also hoped that it will be possible to produce materials with less waste.

The synthesis and control of micro- and nano scale structures may yield unprecedented control of meso-and macro-scale properties in functional materials for use in applications of direct relevance to industry. It has been predicted (Chemical Industry 2003) that over the course of the century, many of the needs of commerce and society may be satisfied through a materials revolution involving synthesis and smart fabrication.

1.7.7 The Information and Communication Technology Industry

Although the future of device fabrication is still centred around the lithographic processes there are other techniques that are increasingly being applied both to on-roadmap developments and to alternative approaches to device materials. Soft lithography techniques where a flexible master is used to stamp out patterns on a range of surfaces have been available for several years. The accuracy demands imposed by the silicon-based industry have, until now, prohibited the use of soft lithographies as the elastic nature of the stamp can cause small, but still unacceptable, physical distortions across a wafer surface. However, for small-area device fabrication and for applications where spatial tolerances are less restrictive, they offer a real alternative to conventional methods, although the fabrication of the master still requires optical or electron beam methods. Soft lithographies can be used for plastic electronics, as can alternative ink-jet based methods which use essentially the same technology as desk-top printers. Although plastic electronics are not truly in the nano-range in terms of critical dimensions, a relatively simple manufacturing technique that can deal with wet chemistries will enable cheap electronic and photonic devices. Such developments, combined with advances in directed self-assembly, may bring the semiconductor, materials and chemical industries closer together, in order to create novel alternative methods for chip production as the end of the roadmap approaches.

1.7.8 Molecular Manufacturing

Molecular manufacturing systems will develop with precision on the atomic scale. These systems will expand our control of systems from the macro to the micro and beyond, while simultaneously reducing the cost associated with manufacturing products. Thus gaining molecular-level control over the structure of matter will bring a wide variety of positive applications.

Molecular machine systems will be able to sense and rearrange patterns of molecules in the human body, providing the tools needed to bring about a state of health, regardless of a disease's cause. Theorists envision creating machines that will be able to travel through the circulatory system, cleaning the arteries as they go; sending out troops to track down and destroy cancer cells and tumors; or repairing injured tissue at the site of the wound, even to the point of replacing missing limbs or damaged organs. The extent of medical repair systems is expected to be quite broad, with the cumulative impact being equally large.

1.8 Safety and Health Risks of Nano Materials

1.8.1 Potential Areas for Concern

Some nano particles, nano spheres, nano tubes, and nano fibers produced via nanotechnologies including adventitious by-products have the potential to raise concerns for humans (public health, consumer safety, and the health and safety of workers) and the environment. As nanomaterials, have a size comparable to biomolecules (e.g. proteins, DNA) there are a number of apprehensions that they may interact with biomolecules in an adverse manner, triggering a toxic effect by passing through protection barriers in cells .In medical applications nanomaterials are used precisely to target infected cells and deliver drug agents locally. They are designed to pass though cell membranes. Further we need to understand

what is their impact on environment ?

What happens when materials containing nanoparticles reach landfills and degrade?

Could this cause harm to ecosystems?

The concerns that nano particles, nano tubes, and nano fibers raise constitute the most significant ones relating to nanotechnologies within the next 3–5 years. They require further studies. In this respect, more often the toxicological, ecotoxicological, and exposure data needed to perform a complete risk analysis which are lacking.

Experts are of the unanimous opinion that the adverse effects of nanoparticles cannot be predicted (or derived) from the known toxicity of material of macroscopic size, which obey the laws of classical physics. This has led the UK Royal Society and the Royal Academy of Engineering to recommend "that chemicals in the form of nanoparticles or nanotubes be treated as new substances under the existing Notification of New Substances (NONS) regulations and in the Registration, Evaluation, Authorisation and Restriction of Chemicals (REACH)" (Chapter 8 "Regulatory issues", Point 18, p. 71.

1.8.2 Environmental Effects of Nanotechnology

Nanotechnology have a number of beneficial environmental effects. On a broad scale, as the technology develops, it is likely the production of green house gases will decline which may lead to an abatement of the global warming process. Because nanotechnology employs a bottom up manufacturing process, there will be a decrease in industrial waste, both in terms of wasted energy and solid waste production.

Environmental Risks

The knowledge of the risks of the particles to the flora and fauna is sparse. The application of nanoparticles to the soil or water sources, for example as fertilizers, may hold many unknown risks. We also know that as surface area increases, the rates of chemical reactions will also increase such that substances which are normally inert in normal forms become explosive as finely divided powders. How will the increased reactivity of nanomaterials affect the environment?

Global warming, not surprisingly, is a major consideration in new technologies.

Nanotechnology may allow the manufacture of materials with less energy consumption and fewer waste gases, but the effect on global warming is yet to be determined. Current evidence suggests that nano technology may cause either warming or cooling of the atmosphere.

1.8.3 Nano Particles Impact on Human Health

Nanoparticles are produced by aeroplane and motor vehicle exhaust emissions, erosion of man-made materials (e.g., tyres), as well as by natural erosion and volcanic activity. Thus humans are exposed to small particles, often in very high concentration, and the mechanisms evolved for defence against microorganisms are also used to defend the body against such particles. So to understand the potential risks to humans from nanoparticles, it is necessary first to consider briefly the body's defenses against particles in general and the properties that particles require to overcome these defences. Access to the human body can occur through the lungs, the skin or the intestinal tract. Each organ presents a barrier to penetration by microorganisms or other particles. Nevertheless, despite the defense mechanisms, certain particles have proved to have toxic effects on humans, just as have certain microorganisms. In general this is a consequence of properties that either allow them to evade or cause damage to defensive mechanisms. An understanding of these mechanisms is of importance to estimate the possible toxic effects of nanoparticles or nanotubes. Three types of particle in particular have provided relevant information: the minerals quartz and asbestos, and the particles associated with air pollution. There is some evidence that established protective measures against ultrafine particles would also be effective against engineered nano particles (ENPs), should these be classified as hazardous. So the question becomes: Are ENPs new hazardous materials and, if so, do these pose a risk for humans and/or the environment. But the risk associated with any material depends on the **exposure route and dose**, Some possible exposure routes for human exposure to nano particles are depicted in Figure 1.38.

In the last few years, a wealth of information has been reported by authoritative research groups on toxicity of ENPs though on animal models or *in vitro*. Another problem is that different testing methods are used in different laboratories, making it difficult to compare results. Research so far has mainly focused on two groups of materials: carbon-based nanomaterials (carbon nanotubes and fullerenes) and metal or metal-oxide nanoparticles (e.g. ultrafine titanium dioxide, TiO_2). Several studies seem to indicate that some forms of carbon nanotubes show pulmonary toxicity and that this depends on the production method and the length and surface properties of the carbon tubes. Similarly, TiO_2 has been reported to cause inflammation in the lungs when inhaled in high doses. The small size of nanoparticles ensures that a high proportion inhaled from the air reaches and is deposited in the deep lung. The size of nanoparticles appears to influence their uptake into cells. Specialized phagocytic cells such as tissue macrophages and leukocytes in the blood generally take up larger particles. This is a mechanism evolved in higher animals for removal of potentially harmful bacteria, and is analogous to the feeding method of unicellular organisms. Nanoparticles in contrast, because of their small size, may pass into cells directly through the cell membrane with the possibility of interfering with important cell functions such as motility and ability to remove bacteria (42).

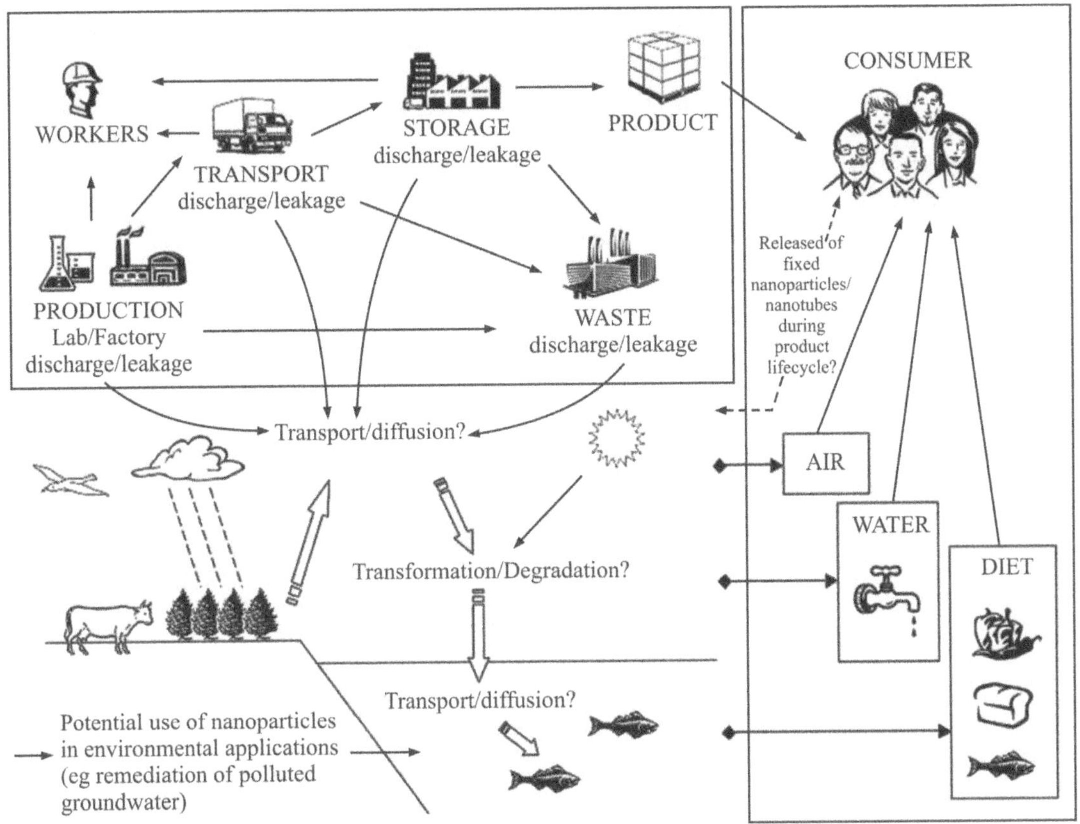

Figure 1.38 Some possible exposure routes for nano particles and nano tubes based on current and potential future applications. (Adapted from National Institute for Resources and Environment, Japan *http://www.nire.go.jp/eco_tec_e/hyouka_e.htm*

Small size alone is not the critical factor in the toxicity of nanoparticles; the overall number and thus the total surface area (essentially the dose) are also important. Although nanoparticles may imply a toxic threat because of their small size and therefore large surface area per unit mass, any toxicity would be expected to be dependent upon inhalation or absorption into the body of a very large number. For nanoparticles with low surface reactivity, potential toxicity to humans and, presumably, other animals, should be considered in relation to the likely dose and route of exposure. From the point of view of the lung, inhalation of small numbers of particles is unlikely to represent a significant risk. Inhalation of very large numbers, as may occur in a manufacturing process, should be controlled by regulation. For specific types of nanoparticle that may be expected to have a more reactive surface, perhaps because of a higher proportion or different combination of transition metals, greater caution would be advised and exposure should be minimised. Toxicological studies, both *in vitro* and *in vivo*, will be required for the investigation of any such substances to which people might be exposed.

Only a few types of manufactured nano particle such as titanium dioxide, carbon black, zinc oxide and iron oxide are currently in industrial production. There is, however, a realistic prospect of industrial production of other nanoparticles for therapeutic or diagnostic use, often based on metals with chemical coatings that confer particular properties with respect to uptake into the body, across natural tissue barriers, and into cells (43). There is also considerable interest in the possible production of nanotubes: some pilot manufacturing plants exist for CNTs, and nanotubes made of carbon and other elements are being extensively investigated in research laboratories. The greatest potential for exposure therefore over the next few years will be in the workplace, both in industry and in universities.

Overall, the scientific community agrees that progress has been made in the toxicological evaluation of nanomaterials. There is still much research to be done, but some key matrices have been identified. For example that surface area is a more important parameter than mass when dealing with engineered nanoparticles, and some targets and common behaviours have been also identified. The question is now how to make a risk assessment framework from these data, how to convert scattered numbers collected in numerous laboratories around the world into a risk management strategy for the safe handling of nanomaterials.

Research into the potential toxic effects of nanomaterials is now a priority in most funding institutions and agencies, as it is clear that the success of nanotechnologies will also depend on how the issue of safety is handled. Further it is also necessary to investigate the social and ethical issues expected to arise from the development of some nanotechnologies.

References

1. http//www.nano/html/facts/the scale of things.html
2. NSTC (2000). National Nanotechnology Initiative: Leading to the Next Industrial Revolution, A Report by the Interagency Working Group on Nano science, Engineering and Technology Committee on Technology, National Science and Technology Council, Washington, D.C. February.
3. M. C. Roco, S. Williams and P. Alivisatos (Editors) (1999). Nanotechnology Research Directions: IWGN Workshop Report - Vision for Nanotechnology R & D in the Next Decade, WTEC, Loyola College in Maryland, September.
4. W. Heisenberg (1958). Physics and Philosophy, Harper and Row, New York, NY.
5. Razavy and Mohsen (2003). Quantum Theory of Tunneling. World Scientific. 4, 462.
6. W Barthlott and C Neinhuis – Planta (1997). Purity of the sacred lotus, or escape from contamination in biological surfaces.

7. Hansen et al. (2005). Proceedings of the National Academy of Science, National Academy of Science, USA. 102 (2), 386-9.
8. A. Kumar et. al (2000). Nature, 405, 681.
9. H.Gleiter (2000). Nanostructured materials: basic concepts and microstructures// Acta Metallurgica, 48, N1, 1-29.
10. R. P. Feynman (1960). There's plenty of room at the bottom - An invitation to enter a new field of physics, Engineering and Science Magazine of Cal. Inst. of Tech., 23, 22.
11. N. Taniguchi (1974). On the Basic Concept of Nano-Technology, Proc. Intl. Conf. Prod. Eng. Tokyo, Part II, Japan Society of Precision Engineering.
12. K. E. Drexler (1992). Nanosystems: Molecular Machinery, Manufacturing, and Computation. John Wiley & Sons, Inc., New York, NY.
13. H. W. Kroto, J. R. Heath, S. C. O'Brien, R. F. Curl and R. E. Smalley (1985). Nature 318, 165.
14. http://serc.carleton.edu/usingdata/nasaimages/index4.html
15. Nano-Scale materials in Chemistry, Wiley 2001.
16. W.H. Qi*, M.P (2004). Wang Size and shape dependent melting temperature of metallic nanoparticles.
17. http://www.physics.hku.hk/~nature/CD/regular_e/lectures/images/chap04/newtonlaw.jpg
18. http://www.antonineeducation.co.uk/Physics_AS/Module_1/Topic_5/em_force.jpg
19. Nanoscience Research Group University of North Carolina (USA) http://www.physics.unc.edu/~rsuper/research
20. Advanced nanomaterials –H.Hofman IMX-HFL Version Sept 2009.
21. K. Lance Kelly, E. Coronado, L.L. Zhao and G.C. Schatz (2003). The Optical Properties of Metal Nanoparticles: The Influence of Size, Shape, and Dielectric Environment, Journal of Physical Chemistry.
22. L Pillppooni. iNano Arhus University Creative common Share Alike 3.0
23. A.P. Alivisatos (1996). Perspectives on the physical chemistry of semiconductor nanocrystals, Journal of Physical Chemistry.
24. R.F. Service Science (2001), 293.
25. J.Garcıa-Otero, M.Parto, J.Rivas and A.Bunde (2000). Phy. Rev. Lett., 84, 167.
26. C.S. Brazel (2009). Pharm. Res. 26, 644.
27. Raul Valenzuela, Frederic Herbst, Souad Ammar and J. Magn. (2012). Mater, 324, 3398.
28. X.T.U. Jeong, Y. Wang, H. Yang and Y. Xia (2007). Adv. Materal. 19, 33.
29. M.A. Zalich (2005). Physical Properties of Magnetic Macromolecule-Metal and Macromolecule-Metal Oxide Nanoparticle Complexes, Virginia Polytechnic Institute and State University.
30. C.S.S.R. Kumar and F. Mohammad (2011). Advanced Drug Delivery Rev.
31. N.T.K. Thanh and L.A.W. (2010). Nano Today 5, 213.
32. F. Montagne, O. Mondain-Monval, C. Pichot, H. Mozzanega and A. ElaIssari (2002). J Magn. Magn. Mater 250, 302.
33. H. Shokrollahi and K. Janghorban (2007). Mater. Sci. Engin. B, 141, 91.

34. S. Sun, H. Zeng, D.B. Robinson, S. Raoux, P.M. Rice, S.X. Wang, G. Li, and J. Ameri (2003). Chem. Soci., 126, 273.
35. Hayatt MA (Ed) (1989). Colloidal Gold, Principles, Methods and Applications. San Diego: Academic Press.
36. Cheng N, Conway JF, Watts NR, Hainfeld JF, Joshi V, Powell RD, Stahl SJ and Wingfield.
37. PE and Steven AC (1999). Tetrairidium, a four-atom cluster is readily visible as a density label in three-dimensional cryo-EM maps of proteins at 10-25 angstrom resolution. J Structural Biology, 127, 169-176.
38. Hainfeld JF and Powell RD (2000). New frontiers in gold labelling, J Histochem & Cytochem, 48, 471-480.
39. Romano EL and Romano M (1977). Staphylococcal protein-a bound to colloidal gold - useful reagent to label antigen-antibody sites in electron-microscopy. Immunochemistry, 14, 711-715.
40. Chan WCW and Nie SM (1998). Quantum dot bioconjugates for ultrasensitive nonisotopic detection. Science, 281 (5385), 2016-2018.
41. Nanotechnology: big prospects for Small Engineering; Roger W. What more FR Eng; Ingenia, Issue, 9. August 2001.
42. Renwick L C, Donaldson K and Clouter A (2001). Impairment of alveolar macrophage phagocytosis by ultrafine particles. Toxicology and Applied Pharmacology 172, 119-127.
43. Borm P J A and Kreyling W (2004). Toxicological hazards of inhaled nanoparticles - potential implications for drug delivery. Journal of Nanoscience and Nanotechnology 4, 1-11.

Recommended Reference Materials Books

1. Introduction to Nanoscience S. M. Lindsay, *Oxford*, 2010.
2. Nanotechnology: Understanding small systems: Rogers, Pennathur & Adams, *CRC press,* 2008.
3. Introductory Nanoscience: Masaru Kuno, *Garland Science,* 2011.
4. Quantum Mechanics for Nanostructures: Vladimir V. Mitin, Dimitry I. Sementsov & Nizami Z. Vagidov, *Cambridge*, 2010.
5. Nanophysics and Nanotechnology: An introduction to the modern concepts in Nanoscience: Edward L. Wolf, *Wiley-VCH*, 2011.
6. Introductory Quantum Mechanics: Richard L. Liboff, *Addison-Wesley*, 1993.
7. Foundations of Nano mechanics: A. N. Cleland, *Springer*, 2003.

Web Sites

Paper by R. Feynman entitled 'There's plenty of room at the bottom':

http://www.zyvex.com/nanotech/feynman.html

Institute of nanotechnology:http://www.nano.org.uk/

European Society for Precision Engineering and Nanotechnology

(EUSPEN): http://www.euspen.org/

Images of quantum corrals:http://www.almaden.ibm.com/almaden/media/image_mirage. html

Molecular abacus:http://www.research.ibm.com/topics/popups/serious/nano/html/show.html

Nanoscience:http://www.research.ibm.com/nanoscience/

Ferroelectrics, microsystems and nanotechnology – the Cranfield University Nanotechnology web site:http://www.nanotek.org/

http://www.nanostart.de/index.php/en/nanotechnology/nanotechnology-information/610-schneller-sparsamer-robuster-nanotechnologie-in-computer-handy-a-co

2

Carbon Nanotube Technologies (CNT)

2.1 Introduction

Graphite, carbon nano tubes, fullerenes and diamond are four allotropes of carbon (Figure 2.1).

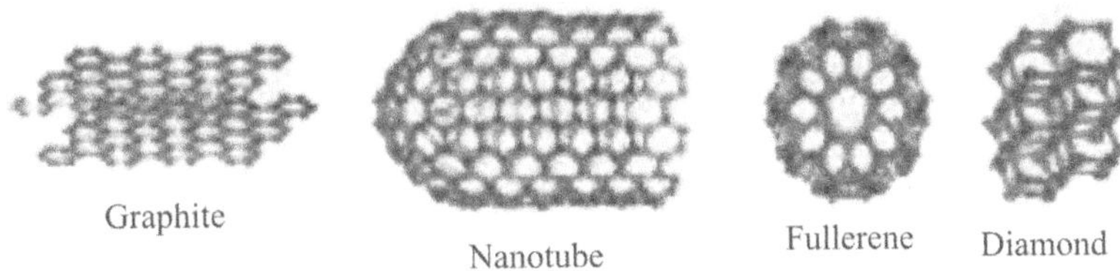

Figure 2.1 The four allotropes of carbon

A carbon nanotube may be described as a graphene sheet rolled up to form a tube, with half-fullerene caps, for closed tubes. The caps are fairly reactive due to curvature strain, and are readily destroyed during processing, yielding open tubes. (1-6).

Nanotubes are formed in a range of diameters and may be single-walled or multi-walled, depending on the reaction conditions.

The single-walled variety exist as a bundle made up of strands, like a rope, in which each strand is a single-walled nanotube (SWNT); each bundle contains typically 10-50 strands and is 5-20 nm in diameter. The diameter of SWNTs is of the order of a nanometer (~ 1.4 nm), and the inter tube spacing within the bundle has been determined to be ~ 0.3 nm, at closest approach.

Multi-walled nanotubes (MWNT) are comprised of several graphene sheets rolled up, resulting in concentric tubes with successively larger radii. Multi-walled nanotubes also aggregate, due to significant van der Waals forces. The diameter of MWNTs is typically 10-20 nm, and the interlayer spacing found to be ~ 0.34 nm.

Fullerenes are large, closed-cage, carbon clusters and have several special properties that were not found in any other compound before and form an interesting class of compounds with great potential for future technologies and applications.

Carbon nanotubes and fullerenes possess unusual photochemical, electronic, thermal and mechanical properties. Single-walled carbon nanotubes (SWCNTs) exhibit properties of metallic, semi-metallic, or semi-conductive one-dimensional objects (7). and their longitudinal thermal conductivity exceeds the in-plane thermal conductivity of graphite (8). Very high tensile strength (~100 times that of steel) of ropes made of SWCNTs proved experimentally (9). When dispersed in another medium, SWCNTs retain their intrinsic mechanical attributes or even augment the structural properties of their medium host (10). exhibit similar electrical conductivity as copper and similar thermal conductivity as diamond.

Thus there is a great deal of interest and activity in the applications for fullerene and carbon nanotube.

2.2 History of Discovery of Fullerenes and Carbon Nano–Tubes and their Salient Features

2.2.1 Fullerenes

A buckyball, or fullerene, is named after Buckminster Fuller, a futurist and global thinker best known for the invention of the geodesic dome. The structure consists of 60 carbon atoms arranged in a sphere with a pattern of hexagons and pentagons similar to a soccer ball (Figure 2.2). The buckyball is hollow, but can be used to carry another material; the outside surface of the buckyball can also be coated. A buckyball is also a semiconductor and small enough to penetrate biomembranes ordinarily thought relatively impenetrable such as cell walls or the blood brain barrier. Given these unique properties, it is not surprising that buckyballs are source of great interest to engineers and medical researchers. Fullerenes were accidentally discovered. in 1985, by Kroto and Smalley (11) who found strange results in mass spectra of evaporated carbon samples. These investigators used laser evaporation of graphite and they found C_n clusters (with n > 20 and even-numbers) of which the most common were found to be C_{60} and C_{70}. For this discovery by Curl, Kroto and Smalley were awarded the 1996 Nobel Prize in Chemistry.

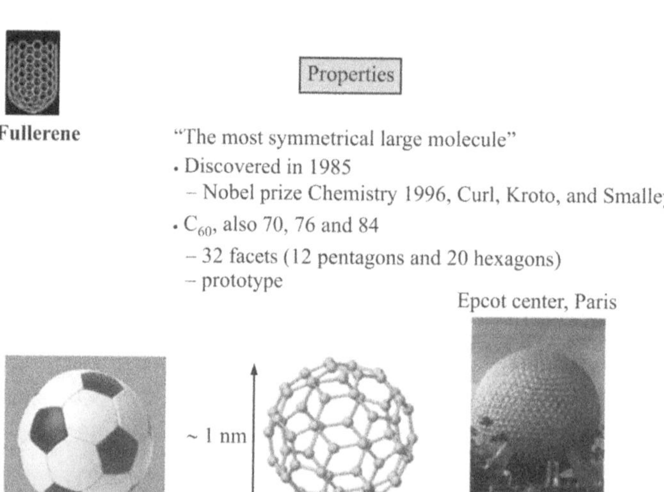

Figure 2.2 Fullerenes-the most symmetrical molecule

Before the first synthesis and detection of the smaller fullerenes C_{60} and C_{70}, it is presumed generally that these large spherical molecules will be unstable. However, some Russian scientists (12, 13) already had calculated that C_{60} in the gas phase will be stable and had a relatively large band gap. Later fullerenes with larger number of carbon atoms (C_{76}, C_{80}, C_{240}, etc.) were also discovered.

Since the time of discovery of fullerenes over a decade and a half ago, a great deal of investigation has gone into these interesting and unique nanostructures. They have found tremendous applications in nanotechnology. In 1990 a more efficient and less expensive method to produce fullerenes was developed by Krätchmer and collaborators (14). Availability of low cost fullerene will paved the way for further research into practical applications of fullerene and its role in nanotechnology.

2.2.2 Carbon Nanotubes

Carbon nanotubes were discovered by lijima in 1991(1) using an electron microscope while studying cathodic material deposition through vaporizing carbon graphite in an electric arc-evaporation reactor under an inert atmosphere during the synthesis of Fullerenes. The nanotubes produced by Lijima appeared to be made up of a perfect network of hexagonal graphite (Figure 2.3).

The nanotube diameter range is from one to several nanometers which is much smaller than its length range which is from one to a few micrometers. Nanotubes are categorized as single-walled nanotubes (SWNTs) and multi-walled nanotubes (MWNTs). Individual nanotubes naturally align themselves into "ropes" held together by van der Waals forces, more specifically, pi-stacking.

2.2.2.1 Structure and Morphology of Carbon Nanotubes

2.2.2.1.1 Single Wall Carbon Nano Tubes (SWCNT)

A single walled carbon nanotube (SWNT) is a seamless cylinder with a diameter in the nanometer range that can be visualized as a rolled up sheet of graphene (i.e., two-dimensional graphite plane). The structure of any given nanotube is determined by its diameter and the relative orientation of the carbon hexagons with respect to the axis of the tube.

Thus Single wall Carbon nano tubes (SWCNTs) can be described as a rolled up graphene sheet that is closed at each end with half of a fullerene. A nanotube is usually characterized by its diameter d_t and the chiral angle θ ($0 \leq |\theta| \leq 30°$) (Figure 2.3). The chiral vector C_h is defined with the two integers (n, m) and the basis vectors of the graphene sheet (15 &16);

$$C_h = n \times a_1 + m \times a_2$$

where a_1 and a_2 are unit vectors in the two-dimensional hexagonal lattice, and n and m are integers. Another important parameter is the chiral angle, which is the angle between C_h and a unit vector (14).

2.2.2.1.2 Classification of Carbon Nanotubes

Carbon nanotubes are classified into three different types:, zigzag, armchair and chiral nanotubes, depending on how the two-dimensional graphene sheet is "rolled up". Fig 2.3.

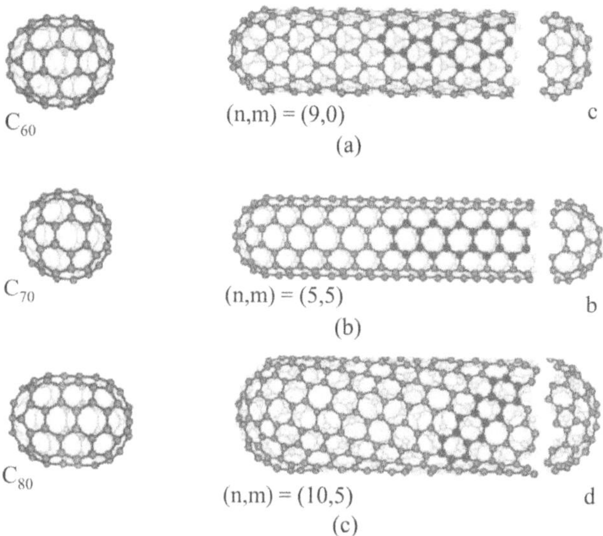

Figure 2.3 (a) Zig-zag single-walled nanotube. Note the zig-zag pattern around circumference and m = 0. (b) Chiral single-walled nanotube. Note twisting of hexagons around tubule body. (c) Armchair single-walled nanotube. Note the chair-like pattern around circumference and n = m.

2.2.2.1.3 Zig-zag Nanotubes

This describes the pattern of hexagons as one move circumferentially around the body of the tubule, when either n or m is zero and the chiral angle is 0° they are called zig-zag nano tubes.

2.2.2.1.4 Armchair Nanotubes

The second of these nanotube structures is termed armchair, which describes one of the two conformers of cyclohexane, a hexagon of carbon atoms, and describes the shape of the hexagons as one move around the body of the tubule.

2.2.2.1.5 Chiral Nanotubes

The third form is known as chiral and is believed to be the most commonly occurring SWNT. The name chiral means handedness and indicates that the tubes may twist in either direction. The geometry of the chiral SWCNT lies between that of the armchair and zigzag SWCNTs.

To discriminate between different types of carbon nanotubes, the chiral angle and vector play an important role in determining the important properties of nanotubes. (Figure 2.3). Thus the structure of a SWNT can be conceptualized by wrapping a one-atom-thick layer of graphite called graphene into a seamless cylinder.

As discussed earlier the way the graphene sheet is wrapped is represented by a pair of indices (*n*, *m*) and the integers *n* and *m* denote the number of unit vectors along two directions in the honeycomb crystal lattice of graphene.

- **Armchair nanotubes:** Are formed when n = m and the chiral angle is 30°.

- **Zigzag nanotubes:** Are formed when either n or m is zero and the chiral angle is 0°.
- **Chiral nano Tubes:** All other nanotubes, with chiral angles intermediate between 0° and 30°, are known as chiral nanotubes.

The properties of nanotubes are also determined by their diameter, which depends on n and m. A nanotube is usually characterized by its diameter dt and the chiral angle θ ($0 \leq |\theta| \leq 30°$). The integers (n, m) determine dt and θ (15-16)

The diameter of an ideal nanotube can be calculated from its (n, m) indices as follows

$$d = \frac{a}{\pi}\sqrt{\left(n^2 + nm + m^2\right)}$$

where a = 0.246 nm.

Thus SWNTs are important variety of carbon nanotube because most of their properties change significantly with the (*n*, *m*) values, and this dependence is non-monotonic. In particular, their band gap can vary from zero to about 2 eV and their electrical conductivity can show metallic or semiconducting behavior. Single-walled nanotubes are likely candidates for miniaturizing electronics.

The unique properties of single-walled carbon nanotubes are a direct result of their distinctive electronic structure, which is composed of C-C bonds more closely related to those in graphite than to those in diamond. That is, while diamond has a coordination number of four with sp^3 hybridization, graphite involves three-coordinated carbons, in which three electrons are in sp^2 hybridization and one is delocalized.

Fullerenes and nanotubes also have carbon bonds with sp^2 hybridization such as graphite, but unlike the graphite structure, which is made up of flat planar honeycomb, the structures of fullerenes and nanotubes involve a high degree of curvature. Thus the electronic band structure of a nanotube can be described by considering the bonding of the carbon atoms arranged in a hexagonal lattice. Each carbon atom ($Z = 6$) is covalently bonded to three neighbor carbons *via* sp^2 molecular orbitals. The fourth valence electron, in the p_z orbital, hybridizes with all the other p_z orbitals to form a delocalized π band. As the unit cell of graphene has two carbon atoms an even number of electrons are contained in the basic nanotube structure, which consequently can be metallic/semiconducting (17).

Mintmire, (18) Hamada, (19) and Saito (20) predicted through tight-binding electronic structure calculations that the relationship between the coefficients (n_1 and n_2) of the translational vector $c_h = n_1a_1 + n_2a_2$, which connects two crystallographically equivalent sites, determines the conducting properties. When $2\,n_1 + n_2$ is an integer multiple of three, the CNT exhibits metallic behavior while non-metallic/semiconducting behavior, with a band gap for s conduction (E_g), is obtained in all other cases.

If all values of the chiral vectors were equally probable, it would be expected that 1/3 of the total SWNTs would be metallic while the remaining 2/3 would be semiconducting (21).

Depending on their geometric structure, they can either be metallic or semiconducting, since slight changes in the geometric configuration can result in significant changes in electronic structure.

This offers materials scientists the opportunity to design at the nanoscale the best material suitable for each specific application.

2.2.2.2 Multi Walled Carbon Nanotubes (MWCNT)

There are two structural models of multi-walled nanotubes.

In the Russian Doll model, a carbon nanotube contains another nanotube inside it (the inner nanotube has a smaller diameter than the outer nanotube).

In the Parchment model, a single graphene sheet is rolled around itself multiple times, resembling a rolled up scroll of paper.

Multi-walled carbon nanotubes will have similar properties to single-walled nanotubes, yet the outer walls on multi-walled nanotubes can protect the inner carbon nanotubes from chemical interactions with outside materials.

Multi-walled nanotubes also have a higher tensile strength than single-walled nanotubes (22).

The typical shapes of parchment model MWCNTs with more than one graphene cylinder nested one into another are shown in Figure 2.4. The spacing of inter shell by using image of TEM with high resolution; spacing found to vary from 0.34, augmenting with the diameter of tube diminishing. The biggest spacing for the smallest diameter is allocated in the high curve, following in a distasteful force augmented, linked to the diameter diminished by the shells of CNTs (23). The value of 0.34 nm spacing in the bulk graphite crystal is approximately that of the CNTs. However, a closer study revealed that the mean value of the interlayer spacing is 0.3444 ± 0.001 nm; the values in CNTs are larger, by a few percent, than those in the bulk graphite crystal. According to theoretical calculations the distance between two layers is $d = 3.39$ Å, slightly bigger than in graphite. Based on TEM images, the interlayer separation of $d = 3.4$ Å is commonly reported for MWCNT (24-25).

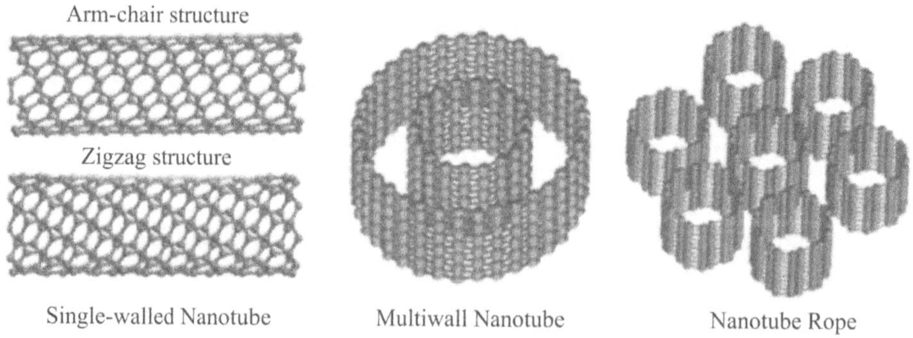

Figure 2.4 Structures of carbon nanotubes (www.iljinnanotech.co.kr , 2002).

Double-walled carbon nanotubes (DWNT) form a special class of nanotubes because their morphology and properties are similar to those of SWNT but their resistance to chemicals is significantly improved. This is especially important when functionalization is required (this means grafting of chemical functions at the surface of the nanotubes) to add new properties to the CNT. In the case of SWNT, covalent functionalization will break some C = C double bonds, leaving "holes" in the structure on the nanotube and, thus, modifying both its mechanical and electrical properties. In the case of DWNT, only the outer wall is modified.

2.2.3 Carbon Nano Buds

Carbon Nano Buds (26) are newly discovered hybrid nano materials where fullerenes are covalently bonded to the outer sidewalls of the underlying nanotube (Fig. 2.5). Consequently, nano Buds exhibit properties of both carbon nanotubes and fullerenes. For instance, the mechanical properties and the electrical conductivity of the Nano Buds are similar to those of corresponding carbon nanotubes, however, because of the higher reactivity of the attached fullerene molecules, the hybrid material can be further functionalized through known fullerene chemistry. Additionally, the attached fullerene molecules can be used as molecular anchors to prevent slipping of the nanotubes in various composite materials, thus improving the composite's mechanical properties.

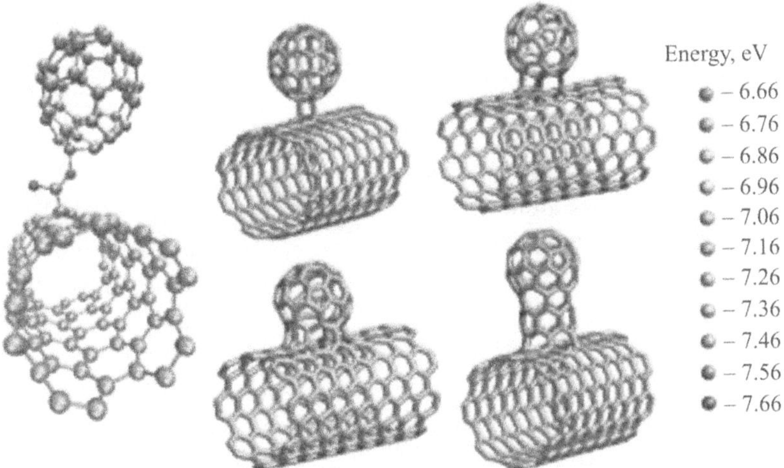

Figure 2.5 Different structures of nanobuds

2.2.4 Carbon Nano Fibers

Carbon nano fibers change from 5 to several hundred microns on length and between 100-300 nm in diameter (27-30). They consist of the graphite sheet completely arranged in various orientations. One of the most outstanding features of these structures is the presence of a plenty of sides which in turn make sites which are accessible to chemical or physical interaction, especially adsorption.

The mechanism of formation of carbon nano fibers (31-32) and the key steps in their formation is depicted in Figure 2.6. When a hydrocarbon is adsorbed on a metal surface (A) and condition exist that favor the scission of a carbon-carbon bond in the molecules, then the resulting atomic species may dissolve in the particle (B), diffuse to the rear faces, and ultimately precipitate at the interface (C) to form a carbon nanostructure. The composition of the reactant gases and the reaction temperature control the degree of crystalline perfection of the deposited fiber (D) dictate chemical nature of the catalyst particle. In this special case, a plate of graphite oriented itself in a "herringbone" classification. Surface science studies have revealed that certain faces prefer precipitation of the carbon in the form of the graphite. The choice of the catalyst, the relation of the hydrocarbon / hydrogen reactant mixture, and reaction conditions, control the morphological qualities, the degree from crystallinity and the orientation of the precipitated graphite crystallites with regard to the fiber axis. It was pointed out that the distance between graphite is 0.34 nm separate from layers one of other one. This spacing can be increased by introducing selected groups between the layers, a process known as intercalation, thereby generating new types of sophisticated molecular sieves. Such unique structural conformations found in carbon nanofibers open up numerous possibilities in the fabrication of new materials.

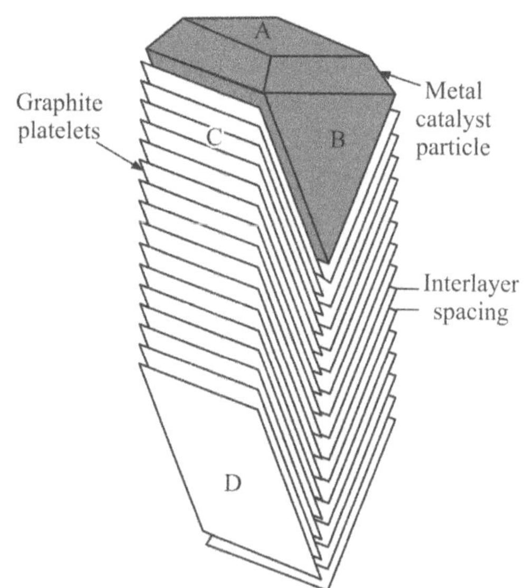

Figure 2.6 Schematic diagram of a catalytically grown carbon nano fiber [www.wtec.org/loyola/nano, 2002].

It is possible to make nano fibers, which will consist completely of plates of graphite, that show double structure depending on crystallographic orientation of catalyst faces, which exist in the metal-carbon interface. Nickel-and iron-based alloys are found to be among the most effective catalysts for the reaction. The "herringbone structures" of the carbon nano fiber are frequently found when alloy catalysts are used in the nano fiber process.

Examination of the high-resolution electron micrographs clearly indicates that the graphite platelets in these two examples are aligned in directions perpendicular (Figure 2.7 a) and parallel (Figure 2.7 b) to the fiber axis. A more detailed appreciation of these structures can be seen from the respective 3-D models, where the darker geometric shapes represent the metal catalyst particles responsible for generating these conformations. The metal catalyst particles (< 0.4%) can be easily removed by acid treatment, thus producing high purity graphite nanofibers.

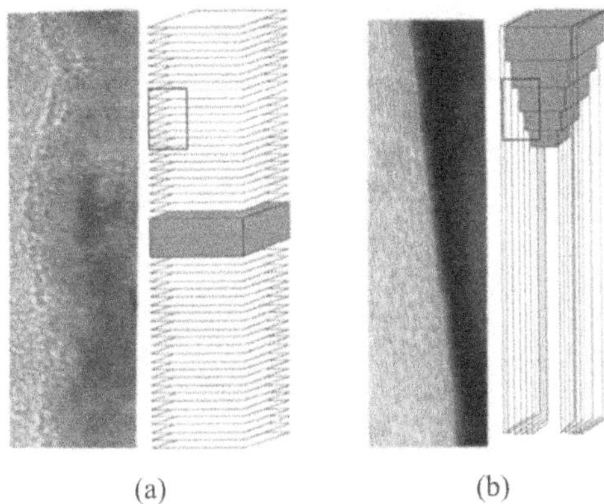

(a) (b)

Figure 2.7 High resolution electron micrographs and schematic representation of carbon nanofibers with their graphite platelets, (a) "perpendicular" and (b) "parallel" to the fiber axis [www.wtec.org, 2002].

2.3 Special Properties of Carbon Nanotubes

The most important electronic, molecular and structural properties of carbon nano materials and their molecular background are presented in the following pages.

2.3.1 Chemical Reactivity

Increased Curvature of CNT surface (which determines pi orbital difference) play important role in imparting them higher chemical reactivity, compared with a graphene sheet. Therefore, a difference must be made between the sidewall and the end caps of a nanotube. For the same reason, a smaller nanotube diameter results in increased reactivity. Covalent chemical modification of either sidewalls or end caps has shown to be possible.

2.3.2 Electrical Conductivity

As presented earlier based on their chiral vector, carbon nanotubes with a small diameter can be either semi-conducting or metallic. Electric conductance depends on Helicity $C_n = n\bar{a}_1 + m\bar{a}_2$

If $C_n = n\bar{a}_1 + m\bar{a}_2$ then metallic or else semiconductor. The differences in conducting properties are caused by the molecular structure that results in a different band structure, and thus a different band gap a The differences in conductivity can easily be derived from the graphene sheet properties .It was found that a (n, m) nanotube is metallic when: $n = m$ or $(n–m) = 3i$, or where i is an integer and n and m are defining the nanotube. The resistance to conduction is determined by quantum mechanical aspects and was proved to be independent of the nanotube length (33).

Thus the electrical properties of carbon nanotubes as discussed above depend on how the hexagons are orientated along the axis of the tube i.e., armchair, zigzag, and chiral (Figure 2.8).

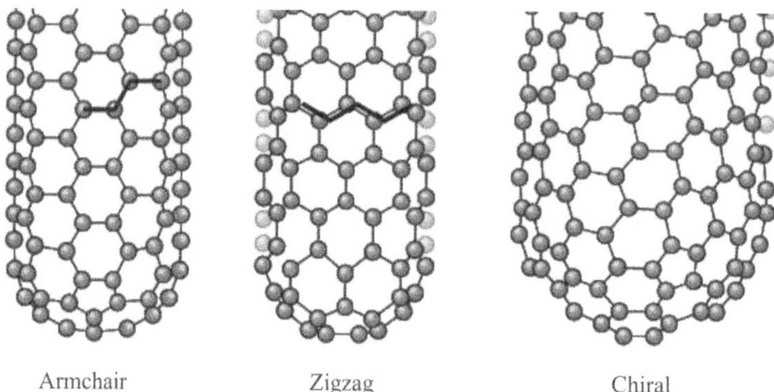

Figure 2.8 Electrical properties depend on the orientation of the hexagon

Arm Chair Carbon nanotubes with the hexagons orientated in the configuration (hexagons are lined up parallel to the axis of the nanotube) exhibit electrical properties similar to metals. When you apply a voltage between two ends of an armchair nanotube, a current will flow. An armchair carbon nanotube is, in fact, a better conductor than the copper normally used in electrical wire, or any other metal.

Methods are being developed to spin carbon nanotubes together to make low-resistance electrical wires that could transform the electrical power grid, as well as reduce the power consumed and weight of wiring in such power- and weight-sensitive uses as spacecraft and airplanes.

Another use that these armchair carbon nanotubes are being considered for is to connect devices in integrated circuits. As devices in integrated circuits become smaller, it may not be possible to pattern narrow enough metal lines, so researchers are considering using armchair carbon nanotubes to replace the metal lines.

The next two possible orientations of hexagons in carbon nanotubes share electrical properties similar to semiconductors. Those with the hexagons oriented in a circle around the nanotube have a configurations labeled zigzag. Those with a twist to the nanotube so the hexagons do not form any line are called chiral. These two configurations of nanotubes will only conduct an electric current when extra energy in the form of light or an electric field is applied to free electrons from the carbon atoms. Semiconducting nanotubes could be useful in building the ever smaller transistors used by the hundreds of millions in integrated circuits for all kinds of electronic devices.

Another interesting property of carbon nanotubes is that their electrical resistance changes significantly when other molecules attach themselves to their carbon atoms. Companies are using this property to develop sensors that can detect chemical vapors such as carbon monoxide or biological molecules.

2.3.3 Thermal Conductivity

Carbon nanotubes also conduct heat and cold really well (they have a high thermal conductivity); some researchers predict a thermal conductivity more than 10 times that of silver.

While metals depend upon the movement of electrons to conduct heat, carbon nanotubes conduct heat by the vibration of the covalent bonds holding the carbon atoms together; the atoms themselves are wiggling around and transmitting the heat through the material. The stiffness of the carbon bond helps transmit this vibration throughout the nanotube, providing very good thermal conductivity.

The strength of the atomic bonds in carbon nanotubes allows them to withstand high temperatures. Because of this, carbon nanotubes have been shown to be very good thermal conductors. When compared to copper wires, which are commonly used as thermal conductors, the carbon nanotubes can transmit over 15 times the amount of watts per meter per Kelvin. The thermal conductivity of carbon nanotubes is dependent on the temperature of the tubes and the outside environment (10). Further nanotubes offer a natural template for exploring the effects of reduced dimensionality towards fabricating the *best* thermoelectric material.

2.3.4 Optical Activity

Carbon nanotubes exhibit specifically optical properties like absorption, photolumine scence (fluorescence), and Raman Spectra. which can be used for their quick and non-destructive characterization. Optical activity of chiral nanotubes disappears if the nanotubes become larger. There is a great demand from industrial point of view for spectral characterization of CNTs which provide a large benefits industrially as numerous parameters of the nanotube synthesis can be changed, to alter the nanotube quality. As discussed below, optical absorption, photoluminescence and Raman spectroscopies allow quick and reliable characterization of "nanotube quality" in terms of non-tubular carbon content, structure (chirality) of the produced nanotubes, and structural defects. Those features determine nearly any other properties such as optical, mechanical, and electrical properties.

Raman spectroscopy is useful for characterizing nanotubes, especially SWNTs. There are three regions of relatively intense peaks: the radial breathing mode (RBM) is found at low wavenumber (~ 150 – 300 cm^{-1}); the D-band (disorder-induced) (~ 1350 cm^{-1}) arises from the presence of amorphous carbon, and is generally used as an indicator of the purity; and the G-band (~ 1500 – 1600 cm^{-1}), the shape of which depends on the conducting nature of the nanotubes. The RBM frequency has been found to be inversely proportional to the diameter according to

$$d(\mathrm{nm}) = \frac{x}{\omega\mathrm{RBM}\left(\mathrm{cm}^{-1}\right)}$$

where x varies from 224 to 248 (depending on whether x is based on theory, or a combination of experiment and theory). The D* band (disorder-induced overtone) (~ 2600 cm^{-1}) has also been used in the investigation of interaction between SWNTs and a polymer matrix or other molecules. The D* band is sensitive to strain as a result of breathing mode restriction.

Although Raman scattering is generally weak, a phenomenon known as resonant Raman scattering occurs with carbon nanotubes. When the Raman laser excitation energy is in resonance with the energy gap between pairs of van Hove singularities (valence → conducting) for example $v_1 \rightarrow c_1$, (Figure 2.9), the Raman scattering is enhanced, and then it will be much more intense.

Figure 2.9 illustrates the relationship between nanotube size (diameter) and band gap; the numerical values correspond to the band gap in eV, and the inverse relationship is evident.

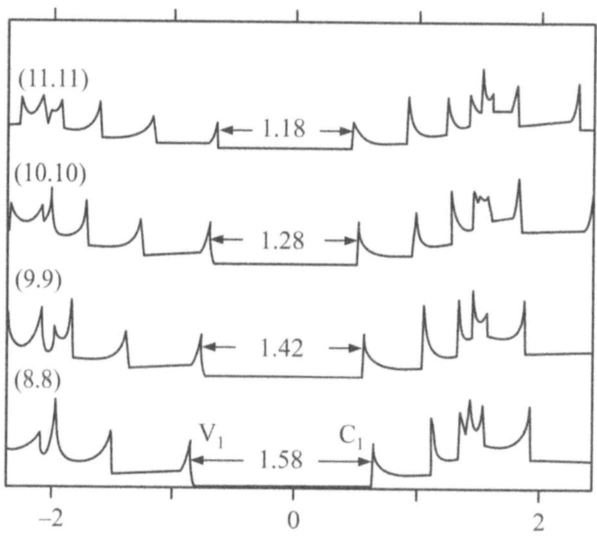

Figure 2.9 Calculated electronic DOS illustrating the nanotube size and band gap relationship for a series of armchair nanotubes

Optical properties of carbon nanotubes derive from electronic transitions within one-dimensional density of states (DOS). A typical feature of one-dimensional crystals is that their DOS is not a continuous function of energy, but it descends gradually and then increases in a discontinuous spike. In contrast, three-dimensional materials have continuous DOS. The sharp peaks found in one-dimensional materials are called Van Hove singularities.

Van Hove singularities result in the following remarkable optical properties of carbon nanotubes:

Optical transitions occur between the $v_1 - c_1$, $v_2 - c_2$, etc., states of semiconducting or metallic nanotubes and are traditionally labeled as S_{11}, S_{22}, M_{11}, etc., or, if the "conductivity" of the tube is unknown or unimportant, as E_{11}, E_{22}, etc. Crossover transitions $c_1 - v_2$, $c_2 - v_1$, etc., are dipole-forbidden and thus are extremely weak, but they were possibly observed using cross-polarized optical geometry (34-38).

The energies between the Van Hove singularities depend on the nanotube structure (Figure 2.10). Thus by varying this structure, one can tune the optoelectronic properties of carbon nanotube. Such fine tuning has been experimentally demonstrated using UV illumination of polymer-dispersed CNTs (39-40).

Optical transitions are rather sharp (~10 meV) and strong. Consequently, it is relatively easy to selectively excite nanotubes having certain (n, m) indices, as well as to detect optical signals from individual nanotubes.

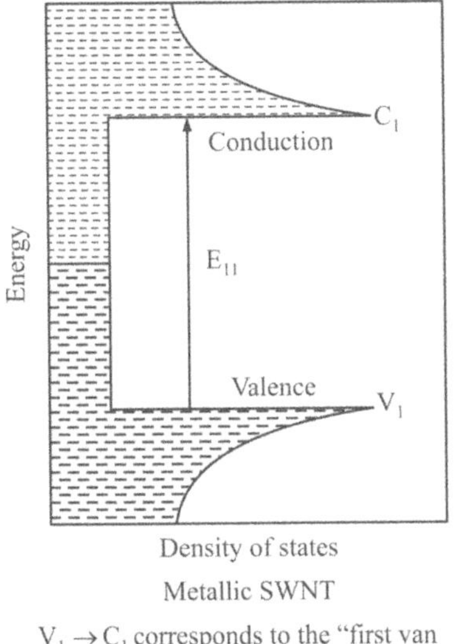

Metallic SWNT

$V_1 \rightarrow C_1$ corresponds to the "first van Hove" optical transition

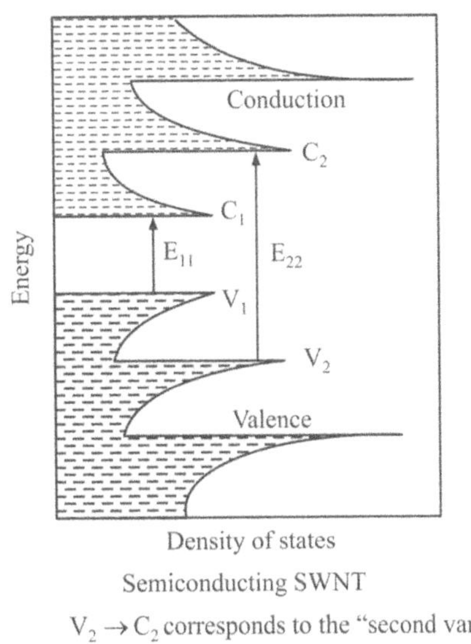

Semiconducting SWNT

$V_2 \rightarrow C_2$ corresponds to the "second van Hove" optical transition

Figure 2.10 Van Hove singularities in metallic SWNT and semi conducting SWNT

2.3.5 Mechanical Strength

Nanotubes are very strong. Tensile strength is a measure of the amount of force an object can withstand without tearing apart. The tensile strength of carbon nanotubes is approximately 100 times greater than that of steel of the same diameter (40). There are two things that account for this strength.

The first is the strength provided by the interlocking carbon-to-carbon covalent bonds. Thus the intrinsic strength of the sp^2 carbon-carbon bonds and the special structure of nanotubes result in high mechanical strength and a very large Young's modulus (values near 1,000 GPa is observed for individual nanotubes, which is around 50 times higher than the modulus of steel).

The second is the fact that each carbon nanotube is one large molecule. This means it doesn't have the weak spots found in other materials, such as the boundaries between the crystalline grains that form steel.

Nanotubes are strong but are also elastic (Table 2.1). This means it takes a lot of force to bend a nanotube, but the little guy will spring right back to its original shape when you release it, just like a rubber band does. Researchers have used atomic force microscopes to physically push nanotubes around and observe their elastic properties. Evaluations with transmission electron microscopes — the kind of microscope sensitive enough to give you a peek at atomic shapes show that the bonds in the atomic lattice don't break when you bend or compress a nanotube.

Young's modulus for carbon nanotubes, a measurement of how much force it takes to bend a material, is about 5 times higher than for steel, so if you were thinking of going out and bending a nanotube, think again. The fact is, there's not another element with a lattice structure in the whole periodic table that bonds to itself with as much strength as carbon atoms. And, since carbon nanotubes have such a perfect structure, they avoid the degradation of strength that you get with other materials.

Table 2.1 Mechanical Properties of CNTs

Young Modulus (stiffness)	
Carbon nanotubes	1250 GPa
Carbon fibers	425 GPa (max.)
High strength steel	200 GPa
Tensile strength (breaking strength)	
Carbon nanotubes	11- 63 GPa
Carbon fibers	3.5 - 6 GPa
High Strength Steel	~2GPa
Elongation to failure	~ 20-30 %
Density:	
Carbon nanotube (SW)	1.33 – 1.40 gram / cm^3
Aluminium	2.7 gram / cm^3

In addition to being strong and elastic, carbon nanotubes are also lightweight, with a density about one quarter that of steel.

Carbon nanotubes have a very large Young modulus in their axial direction. The nanotube as a whole is very flexible because of the great length. Therefore, these compounds are potentially suitable for applications in composite materials that need anisotropic properties.

2.4 Carbon Nano Tubes Growth Mechanism

2.4.1 Multiple Growth Mechanisms in Operation (37-41)

Though there are uncertainties in the growth mechanisms of CNTs it is generally believed the growth involves three main steps (Figure 2.11).

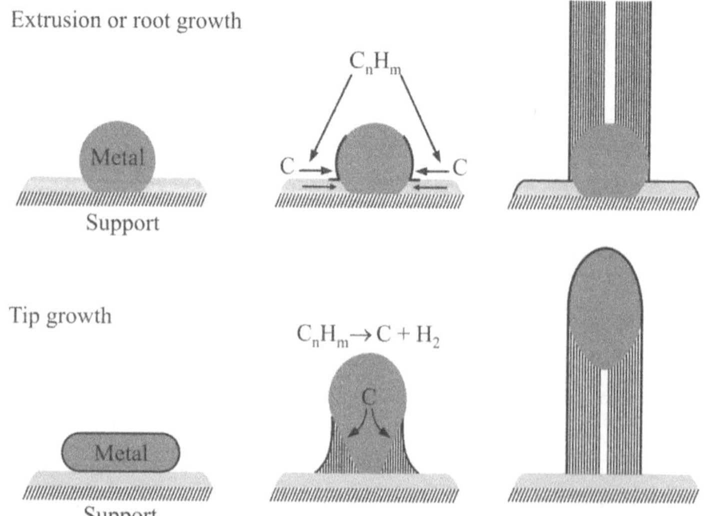

Figure 2.11 Visualization of a possible carbon nanotube growth mechanism

1. In the first step a precursor to the formation of nanotubes and fullerenes, C_2, will be formed on the surface of the metal catalyst particle.
2. From this metastable carbide particle, a rodlike carbon is formed rapidly followed by a slow graphitization of its wall.
3. In the third step filaments can form either by 'extrusion (also known as base growth)' in which the nanotube grows upwards from the metal particles that remain attached to the substrate, or the particles detach and move at the head of the growing nanotube.

This mechanism is based on *in-situ* TEM observations. The exact atmospheric conditions depend on the technique used. The actual growth of the nanotube seems to be the same for all techniques mentioned.

Some other theories on the growth mechanism for nanotubes (42-43) propose that the precursors of nano tubes will be deposited on a metal catalyst particles (which are spherical in shape). As the catalyst particles are spherical or pear-shaped, the deposition will take place on only one half of the surface (this is the lower curvature side for the pear shaped particles). The carbon diffuses along the concentration gradient and precipitates on the opposite half, around and below the bisecting diameter. However, it does not precipitate from the apex of the hemisphere, which accounts for the hollow core that is characteristic of these filaments. For supported metals, filaments can form either by 'extrusion (also known as base growth)' in which the nanotube grows upwards from the metal particles that remain attached to the substrate, or the particles detach and move at the head of the growing nanotube, labelled 'tip-growth'. Depending on the size of the catalyst particles, SWNT or MWNT are grown. In arc discharge, if no catalyst is present in the graphite, MWNT will be grown on the C2-particles that are formed in the plasma.

2.4.2 Role of Hydrogen Flow Rate on the Growth of CNT

It is important to understand the effect of hydrogen flow rate on the formation of carbon nano and microstructure material because hydrogen is frequently present in the hydrocarbon processing system. The effect of hydrogen can be both acceleration and suppression. The effect of hydrogen acceleration on carbon formation may be interpreted in two ways (Figure 2.12). The first interpretation suggested that, hydrogen decompose inactive metal carbides to form catalytically active metal. The other interpretation pertains to the removal, by hydrogen, of the surface carbon and precursors of carbon, which block the active site. The suppressing effect has also been reported to be due to the surface hydrogenation reactions to form methane.

All these theories together with experimental data leads to the following mechanism for the deposition of graphitic carbon on the metal surface. The promotional effect of hydrogen on carbon nanotubes formation from the metal catalyzed decomposition of carbon-containing gas molecules has been attributed to its ability to convert inactive metal carbides into the catalytically active metallic state as well as to prevent the formation of graphitic overlayers on the particle surface.

Thus the catalytic decom-position of hydrocarbon is highly sensitive to substrate catalyst, while the hydrogenation of carbon is relatively less sensitive to catalyst. For the catalyst, which is not highly active for decomposition, the hydrogenation reaction becomes important and the net carbon deposition rate is lowered by hydrogen gas.

2.4.3 The Role of Reaction Temperature

Thermodynamically, higher temperatures favor the surface decomposition of hydrocarbon rather than the hydrogenation reactions. It is generally accepted that carbon materials are formed by carbon atom dissolving, diffusing, and precipitating through the catalyst droplets in Chemical Vapour Dissolves CVD process. The dissolving, diffusing and precipitating rates of the carbon atoms are affected by both the carbon atoms concentration and the temperature. The carbon precipitation region on the Fe catalyst droplets can be distinguished into two areas, surface area and internal area. At low temperature, the dissolving and diffusing rates are limited by the low concentration of carbon atoms so that carbon atoms can only precipitate on the surface area of the catalyst droplets to form completely hollow CNTs.

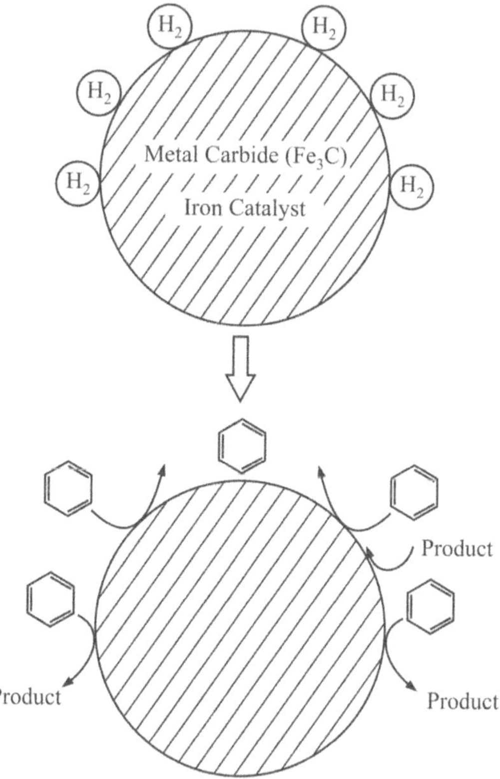

Figure 2.12 Promotional effect of hydrogen on carbon nanotubes formation through metal carbide formation

The diameter of CNTs gets bigger with the increase in temperature. This can probably be attributed to small catalyst droplets agglomerate at high temperature to form bigger catalyst particle which will form big CNTs. High reaction temperature will promote the decomposition of hydrocarbon to increase the concentration of carbon atoms, which will increase the growth rate of CNTs to form bigger CNTs. With the increase of the temperature, the dissolving and diffusing rates of carbon atoms will increase, and carbon atoms can get to the internal area of the catalyst droplet to form CNFs.

2.5 Production Methods for Carbon Nanotubes and Carbon Nanofibers

2.5.1 Introduction

There are three main techniques used for the production of carbon nano tubes i.e., (i) arc discharge (ii) laser ablation and (iii) chemical vapour deposition.

In the arc discharge a vapor is created by an arc discharge between two carbon electrodes with or without catalyst, while in the laser ablation technology laser beam provides carbon, containing feedstock gas (methane or carbon monoxide). However laser ablation produces a very small amount of pure nanotubes, while an arc-discharge method produces in general large amounts of the impure material while CVD is better option for industrial scale-up due to the relatively low growth temperature, high yields and high purities that can be achieved. The details of these techniques are discussed in the following pages.

2.5.2 Arc Discharge

In this arc discharge method, vaporization of two carbon rods placed end to end, separated by approximately 1mm, in an enclosure filled with inert gas (helium, argon) at low pressure (between 50 and 700 mbar) will produce carbon nano tubes. A direct current of 50 to 100 A driven by approximately 20 V creates a high temperature discharge between the two electrodes. The discharge vaporises one of the carbon rods and forms a small rod shaped deposit on the other rod. Producing nanotubes in high yield depends on the uniformity of the plasma arc and the temperature of the deposit form on the carbon electrode (41) (Figure 2.13). This method is simple and initially used for producing C_{60} fullerenes. One of the drawback of this method is mixture of components are produced and requires separation of nanotubes from the soot and the catalytic metals present in the crude product.

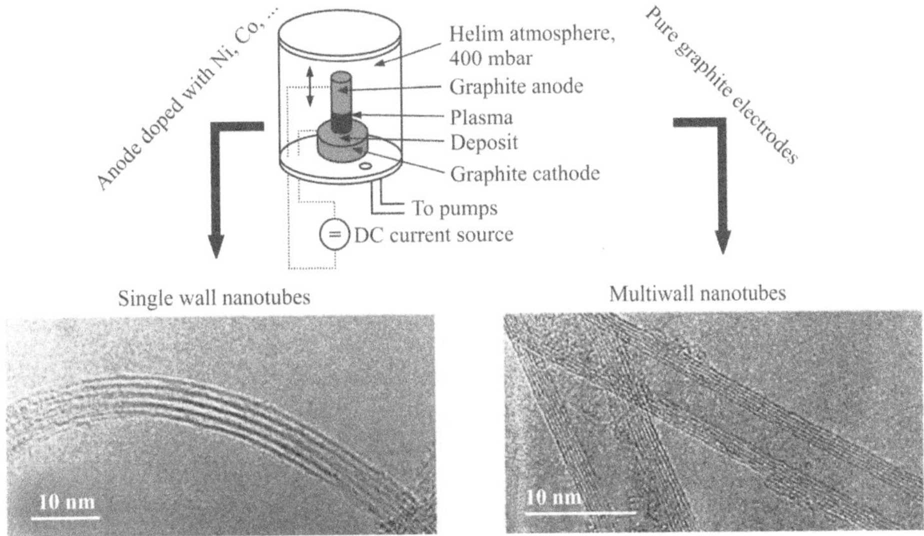

Figure 2.13 Experimental set-up of an arc discharge apparatus

Studies on the growth mechanism have revealed that different diameter distributions will occur depending on the mixture of helium and argon. These mixtures have different diffusions coefficients and thermal conductivities which affect the speed with which the carbon and catalyst molecules diffuse and cool influencing nanotube diameter in the arc process. Thus single-layer tubules nucleate and grow on metal particles in different sizes depending on the quenching rate in the plasma which suggests that temperature and carbon and metal catalyst densities affect the diameter distribution of nanotubes (43). Based on exact technique implemented it is possible to selectively grow SWNTs or MWNTs.

2.5.2.1 Production of SWNT

For producing SWNTs the anode need to be doped with metal catalyst, such as Fe, Co, Ni, Y Mo. A lot of elements and mixtures of elements have been tested by various authors (44) and it is noted that the results vary a lot, even though they use the same elements. This is not surprising as experimental conditions differ. The quantity and quality of the nanotubes obtained depend on various parameters such as the metal concentration, inert gas pressure, kind of gas, the current and system geometry. Usually the diameter is in the range of 1.2 to 1.4 nm. A number of ways to improve the process of arc discharge are stated below.

(a) **Influence of inert gas medium:** The most common problems with SWNT synthesis are that the product contains a lot of metal catalyst, SWNTs have defects and purification is hard to perform. On the other hand, an advantage is that the diameter can slightly be controlled by changing thermal transfer and diffusion, and hence condensation of atomic carbon and metals between the plasma and the vicinity of the cathode can control nanotube diameter in the arc process (44). Argon, with a lower thermal conductivity and diffusion coefficient, gave SWNTs with a smaller diameter of approximately 1.2 nm. A linear fit of the average nanotube diameter showed a 0.2 nm diameter decrease per 10 % increase in argon helium ratio, when nickel/yttrium is used (C/Ni/Y was 94.8:4.2:1) as catalyst.

(b) **Plasma control:** Plasma can be controlled by changing the anode to cathode distance (ACD) which can be adjusted in-order to obtain strong visible vortices around the cathode. This enhances anode vaporisation, which improves nanotubes formation. Combined with controlling the argon-helium mixture, one can simultaneously control the macroscopic and microscopic parameters of the nanotubes formed (45). With a nickel and yttrium catalyst (C/Ni/Y is 94.8:4.2:1) the optimum nanotube yield was found at a pressure of 660 mbar for pure helium and 100 mbar for pure argon. The nanotube diameter ranges from 1.27 to 1.37 nanometre.

(c) **Influence of catalyst in arc discharge method:** Chemical vapour deposition (CVD) produces SWNTs with a diameter of 0.6-2 nm in presence of catalyst. Though all the catalysts could not give positive results in the production of nanotubes, a correlation of diameter of SWNTs synthesised by CVD and arc discharge is observed. Thus the diameter can be controllably lowered to a range of 0.6-1.2 nm with arc-discharge. Mixture of Co and Mo catalysts produced high concentrations of SWNTs with Diameters smaller than 1.2-1.4 nm (44), which is the usual size produced from arc-discharge (46).

(d) **Oxidation resistance:** A strong oxidation resistance is needed for the nanotubes if they have to be used for applications such as field emission displays. Oxidation resistance SWCNTs is due to the defects present in nanotubes. Modified arc-discharge method using a bowl-like cathode (Figure 2.14) is found to, decrease the defects and produce pure nanotubes due to increased oxidation resistance (47).

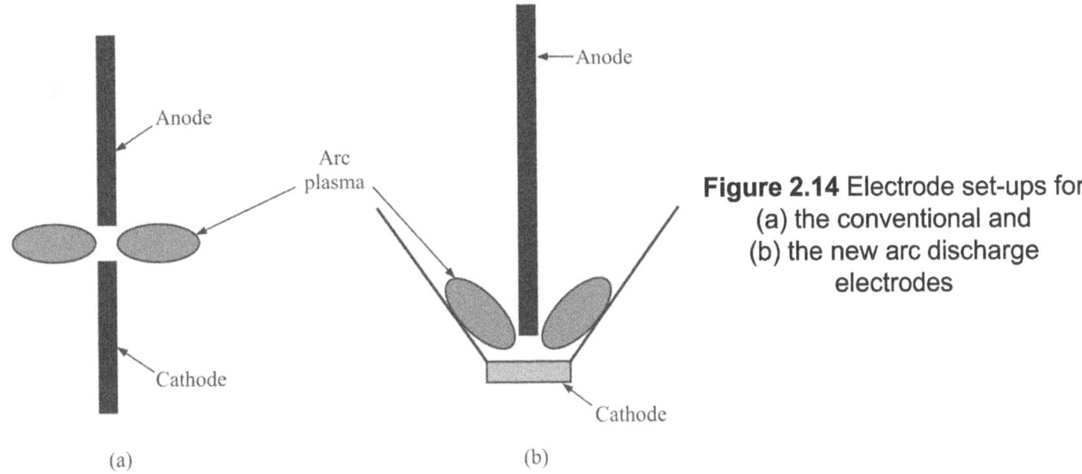

Figure 2.14 Electrode set-ups for (a) the conventional and (b) the new arc discharge electrodes

The Nanotubes produced by modified arc discharge method are found to be more pure and less defective (based on their Raman Spectra) compared with those synthesized by conventional methods.

2.5.2.2 Synthesis of MWNT

Multi Walled Carbon Nao Tubes (MWNT)s (with inner diameter 1-3 nm, outer diameter 10 nm) can be produced by choosing graphite for both electrodes. However a number of additional components like fullerenes, amorphous carbon, and some graphite sheets are also produced in this process which need purification due to which some losses may be there due to structure losses and walls disorders. Modified highly effective single step process without catalyst are being developed for which need no further purification avoiding any loss of structure of MWNT and any defects.

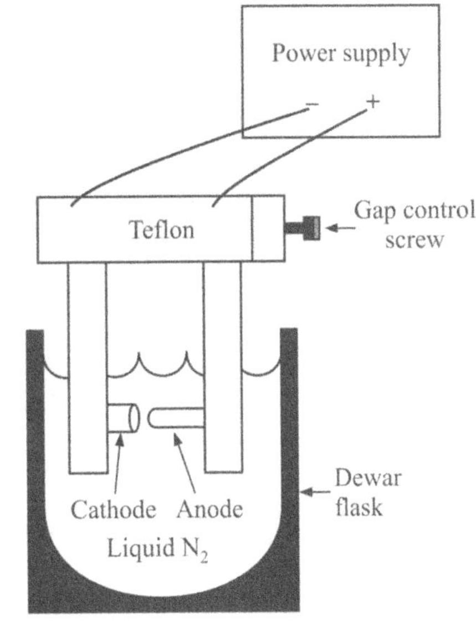

Figure 2.15 Arc discharge apparatus in liquid nitrogen

(a) Production in liquid nitrogen: Arc discharge in liquid nitrogen (48) is a commercially viable mass production (70% yield) method for production of highly crystalline MWNT (Figure 2.15) and low cost process as this process does not require low pressures and expensive inert gas medium.

(b) Magnetic field synthesis: MWNTs production in a magnetic field (49-51) produces defect-free and high purity MWNTs which can be controlled by regulating magnetic field around arc plasma (Figure 2.16). These MWNT can be applied as nano sized electric wires for device fabrication.

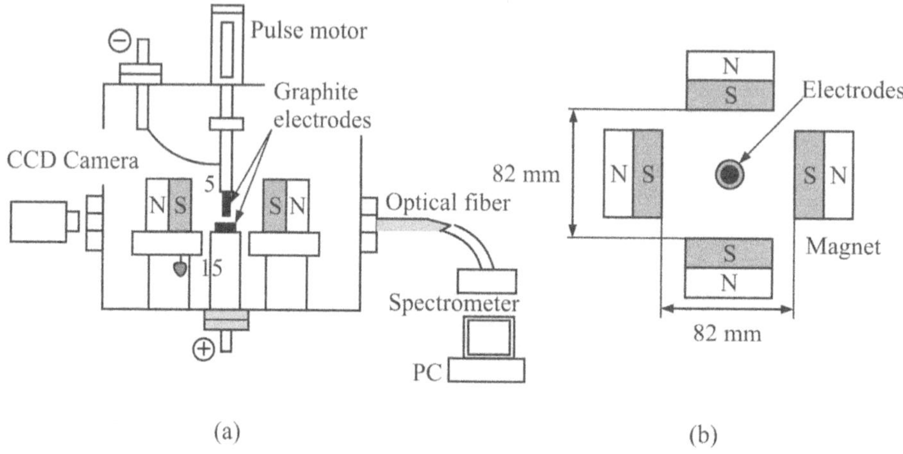

Figure 2.16 Production of MWNTs in a magnetic field.

MWNTs of high purity (purity > 95%) which does not need further purification can be produced using extremely pure graphite rods (purity > 99.999 %) as electrodes (Fig. 2.17).

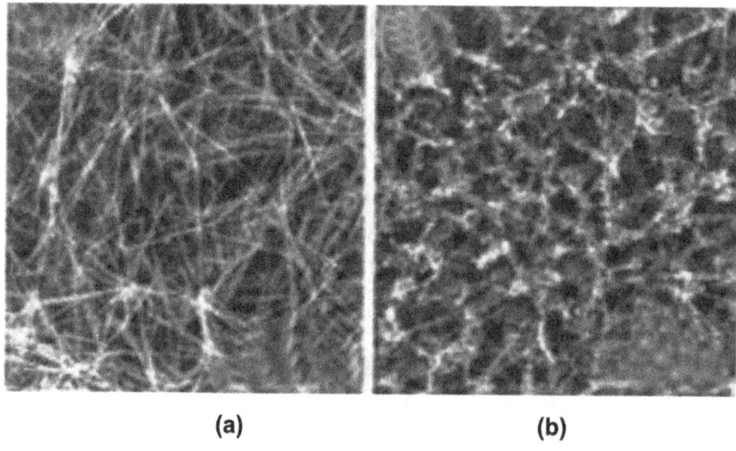

(a) (b)

Figure 2.17 SEM images of MWNTs synthesized with (a) and without (b) the magnetic field.

(c) **Plasma rotating arc discharge method:** Plasma rotating arc discharge technique (52) is one of the cost effective mass production method for MWNT. The centrifugal force caused by the rotation generates turbulence and accelerates the carbon vapour perpendicular to the anode. In addition, the rotation distributes the micro discharges uniformly and generates a stable plasma. Consequently, it increases the plasma volume and raises the plasma temperature (Figure 2.18).

At a rotation speed of 5000 rpm a yield of 60 % was found at a formation temperature of 1025 °C without the use of a catalyst. The yield increases up to 90% after purification if the rotation speed is increased and the temperature is enlarged to 1150 °C.

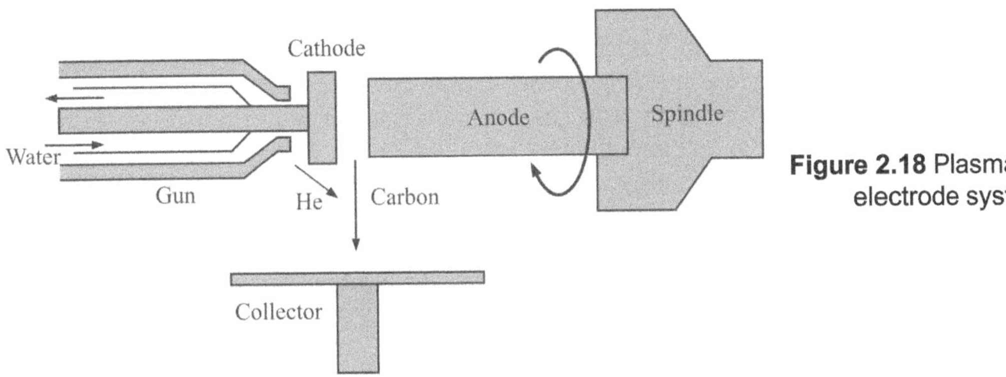

Figure 2.18 Plasma rotating electrode system

2.5.3 Laser Ablation

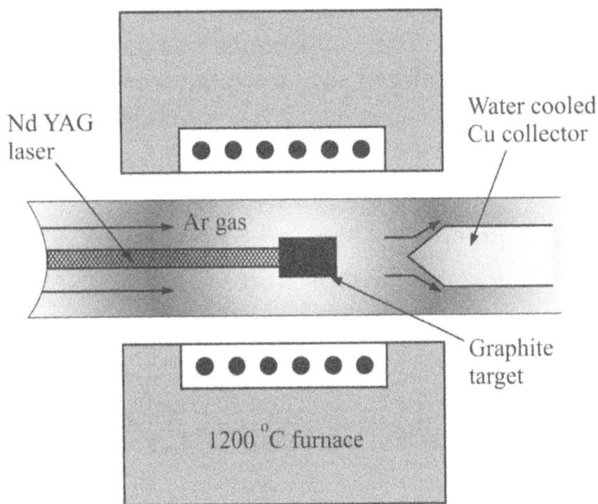

Figure 2.19 Laser ablation apparatus

Production of carbon nanotubes by lasers vaporization was reported by Smalley (53) at Rice University in 1995 and the corresponding laser vaporisation apparatus used by Smalley's group is shown in Figure 2.19. A pulsed, or continuous (53, 54) laser is used to vaporise a graphite target in an oven at 1200 °C. The main difference between continuous and pulsed laser, is that the pulsed laser demands a much higher light intensity (100 kW/cm^2 compared with 12 kW/cm^2). The oven is filled with helium or argon gas in order to keep the pressure at 500 Torr. A very hot vapour plume forms, then expands and cools rapidly. As the vaporised species cool, small carbon molecules and atoms quickly condense to form larger clusters, possibly including fullerenes. The catalysts also begin to condense, but more slowly at first, and attach to carbon clusters and prevent their closing into cage structures (55). Catalysts may even open cage structures when they attach to them. From these initial clusters, tubular molecules grow into single-wall carbon nanotubes until the catalyst particles become too large, or until conditions have cooled sufficiently that carbon no longer can diffuse through or over the surface of the catalyst particles. It is also possible that the particles become that much coated with a carbon layer that they cannot absorb more and the nanotube stops growing. The SWNTs formed in this case are bundled together by van der Waals forces (56).

Laser ablation is almost similar to arc discharge, since the optimum background gas and catalyst mix is the same as in the arc discharge process. This might be due to very similar reaction conditions needed, and the reactions probably occur with the same mechanism.

2.5.4 Chemical Vapour Deposition (CVD)

Chemical vapour deposition (CVD) synthesis involves combination of a carbon source like methane/carbon monoxide/acetylene in the gas phase and an energy source, such as plasma or a resistively heated coil, to transfer energy to a gaseous carbon molecule. The energy source is used to "crack" the molecule into reactive atomic carbon which diffuses towards the substrate, heated and coated with a catalyst (usually a first row transition metal such as Ni, Fe or Co) where it will bind. By properly maintaining experimental conditions Carbon nanotubes with excellent alignment and tailored nano scale diameter can be produced by CVD process (57) will be formed if the proper parameters are maintained. The appropriate metal catalyst can preferentially grow single rather than multi-walled nanotubes (58).

CVD carbon nanotube synthesis is essentially a two-step process consisting of a catalyst preparation step followed by the actual synthesis of the nanotube. The catalyst is generally prepared by sputtering a transition metal onto a substrate and then using either chemical etching or thermal annealing to induce catalyst particle nucleation. Thermal annealing results in cluster formation on the substrate, from which the nanotubes will grow. Ammonia may be used as the Etchant (58-61). The temperatures for the synthesis of nanotubes by CVD are generally within the 650-900 °C with typical yields for CVD are approximately 30%.

2.5.4.1 Plasma Enhanced Chemical Vapour Deposition

A glow discharge will be produced by applying high frequency voltage in between two electrodes in reaction chamber or furnace in the plasma enhanced CVD Process (Figure 2.20).

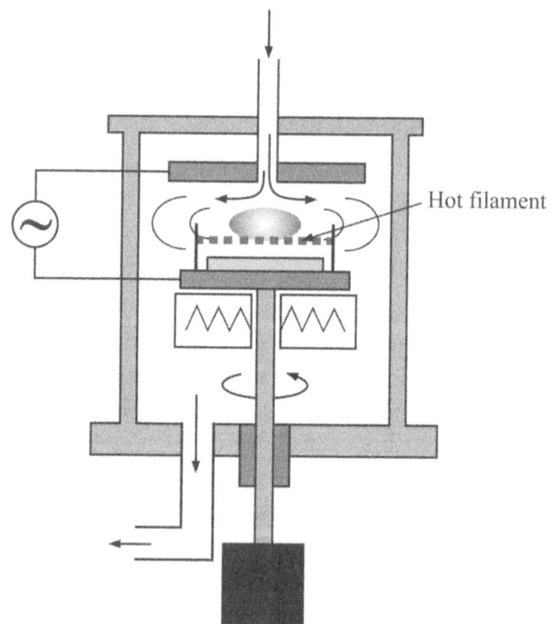

Figure 2.20 Plasma CVD apparatus

A substrate is placed on the grounded electrode. In order to form a uniform film, the reaction gas is supplied from the opposite plate. Catalytic metal, such as Fe, Ni and Co are used on for example a Si, SiO_2, or glass substrate using thermal CVD or sputtering. After nanoscopic fine metal particles are formed, carbon nanotubes will be grown on the metal particles on the substrate by glow discharge generated from high frequency power. A carbon containing reaction gas, such as C_2H_2, CH_4, C_2H_4, C_2H_6, CO is supplied to the chamber during the discharge (61).

The catalyst has a strong effect on the nanotube diameter, growth rate, wall thickness, morphology and microstructure. Ni seems to be the most suitable pure-metal catalyst for the growth of aligned multi-walled carbon nanotubes (MWNTs)(62). The diameter of the MWNTs is approximately 15 nm. The highest yield of carbon nanotubes achieved was about 50% and was obtained at relatively low temperatures (below 330 °C) (63).

2.5.4.2 Thermal Chemical Vapour Deposition

Figure 2.21. shows a schematic diagram of thermal CVD apparatus in the synthesis of carbon nanotubes. Initially Fe, Ni, Co or an alloy of the three catalytic metals will be deposited on a substrate which is etched in a diluted HF solution with distilled water and the specimen is then placed in a quartz boat. The boat will be positioned in a CVD reaction furnace, and nanometre-sized catalytic metal particles are formed after an additional etching of the catalytic metal film using NH_3 gas at a temperature of 750 to 1050 °C. When growing carbon nanotubes on a Fe catalytic film by thermal CVD, the diameter range of the carbon nanotubes depends on the thickness of the catalytic film. As carbon nanotubes are grown on these fine catalytic metal particles in CVD synthesis, forming these fine catalytic metal particles is the most important process.

By using a thickness of 13 nm, the diameter distribution lies between 30 and 40 nm. When a thickness of 27 nm is used, the diameter range is between 100 and 200 nm. The carbon nanotubes formed are multi walled (63).

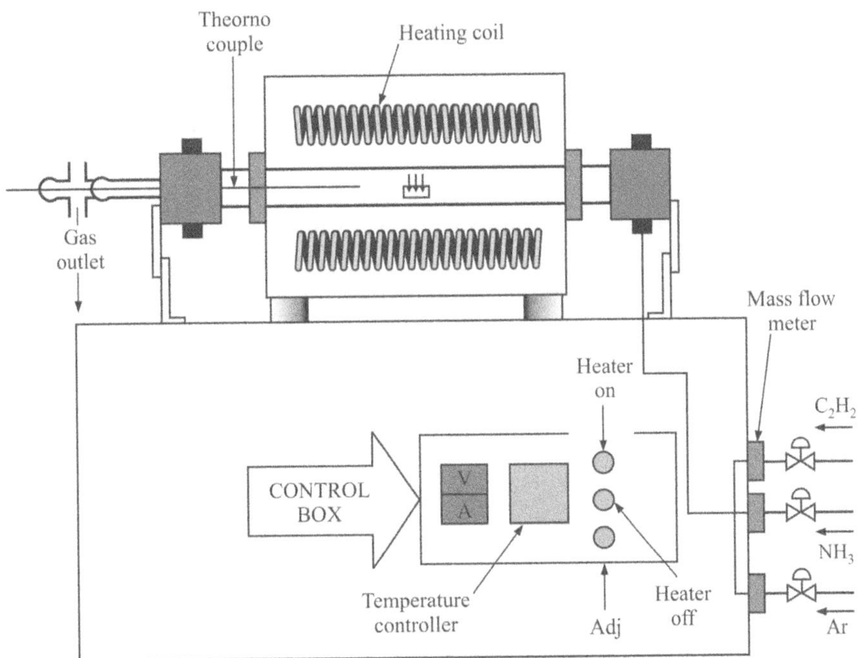

Figure 2.21 Thermal CVD apparatus

2.5.4.3 Alcohol Catalytic Chemical Vapour Deposition (ACCVD)

In this technique, evaporated alcohols, methanol and ethanol will be decomposed over iron and cobalt catalytic metal particles supported with zeolite. SWNTs production will occur at a relatively low minimum temperature of about 550 °C. It seems that hydroxyl radicals, which come from reacting alcohol on catalytic metal particles, remove carbon atoms with dangling bonds, which are obstacles in creating high-purity SWNTs. The diameter of the SWNTs is about 1 nm. Figure 2.22 shows the ACCVD experimental apparatus.

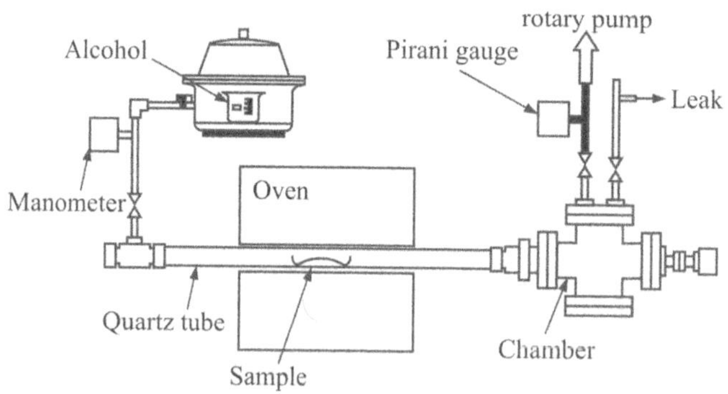

Figure 2.22 ACCVD experimental apparatus

The lower reaction temperature and the high-purity features of this ACCVD technique guarantee an easy possibility to scale production up at low cost. Furthermore, the reaction temperature, which is lower than 600 °C, ensures that this technique is easily applicable for the direct growth of SWNTs on semiconductor devices already patterned with aluminium (64).

2.5.4.4 Vapor Phase Growth

Figure 2.23 shows a schematic diagram of a vapour phase growth apparatus. Two furnaces are placed in the reaction chamber. Ferrocene is used as catalyst. In the first furnace, vaporisation of catalytic carbon is maintained at a relatively low temperature. Fine catalytic particles are formed and when they reach the second furnace, decomposed carbons are absorbed and diffused to the catalytic metal particles. Here, they are synthesised as carbon nanotubes. The diameter of the carbon nanotubes by using vapour phase growth are in the range of 2-4 nm for SWNTs and between 70 and 100 nm for MWNTs (65-67).

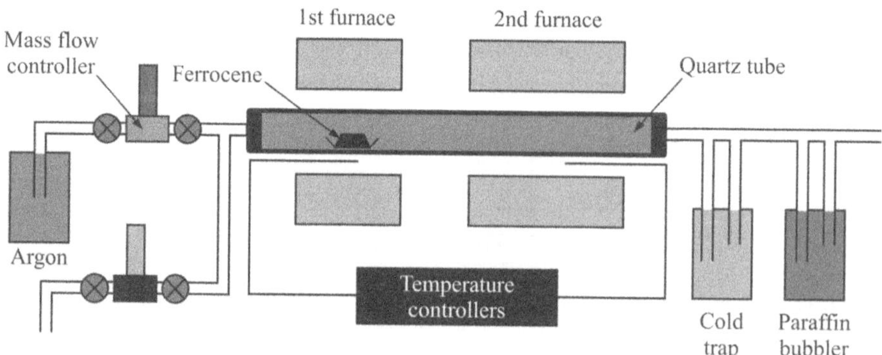

Figure 2.23 Schematic diagram of a vapour phase growth apparatus

2.5.4.5 CoMoCAT Process

The strategy used in the CoMoCAT method is to keep the active Cobalt species (Co) stabilized in a non-metallic state by interaction with Molybdenum oxide (MoO_3) before it is reduced by the carbon-containing compound (CO).

When exposed to carbon monoxide, the Co-Mo dual oxide is carburized, producing Molybdenum carbide and small metallic Co clusters, which remain in a high state of dispersion and result in high selectivity towards SWNT of very small diameter.

Lower temperature syntheses and stabilization of small metal clusters yield a CoMoCAT (66-68) (nanotube product with a smaller average diameter and a narrower distribution of structures compared to other synthesis methods.

The CoMoCAT process utilizes fluidized bed reactors to maintain precise control of the temperature and flow rates, resulting in high (n, m) selectivity.

During the SWNT reaction, cobalt is progressively reduced from the oxidic state to the metallic form. Simultaneously Molybdenum is converted to the carbidic form (Mo_2C). Co acts as the active species in the activation of CO, while the role of the Mo is possibly dual. It would stabilise Co as a well-dispersed CO^{2+} avoiding its reduction and would act as a carbon sink to moderate to the growth of carbon inhibiting the formation of undesirable forms of carbon.

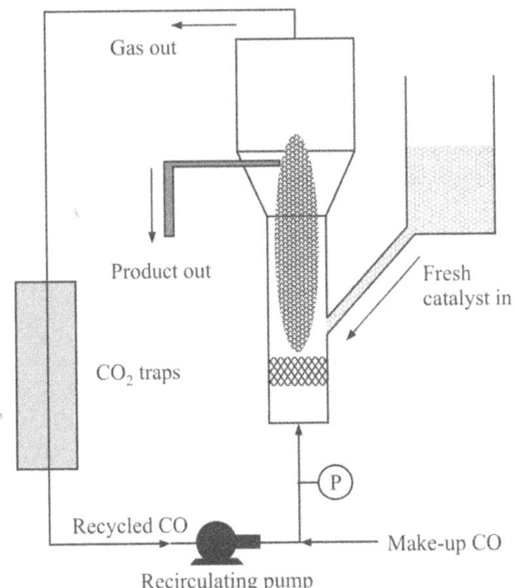

Figure 2.24 CoMoCAT apparatus

It is found that one of the critical conditions for an effective reactor operation is that the space velocity has to be high enough to keep the CO conversion as low as possible. Figure 2.24 shows a fluidized bed reactor for a CoMoCat process. The most important advantage of fluidized bed reactors is that they permit continuous addition and removal of solid particles from the reactor, without stopping the operation.

This method can be scaled-up without losses in SWNT quality. By varying the operation conditions, SWNTs can be produced with different diameter ranges. In Table 2.2 several diameter ranges at different temperatures are given.

Table 2.2 Diameter ranges of SWCNT at different temperatures

Temperature (°C)	Diameter range (nm)
750	0.9 ± 0.05
850	0.9 – 1.25
950	1.00 – 1.40

2.6 Characterization Techniques for Nanotubes

The scanning electron microscopy (SEM) is used to characterize the morphology (diameter and the length) of the deposited powder, while the structure is characterized using the transmission electron microscopy (TEM) for the selected sample. The SEM gives good results in characterizing the length and the diameter of the CNT and the CNF because it is used to scan only the surface of the sample, but it cannot be used to distinguish between the structure of the CNT and CNF, while the TEM is the best way to distinguish between them.

These characterization tools are discussed in detail in Chapter 4.

2.7 Purification Techniques of CNT

A large problem with nanotube application is next to large-scale synthesis also the purification. The produced SWNT soot contains a lot of impurities. The main impurities in the soot are graphite (wrapped up) sheets, amorphous carbon, metal catalyst and the smaller fullerenes.

The purification techniques can be divided into two mainstreams, structure selective and size selective separations. The first one will separate the SWNTs from the impurities; the second one will give a more homogeneous diameter or size distribution.

2.7.1 Oxidation

Carbonaceous materials can be effectively removed by Oxidation (68-74). The main disadvantages of oxidation are that not only the impurities are oxidised, but also the SWNTs. Luckily the damage to SWNTs is less than the damage to the impurities. These impurities have relatively more defects or a more open structure. Another reason why impurity oxidation is preferred is that these impurities are most commonly attached to the metal catalyst, which also acts as oxidising catalyst. Altogether, the efficiency and the yield of the procedure are highly dependable on a lot of factors, such as metal content, oxidation time, environment, oxidising agent and temperature.

The fact that metal acts as oxidising catalyst, the metal content should certainly be taken into consideration, when looking at the oxidising time. For example, when the temperature is raised above 600 °C, SWNTs will also oxidise, even without catalyst. This is the case with thermal, fixed air and pure oxygen oxidations. These can easily oxidise all the components, so the temperature and the time should be in good control.

2.7.2 Acid Treatment

In general the acid treatment will remove the metal catalyst, the surface of the metal need to be exposed by oxidation or sonication and the metal catalyst will be then exposed to acid and solvated. SWNTs purification using HNO_3 will have only effect on the metal catalyst and no effect on the SWNTs and other carbon Particles (75-77). If HCl is used, the acid will also have a little effect on the SWNTs and other carbon particles. The mild acid treatment (4 M HCl reflux) is basically the same as the HNO_3 reflux, but here the metal has to be totally exposed to the acid to solvate it.

2.7.3 Ultra Sonication

In this technique particles will be separated due to ultrasonic vibrations resulting nano particle agglomerates to vibrate and be more dispersed. The separation of the particles is highly dependable on the surfactant, solvent and reagent used. The solvent influences the stability of the dispersed tubes in the system. In poor solvents the SWNTs are more stable if they are still attached to the metal while in alcohols, mono-dispersed particles will be relatively stable (77-79). When an acid is used, the purity of the SWNTs depends on the exposure time. When the tubes are exposed to the acid for a short time, only the metal solvates, but for a longer exposure time, the tubes will also be chemically cut.

2.7.4 Magnetic Purification

In this process the SWNT suspensions will be mixed with ferromagnetic catalytic particles like $ZrO_2/CaCO_3$ in ultrasonic bath (Figure 2.25) and then these particles will be removed mechanically from graphitic shells by trapping with permanent magnetic poles. High pure SWNT can be obtained after subsequent chemical treatment (78).

This process does not require large equipment and enables the production of laboratory-sized quantities of SWNTs containing no magnetic impurities.

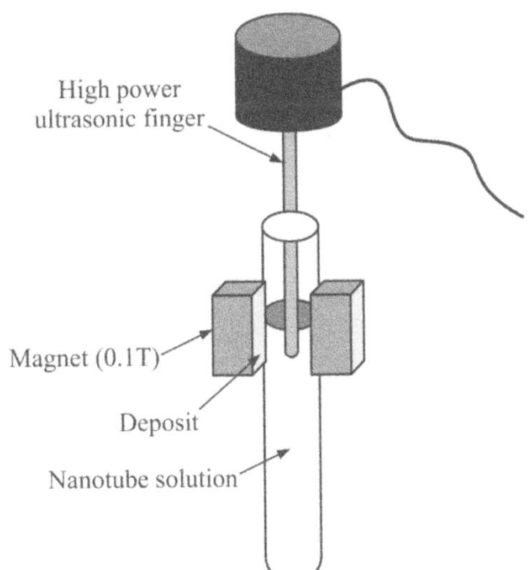

Figure 2.25 Ultrasonic Magnetic purification Apparatus

2.7.5 Micro Filtration

In micro filtration the SWNTs will be removed based on size separation by cross flow filtration through a hallow nano fiber filter. Initially by soaking in CS_2 the soluble graphite will be separated by filtering the insolubles in CS_2 through the nano fiber filter. Figure 2.26 Shows a micro filtration cell (79).

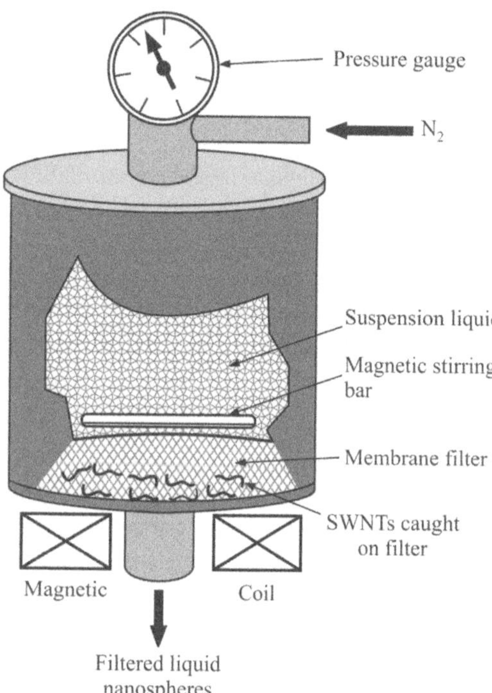

Figure 2.26 Micro filtration

In cross flow filtration the filtrate is pumped down the bore of the fibre at some head pressure from a reservoir and the major fraction of the fast flowing solution which does not permeate out the sides of the fibre is fed back into the same reservoir to be cycled through the fibre repeatedly. A fast hydrodynamic flow down the fibre bore (crossflow) sweeps the membrane surface preventing the build-up of a filter cake (79).

2.7.6 Cutting

An ultra-sonical induced cutting in acid solution includes a combination of mechanical and chemical cutting of SWNTs (80). SWNTs can be chemically cut by partially functionalising the tubes, for example with fluoro groups and then, the fluorated carbon will be driven off the sidewall with pyrolisation in the form of CF_4 or COF_2. This will leave behind the chemically cut nanotubes. Mechanical cutting of the nanotubes can be induced by ball-milling. Here, the bonds will break due to the high friction between the nanoparticles and the nanotubes will be disordered. Then, in combination with acid the nanotubes will rupture at the defect sites.

2.7.7 Functionalisation

In this process SWNTs will be made more soluble than the impurities by attaching other groups to the tubes which will be separated from impurities like metals by filtration (81-82).

Another functionalisation technique also leaves the SWNT structure intact and makes them soluble for chromatographic size separation. For recovery of the purified SWNTs, the functional groups can be simply removed by thermal treatment, such as annealing.

2.7.8 Chromatography

SWNTs can be run through chromatographic columns like Gel Permeation Chromotography (GPC) and HPLC -SEC High Performance Liquid Chromatography –Size. Exclusion Chromatography) columns and SWNTs will flow through depending on their size.

Larger sizes come first and smaller ones comes last at the end of the column. Thus the pore size will control what size distribution can be separated. The main hurdle is that it is necessary to keep SWNTs in dispersed or soluble form which can be achieved by pretreatment with ultrasonication or functionalization with soluble groups (82-83).

2.8 Important Industrial and Bio-Medical Applications of Carbon Nanotubes (CNTs)

The various industrial applications related to the specific properties where carbon nanostructure present a distinctive advantage are depicted in Figure 2.27 of which some are at full scale commercialization stage while some are at development stage.

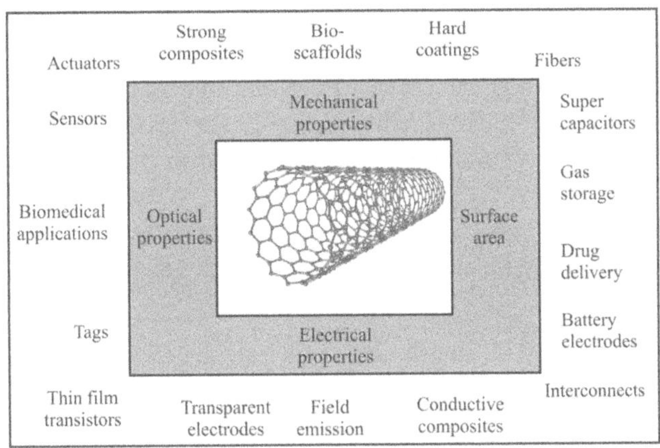

Figure 2.27 Various applications based on extraordinary properties of carbon nanotubes

The cost and availability of nanotubes of consistent quality are becoming main barriers for their large scale applications.

Some of the most widely investigated applications include 1. Composites (conductive and high-strength nanotube/polymer composites, 2. field Emitters/Emissions, field emission displays and radiation sources, 3. Electro Chemistry (Special Electrodes, transparent electrodes, sensors and nano electromechanical devices, Energy Production and Storage-batteries - solar Cells, 4. Electronics semiconductor devices (e.g., transistors) and interconnects.

2.8.1 Composites

Strength: Carbon nanotubes have a higher tensile strength than steel and Kevlar due to the sp^2 bonds between the individual carbon atoms which is even stronger than the sp^3 bond found in diamond. Under high pressure, individual nanotubes can bond together, trading some sp^2 bonds for sp^3 bonds. This gives the possibility of producing long nanotube wires. Carbon nanotubes are not only strong, they are also elastic. You can press on the tip of a nanotube and cause it to bend without damaging to the nanotube, and the nanotube will return to its original shape when the force is removed. A nanotube's elasticity does have a limit, and under very strong forces, it is possible to permanently deform to shape of a nanotube. A nanotube's strength can be weakened by defects in the structure of the nanotube. Defects occur from atomic vacancies or a rearrangement of the carbon bonds. Defects in the structure can cause a small segment of the nanotube to become weaker, which in turn causes the tensile strength of the entire nanotube to weaken. The tensile strength of a nanotube depends on the strength of the weakest segment in the tube similar to the way the strength of a chain depends on the weakest link in the chain.

The high aspect ratio (length to radius ratio) and high conductivity of CNTs makes them excellent for conducting composites. A mixture of conducting and insulating phases becomes conducting when the volume fraction of conducting phase exceeds a 'percolation threshold' of 16%, the minimum amount to give a continuous path across the whole sample. This threshold is independent of the size and shape of the conducting phase, as long as its particles are equiaxed. When the conducting phase consists of long thin particles, the chance of contact increases, which reduces the percolation threshold so that conduction occurs at much lower loading.

Existing conducting composites are made of carbon black in polymers. This should require about 16% of carbon, but it has been reduced by clever processing. Figure 2.28 shows an example of composites of MWNTs in epoxy resin achieving percolation at loadings of 0.01% and even 0.004% with careful processing (84-85). These are extremely low loadings. The first use of such composites was by Hyperion for electrostatically applying paint onto car components (90). Another use of nanotube composites is as antistatic shielding, on airplane wings and fuselages.

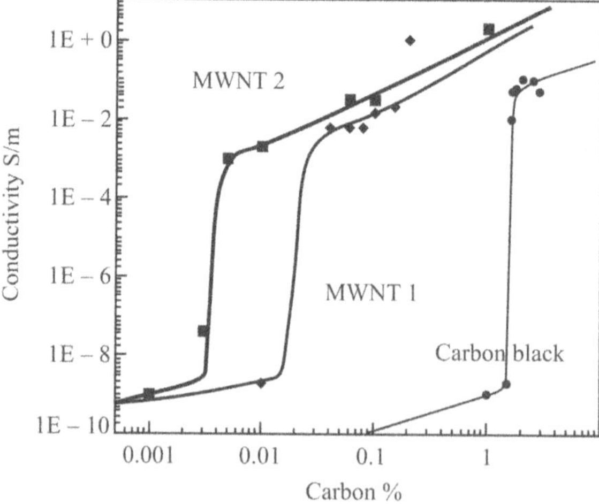

Figure 2.28 Conductivity versus carbon content for three different carbon/epoxy composite systems showing the rapid increase in conductivity corresponding to the percolation threshold. Very low threshold values are possible for MWNTs (*Source:*92)

A future use is for shielding of electromagnetic interference, a critical application for many industrial sectors. This would require a composite with conductivity of 1 S•cm. Necessity that the carbon phase should be highly conducting and is probably only achievable if SWNTs are used as the loading.

Another application of conducting composites is as a transparent conductor. There is a huge market for transparent electronic conductors such as indium tin oxide (ITO) in displays. In this field, there is a drive towards flexible displays on plastic substrates. ITO is less good for this situation as it is brittle and has poor adhesion to plastic. Conducting composites of SWNTs can be transparent if thin enough (86). They have the huge advantage of being truly flexible and compatible with polymer substrates. This is a large potential market with few competitors.

SWNTs have the highest Young's modulus of about 1 TPa if normalized to their diameter and, therefore, all types of CNTs have attracted much interest for low weight structural composites and many times their performance might eventually exceed that of carbon fiber reinforced composites.

2.8.2 Field Emission

CNTs are ideal field-emission (FE) materials (87-90) due to their high aspect ratio. FE is the emission of electrons from a solid under an intense electric field. The simplest way to create such a field is by field enhancement at the tip of a sharp object. Si or W tips were initially used, made by anisotropic etching or deposition. CNTs have an advantage over Si or W tips in that their strong, covalent bonding means they are physically inert to sputtering, chemically inert to poisoning, and can carry a huge current density of 10^9 A/cm^2 before electromigration. In addition, when driven to high currents, their resistivity decreases, so that they do not tend to electric-field-induced sharpening, which causes instabilities in metal tip field emitters. CNTs have better FE performance than other forms of carbon such as diamond and diamond-like carbon. Both SWNTs and MWNTs exhibit emission at low threshold electric fields of 1 to 2 V/μm, indicating that their sharp tips geometrically enhance local electric fields by factors of 1000 or more. The tolerance of nanotubes for extraordinary current densities further allows these emitters to operate at high power densities. Individual field emitters constitute a small commercial market, but arrays of emitters can be used for general lighting, flat panel displays, and high power electronics. In each of these applications, the nanotube emission exceeds the minimum performance specifications to be competitive; the primary barriers to commercialization involve other issues such as fabrication, compatibility, and product complexity. In lighting, these barriers are relatively small and high-brightness, high-efficiency lighting products exist.

2.8.2.1 Chemical and Biological Sensors

Chemical sensor can exploit one of the important features of SWNT circuits, namely their sensitivity to chemical perturbations. SWNTs are ideal for electronic chemical sensors as charge carriers are exclusively confined to the conductor's surface and the measured resistance can be sensitive to individual scattering sites. Even though applications requiring large numbers of identical FETs remain out of reach for SWNT electronics, a chemical sensor demands only a few individual circuits, By doping individual SWNTs exhibit unexpected sensitivities to air exposure to reagents like NH_3 and NO_2.

2.8.3 Electrochemistry

Large surface area and low resistivity of CNTs makes them of great interest in electrochemistry. At any electrode in contact with an electrolyte, there exists a dipole layer, the so-called electrochemical double layer (ECDL) (Figure 2.29). Typically ~0.4 nm thick, the ECDL acts as a capacitor if the potential difference across it is less than that needed to dissociate the electrolyte (into hydrogen or oxygen in the case of water). The capacitance is given by $C = \varepsilon A/d$, where ε is the dielectric constant, A is the electrode surface area, and d is the ECDL thickness.

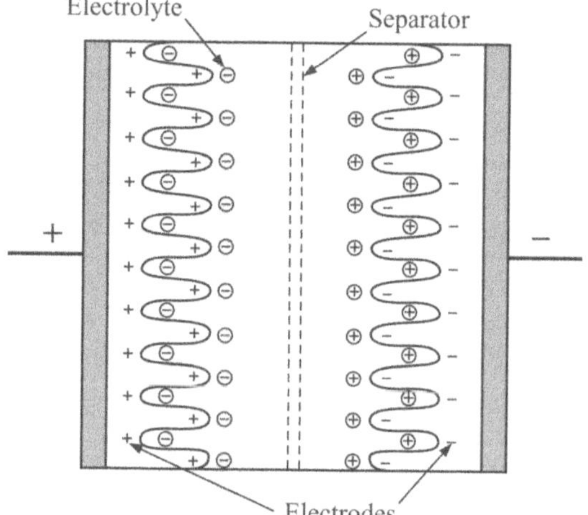

Figure 2.29 Schematic of Super Capacitor

This capacitance can be huge for cases such as porous carbon, because of the large A and small d values. This has led to the use of porous carbons such as activated carbon for the electrodes in so-called super capacitors (91-93).

Super capacitors consist of two carbon-based electrodes on a mylar backing, an ion-permeable membrane, and an electrolyte, as shown in Fig 2.29. If the electrolyte is water, this restricts the voltage to 1.23 V, but higher voltages of 2.3 V are possible by using a non-aqueous electrolyte such as tetra- ethyl ammonium fluoroborate in propylene carbonate. 1000 F super capacitors are typically 12 cm × 2.5 cm in size and there is already a substantial market in inverters. The electrodes are presently activated carbon, but CNTs are a possible replacement given their large and controllable surface area.

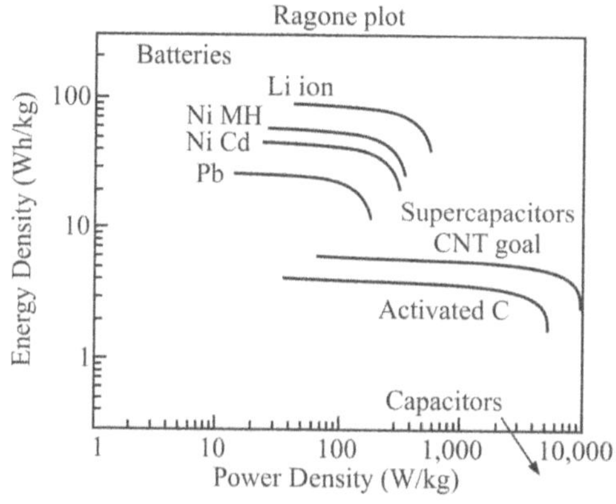

Figure 2.30 The power density versus energy density plot distinguishes various electricity storage systems (Source 95)

The performance of super capacitors is judged in terms of their energy density (per unit volume) or $\frac{1}{2}CV^2$ and power density or V^2/R, where R is their series discharge resistance. The performance can be plotted on a 'Ragone plot (Figure 2.30). The energy density of super capacitors is proportional to C and to the surface area per unit volume of carbon. SWNTs have the largest surface area to volume ratio of any carbon material, 3000 m^2/g, as all their atoms are on the surface, and should make the ultimate electrode material. In fact, the batteries are high energy density but low power density, whereas normal capacitors are low energy density but high power density. Li^+ batteries have the highest energy density of any battery. The advantage of CNTs is that they have a much lower resistance than activated carbon, which could greatly increase their power density. The limiting factor is contacting CNTs to the electrode polymer backing, the ability to do this at low temperatures, and at a low cost, as activated carbon is very cheap.

2.8.3.1 Energy Storage - Li-Batteries

Electrodes made of carbon nanotubes can be ten times thinner and lighter than amorphous carbon electrodes and their conductivity is more than one thousand times greater. In some cases, such as electric vehicles, the reduction in weight can make a significant reduction in battery power requirements. Carbon nanotubes have been used in super capacitors producing a power density of 30 kW/kg (compared to 4 kW/kg for commercially available devices). Such super capacitors could drastically reduce the time it takes to recharge devices such as laptops and cell phones.

Ultra-thin flexible batteries have been made with CNT infused paper (93). Ionic liquid is soaked into the paper as the battery's electrolyte. Electrolytes in human blood, sweat, and urine can also help to power the battery which may be useful in implantable medical devices. These batteries can be rolled, folded, or cut without loss of efficiency. They can also be stacked to boost their output power. Although the materials used in the batteries, which are over 90% cellulose, are very inexpensive an expensive method of mass-production has not yet been developed.

2.8.3.2 Solar Cells

Solar cell made from carbon nanotubes (CNTs) are twice as good at converting sunlight into power than the best previous such cells has been unveiled by a team of researchers in the US (Figure 2.31). The National Renewable Energy Laboratory (NREL) has already independently certified the performance of the device – a first for a CNT-based solar cell (94).

Thin-film photovoltaic materials are better than conventional solar-cell materials (such as silicon) because they are lighter, more flexible and cheaper to make. They work by absorbing photons from sunlight and converting these into electron–hole pairs (or excitons). To generate electric current, an electron and hole must be rapidly separated before the two particles have a chance to come back together and be reabsorbed into the material. In solar cells, the exciton must quickly travel to another layer in the device (where the charge separation will occur) for the best light-absorption efficiencies.

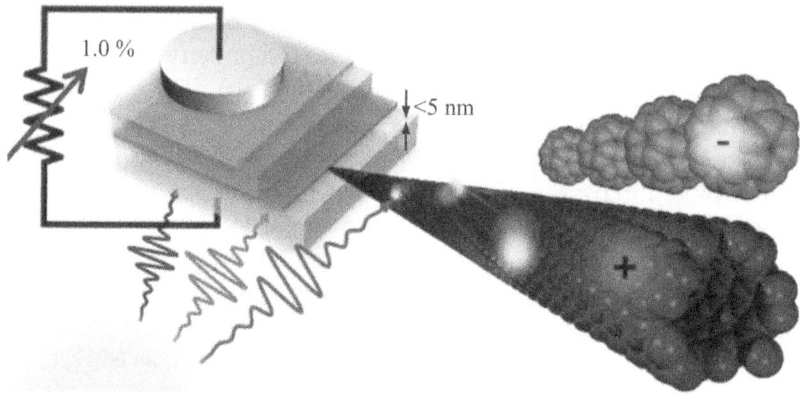

Figure 2.31 Light from the sun creates charges in an ultra thin film of carbon nanotubes (blue), which are extracted by fullerene C60 (brown) in this schematic of the ground breaking proof-of-concept solar cell with greater than 1 percent efficiency (94)

Single-walled carbon nanotubes (SWCNTs) are ideal as thin-film photovoltaics because they absorb light across a wide range of wavelengths from the visible to the near-infrared and possess charge carriers (electrons and holes) that move quickly. However, most thin-film cells containing SWCNTs have so far suffered from limited current and voltage, and therefore poor power-conversion efficiencies.

2.8.3.3 CNTs for Hydrogen Storage for Clean Energy

Two of the major challenges for humanity in the next 20 years are the shrinking availability of fossil fuels and the global warming and potential climate changes that result from their ever-increasing use. One possible solution to these problems is to use an energy carrier such as hydrogen, and ways to produce and store hydrogen in electric power plants and vehicles is a major research focus for materials scientists and chemists. To realize hydrogen-powered transport, for example, it is necessary to find ways to store hydrogen onboard vehicles efficiently and safely. Carbon nanotubes are one of the most promising materials, and hydrogen storage through both physisorption and chemisorption mechanisms.

The large surface area and tubular structure of CNTs suggest that capillarity effects are important. Thus, CNTs could be very useful for the storage of hydrogen, a key aspect of the clean energy economy. A storage capacity of 10%, if readily cyclable, would be an extremely valuable property compared to other systems. Initial studies suggested a storage capacity of 60% by weight, but more realistic estimates were 8-10 wt% based on extrapolations to 100% nanotube content.

2.8.4 Electronics

CNTs can carry the highest current density of any metal, 10^9-10^{10} A/cm^2, over 1000 times that for Cu, before failing as a result of electromigration (self-electrolysis). It is well known that feature sizes in Si integrated circuits are continually reducing. This forces the interconnects between each transistor to carry ever larger current densities. Graham et.al., have (95) shown how the vertical interconnects known as vias could be replaced by CNTs and thus satisfy this requirement. This application is relatively easy, particularly as it requires vertical nanotube growth. Multiple CNTs in parallel can be used to lower the overall resistance (of any Schottky barrier). However, the greatest demand is not for vias but for horizontal interconnects. This application is much more difficult to satisfy, as horizontally-directed nanotube growth can be done, but not with high reliability and yield.

CNTs have also been made into field-effect transistors (FETs). The first examples were made by dispersing SWNTs on SiO_2-covered Si wafers and then making contacts. The Si substrate is used as the bottom gate electrode. This is not suitable for a real device, as the gate would be common to all FETs on the wafer. The second generation FET requires a separate 'top gate' over each individual nanotube. Wind *et al.* (96) have achieved FETs with very good performance figures. Javey *et al.* (97) subsequently used a ZrO_2 gate oxide to achieve the best gate turn-on performance of any nanotube device.

A significant problem for electronic applications is that the band gap of a SWNT depends on its chirality. The cohesive energy depends only weakly on diameter and very weakly on chirality, so that a mixture will contain both metallic and semiconducting nanotubes. This is a huge disadvantage in electronics. It is necessary to be able to grow a specific type of tube at a defined position in a defined direction with near 100% yield. Control of position and direction is achievable now. It is hoped to be able to gain some control of chirality in the CVD process by the design of the catalyst (98-100).

CNTs show a large nonlinear absorption of light (100). This arises because of their one-dimensional band structure with singularities in the electronic density of states. Optical absorption creates electron-hole pairs, but these rapidly fill up the first bands and are then forced to occupy the next bands at higher energy. This creates a nonlinear absorption coefficient. The interesting aspect is that random mixtures of CNTs contain some with band gap singularities covering the whole optical spectrum, so that selection is not necessary. In addition, excited electrons decay rapidly in picoseconds. Thus fast saturable absorber, limiter, and optical switches are possible, without requiring excessive processing of the CNTs.

2.8.5 Recent Research on Some Special Applications of CNTs

2.8.5.1 Carbon Nanotube Cupcakes to Measure Terahertz Laser Light

CNT cup cakes are made up of vertically aligned carbon nanotube arrays (VACNTAs) that are grown on silicon. Using a razor blade, visible chunks of these densely packed arrays can be sliced off and placed on top of a laser power detector to detect terahertz radiation. Terahertz radiation can penetrate materials like plastic, clothing, paper and some biological tissues, making it an attractive candidate for applications such as concealed weapons detection, package inspection and imaging skin tumors. Until now, there wasn't a standard method for measuring the output power of terahertz lasers, but researchers at the National Institute of Standards and Technology (NIST) claim they are fore runners with these nano-cakes.

2.8.5.2 Carbon Nanotubes for Cleaning Polluted Water

Scientists found that by using filters made of carbon nanotubes, pollutants could be removed more effectively from contaminated water as compared to common charcoal filters. CNTs have a very large surface area (e.g., 500 m^2 per gram of nanotube) that gives them a high capacity to retain pollutants such as water soluble drugs. A team at the University of Vienna found that at concentrations likely to occur in the environment, the tubes removed 13 tested Polycyclic Aromatic Hydrocarbons (PAHs) from contaminated water.

2.8.5.3 Boosting Solar Energy Storage by a Factor of 10,000

The business of storing solar energy in molecules that change state in response to light could be entirely transformed by carbon nanotubes. Researchers at the Massachusetts Institute of technology (MIT) have announced a new solar thermal fuel that could store up to 10,000 times more energy than previous systems. The fuel consists of carbon nanotubes modified with azobenzene, a mix that is expected to provide the same energy storage per volume as lithium-ion batteries and can store solar energy almost indefinitely. It can also be recharged by simply exposing it to sunlight – no electricity required.

Pure metallic carbon nanotubes could be the key to overhauling the electrical power grid with more efficient transmission lines -- but only if they could be made in huge quantities and uniformly. Through a refined version of a technique called "amplification," researchers at Rice University plan to make long and highly conductive nanotube fibers they dubbed "armchair quantum wire" that could be woven into more efficient electrical transmission lines.

2.8.5.4 Molecular Syringes, Probes, and Bioelectronic Noses

Medical researchers are eying carbon nanotubes as well, such as potential needles for injecting drugs or genes into sick cells. Size and shape are only half the reason why. As probes, their physical properties, including their great electrical and thermal conductivity, make them particularly suited for exchanging information between the inside and outside of the cell. Nano tube probes may be used to test for certain substances and test certain processes beyond cell membranes.

2.8.5.5 Bio-Medical Applications of CNTs

Carbon Nano Tubes (CNTs) possess unique structures and exceptional optical, electrical, magnetic, mechanical, thermal properties which can be exploited for bio medical applications. Further as they possess high surface reactivity which facilitates their functionalization with different organic functional groups to carry simultaneously several moieties for targeting, imaging, and therapy. Ultra-sensitive detection of biological species with carbon nanotubes can be realized after surface passivation to inhibit the non-specific binding of bio-molecules on the hydrophobic nanotube surface. Electrical nanosensors based on nanotubes provide a label-free approach to biological detections. Surface enhanced Raman spectroscopy of carbon nanotubes opens up a method of protein microarray with down to 1 fM detection sensitivity.

Functionalized CNTs are promising novel materials for a variety of biomedical applications. These potential applications are particularly enhanced by their ability to penetrate biological membranes with relatively low cytotoxicity.

Functionalization of carbon nanotubes for bio-medical applications

The overall objective of functionalizing CNTs for biomedical applications is to increase their solubility or dispersion in biocompatible (aqueous) media, thereby reducing toxic effects. Surface functionalization of carbon nanotubes may be covalent or non-covalent (Figure 2.32 a). Chemical reactions forming bonds with nanotube sidewalls are carried out in the covalent functionalization case, while non-covalent functionalization exploits favorable interactions between the hydrophobic domain of an amphiphilic molecule and the CNT surface, affording aqueous nanotubes wrapped by surfactant (Fig 2.32 b)

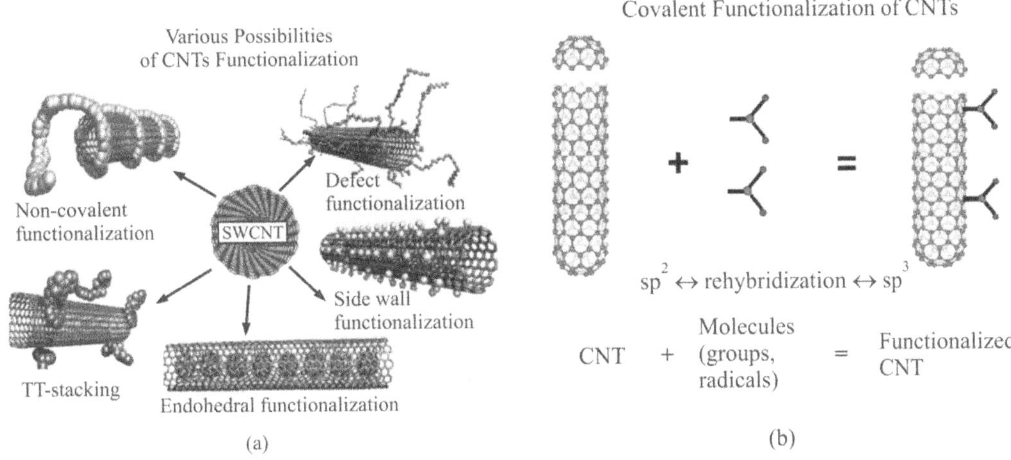

Figure 2.32 (a) Various options for CNT functionalization
(b) Covalent functionalization of CNTs

Covalent functionalization of carbon nanotubes

Various covalent reactions have been developed to functionalize carbon nanotubes. Among them oxidization is one of the most common. CNT oxidization is carried out with oxidizing agents such as nitric acid (101). During the process, carboxyl groups are formed at the ends of tubes as well as at the defects on the sidewalls. Zeng et al., (102) observed sp^3 carbons on SWNTs after oxidization and further covalent conjugation with amino acid Figure 2.33 (103). However, although oxidized CNTs are soluble in water, they will aggregate in the presence of salt due to charge screening effects, and thus cannot be directly used for biological applications due to the high salt content of most biological solutions. Further

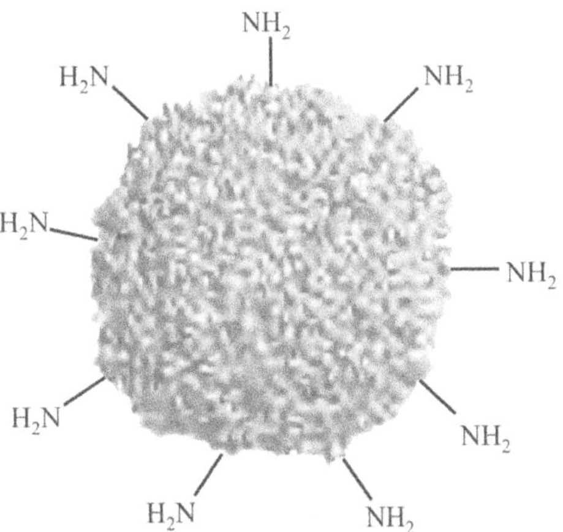

Figure 2.33 Covalent conjugation with amino acid

modification can be achieved by attaching oxidized CNTs with hydrophilic polymers such as poly ethylene glycol (PEG), yielding CNT-polymer conjugates stable in biological environments]. Another type of widely used type of covalent reaction to functionalize CNTs is the cylcoaddition reaction, which occurs on the aromatic sidewall, instead of nanotube ends.

It has been reported that after modifications, functionalized carbon nano tubes (fCNT) solubility increased significantly. Many studies have shown that increased solubility (or dispersion) of fCNTs improves their performance and lowers their toxicity (104-106). These fCNTs have excellent electro-optical properties, high tensile strength, and a high surface-area-to-volume ratio that facilitates surface functionalization. The addition of a layer of biocompatible material can be used to annihilate the toxicity of pristine CNT aggregates by making them more dispersible in aqueous solutions. Functionalization of CNTs can be achieved by using oligonucleotides, biomolecules, surfactants, and polymers, thus increasing the dispersibility of CNTs and decreasing their cytotoxicity.

In vitro and *in vivo* toxicity studies reveal that well water soluble and serum stable functionalized nanotubes are biocompatible, non-toxic and potentially useful for a variety of biomedical applications.

2.8.6 Current Hurdles on the Application of CNTs

There are several obstacles, technical and non-technical to CNT success.

Electronic Heterogeneity

One problem with nano tube production for electronics is that batches of nano tubes are heterogeneous mixtures of metallic and semiconducting tube types. Electrical devices typically require these types to be separated, but so far it has been difficult to tune production in this regard. There also many issues with doping, or tuning conductivity, and electrical behavior at contact points.

2.8.6.1 Orientation

A problem with nano tubes where recent progress has been made is controlling their orientation. Nano tubes are commonly grown in a chaotic organization (affectionately known as a "rat's nest"), which are difficult to use in microprocessors. Recently John Rogers and his team at the University of Illinois, Urbana Champagne, discovered that carefully growing nano tubes on quartz wafers can lead to a highly organized configuration.

2.8.6.2 Size and Density

The size of manufactured nano tubes typically varies widely. For commercial use, nano tube manufacturers will need to make size more consistent. Though nano tubes are very narrow, nano tube matrices typically have quite large (100 nm) spacing between tubes.

2.8.6.3 Export Policy

As with other multi-use technologies, nano tubes may be subject to export controls. Finding information about this was difficult, multiple sources were unaware of restrictions but in at least one case a foreign researcher was denied access to nano tubes. Adelina Santos, a Brazilian nuclear scientist, says a U.S. based supplier refused to ship him nano tubes due to federal regulations. However, restrictions seem difficult to enforce; Santos had a friend smuggle a gram of nano tubes to him. Current customs protocols probably do not place a priority on detecting nano tubes.

Export restrictions could slow adoption of nano tube technologies and prevent standardization. Regulation of commerce in nano tube technology will increase costs.

2.8.6.4 Environmental Concerns

The environmental risks of nano tubes are still unclear. Naturally occurring carbon is fairly benign, and is largely unregulated, but nano tubes interact with the environment differently. Though there have been several studies performed to test the effects of carbon nano tubes on living systems still the effects are not clear.

In some situations, coatings applied to the nanotubes, rather than the nanotubes themselves, may become environmental culprits. The solar cells are coated with a cadmium-telluride mix, which would be too toxic for widespread use. Perhaps most sobering to consider is that some forms of nanotubes biodegrade slowly, tubes released into the environment may make their way into our food supply, and from there, throughout our bodies.

References

1. S. Iijima (1991). Nature 354, 56.
2. P. J. F. Harris (2009). Carbon Nanotube Science - Synthesis, Properties, and Applications. Cambridge Univ. Press, Cambridge.
3. B. Peng et al. (2008). Nat. Nanotechnol. 3, 626.
4. B. Q. Wei, R. Vajtai and P. M. Ajayan (2001). Appl. Phys. Lett. 79, 1172.
5. E. Pop, D. Mann, Q. Wang, K. Goodson and H. J. Dai (2006). Nano Lett. 6, 96.
6. A. Y. Cao, P. L. Dickrell, W. G. Sawyer, M. N. Ghasemi-Nejhad and M. Ajayan (2005). Science 310, 1307.
7. Ren, Z. F., Huang, Z. P., Wang, D. Z., Wen, J. G., Xu, J. W., Wang, J. H., Calvet, L. E., Chen, J., Klemic, J. F., and Reed, M. A. (1999). Applied Physics Letters, 75, (8).
8. M. Yudasaka, Masako, Kikuchi, Rie, Matsui, Takeo, Ohki, Yoshimasa, Yoshimura, Susumu, and Ota. Etsuro, Applied Physics Letters, 67, (17).
9. Yudasaka, Masako, Kikuchi, Rie, Ohki, Yoshimasa, Ota, Etsuro, and Yoshimura, Susumu, Applied Physics Letters, 70, (14), 1997.
10. Chen, M., Chen, C. M., and Chen, C. F. (2002). Journal of Materials Science, 37, (17), 3561-3567.
11. H.W. Kroto, J.R. Heath, H.C. Obrien, R.F. Curl, and R.E. Smalley (1985). Nature 318, 162.
12. D.A. Bochvar and E.G. Gal'pern (1973). Dokl. Akad. Nauk. USSR, 209, 610.
13. I.V. Stankevich, M.V. Nikerov, and D.A. Bochvar (1984). Russ. Chem. Rev., 53, 640.
14. W. Krätchmer, L.D. Lamb, K. Fostiropoulos and D.R. Huffman (1990). Nature, 347-354.

15. Dresselhaus, M. S., Dresselhaus, G., Jorio, A., Souza Filho, A. G., Pimenta, M. A., and Saito, R. (2002). Accounts of Chemical Research, 35, 1070-1078.
16. S. Saito and A. Oshiyama, Phys. Rev. Lett. 66 2637 (1991). aus, M. S., Physical Review Letters, 86, (6), 1118-1121, 2001.
17. Kittel (1995). Introduction to Solid state Physics, 7th edn, John Wiley, New York.
18. J. W. Mintmire, B. I. Dunlap, and C. T. White (1992). Phys. Rev. Lett. 68, 631.
19. N. Hamada, S. Sawada and A. Oshiyama (1992). Phys. Rev. Lett. 68, 1579.
20. R. Saito, M. Fujita, G. Dresselhaus, and M. S. Dresselhaus (1992). Appl. Phys. Lett. 60, 2204.
21. P. M. Ajayan (1999). Chem. Rev. 99, 1787.
22. Carbon nanotube. (2008, April 24). In Wikipedia, The Free Encyclopedia. Retrieved 03:04, April 28, 2008, from http://en.wikipedia.org/w/index.php?title=Carbon_nanotube&oldid=207910667.
23. Y. Saito et. al. (1993) J. Physics. Chem. Solids 54, 1849.
24. P. M. Ajayany and T. W. Ebbesenz (1997). Nanometre-Size Tubes of Carbon, Rep. Prog. Phys., 60, 1025–1062.
25. Ajayan, P. M. and Ebbesen, T. W. (2003). Rep. Prog. Phys., 60, 1025-1065.
26. Nasibulin, Albert G. et al. (2007). A novel hybrid carbon material. Nature Nanotechnology 2 (3): 156–161. doi:10.1038/nnano.2007.37.PMID 18654245.
27. Morgan, P. (2005). Carbon Fibers and their Composites, Taylor & Francis Group, CRC Press, Boca Raton, FL.
28. J Tibbetts, G. Lake, M. Strong, K. and Rice, B. (2007). A review of the fabrication and properties of vapor-grown carbon nanofiber/polymer composites. Composites Science and Technology 67 (7–8): 1709.doi:10.1016/j.compscitech.2006.06.015. edit.
29. Hammel E, Tang X, Trampert M, Schmitt T, Mauthner K, Eder A and Pötschke P. (2004). Carbon nanofibers for composite applications. Carbon 42 (5–6): 1153.doi:10.1016/j.carbon.2003.12.043. edit.
30. Burchell, T.D. (1999). Carbon Materials for Advanced Technologies, Pergamon (Elsevier Science Ltd.), Oxford, UK.
31. Prabhakar R. Bandaru (2007). Electrical Properties and Applications of Carbon Nanotube Structures. Journal of Nanoscience and Nanotechnology, Vol. 7, 1–29.
32. Wang, X.; Li, Qunqing; Xie, Jing; Jin, Zhong; Wang, Jinyong; Li, Yan; Jiang, Kaili and Fan, Shoushan (2009). Fabrication of Ultralong and Electrically Uniform Single-Walled Carbon Nanotubes on Clean Substrates. Nano Letters 9 (9): 3137–3141.
33. Charlier, J.C, Blasé, X & Roche, S. (2007). Electronic and transport properties of nano tubes, Rev. Mod. Phys, 79, 677(732).
34. S. Bellucci (ed.) (2011). Physical Properties of Ceramic and Carbon Nanoscale Structures, 45 Lecture Notes in Nanoscale Science and Technology 11, DOI 10.1007/978-3-642-15778-3_2, C Springer-Verlag Berlin Heidelberg.
35. J. A. Misewich et al. (2003). Electrically Induced Optical Emission from a Carbon Nanotube FET. Science 300 (5620): 783–786.BibH.

36. Kataura et al. (1999). Optical Properties of Single-Wall Carbon Nanotubes. Synthetic Metals 103 (1–3): 2555–2558. doi:10.1016/S0379-779(98)00278code:2003 Sci...300.. 783M. doi:10.1126/science.1081294.PMID 12730598.
37. Ajayan, P. M. and Zhou, O. Z. (2001). Carbon Nanotubes, 80, 391-425.
38. Damnjanovic, M., Milosevic, I., Vukovic, T., and Sredanovic, R. (1999). Physical Review B, 60, (4), 2728-2739.
39. Yasuda, Ayumu, Kawase, Noboru, and Mizutani, Wataru (2002). Journal of Physical Chemistry B, 106, 51.
40. Sinnot, S. B., Andrews, R., Qian, D., Rao, A. M., Mao, Z., Dickey, E. C., and Derbyshire, F. (1999). Chem. Phys. Lett., 315, 25-30.
41. Wal, R. L. V., Berger, G. M., and Hall, L. J. (2002). Journal of Physical Chemistry B, 106(14), 3564-3567.
42. Sinnot, S. B., Andrews, R., Qian, D., Rao, A. M., Mao, Z., Dickey, E. C., and Derbyshire, F. (1999). Chem. Phys. Lett., 315, (25-30).
43. Jung, S. H., Kim, M. R., Jeong, S. H., Kim, S. U., Lee, O. J., Lee, K. H., Suh, J. H., and Park, C. K. (2003). Applied Physics A-Materials Science & Processing, 76(2), 285-286.
44. Ebbesen, T. W. and Ajayan, P. M. (1992). Nature, 358, 220-222. Journet, C. and Bernier, P. (1998). Applied Physics A-Materials Science & Processing, 67(1), 1-9.
45. Farhat, S., La Chapelle, M. L., Loiseau, A., Scott, C. D., Lefrant, S., Journet, C., and Bernier, P. (2001). Journal of Chemical Physics, 115(14), 6752-6759.
46. Farhat, S., Hinkov, I., Chapelle, D. I., Fan, S. S., Li, G. H., and Scott, C. D. (2001). NASA Conference Publication.
47. Costa, Pedro, Xu, Cigang, Coleman, Karl, Sloan, Jeremy, and Green, Malcolm L H, (2002).
48. Huang, H. J., Marie, J., Kajiura, H., and Ata, M. (2002). Nano Letters, 2(10), 1117-1119.
49. Takikawa, H., Ikeda, M., Hirahara, K., Hibi, Y., Tao, Y., Ruiz, P. A., Sakakibara, T., Itoh, S., and Iijima, S. (2002). Physica B: Condensed Matter (Amsterdam, Netherlands), 323, (1-4).
50. Anazawa, Kazunori, Shimotani, Kei, Manabe, Chikara, Watanabe, Hiroyuki, and Shimizu, Masaaki (2002). Applied Physics Letters, 81, (4).
51. Lee, Seung Jong, Baik, Hong Koo, Yoo, Jae eun, and Han, Jong hoon (2002). Diamond and Related Materials, 11, (3-6).
52. Guo, T., Nikolaev, P., Thess, A., Colbert, D. T., and Smalley, R. E. (1995). Chemical Physics Letters, 243, (1,2).
53. Yudasaka, M., Yamada, R., Sensui, N., Wilkins, T., Ichihashi, T., and Iijima, S. (1999). Journal of Physical Chemistry B, 103, (30).
54. Eklund, P. C., Pradhan, B. K., Kim, U. J., Xiong, Q., Fischer, J. E., Friedman, A. D., Scott, C. D., Arepalli, S., Nikolaev, P., and Smalley, R. E. (2001). Applied Physics A: Materials Science & Processing, 72, (5).
55. Scott, CD, Arepalli S, Nikolaev P, Smalley RE. (2001). Growth mechanisms for single-wall carbon nanotubes in a laser-ablation process. Applied Physics A: Materials Science & Processing 72(5), 573-580.

56. Ren, Z. F., Huang, Z. P., Xu, J. W., Wang, J. H., Bush, P., Siegel, M. P., and Provencio, P.N. (1998). Science (Washington, D.C.), 282,(5391).

57. Yasuda, Ayumu, Kawase, Noboru, and Mizutani, Wataru (2002). Journal of Physical Chemistry B, 106, (51).

58. Ren, Z. F., Huang, Z. P., Wang, D. Z., Wen, J. G., Xu, J. W., Wang, J. H., Calvet, M. E., Chen, J., Klemic, J. F., and Reed, M. A. (1999). Applied Physics Letters, 75, (8).

59. Chen, M., Chen, C. M., and Chen, C. F. (2002). Journal of Materials Science, 37(17), 3561-3567.

60. Huang, Z. P., Wang, D. Z., Wen, J. G., Sennett, M., Gibson, H., and Ren, Z. F. (2002). Applied Physics A-Materials Science & Processing, 74(3), 387-391.

61. Park, J. B., Choi, G. S., Cho, Y. S., Hong, S. Y., Kim, D., Choi, S. Y., Lee, J. H., and Cho, K. I. (2002). Journal of Crystal Growth, 244(2), 211-217.

62. Maruyama, S., Chiashi, S., and Miyauch, Y. (2003). Thermal Engineering Joint Conference, 6.

63. Lee, Cheol Jin, Lyu, Seung Chul, Kim, Hyoun Woo, Park, Chong Yun, and Yang, Cheol Woong (2002). Chemical Physics Letters, 359, (1,2).

64. Lolli, G. Zhang, L. A. Balzano, L. Sakulchaicharoen, N. Tan, Y. Q. and Resasco, D. E. (2006). Tailoring (n,m) structure of single-walled carbon nanotubes by modifying reaction conditions and the nature of the support of CoMo catalysts, Phys. Chem. B., 110, 2108-2115.

65. Resasco D. E.; Alvarez, W. E.; Pompeo, F.; Balzano, L.; Herrera, J. E.; Kitiyanan, B.; Borgna, A. J. (2002). A scalable process for production of single-walled carbon nanotubes (SWNTs) by catalytic disproportionation of CO on a solid catalyst, Nanopart. Res. 4, 131-136.

66. Goto, Hajime, Furuta, Terumi, Fujiwara, Yoshiya, and Ohashi, Toshiyuki., (2002). 166-386, (2003007924).

67. Borowiak-Palen, E., Pichler, T., Liu, X., Knupfer, M., Graff, A., Jost, O., Pompe, W., Kalenczuk, R. J., and Fink, J. (2002). Chemical Physics Letters, 363(5-6), 567-572.

68. Huang, S. M. and Dai, L. M. (2002). Journal of Physical Chemistry B, 106(14), 3543-3545.

69. Chiang, I. W., Brinson, B. E., Smalley, R. E., Margrave, J. L., and Hauge, R. H. (2001). Journal of Physical Chemistry B, 105(6), 1157-1161.

70. Harutyunyan, A. R., Pradhan, B. K., Chang, J. P., Chen, G. G., and Eklund, P. C. (2002). Journal of Physical Chemistry B, 106(34), 8671-8675.

71. Farkas, E., Anderson, M. E., Chen, Z. H., and Rinzler, A. G. (2002). Chemical Physics Letters, 363(1-2), 111-116.

72. Yang, F. H. and Yang, R. T. (2002). Carbon, 40(3), 437-444.

73. Bauschlicher and Charles W., Jr. (2001). Nano Letters, 1(5).

74. Hou, Peng Xiang, Liu, C., Tong, Y, Liu, M., and Cheng, H. M. (2001). Journal of Materials Research, 16(9), 2526-2529.

75. Huang, Houjin, Shiraishi, Masashi, Yamada, Atsuo, Kajiura, Hisashi, and Ata, Masafumi. 2001JP10713, (0245812).

76. Thien-Nga, L., Hernadi, K., Ljubovic, E., Garaj, S., and Forro, L. (2002). Nano Letters, 2, (12), 1349-1352.

77. Shelimov, K. B., Esenaliev, R. O, Rinzler, A. G., Huffman, C. B., and Smalley, R. E. (1998). Chem. Phys. Lett., 282, (429-434).
78. S. Bandow, A. M. Rao, K.A. Williams (1997). The Journal of physical chemistry B. 101 (44) 8839-42.
79. Gu, Z., Peng, H., Hauge, R. H., Smalley, R. E., and Margrave, J. L. (2002). Nano Letters, 2, (9), 1009-1013.
80. Gao, B., Bower, C., Lorentzen, J. D., Fleming, L., Kleinhammes, A., Tang, X. P., Mcneil, L. E., Wu, Y., and Zhou, O. (2000). Chemical Physics Letters, 327(1-2), 69-75.
81. Niyogi, S., Hu, H., Hamon, M. A., Bhowmik, P., Zhao, B., Rozenzhak, S. M., Chen, J., Itkis, M. E., Meier, M. S., and Haddon, R. C. (2001). Journal of the American Chemical Society, 123(4), 733-734.
82. Zhao, B., Hu, H., Niyogi, S., Itkis, M. E., Hamon, M. A., Bhowmik, P., Meier, M. S., and Haddon, R. C. (2001). Journal of the American Chemical Society, 123(47), 11673-11677.
83. Georgakilas, Vasilios, Voulgaris, Dimitrios, Vazquez, Ester, Prato, Maurizio, Guldi, Dirk M., Kukovecz, Akos, and Kuzmany, Hans (2002). Journal of the American Chemical Society, 124, (48).
84. Farkas, E., Anderson, M. E., Chen, Z. H., and Rinzler, A. G. (2002). Chemical Physics Letters, 363(1-2), 111-116.
85. J.K.W. Sandler, et al (2003). Polymer, 44, 5893.
86. R.H. Baughman, et al. (2002). Science, 297, 787.
87. C. Park, et al. (2002). Chem. Phys. Lett., 364, 303 www.eikos.com
88. W.I. Milne, et al. (2004). J. Mater. Chem., 14, 933.
89. N. de Jonge, et al. (2002). Nature, 420, 393.
90. Ch. Emmenegger, et al. (2003). J. Power Sources, 124, 321.
91. M. Endo, et al. (2001). Carbon, 39, 1287.
92. Rechargeable Batteries with High Energy Storage Activated by *In-situ* Induced Fluorination of Carbon Nanotube Cathode; Xinwei Cui, Jian Chen, Tianfei Wang & Weixing Chen Scientific Reports, 4, Article number:5310 (2014).
93. Polychiral Semiconducting Carbon Nanotube-Fullerene Solar Cells in the August 7 issue of Nano Letters. (2014).
94. A.P. Graham, et al. (2004). Diamond Relat. Mater., 13, 1296.
95. Wind, S. J., et al. (2002). Appl. Phys. Lett. 80, 3817; Avouris, P., et al. (2003). Proc. IEEE 91, 1772.
96. Javey, A., et al. (2002). Nat. Mater. 1, 241; Nano Lett. (2004) 4, 447.
97. Bachilo, S. M., et al., J. Am. Chem. Soc. (2003) 125, 11186; Li, Y., et al., Nano Lett. (2004) 4, 31
98. Y.C. Chen, et al. (2002). Appl. Phys. Lett., 81, 975.
99. Niyogi, S.; Hamon, M. A.; Hu, H.; Zhao, B.; Bhowmik, P.; Sen, R.; Itkis, M. E. and Haddon, R. C.(2002). Chemistry of single-walled carbon nanotubes. Acc. Chem. Res., 35, 1105-1113.

100. Rosca, I. D. Watari, F. Uo, M. and Akaska, T. (2005). Oxidation of multiwalled carbon nanotubes by nitric acid. Carbon, 43, 3124-3131.

101. Zeng, L.; Alemany, L. B.; Edwards, C. L. and Barron, A. R. (2008). Demonstration of Covalent Sidewall Functionalization of Single Wall Carbon Nanotubes by NMR Spectroscopy: Side Chain Length Dependence on the Observation of the Sidewall sp3 Carbons. Nano Res. 1, 72-88.

102. Liu, Z.; Winters, M.; Holodniy, M. and Dai, H. J. (2007). siRNA delivery into human T cells and primary cells with carbon-nanotube transporters. Angew. Chem. Int. Ed., 46, 2023-2027.

103. Wu, W. Wieckowski, S. Pastorin, G. Benincasa, M. Klumpp, C. Briand, J. P. Gennaro, R. Prato, M. and Bianco, A. (2005). Targeted delivery of amphotericin B to cells by using functionalized carbon nanotubes. Angew. Chem. Int. Ed., 44, 6358-6362.

104. Dumortier, H. Lacotte, S. Pastorin, G. Marega, R. Wu, W. Bonifazi, D. Briand, J. P.; Prato, M. Muller, S. and Bianco, A. (2006). Functionalized carbon nanotubes are non-cytotoxic and preserve the functionality of primary immune cells. Nano Lett., 6, 3003.

105. Chen, X. Lee, G. S. Zettl, A. and Bertozzi, C. R. (2004). Biomimetic engineering of carbon nanotubes by using cell surface mucin mimics. Angew. Chem. Int. Ed., 43, 6111-6116;

106. Chin, S. F. Baughman, R. H. Dalton, A. B. Dieckmann, G. R. Draper, R. K. Mikoryak, C. Musselman, I. H. Poenitzsch, V. Z. Xie, H. and Pantano, P. (2007). Amphiphilic helical peptide enhances the uptake of single-walled carbon nanotubes by living cells. Exper. Biol. Med., 232, 1236-1244.

3

Nano/Micro Fabrication Technologies

3.1 Nano/Micro Fabrication Techniques-Introduction

Nano fabrication techniques are extension of Micro fabrication techniques which were developed for applications in the semiconductor industry. Both micro- and nanofabrication offer a number of possibilities for the study of chemical, biological, and physical processes at the cellular and molecular scale, and for the design of synthetic devices capable of interacting with biological systems at these levels.

Some of the advantages of micro- and nanofabricated devices include

(a) ability to control the features to the nanometer scale for reproducible mass production of structures and devices

(b) ability to miniaturize already-existing systems for the study of cellular or molecular processes, the capacity of including electronics within structural devices through the use of the well-developed semiconductor techniques, and

(c) high throughput possible with some of the micro- and nanofabrication methods.

However some of the challenges to be addressed by nano/micro fabrication techniques are to

(i) overcome the huge surface energy, a result of enormous surface area or large surface to volume ratio,

(ii) to produce nanomaterials with desired size, uniform size distribution, morphology, crystallinity, chemical composition, and microstructure. Sometimes it is not only the size but also the position of various groups need to be maintained and

(iii) to prevent nanomaterials and nanostructures from coarsening through either Ostwald ripening or agglomeration as time evolutes.

The following are three basic characteristics of nano materials to be achieved by any nano fabrication method.

(i) to produce nano particles/materials of Identical size of all particles (also called monosized or with uniform size distribution),

(ii) to produce nanoparticles/materials of identical shape or morphology, and

(iii) to produce nano particles/materials of identical chemical composition and crystal structure that are desired among different particles and within individual particles, such as core and surface composition must be the same.

The integration of the knowledge gained from micro- and nano-fabrication can lead to design principles for nano devices that can detect substances, analyze their environment, and perform tasks such as the release of a specific molecule.

The number of nano products is rapidly growing since more and more nano engineered materials are reaching the global market. The continuous revolution in nanotechnology will result in the fabrication of nanomaterials with properties and functionalities which are going to have positive changes in the lives of our citizens, be it in health, environment, electronics or any other field. However this demands innovations in nano fabrication techniques to meet the product specifications.

3.2 Nano/Micro Fabrication Techniques

As discussed in Chapter I, in general, nano materials can be fabricated basically by two approaches

(i) top – down and

(ii) bottom-up methods.

In **Top-down** methods nano materials are derived from a bulk substrate and obtained by progressive removal of material, until the desired nano material is obtained.

In **Top-Down** methods starting with macroscopic structures nano/structures materials will be obtained by process like etching through the mask, ball milling, etc., which can be externally-controlled.

Top-down methods begin with a pattern generated on a larger scale, then reduced to nanoscale. By nature these methods are slow and costly and not suitable for large scale manufacture. Some of the common problems associated with Top-Down approach are that they introduce internal stress, Surface defects (i.e., imperfections), non homogeneous nano materials and may carry some contaminants.

Bottom-up methods work in the opposite direction: the nanomaterial, such as a nano coating, is obtained starting from the atomic or molecular precursors and gradually assembling it until the desired structure is formed. These approaches include the miniaturization of materials components (up to atomic level) with further self-assembly process leading to formation of nano structures. During self-assembly the physical forces operating at nanoscale are used to combine basic units into larger stable structures. Typical examples are quantum dot formation during epitaxial growth and formation of nano particles from colloidal dispersion. Thus these methods start with atoms or molecules and build up to nanostructures. Fabrication by bottom-up is much less expensive. As Bottom-up methods are driven mainly by the reduction of Gibbs free energy, the product will have a closer to a thermodynamic equilibrium state. So the nano materials produced by bottom-up approach will have less surface defects, homogeneous chemical composition and better long range and short range ordering.

In both top –down and bottom-up methods:

(a) control of the fabrication conditions (e.g., energy of the electron beam) and

(b) control of the environment conditions (presence of dust, contaminants, etc). are essential.

For these reasons, nanotechnologies use highly sophisticated fabrication tools that are mostly operated in a vacuum in clean-room laboratories.

3.2.1 Top-Down Approach

A number of top-down fabrication methods used in nanotechnologies are derived from the fabrication methods used in the semiconductor industry to fabricate the various elements of computer chips (integrated circuits). These methods are collectively called **lithography** and use a light or electron beam to selectively remove micron-scale structures from a precursor material called **resist**. In recent years there has been a tremendous push to reduce the size of electronic devices and integrate more functions into them, which is due to the advances in lithographic fabrication methods. Nowadays it is possible to obtain single features below 100 nm (the transistors in latest generation processors are about 45 nm). **Therefore in the semiconductor industry nanostructures are routinely fabricated**.

3.2.1.1 Lithography

Lithography is the technique used to transfer a computer generated pattern on to a substrate (Si, GaAs etc.). This pattern is subsequently used to each an underlying thin film (oxide, nitride etc.) for various purposes (doping, etching etc.). A typical lithographic process consists of three successive steps:

(i) coating a substrate (Si wafer or glass) with a sensitive polymer layer (called resist),

(ii) exposing the resist to light, electrons or ion beams,

(iii) developing the resist image with a suitable chemical (developer), which reveals a positive or negative image on the substrate depending on the type of resist used (i.e., positive tone or negative tone resist). In conventional micro fabrication used in the semiconductor industry, the next step after lithography is the pattern transfer from the resist to the underlying substrate. This is achieved through a number of transfer techniques, such as chemical etching and dry plasma etching.

Lithographic techniques can be broadly divided in two main groups:

1. Methods that use a **physical mask**, where the resist is irradiated through the mask which is in contact or in proximity with the resist surface. These methods are collectively called mask lithography, among which photolithography is the most used.
2. Methods that use **software mask**. Here, a scanning beam irradiates the surface of the resist sequentially, point by point, through a computer-controlled program where the mask pattern is defined. These methods are collectively called scanning lithography.

The main difference between mask and scanning lithography is speed: whereas mask lithography is a parallel, fast technique, scanning lithography is a slow, serial technique. Another important difference is resolution which, in general terms, is higher for scanning methods. The price paid for higher resolution is the use of more energetic radiation sources, which entails expensive equipment. The **resolution of contact mode lithography is typically 0.5-0.8 μm when UV light (360-460 nm) is used**. Higher resolutions cannot be achieved due to the inability to reduce the gap between the mask and the flat substrate below ~1 μm, even when elaborate vacuum systems are used to hold the two parts together. To produce patterns with higher resolution, projection photolithography or "next-generation photolithography" techniques (i.e., extreme UV and X-ray photolithography) need to be employed. These technologies use very expensive equipment and therefore their use is limited to selected applications (such as photo mask fabrication). The equipment needed is available only in specialized laboratories.

3.2.1.2 Photolithography

Photolithography is the photographic technique to transfer copies of a master pattern, usually a circuit layout in IC applications, onto the surface of a substrate of some photo sensitive material (usually a silicon wafer) by selective exposure to a radiation source such as light. A photosensitive material is a material that experiences a change in its physical properties when exposed to a radiation source. If a photosensitive material is selectively exposed to radiation (e.g., by masking some of the radiation) the pattern of the radiation on the material is transferred to the material exposed, as the properties of the exposed and unexposed regions differs. This exposed region can then be removed or treated providing a mask for the underlying substrate.

Photolithography is typically used with metal or other thin film deposition, wet and dry etching. The starting point subsequent to the creation of the computer layout for a specific fabrication sequence is the generation of a photomask. This involves a sequence of photographic processes (using optical or e-beam pattern generators), which results in a glass plate having the desired pattern in the form of a thin (≈ 100 nm) chromium layer. Following the generation of the photomask, the lithography process can proceed as shown in Fig.3.1.

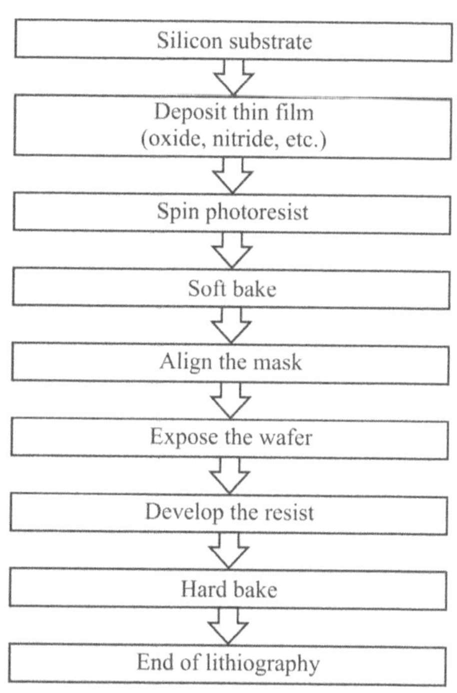

Figure 3.1 Lithography process flow

As can be seen the steps involved in the photolithographic process are

(i) wafer cleaning;
(ii) barrier layer formation;
(iii) photoresist application;
(iv) soft baking;
(v) mask alignment;

(vi) exposure and development; and
(vii) hard-baking.

This sequence demonstrates the pattern transfer onto a substrate coated with silicon dioxide; however, the same technique is applicable to other materials. After depositing the desired material on the substrate, the photolithography process starts with spin-coating the substrate with a photoresist. Photoresist is composed of the base resin, solvent and light sensitive material, called photoinitiator. The base resin determines the mechanical and thermal properties and photoinitiator the sensitivity to radiation. Photoinitiator is a molecule (usually organic), which absorbs light and forms reactive initiating species like anion radicals, H^+, or cation radicals. Solvents are used to control the viscosity of the resist. Viscosity has direct impact on the achievable layer thickness. Depending on the type of the resist, either positive or negative, the areas that are exposed to light or the areas that are not exposed to light will be removed in development, respectively. In standard optical lithography a photomask, made of glass or quartz with light blocking metal layer, usually chromium is used.

An adhesion promoter such as hexamethyl disilazane (HMDS) is used prior to the application of the resist). The spinning speed and photoresist viscosity will determine the final resist thickness, which is typically in the range 0.5–2.5 μm. Two different kinds of photoresist are available: positive and negative. With positive resist, UV-exposed areas will be dissolved in the subsequent development stage, whereas with negative photoresist, the exposed areas will remain intact after UV development. Due to its better performance with regard to process control in small geometries, positive resist is the most extensively used in the VLSI processes. After spinning the photoresist onto the wafer, the substrate is soft-baked (5–30 min at 60–100 °C) in order to remove the solvents from the resist and improve adhesion. Subsequently, the mask is aligned to the wafer and the photoresist is exposed to a UV source. The schematic of UV photo lithography operations is shown in Fig. 3.2.

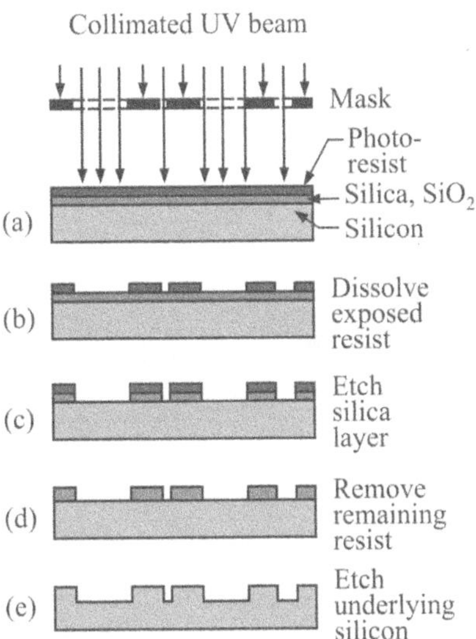

Figure 3.2 Schematic of UV photo lithography

A beam of UV light activates the photo resist transferring the pattern from the mask to sample.

Depending on the separation between the mask and the wafer, three different exposure systems are available: (a) contact (b) proximity and (c) projection.

Fig. 3.2 shows how an oxidized silicon wafer will be coated with 1 um thick photo resist layer. After exposure to UV light the photo resist undergoes a photo chemical reaction which breaks down the polymer by rupturing the polymer chains. When the wafer is rinsed in a developing solution the exposed areas are removed.

Although contact printing gives better resolution compared with the proximity technique, the constant contact of the mask with the photoresist reduces the process yield and can damage the mask. Projection printing uses a dual-lens optical system to project the mask image onto the wafer. Since only one die at a time can be exposed, this requires a step-and-repeat system to cover the whole wafer area. Projection printing is by far the most widely used system in microfabrication adhesion. The hard bake step concludes the photolithography sequence by creating the desired pattern on the wafer. Next, the underlying thin film is etched and the photoresist is stripped using acetone or other organic removal solvents.

3.2.1.2.1 Problems and Advantages in Lithography

Though the concept of photo lithography are simple, in practice the implementation is very complex and expensive as

(i) Nano structures significantly smaller than 100 nm are difficult to produce due to diffraction effects

(ii) Masks need to be perfectly aligned with the pattern on the wafer

(iii) Identity of defects needs to be carefully controlled and

(iv) Photolithographic tools are very expensive ranging from 10-100M$.

Advantages of Lithography

Once the master template has been made no special equipment is required.

Soft Lithographic methods are capable of producing nano structures in a wide range of materials and they can print or mold on curved as well as planar surfaces.

3.2.1.3 Scanning Lithography

Although photo lithography, i.e., lithography using an ultraviolet (UV) light source, is by far the most widely used lithography technique in the microelectronic fabrication, electron-beam (e-beam) and x-ray lithography are two other alternatives which have attracted considerable attention in nanofabrication areas.

Energetic particles such as **electrons and ions can be used to pattern appropriate resist films** leading to features with nano metre resolution. When using electrons, the technology is called electron beam lithography (**e-beam**), whereas when using ions the technology is called **focused ion beam lithography**.

3.2.1.3.1 E-beam Lithography

Electron beam lithography (EBL) is a specialized technique for creating the extremely fine patterns (much smaller than can be seen by the naked eye) required by the modern electronics industry for integrated circuits. Derived from the early scanning electron microscopes, the technique in brief consists of scanning a beam of electrons across a surface covered with a resist film sensitive to those electrons, thus depositing energy in the desired pattern in the resist film thus e-beam gun is usually part of a scanning electron microscope (SEM), although transmission electron microscopes (TEM) can also be used. Thus in a typical e-beam lithography process, a tightly focused beam of electrons scans across the surface of an electron-sensitive resist film, such as poly(methyl methacrylate) (PMMA).

The main attributes of the EBL technology are

1. it is capable of very high resolution, almost to the atomic level;
2. it is a flexible technique that can work with a variety of materials and an almost infinite number of patterns;
3. it is slow, being one or more orders of magnitude slower than optical lithography; and
4. it is expensive and complicated - electron beam lithography tools can cost many millions of dollars and require frequent service to stay properly maintained.

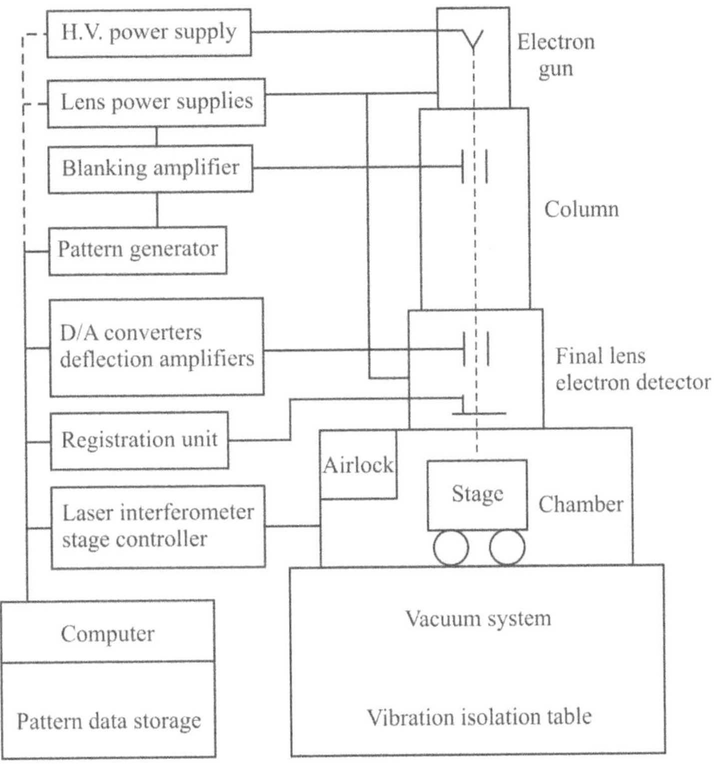

Figure 3.3 Block diagram showing the major components of a typical electron beam lithography system

The first electron beam lithography machines, based on the scanning electron microscope (SEM), were developed in the late 1960s. Shortly thereafter came the discovery that the common polymer PMMA (polymethyl methacrylate) made an excellent electron beam resist. It is remarkable that even today, despite sweeping technological advances, extensive development of commercial EBL, and a myriad of positive and negative tone resists, much work continues to be done with PMMA resist on converted SEMs. Fig. 3.3 shows a block diagram of a typical electron beam lithography tool. The column is responsible for forming and controlling the electron beam.

Underneath the column is a chamber containing a stage for moving the sample around and facilities for loading and unloading it. Associated with the chamber is a vacuum system needed to maintain an appropriate vacuum level throughout the machine and also during the load and unload cycles. A set of control electronics supplies power and signals to the various parts of the machine. Finally, the system is controlled by a computer, which may be anything from a personal computer to a mainframe. The computer handles such diverse functions as setting up an exposure job, loading and unloading the sample, aligning and focusing the electron beam, and sending pattern data to the pattern generator.

The main advantage of e-beam lithography over photolithography is its high resolution: patterns with **features as small as 50 nm** can be routinely generated. Since diffraction effects are largely reduced due to the wavelength of electrons there will not be any blurring of features and thus the resolution is greatly improved. The resolution of this technology is mainly determined by the scattering of the electrons in the resist film and the substrate. When using particles with a mass higher than electrons, however, this effect is largely reduced.

Electron Beam Write Time

The minimum time to expose a given area for a given dose is given by the following formula:

$$D.\ A = T.I$$

where T is the time to expose the object (can be divided into exposure time/step size), I is the beam current, D is the dose and A is the area exposed(1).

For example, assuming an exposure area of 1 cm^2, a dose of 10^{-3} coulombs/cm^2, and a beam current of 10^{-9} amperes, the resulting minimum write time would be 10^6 seconds (about 12 days). This minimum write time does not include time for the stage to move back and forth, as well as time for the beam to be blanked (blocked from the wafer during deflection), as well as time for other possible beam corrections and adjustments in the middle of writing. To cover the 700 cm^2 surface area of a 300 mm silicon wafer, the minimum write time would extend to 7×10^8 seconds, about 22 years. This is a factor of about 10 million times slower than current optical lithography tools. It is clear that throughput is a serious limitation for electron beam lithography, especially when writing dense patterns over a large area.

E-beam lithography is not suitable for high-volume manufacturing because of its limited throughput. The smaller field of electron beam writing makes for very slow pattern generation compared with photolithography (the current standard) because more exposure fields must be scanned to form the final pattern area ($\leq$ mm^2 for electron beam vs. $\geq$40 mm^2 for an optical mask projection scanner). The stage moves in between field scans. The electron beam field is small enough that a rastering or serpentine stage motion is needed to pattern a 26 mm × 33 mm area for example, whereas in a photolithography scanner only a one-dimensional motion of a 26 mm × 2 mm slit field would be required.

Thus the primary advantage of electron beam lithography is that it is one of the ways to beat the diffraction limit of light and make features in the nanometer regime. This form of maskless lithography has found wide usage in photomask-making used in photolithography, low-volume production of semiconductor components, and research and development. Although electron wavelengths of the order of 1 Å can be easily achieved, electron scattering in the resist limits the attainable resolutions to >10 nm. Beam control and pattern generation is achieved through a computer interface. It was developed for manufacturing integrated circuits, and is also used for creating nanotechnology architectures.

Disadvantages

Defects in electron-beam lithography.

Despite the high resolution of electron-beam lithography, the generation of defects during electron-beam lithography is often not considered by users. Defects may be classified into two categories: data-related defects, and physical defects.

Data-related defects may be classified further into two sub-categories. Blanking or deflection errors occur when the electron beam is not deflected properly when it is supposed to, while shaping errors occur in variable-shaped beam systems when the wrong shape is projected onto the sample. These errors can originate either from the electron optical control hardware or the input data that was taped out. As might be expected, larger data files are more susceptible to data-related defects.

Physical defects are more varied, and can include sample charging (either negative or positive), backscattering calculation errors, dose errors, fogging (long-range reflection of backscattered electrons), outgassing, contamination, beam drift and particles. Since the write time for electron beam lithography can easily exceed a day, "randomly occurring" defects are more likely to occur. Here again, larger data files can present more opportunities for defects.

Photo mask defects largely originate during the electron beam lithography used for pattern definition

The electron beam technique **is very expensive and very slow**:

E-beam lithography is serial and hence has **low throughput**. Although this is not a major concern in fabricating devices used in studying fundamental microphysics, it severely **limits large-scale nanofabrication**.

As it takes very long time to expose an entire silicon wafer or glass substrate it leaves the **user vulnerable to beam drift or instability** which may occur during the exposure. Also, the turn-around time for reworking or re-design is lengthened unnecessarily if the pattern is not being changed the second time.

3.2.1.3.2 Ion Beam Lithography

Ion beam lithography is the practice of scanning a focused beam of ions in a patterned fashion across a surface in order to create very small structures such as integrated circuits or other nanostructures. Ion beam (focus ion beam) lithography has high potential to play an important role in nanometer technology because of the lack of backscattered electrons. Focused ion beam lithography works on the same principle of e-beam lithography, but ions such as H^+, He^{++}, Li^+ and Be^{++} are used. Both techniques provide a resolution much higher than photolithography but share a main disadvantage: both are serial techniques, very slow in process, so their use is mostly limited to producing photo masks in optical lithograph. It is known that focused-ion-beam lithography has the capability of writing extremely fine lines (less than 50 nm line and space has been achieved) without proximity effect. However, because the writing field in ion-beam lithography is quite small, large area patterns must be created by stitching together the small fields. Common operational energy levels will be in the range of 10 to 200 keV. Focused ion beam lithography can also be used to pattern features directly to a substrate without the need for a photo mask by either selective material removal or deposition. Compared with electron ion beam lithography, the patterning speed offered by this technique is significantly slower because of lower achievable ion current density, when compared with that of electrons. Common feature resolutions are in the range of 20 nm, with a minimal 5 nm lateral feature size. Figure 3.4 depicts the schematic of Ion beam lithography.

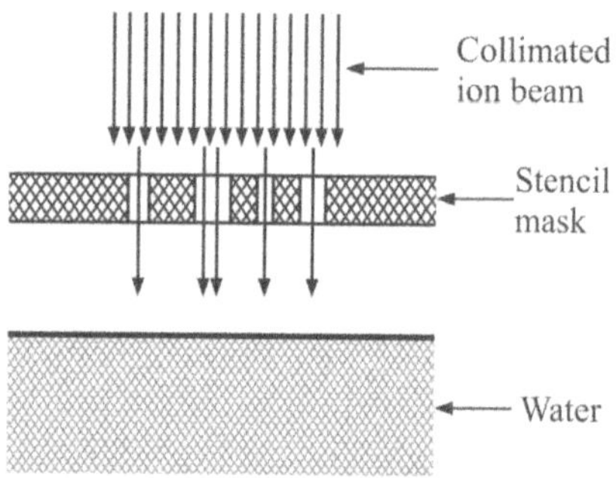

Figure 3.4 Schematic of Ion beam lithography(2)

Advantages of Ion Beam Lithography

(i) Ions have heavy mass than electrons,

(ii) less proximity effects (due to back scattering),

(iii) less scattering effect,

(iv) offers higher resolution patterning than UV, X-ray, or electron beam lithography because these heavier particles have more momentum.

(v) this gives the ion beam a smaller wavelength than even an e-beam and therefore almost no diffraction.

Ion beam sources: Gas source: Hydrogen, helium;

Metal sources: The most common used source metal is Ga; Au/Si and Au/Si/Be alloys have also been used for lithography because these sources can supply lighter mass ions.

Resists: Polymethylmethacrylate (PMMA) is most common, Poly(butene-1-sulfone), Poly(2,2,2-trifluoroethyl-α-chloroacrylate).

Features: Penetration of the particle beam is small, compared to electron beam

Reduce blurring resulting from beam scattering (reduce proximity effect; increase localization, precision)

Two types of ion beam columns are in use for different applications (Fig. 3.5).

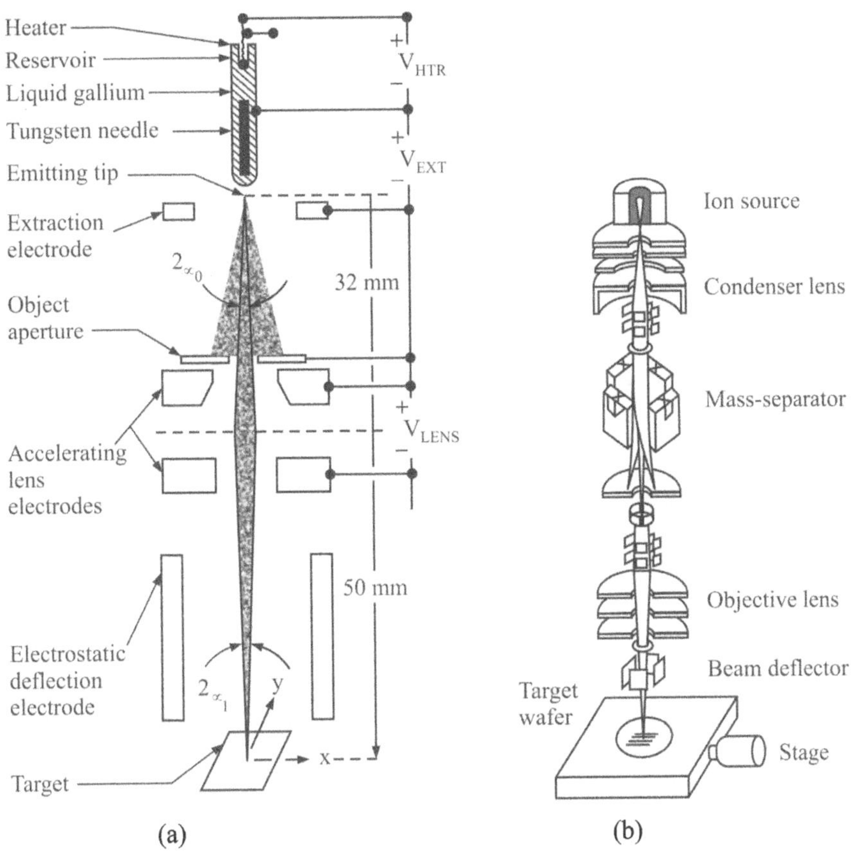

Figure 3.5 (a) Low acceleration Voltage (<50 KV) focusing columns without mass separator, (b) High acceleration voltage (> 100 KV) focusing column with separator,3.4)

(i) Low acceleration voltage (<50 kV) columns without mass separation (designed for applications such as mask repair, microcircuit modification, and scanning ion microscopes using a Ga LMIS).

(ii) High acceleration voltage (> 100 kV) columns with mass separators. (designed for applications such as ion implantation and lithography with an alloy LMIS).

3.2.1.3.3 Ion Track Technology

Ion track technology is a deep cutting tool with a resolution limit around 8 nm applicable to radiation resistant minerals, glasses and polymers. It is capable to generate holes in thin films without any development process. Structural depth can be defined either by ion range or by material thickness. Aspect ratios up to several 10^4 can be reached. The technique can shape and texture materials at a defined inclination angle. Random pattern, single-ion track structures and aimed pattern consisting of individual single tracks can be generated.

3.2.1.3.4 X-ray Lithography

X-ray lithography, is a process used in electronic industry to selectively remove parts of a thin film. It uses X-rays to transfer a geometric pattern from a mask to a light-sensitive chemical photoresist, or simply "resist," on the substrate. A series of chemical treatments then engraves the produced pattern into the material underneath the photo resist.

This technique employs electromagnetic radiation of wavelengths in the range of 0.5 to 4 nm (soft X-rays) for pattern transfer from a mask to a substrate material (3).

Common sources of soft X-rays are synchrotrons or laser-induced plasma generators.

X-ray masks are usually in the order of a few microns in thickness, and are usually composed of silicon carbide. Absorber patterns, analogous to the opaque regions of a photolithographic mask, will be created with heavy metals such as gold, tungsten or tantalum (5).

In this technique, it is possible to perform lithography with a mask-to-wafer distance of several microns. However, larger gap distances proportionally reduce the achievable resolution (5, 6). Additional parameters that influence the resolution offered by this technique are Fresnel diffraction, also known as near-field diffraction, and photoelectron diffusion into resist films, or photoelectron blur.

Unfortunately, the Fresnel diffraction is proportional to the wavelength of energy used, while the photoelectron blur is inversely proportional. Consequently, to achieve patterns with small resolutions it is necessary to balance these two variables. Resolutions of less than 30 nm have been reported (5).

3.2.1.4 Soft Lithography

Soft lithography is a name for a number of techniques that fabricate and use a soft mold prepared by casting a liquid polymer precursor against a rigid master. These methods have been developed specifically for making large-scale micro and nanostructures with equipment that is easier to operate compared to those used in "conventional" lithography, cheaper and also available in biological laboratories.

The resolution of soft lithography is mainly determined by vander Waals interactions, by wetting and by kinetic factors such as filling the capillaries on the surface of the master, but not by optical diffraction. This is an important advantage over "conventional" lithographic techniques. The master is normally fabricated via a conventional lithographic method.

Thus soft lithography (Fig. 3.6) is the collective name for a set of lithographic techniques like

(i) replica molding (REM),

(ii) micro contact printing (μCP),

(iii) micro molding in capillaries (MIMIC),

(iv) micro transfer molding (μTM),

(v) solvent-assisted micro molding (SAMIM), and

(vi) near-field conformal photolithography using an elastomeric phase-shifting masks that has been developed as an alternative to photolithography and a replication technology for micro- and nanofabrication. These techniques use a patterned elastomer (usually poly dimethylsiloxane (PDMS) as the mold, stamp, or mask to generate or transfer the pattern. Soft lithography offers immediate advantages over photolithography and other conventional micro fabrication techniques for applications in which patterning of non planar substrates, unusual materials, or large area patterning are the major concerns. It is especially promising for micro-fabrication of relatively simple, single-layer structures for uses in cell culture, as sensors, as micro analytical systems, in MEMS, and in applied optics. Positive Photoresist as photolithography (7) with PDMS has been readily used in the biomedical and pharmaceutical fields because of its biocompatibility, and good thermal, mechanical and optical properties. The main advantage of soft lithography is that once the reusable mold is made, none of the other steps require clean room manipulation. As a result, it is a less expensive technique that provides great resolution through a simpler process (7, 8). There are three main soft lithography processes: micro-stamping, stencil patterning, and microfluidic patterning.

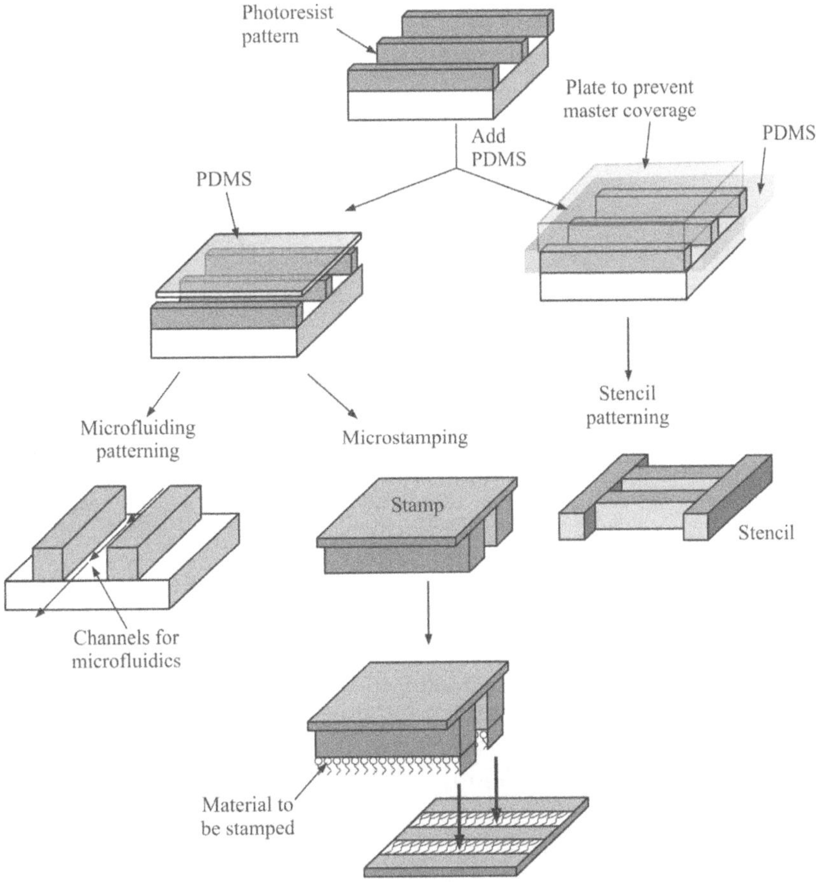

Figure 3.6 Overall scheme of soft lithography

3.2.1.4.1 Replica Molding (REM)/Micro Stamping

The capability and feasibility of replica molding can be greatly enhanced by incorporating elastomeric materials into the procedure, at some cost in reproducibility. In REM, an elastomer (for example, PDMS) is cast against a rigid master and the elastomeric replica is subsequently used as a mold against which new replicas are produced in organic polymers. The introduction of elastomers into replica molding makes it easier to preserve small, fragile structures on the original master; the low surface energy (~21.6 dyn/ cm) of PDMS, in particular, also allows the replica to be separated from the mold easily. An elastomeric mold also offers the opportunity to manipulate the size and shape of structures on the mold by mechanical or thermal deformation. For example, we have been able to reduce the lateral dimension of nanostructures from ~50 to ~30 nm by replica molding against a PDMS mold which is compressed mechanically

3.2.1.4.2 Stencil Patterning

The second soft lithography technique, stencil patterning, creates templates by preventing PDMS from covering the master template, as can be seen in Fig. 3.7.

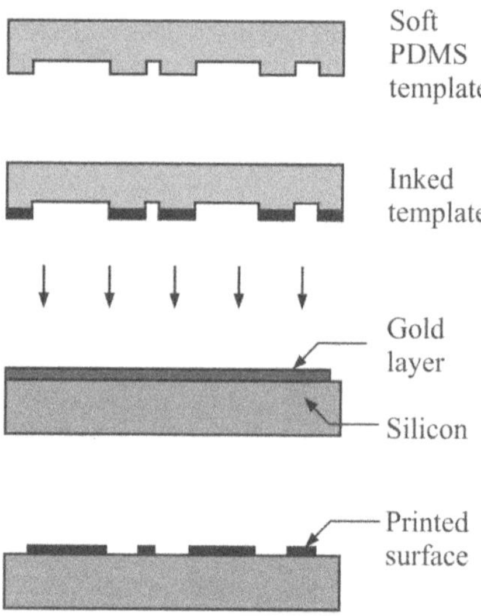

Figure 3.7 Stencil patterning-templates creation

The end result of this process is a PDMS model with holes in the pattern of the master. Different methods can be used to prevent PDMS from covering the master features, such as placing plates against the master following PDMS addition, or adding the PDMS to a thickness smaller than that of the master features (9). One of the major challenges for μCP has been to achieve the capability to print with high resolution, i.e., with lateral dimension lower than 100 nm. This has been achieved by improving the stability of the PDMS, which, being soft and highly compressible, has a tendency to deform and collapse. One way to improve the stability of the patterns is to affix a stiff backplane to the stamp or to change the chemical formulation of the stamp itself, in order to obtain a harder polymer. With these modifications, it is now possible to print features as small as 50 nm. This printing method, which uses harder stamps, is called nano contact printing (nCP).

3.2.1.4.3 Microfluidic Patterning

This soft lithographic technique utilizes a PDMS mold to create micro channels against a substrate. These micro channels can then be used to pattern fluid materials onto a substrate (9). This technique has been utilized for the patterning of cells in tissue engineering applications (10). One important feature of these micro channels is their ability to maintain separate fluid streams through a single channel because of laminar flow (11). This characteristic can be utilized for the study of cellular response to the stimuli of different materials.

3.2.1.4.4 Polymers for Molding in Soft Lithograhy

Various polymers (*e.g.*, polyurethanes, epoxides and polyimides) can be used for moulding;

3.2.1.4.4.1 PDMS

Most commonly, the elastomer poly (dimethylsiloxane) (PDMS) is used. The chemical structure of PDMS is presented in Figure 3.8.

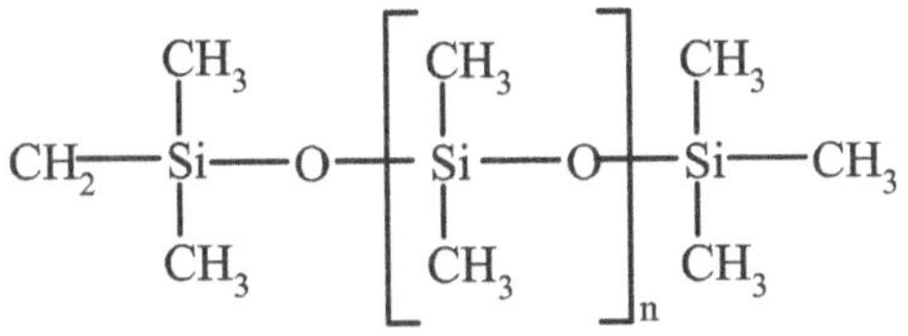

$$CH_2-\underset{CH_3}{\overset{CH_3}{Si}}-O-\left[\underset{CH_3}{\overset{CH_3}{Si}}-O\right]_n-\underset{CH_3}{\overset{CH_3}{Si}}-CH_3$$

Figure 3.8 Chemical structure of PDMS

PDMS is a silicone elastomer and it has several desirable properties such as optical transparency in UV-VIS regions and low cost. It is also non-fluorescent, biocompatible and non-toxic. Young's modulus of PDMS can be tailored by changing the base curing agent ratio. The simple fabrication of PDMS structures by casting method is one of the main reasons why PDMS has become the most used material in microfluidic applications. The bonding of PDMS to PDMS is also easy which makes the fabrication of closed structures possible PDMS being non-toxic can be used safely with biological materials, including live cells. This is a big advantage in devices that aim to integrate nanostructures with biological systems.

A PDMS mould is fabricated by pouring its liquid precursor over a lithographically made master (e.g., a photoresist or silicon master), cured to induce cross-linking, and then peeled off from it. The stamp can then be used either for printing a desired material (the "ink") from the stamp to a suitable surface (micro contact printing, μCP) or, when in contact with a flat or curved surface, to define physical constraints where a liquid can be confined.

Disadvantages of PDMS

(i) One of the main disadvantages of PDMS is its poor nonpolar solvent compatibility. PDMS swells when in contact with the solvents like hydrocarbons, toluene and dichloromethane. This problem was overcome by coating the surface of the PDMS micro channel. By coating PDMS micro channels with a borosilicate glass coatings it was found to solve this problem. The coating was found to prevent swelling of the PDMS when exposed to toluene. The high gas permeability of PDMS can also be an issue although it has also been used as an advantage in applications, like cell culture devices, where oxygen can be supplied through the material.

(ii) The other main issue is the hydrophobicity of PDMS. Hydrophobic nature of PDMS makes its use in contact with biomolecules such as proteins and peptides problematic due to hydrophobic interactions. This problem can be overcome by oxygen plasma treatment of the PDMS. By using oxygen plasma, the water contact angle of PDMS will decrease from about 108° to 30° which can be explained by the conversion of $-OSi(CH_3)_2O-$ groups at the surface to On $Si(OH)_4{-}n$. The decrease of the contact angle has several benefits. The oxidized PDMS channels fill much more easily with liquids with high surface tension, such as water. Also the bubble formation inside the channel is reduced. The oxidized PDMS channels also support electro-osmotic flow (EOF) which can be used in biomolecule separation. The interactions between channel surface and biomolecules is also reduced which is also important in separation devices. The surface of the PDMS will however regenerate back to hydrophobic when in contact with air which is due to a diffusion of the hydrophobic residues into a surface. When PDMS is stored in water, the regeneration is not as fast. It has also been reported that by thermal aging of PDMS the hydrophobic recovery can be reduced. Despite all the above mentioned problems associated with PDMS, it has been and probably will be the most used material in research laboratories for early stage testing of new concepts of hard PDMS.

Another siloxane polymer, called hard-PDMS (h-PDMS), was first developed for stamp material for micro contact printing where the elasticity of common PDMS (Sylgard 184) causes problems. The Young's modulus of h-PDMS and 184-PDMS are 8.2 MPa and 1.8 MPa, respectively. The stamps fabricated of h-PDMS are composite stamps where h-PDMS is used for patterns and common PDMS as support since h-PDMS would be too brittle to be used alone. The use of h-PDMS/PDMS composite stamp enables the fabrication of nanosized patterns with high precision.

3.2.1.4.4.2 Thermoplasts

Thermoplast is a polymer that softens when temperature rises above its glass transition temperature. Thermoplasts return to their original state after cooling. Polymethylmethacrylate (PMMA) and polycarbonate (PC) are two of the most used thermoplastic polymers in microfluidic applications. The chemical structures of both polymers are presented in Fig. 3.9.

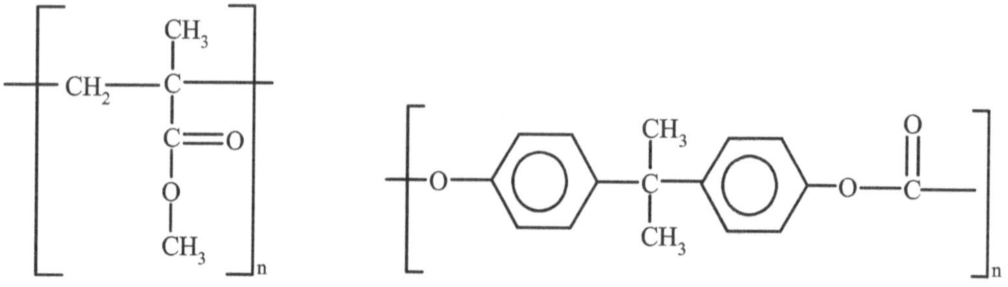

$$\left[-CH_2-\underset{\underset{\underset{CH_3}{|}}{\underset{O}{|}}}{\underset{C=O}{|}}{\overset{CH_3}{\overset{|}{C}}}-\right]_n$$

(a)

$$\left[-O-C_6H_4-\overset{CH_3}{\underset{CH_3}{C}}-C_6H_4-O-\overset{O}{\overset{\|}{C}}-\right]_n$$

(b)

Figure 3.9 Chemical structure (a) (PMMA) (b) polycarbonate (PC)

PMMA has a glass transition temperature of 100-122 °C and it starts to decompose at over 200 °C. It is chemically compatible with diluted acids and bases but it will dissolve into aromatic and chlorine-based solvents. The transparency of PMMA is relatively high. For microfluidic applications PMMA is usually hot embossed but injection molding and laser ablation are also possible. PMMA is also generally used as a resist for e-beam lithography. Polycarbonate has a glass transition temperature of 145-148 °C and it starts to de-compose at 230 °C. Polycarbonate can be processed by hot embossing, injection molding and laser ablation.

The UV-irradiation of polycarbonate was reported to decrease the water contact angle of polycarbonate as well as increase the electro osmotic mobility. Topas® is a cyclic olefin copolymer (COC) and has excellent physical and chemical properties for lab-on-a-chip applications. It is chemically resistant, has low water absorption and is transparent to UV-light. Glass transition temperature of the material can be tailored between 78 and 178 by changing the content of the norbornene between 60 and 85 wt.%, respectively. Topas has been structured by hot embossing, injection molding and nanoimprint lithography. The main disadvantage of Topas® is its hydrophobicity (contact angle 100°) although preliminary studies show that it can be reduced to less.

3.2.1.4.4.3 Fabricating enclosed Polymer Channels

Most of the micro fabrication methods for polymers produce open microfluidic structures. However, in many applications enclosed channels are needed. In microfluidic applications, the situation where all the walls of the micro channel are of the same material is usually optimum, especially when electro-osmotic flow is used to drive the fluidic in the device. Enclosed polymer micro channels can be fabricated with several different techniques depending on the used polymer.

3.2.1.4.4.4 Bonding of PDMS

One major advantage of PDMS is easy bonding. PDMS can be bonded irreversible or reversible not just to it itself but also to many other materials. Reversible bonding can be done to basically any clean surface and even though the bond does not tolerate high pressures this method can be used in many applications. By using reversible bonding, it is also easier to clean and reuse the channels. However, if stronger bonding is needed, PDMS can be activated by oxygen plasma treatment (Fig. 3.10). The formation of bridging, covalent siloxane (Si-O-Si).

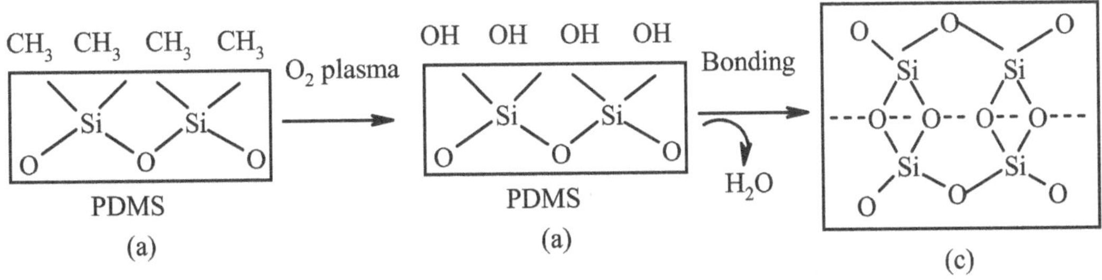

Figure 3.10 (a) native PDMS surface (b) PDMS surface after oxygen plasma treatment (simplified image) (c) PDMS to PDMS bonding

bonds by a condensation reaction will form the irreversible bonding between two PDMS substrates. Formation of irreversible bonding between several plastics and PDMS has also been presented. The surface of the PDMS will however regenerate back to hydrophobic when in contact with air which is due to a diffusion of the hydrophobic residues into the surface. For this reason the bonding of the PDMS have to be done less than one hour after the plasma exposure.

3.2.1.4.4.5 Nano/Micro Contact Printing /Embossing

Micro/nano contact printing (nCP) is useful for patterning features with lateral dimension of 500 nm or larger. Embossing, sometimes also called imprinting, is a process where structures of the stamp are transferred to a substrate by pressing them together. The embossing process can be roughly divided into two category; UV- and thermal embossing, depending on the hardening method used. Both micro- and nanosized structures can be fabricated with these methods. All embossing proccsses are inherently high throughput, because of parallel imprinting. By utilizing embossing methods, high throughput fabrication of nanostructures is thus also possible.

A typical embossing process can be divided into three steps:

- Stamp fabrication
- Replication
- Demolding

One of the main advantages of any embossing technique compared to photolithography, is that it is non-optical and the resolution is thus limited only by the structures that can be fabricated on the stamp. Depending on the pattern size either UV-lithography (for micro scale and larger) or electron beam lithography (for nanosized) can be used. The selection of stamp material depends on the used embossing method. In the embossing step, the ability of the embossed material to flow is important, so that the embossing process is completed within a practical time frame. Demolding step is commonly the most difficult part of the embossing process. The stamps usually contain some nanoscale roughness on the surface which increases the surface area that contacts the embossed polymer. This will lead to increased adhesion of the embossed polymer to the stamp. Typically stamps are coated with some anti adhesion layer prior embossing to prevent this problem.

This is based on the construction of a stamp that has the desired pattern with the PDMS mold. Molecules to be transferred are placed on the surface of the stamp and "printed" on a receiving surface upon stamping, thus forming a self-assembled monolayer (SAM) (5, 9). Depending on the application, peptides, proteins, polysaccharides and other molecules can be stamped. The stamped layer can protect the substrate during etching or deposition procedures, which will be described in more detail later in this section. One great advantage of this micro fabrication method is that the stamp can be reused to make pattern replicas.

3.2.1.4.4.6 Nano-Imprint Lithography

The concept of nano-imprint lithography is to use a hard master with a 3D nanostructure to mould another material, which assumes its reverse 3D structure. Since the master has a fine nano-structure, to be successful the process must be done under pressure, a coating must first be placed on the master to avoid catastrophic adhesion to the mould, and the mould must be heated (above its Tg temperature) in order to be soft enough to completely enter the fine master nanostructure and be effectively replicated. The method is the equivalent of embossing at the nanoscale. It requires specialised equipment.

3.2.1.4.4.7 Nano Sphere Lithography

In nanosphere lithography an ensemble of nanospheres ordered on a surface are used as a mask (Fig. 3.11). The nanospheres are dispersed in a liquid (i.e., a colloid) and a droplet placed on a surface and left to dry. Depending on the surface properties (e.g., charge) and media used in the colloid (e.g., presence of electrolytes) the nanosphere will self-assemble in an ordered pattern.

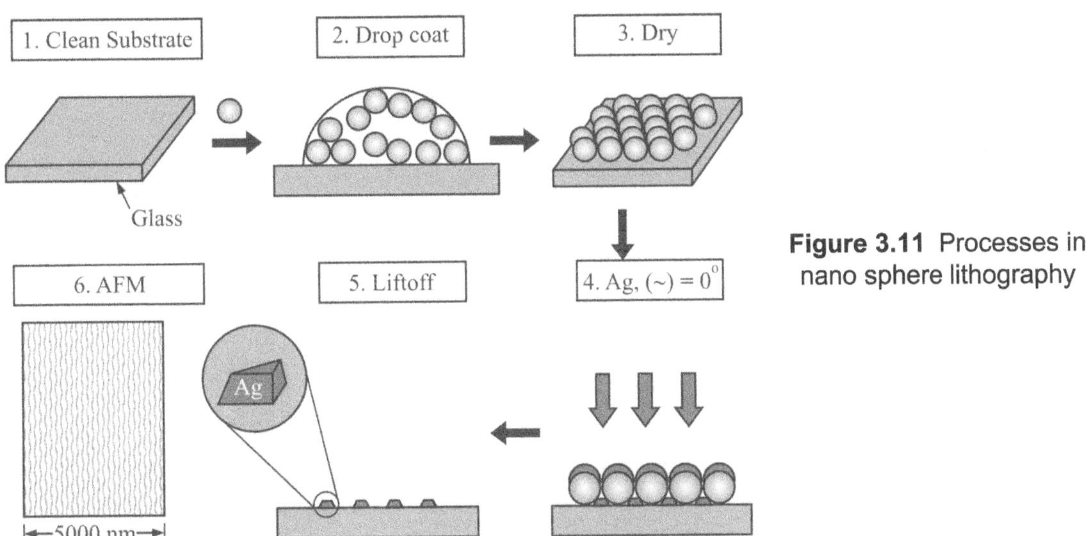

Figure 3.11 Processes in nano sphere lithography

In some conditions, a colloidal crystal is obtained: each nanoparticle is surrounded by six other nanospheres. This regular arrangement (which is a **2D colloidal crystal**) can be used to create ordered structures on surfaces. In the regular arrangement of nano spheres there will be an empty space between them, which is regularly repeated in the entire surface. In the most simple method, this space is employed to create relatively flat nanopatterns on the surface. The nano sphere pattern is used as a mask, and a material (e.g., gold, silver, etc) sputtered on top of it. Once the nano spheres are removed, a regular pattern is left of "dots" each shaped like a triangle but with concave sides. Nano sphere lithography has now evolved into a method that allows the fabrication of very complex arrays of nanostructures, including 3D features with small holes in them.

3.2.1.4.4.8 Colloidal Lithography

Colloid lithography shares the same principle of nano sphere lithography of using a colloid as a mask for the fabrication of nanostructures on surfaces. In this method electrostatic forces are employed to obtain short-range ordered arrays of nano spheres on the surface. The array can then be used to create a number of different nanostructures, through various processes like etching, liftoff, etc. (Fig. 3.12).

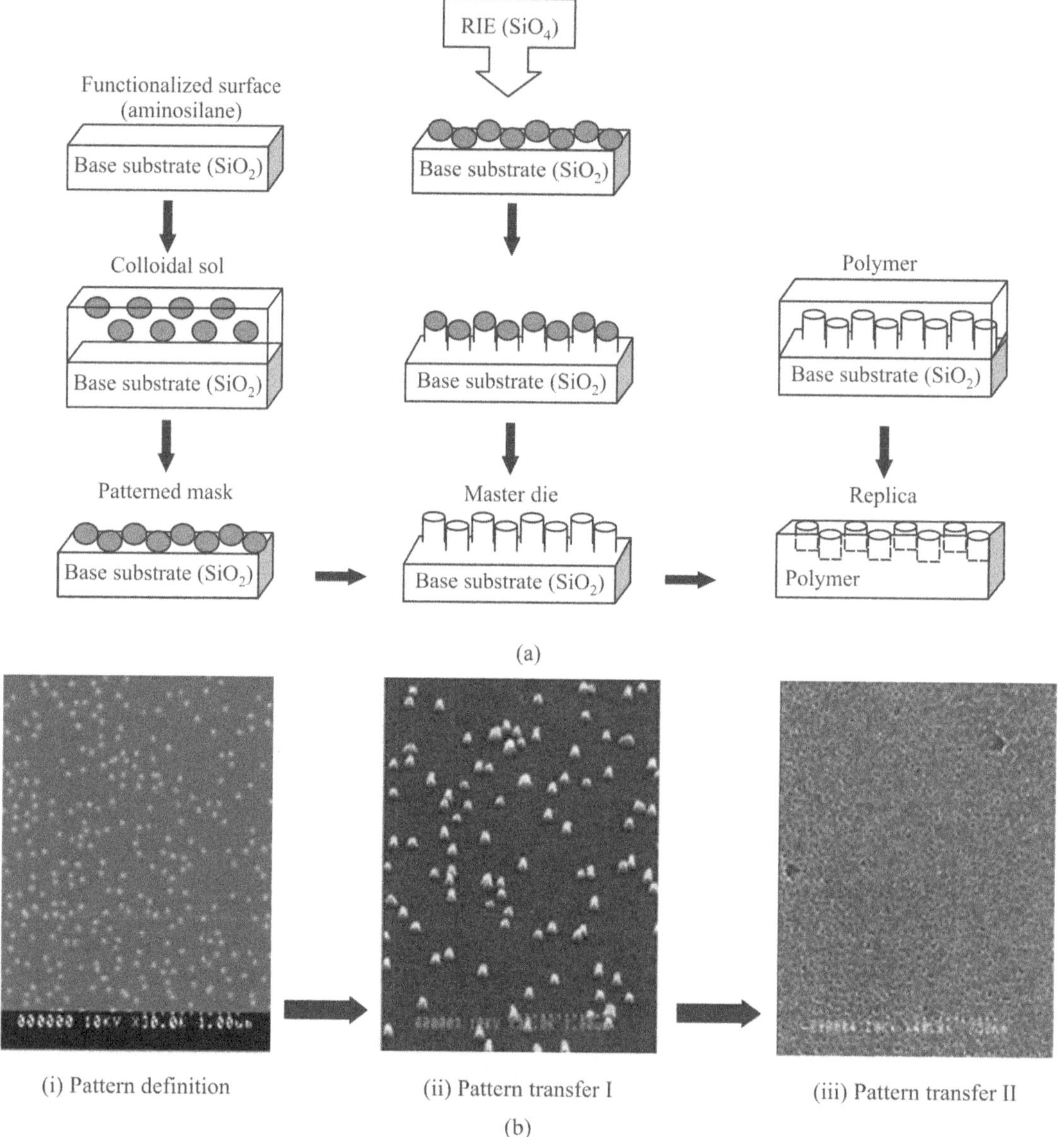

Figure 3.12 Steps in colloidal nano lithography

(Source: M.A. Wood,; Interface, Royal Society Publishing DOI,10.1098,rsif 2006 -0419)

This lithographic method utilizes self-organized one- or two-dimensional colloidal systems as layers for nano fabrication. This technique is an economical alternative to the common electron or ionic lithographic methods, but is still able to produce patterns at the nanoscale.

Colloidal mono-layers can be generated through a number of self-assembly processes. Colloidal particles can be deposited on the surface of a substrate in solution prior to solvent evaporation, through spin coating, or through electrophoresis (12).

If the size and geometry of the colloidal particles is precisely controlled, it is in theory possible to also control the spatial distribution of the colloids. For example, mono-dispersed spherical particles can be close-packed into hexagonally arranged monolayers, i.e., the conformation of highest density for this geometry (12).

However a number of parameters including colloid concentration, solvent evaporation rate, wetting characteristics of the substrate, and competition with multiple-layer formation influence the resulting array.

It has been reported that the achievable resolution offered by this technique, when the colloidal mono-layer is utilized as a resist, can be as low as 5 nm in all three dimensions (13).

Fig. 3.12 gives Overview of the main nanolithography steps used during fabrication, where colloids are utilized as a model system.

A schematic overview is provided in (a) with scanning electron microscope (Hitachi S-900) images of the final patterns provided in (b) Pattern definition (i) requires a mask to be produced, in this instance by functionalizing a base substrate of silica and immersing in a colloidal sol, producing a patterned mask following sol evaporation (50 nm diameter colloidal gold particles in (b)).

Pattern transfer I (ii) uses additive or subtractive techniques to produce a master die, for example, colloids can be used as a mask during reactive ion etching (RIE), where ions etch around the mask, producing, in this instance, a pillared topography (50 nm diameter, 200 nm high pillars, (b) with colloids removed post-etch).

Colloids can subsequently be removed using a gold etch, resulting in a pillared master die with uniform chemistry.

Pattern transfer II (iii), a process often termed replication technologies, uses a polymeric material to produce many copies of the master die topography, inversely (20 nm diameter, 200 nm deep nanopits in polystyrene, (b)).

These colloid monolayers can be used as a protective barrier against etching and, consequently, transfer their semi-random distribution pattern to the substrate material. More specifically, these etch masks would result in the removal of the substrate material located below the colloid interstices.

The same analogy can be made for applications in selective film deposition. Depending on the application, the colloidal particles can be either removed or left in place.

3.2.1.5 Electrically Induced Nano Patterning

Electrically induced nano patterning techniques utilize electrostatic interactions between a thin dielectric material liquid film and an electric field gradient to produce lateral patterns and structures at the nanometer scale (14). The nanofabrication system is analogous to a capacitor because it consists of two parallel electrodes separated by an air gap of less than a micron, as shown in Fig.3.13.

A polymer, such as polystyrene, is applied to one of the electrodes by spin coating. Upon exposure to high temperature (above the glass transition temperature of the polymer) and to an electric field generated by the voltage across the electrodes, electrostatic forces develop and result in the destabilization of the polymeric film.

Since the instabilities have a defined periodic undulation pattern, this system ultimately results in the formation of polymeric columns at the location of the peaks of the polymer "wave", i.e., those locations with the highest polymer thickness or smallest air gap between the polymer and the top electrode (14,15). Lateral column density and order can be altered by changing parameters such as the initial polymer film thickness, or the inter-electrode spacing (14).

If in addition the electrode opposite to the polymer possesses microstructured patterns, the instabilities will first develop at the locations where the distance between the electrodes is minimal because of the increased electro-static driving force. The result of these instabilities is that the polymer extends across the air gap toward the opposite electrode, thus generating nanometer scale "columns" in the specific pattern of the top electrode (16), as can be seen in Fig. 3.13.

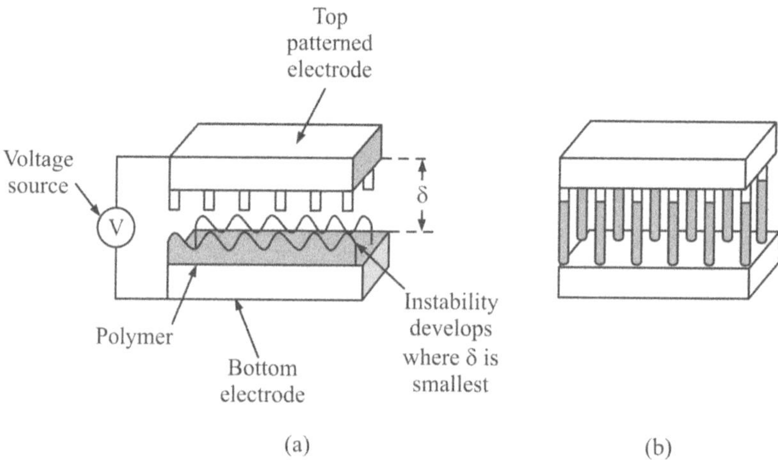

Figure 3.13 Schematic of electrically-induced nanopatterning process. (a) The system utilized for electrically induced micropatterning consists of two electrodes separated by an air gap of thickness δ (b) A thin film of a polymer to be molded is applied to the substrate.

These columns will be then solidified by cooling below the polymer's glass transition temperature. This technique has further been utilized to generate porous templates for the fabrication of nanowire arrays with very high densities (16).

3.2.1.6 Etching Processes

3.2.1.6.1 Etching and Substrate Removal

Thin film and bulk substrate etching is another fabrication step that is of fundamental importance to micro/nanofabrication as various conducting and dielectric thin films deposited for passivation or masking purposes need to be removed at some point or another. In micro/nanofabrication, in addition to thin-film etching, very often the substrate (silicon, glass, GaAs, etc.) also needs to be removed in order to create various mechanical micro/nanostructures (beams, plates, etc.). Two important things for any etching process are selectivity and directionality. Selectivity is the degree to which the etchant can differentiate between the masking layer and the layer to be etched. Directionality has to do with the etch profile under the mask.

Etching processes can be divided into different classes depending on the point of view. Sometimes they are divided into wet and dry etching according to the phase of the etchant. Sometimes division to isotropic and anisotropic etching is used. Isotropic etching means that etching propagates on all directions with the same etch rate. This can be a problem when the structures to be etched are on the same size range as the etch depth. When this is the case, anisotropic etching should be used. There etching propagates faster in some direction than others. All the wet etching processes are not isotropic nor are all the dry etching methods anisotropic or vice versa. Dry etching is also many times called plasma etching since plasma is used to excite the etchants.

In wet etching the material is dissolved when immersed in a chemical solution while in the dry etching , the material is sputtered or dissolved using reactive ions or a vapor phase etchant.

3.2.1.6.2 Wet Etching

Wet etchants are by and large isotropic and show superior selectivity to the masking layer as compared with various dry techniques. In addition, due to the lateral undercut, the minimum feature size achievable with wet etchants is limited to > 3 μm. Silicon dioxide is commonly etched in dilute (6 : 1, 10 : 1, or 20 : 1 by volume) or buffered HF (BHF: HF + NH_4F) solutions (z of ≈ 1000 Å/min in BHF).

Photoresist and silicon nitride are the two most common masking materials for the wet oxide etch. The wet etchant for silicon nitride is hot (140-200 °C) phosphoric acid with silicon oxide as the masking material.

Nitride wet etch is not very common (except for blanket etch) due to the masking difficulty and non repeatable etch rates.

Metals can be etched using various combinations of acid and base solutions. There are also many commercially available etchant formulations for aluminum, chromium, and gold which can easily be used (17).

3.2.1.6.3 Isotropic Etching

Isotropic etchants etch the material at the same rate in all directions, and consequently remove material under the etch masks at the same rate as they etch through the material; this is known as undercutting (Fig. 3.14 (a) and (b)). Etch rates can slow down and in some cases (for example, in deep and narrow channels) they can stop due to diffusion limiting factors. However, this effect can be minimized by of the etchant, resulting in structures with near perfect and rounded surfaces (Fig. 13.14 b). Isotropic etchants are limited by the geometry of the structure to be etched. The surface of these grooves can be atomically smooth if the etch is carried out correctly, with dimensions and angles being extremely accurate.

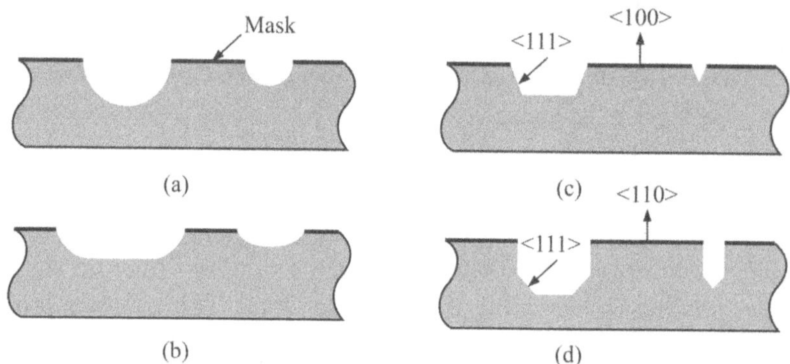

Figure 3.14 Isotropic etching with (a) and without (b) agitation, and anisotropic etching wet etching of (100) and (110) silicon (c and d)

In isotropic etching of silicon HF/HNO_3 /CH_3 COOH (various different formulations have been used) will be used to thin down the silicon wafer. The etch mechanism for this combination involves HNO_3 will oxidize the silicon, which is subsequently dissolved away in the HF. The acetic acid used will prevent the dissociation of HNO_3 (the etch works as well without the acetic acid). For short etch times, silicon dioxide can be used as the masking material; however, one needs to use silicon nitride if a longer etch time is desired. This etch also shows dopant selectivity, with the etch rate dropping at lower doping concentrations (< $10^{17} cm^{-3}$ n-or p-type). Although this effect can potentially be used as an etch-stop mechanism in order to fabricate microstructures, the difficulty in masking has prevented widespread application of this approach. Glass can also be isotropically etched using the HF/HNO_3 combination with the etch surfaces showing considerable roughness. This has been extensively used in fabricating microfluidic components (mainly channels). Although Cr/Au is usually used as the masking layer, long etch times require a more robust mask (bonded silicon has used for this purpose).

3.2.1.6.4 Anisotropic Etching

Silicon anisotropic wet etch constitutes an important technique in bulk micromachining The three most important silicon etchants in this category are potassium hydroxide (KOH), ethylene diamine pyrochatechol (EDP), and tetramethyl ammonium hydroxide (TMAH). These are all anisotropic etchants which attack silicon along preferred crystallographic direction at some point or another. Some single crystal materials, such as silicon, will have different etching rates depending on the crystallographic orientation of the substrate (where Si <111> planes etch approximately 100 times slower than other planes crystallographic orientations). Therefore, etching a rectangular hole in a (100)-Si wafer resulting a pyramid shaped etch pit with 54.7° walls, instead of a hole with curved sidewalls as with isotropic etching.

3.2.1.6.5 HF Etching

Hydrofluoric acid is commonly used as an aqueous etchant for silicon dioxide (SiO_2, also known as BOX for SOI), usually in 49% concentrated form, 5:1, 10:1 or 20:1 BOE (buffered oxide etchant) or BHF (Buffered HF). They were first used in medieval times for glass etching. It was used in IC fabrication for patterning the gate oxide until the process step was replaced by RIE. Hydrofluoric acid is considered one of the more dangerous acids in the clean room. It penetrates the skin upon contact and it diffuses straight to the bone. Therefore the damage is not felt until it is too late.

3.2.1.6.6 Electrochemical Etching

Electrochemical etching (ECE) for dopant selective removal of silicon is a common method to automate and to selectively control etching. An active p-n diode junction is required, and either type of dopant can be the etch-resistant ("etch-stop") material. Boron is the most common etch-stop dopant. In combination with wet anisotropic etching as described above, ECE has been used successfully for controlling silicon diaphragm thickness in commercial piezo resistive silicon pressure sensors. Selectively doped regions can be created either by implantation, diffusion, or epitaxial deposition of silicon.

3.2.1.6.7 Dry Etching

Dry etching relies on vapour phase or plasma-based methods of etching using suitably reactive gases or vapours usually at high temperatures.

Some typical vapor phase etching processes are given below

Vapor Phase etching process

(i) Halogen etching for Si, W, Mo, Ti

– $W + 6F \rightarrow WF6$

(ii) Fluorocarbon (CF_4, C_2F_6, CHF_3)

for SiO_2, Si_3N_4

$- e^- + CF_4 \rightarrow CF_3 + F + e^-$

$-12F + Si_3N_4 \rightarrow 3SiF_4 + 2N_2$

(iii) $CF_4 - H_2$ mixture: increasing HF, reducing F concentration

- Increasing SiO_2 etch, decreasing Si and photoresist etch

(iv) Etching of Aluminum by BCl_3, CCl_4

The two most common special dry etches are:

(a) Xenon difluoride (XeF_2) etching and

(b) Reactive ion etching (RIE)

(a) **Xenon difluoride (XeF_2) ethching:** It is an isotropic silicon etch process and has a strong selectivity for silicon above Al, SiO_2, Si_3N_4 and photoresist. The typical etch rates are 1 to 3 m/min and it is commonly used for release etch.

(b) Reactive ion etching (RIE): Reactive Ion Etching (RIE) utilizes additional energy in the form of radio frequency (RF) power to drive the chemical reaction. Energetic ions are accelerated towards the material to be etched supplying the additional energy needed for the reaction; as a result the etching can occur at much lower temperatures (typically 150°-250 °C, sometimes room temperature or even lower) than those usually needed (above 500 °C). RIE is not limited by the crystal planes in the silicon, and as a result, deep trenches and pits, or arbitrary shapes with vertical walls can be etched.

This is known as the chemical part of reactive ion etching. There is also a physical part, which is similar to the sputtering deposition process. If the ions have high enough energy, they can knock atoms out of the material to be etched without a chemical reaction. It is a very complex task to develop dry etch processes that balance chemical and physical etching, since there are many parameters to adjust. By changing the balance it is possible to influence the anisotropy of the etching, since the chemical part is isotropic and the physical part highly anisotropic the combination can form sidewalls that have shapes from rounded to vertical.

RIE can be deep (Deep RIE or deep reactive ion etching (DRIE)).

Deep RIE (DRIE) is a special subclass of RIE that is growing in popularity. The MEMS process Deep Reactive Ion Etching (DRIE) is a much higher-aspect-ratio etching method. It involves an alternating process of high-density plasma etching (as in RIE) and protective polymer deposition to achieve greater aspect ratios.

Etch rates depend on time, concentration, temperature and material to be etched. Maintaining a high level of performance in all these areas simultaneously is often a balancing act. Companies are doing 20:1 aspect ratios in production, research facilities have demonstrated 50:1 ratios. Selectivity to photoresist or oxides in excess of 100:1 is possible. Etch rates are routinely 20 μm/min and as high as 40 μm/min. With very high etch rates, surface roughness increases and, with a large etched silicon area, the in homogeneity of the etch depth will increase.

3.2.1.6.8 Plasma Etching

Plasma etchers can operate in several modes by adjusting the parameters of the plasma (Fig. 3.15).

Ordinary plasma etching operates between 0.1 and 5 Torr. (This unit of pressure, commonly used in vacuum engineering, equals approximately 133.3 pascals.). The plasma produces energetic free radicals, neutrally charged, that react at the surface of the wafer. Since neutral particles attack the wafer from all angles, this process is isotropic.

Plasma etching can be isotropic, i.e., exhibiting a lateral undercut rate on a patterned surface approximately the same as its downward etch rate, or can be anisotropic, i.e., exhibiting a smaller lateral undercut rate than its downward etch rate. Such anisotropy is maximized in deep reactive ion etching. The use of the term anisotropy for plasma etching should not be conflated with the use of the same term when referring to orientation-dependent etching.

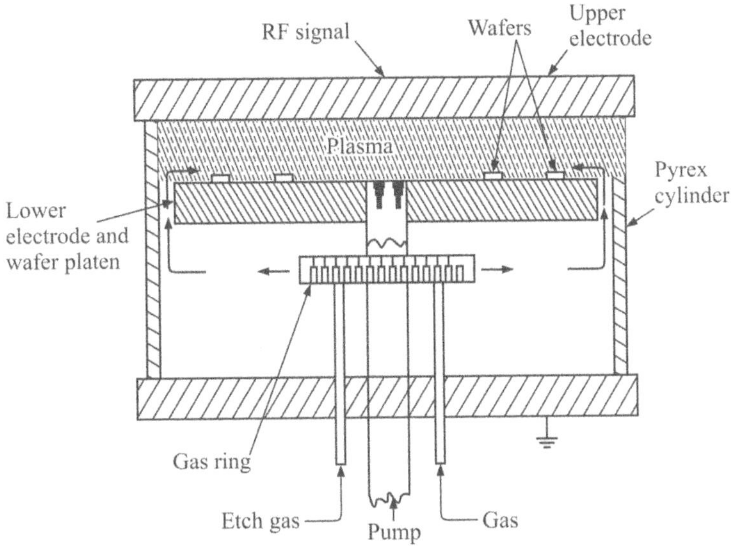

Figure 3.15 Plasma based dry etching process

The source gas for the plasma usually contains small molecules rich in chlorine or fluorine. For instance, carbon tetrachloride (CCl_4) etches silicon and aluminium, and trifluoro methane etches silicon dioxide and silicon nitride. A plasma containing oxygen is used to oxidize ("ash") photoresist and facilitate its removal.

Ion sputter etching, uses lower pressures, often as low as 10^{-4} Torr (10 mPa). It bombards the wafer with energetic ions of noble gases, often Ar^+, which knock atoms from the substrate by transferring momentum. Because the etching is performed by ions, which approach the wafer approximately from one direction, this process is highly anisotropic. On the other hand, it tends to display poor selectivity. Reactive-ion etching (RIE) operates under conditions intermediate between sputter and plasma etching (between 10^{-3} and 10^{-1} Torr). Deep reactive-ion etching (DRIE) modifies the RIE technique to produce deep, narrow features. The main motivation to use dry etching are:

(a) accuracy of processing,
(b) enabling smaller features of sizes (insurface micro machining) and
(c) CMOS compatibility milling.

Present DRIE equipment suppliers are increasing their etching rates.

3.2.1.6.9 Over-view of General Protocols of Lithographic Techniques for Nano Fabrication/Manufacturing

The general protocols for Lithographic nano manufacturing require high resolution and low cost for fabricating devices (18). This may include the high-cost fabrication of a master, with structures replicated in great quantity in an inexpensive process, resulting in the initial outlay of the master being an insignificant part of the overall cost of the device.

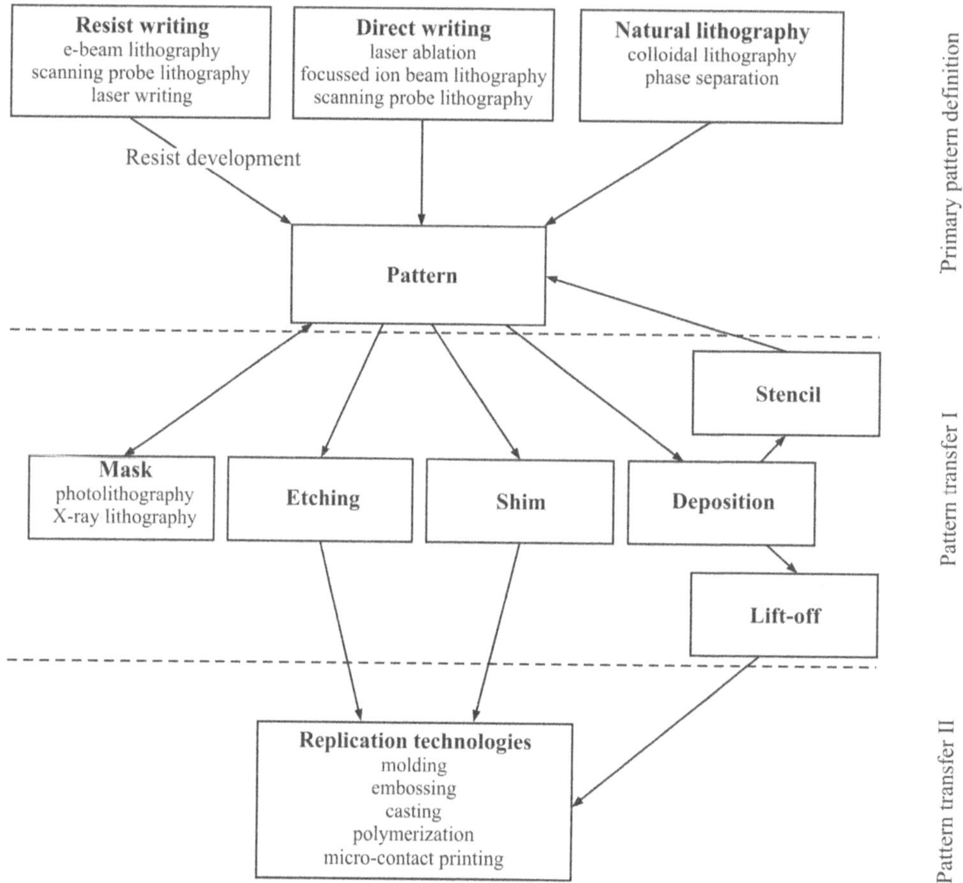

Figure 3.16 Overview of protocols for nano patterning technologies

The definition of a deliberate pattern is by its very nature a serial process, where every picture element has to be defined one by one. If utilized in the production of replicas, the structures, having undergone the writing process, are referred to as master dies (19).

The economical and convenient route to multiple copies of a master is termed replication, where information is transferred to a functional material, usually of a polymeric nature, from the master, rapidly, in a single step and with high precision (8).

Thus, nanopatterning can be separated into three main categories: primary pattern definition or writing, pattern transfer and mechanical transfer technique (Fig. 3.16).

3.2.2 Bottom-up Methods for Nano Fabrication

Bottom-up, or self-assembly, approaches to nanofabrication use chemical or physical forces operating at the nanoscale to assemble basic units into larger structures. As component size decreases in nanofabrication, bottom-up approaches provide an increasingly important complement to top-down techniques. Inspiration for bottom-up approaches comes from biological systems, where nature has harnessed chemical forces to create essentially all the structures needed by life.

A number of bottom-up approaches have been developed for producing nanoparticles, ranging from condensation of atomic vapours on surfaces to coalescence of atoms in liquids. For example, liquid-phase techniques based on inverse micelles have been developed to produce size-selected nanoparticles of semiconductor, magnetic, and other materials. An example of self-assembly that achieves a limited degree of control over both formation and organization is the growth of quantum dots.

Indium gallium arsenide (InGaAs) dots can be formed by growing thin layers of InGaAs on GaAs in such a manner that repulsive forces caused by compressive strain in the InGaAs layer results in the formation of isolated quantum dots. After the growth of multiple layer pairs, a fairly uniform spacing of the dots can be achieved. Another example of self-assembly of an intricate structure is the formation of carbon nanotubes under the right set of chemical and temperature conditions.

3.2.2.1 Deposition

One of the basic building blocks in nano material production is the ability to deposit thin films of material with a thickness anywhere between a few nanometres to about 100 micro metres.

An important process step in most micro fabrication schemes is thin film deposition. Thin films can be deposited with variety of techniques, such as sputtering, evaporation, thermal chemical vapor deposition (thermal CVD), plasma enhanced chemical vapor deposition (PECVD) and atomic layer deposition (ALD). The first two are called physical vapor deposition techniques (PVD) whereas the others are called CVD techniques. This is due to a difference in the deposition method. CVD involves a chemical reaction at the surface to be coated, whereas PVD process is purely physical condensation. Depending on the application, there can be many requirements for the thin films (such as thickness uniformity, good step coverage and fast deposition) and not any of the above mentioned methods fulfill all of them. For example the step coverage and thickness uniformity is excellent in atomic layer deposition (ALD) but the method is very slow and the thickness of the thin films is limited to few hundred nanometers and in some cases only tens of nanometers. ALD is mostly used to deposit oxides, nitrides and sulphides, although some metal depositions have also been presented. The evaporation on the other hand is relatively simple method and almost all metals can be deposited, but it does not work well for compound deposition. It also suffers from poor adhesion and poor step coverage.

3.2.2.1.1 Epitaxial Growth of Thin Films

Epitaxy means the growth of a single crystal film on top of a crystalline substrate. For most thin film applications (hard and soft coatings, optical coatings, protective coatings) it is of little importance. However, for semiconductor thin film technology it is crucial.

Types of Epitaxy

Homo epitaxy, In this category the film and the substrate are the same material. Often used in Si on Si growth. Epitaxially grown layers are purer than the substrate and can be doped independently of it.

Hetero epitaxy, In this film and substrate are different materials. *E.g.*, AlAs on GaAs growth. This allows for optoelectronic structures and band gap engineered devices.

Growing a layer of a different material on top of a substrate leads to unmatched lattice parameters. This will cause strained or relaxed growth and can lead to interfacial defects. Such deviations from normal would lead to changes in the electronic, optic, thermal and mechanical properties of the films.

Lattice Strains: For many applications nearly matched lattices are desired to minimize defects and matched increase electron mobility. As the mismatch gets larger, the film substrate material may strain to accommodate the lattice structure of the substrate. This is the case during the early stages of film formation (pseudomorphic growth) and with film materials of the same lattice structure.

There are two types of deposition processes such as Chemical Deposition and Physical Deposition.

3.2.2.1.2 Chemical Deposition

3.2.2.1.2.1 Chemical Vapour Deposition

Chemical Vapour Deposition (CVD) is a chemical process used to produce high-purity, high-performance solid materials. In a typical CVD process, the substrate is exposed to one or more volatile precursors, which react and/or decompose on the substrate surface to produce the desired deposit. Frequently, volatile by-products are also produced, which are removed by gas flow through the reaction chamber. The energy needed for the chemical reaction to occur is usually supplied by maintaining the substrate at elevated temperature (Fig.3.17). Alternative energy sources such as plasma or optical excitation are also used, with the advantage of requiring a lower temperature at the substrate. The most common CVD processes in microfabrication are low-pressure CVD (LPCVD) and plasma-enhanced VD (PECVD).

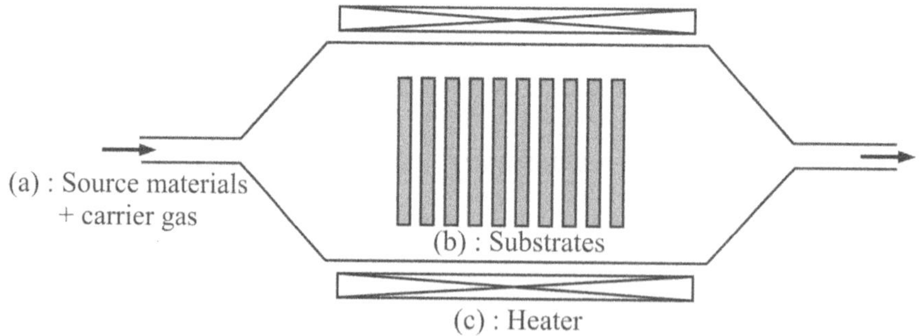

Figure 3.17 General process of CVD

3.2.2.1.2.2 Low Pressure CVD (LPCVD)

The LPCVD process is typically carried out in electrically heated tubes, similar to oxidation tubes, equipped with pumping capabilities to achieve the low pressures required (0.1-1.0 Torr). Large numbers of wafers can be processed simultaneously and the material is deposited on both sides of the wafers. The process temperature depends on the material to be deposited, but generally is in the range 550-900 °C. As in oxidation, high temperatures and contamination issues can restrict the type of processes used previous to the LPCVD. Typical materials deposited by LPCVD include silicon oxide (*e.g.*, $SiCl_2\ H_2 + 2N_2\ O \Rightarrow SiO_2 + 2N_2 + 2HCL$ at 900 °C), silicon nitride (*e.g.*, $3SiH_4 + 4NH_3 \Rightarrow Si_3\ N_4 + 12H_2$ at 700 – 900 °C), and polysilicon (*e.g.*, $SiH_4 \Rightarrow Si + 2H_2$ at 600 °C). Due to its faster etch rate in HF, *in situ* phosphorous-doped LPCVD oxide (phosphosilicate glass, PSG) is extensively used in surface micromachining as the sacrificial layer. Conformality in this process is excellent, even for very high-aspect-ratio structures. Mechanical properties of LPCVD materials are good compared with others such as PECVD, and are often used as structural materials in micro fabricated devices.

3.2.2.1.2.3 Plasma Enhanced Chemical Vapour Deposition (PECVD)

PECVD is a single wafer process, while Low Pressure Chemical Vapour Deposition (LPCVD) deposit very uniform layers on a batch of wafers simultaneously. Typical LPCVD process temperatures are above 600 °C, limiting its use in general to the very beginning of the processing (Fig. 3.18). In nano fabrication it is often used to deposit polysilicon (as mechanical layer in surface micromachining) and low stress silicon nitride as a membrane layer or to be used as a masking material in bulk micromachining.

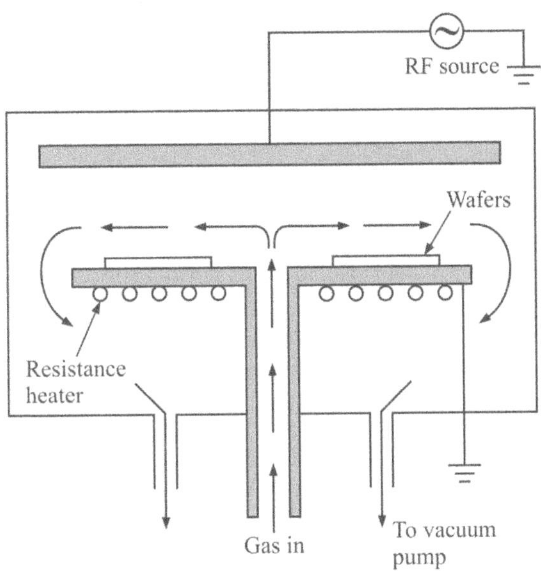

Figure 3.18 Schematic representation of a typical PECVD system

In PECVD stress in the deposited layers depends on the material, deposition conditions, and subsequent thermal history (*e.g.*, post deposition annealing). Typical values are: 100-300 MPa (compressive) for oxide, ≈ 1 GPa (tensile) for stoichiometric nitride, and ≈ 200-300 MPa (tensile) for polysilicon. The stress in nitride layers can be reduced to nearly zero by using a silicon-rich composition. Since the stress values can vary over a wide range, one has to measure and characterize the internal stress of deposited thin films for any specific equipment and deposition conditions.

The PECVD process is performed in plasma systems such as that represented in Fig. (3.18). The use of radiofrequency (RF) energy to create highly reactive species in the plasma allows for the use of lower temperature at the substrate (150-350 °C). Parallel-plate plasma reactors normally used in micro fabrication can only process a limited number of wafers per batch. The wafers are positioned horizontally on top of the lower electrode so only one side gets deposited. Typical materials deposited with PECVD include silicon oxide nitride, and amorphous silicon. Conformality is good for low-aspect-ratio structures, but becomes very poor for deep trenches (20% of the surface thickness inside through-wafer holes with an aspect ratio of 10). Stress depends on deposition parameters and can be either compressive or tensile. PECVD nitrides are typically non stoichiometric (Si_x N_y) and are much less resistant to etchants in masking applications.

Another interesting type of CVD is epitaxial growth. In this process, a single-crystalline material is grown as an extension of the crystal structure of the substrate. It is possible to grow dissimilar materials if the crystal structures are somehow similar (lattice matched). Silicon-on-sapphire (SoS) substrates and some hetero structures are fabricated in this way. However, most common in micro fabrication is the growth of silicon on another silicon substrate. Of particular interest for the formation of microstructures is selective epitaxial growth. In this process the silicon crystal is allowed to grow only in windows patterned on a masking material. Many CVD techniques have been used to produce epitaxial growth. The most common for silicon is thermal chemical vapor-deposition or vapour-phase epitaxy (VPE). Metal organic chemical vapor deposition (MOCVD) and molecular-beam epitaxy (MBE) are the most common for growing high-quality III–V compound layers with nearly atomic abrupt interfaces. The former uses vapors of organic compounds with group III atoms such as trimethyl gallium ($Ga(CH_3)_3$) and group V hydrides such as AsH_3 in a CVD chamber with fast gas switching capabilities. The latter typically uses molecular beams from thermally evaporated elemental sources aiming at the substrate in an ultrahigh-vacuum chamber. In this case, rapid on/off control of the beams is achieved by using shutters in front of the sources. Many metals (molybdenum, tantalum, titanium, and tungsten) can also be deposited using LPCVD. These are attractive for their low resistivities and their ability to form silicides with silicon. Due to its application in new interconnect technologies, copper CVD is an active area of research.

3.2.2.1.3 Physical Deposition

3.2.2.1.3.1 Physical Vapour Deposition

Physical Vapour Deposition (PVD) is a technique used to deposit thin films one atom (or molecule) at a time onto various surfaces (e.g., onto semiconductor wafers). The coating source is physical (i.e., solid or liquid) rather than chemical as in chemical vapour deposition. Evaporation (in which a material is evaporated from a target using either heat (thermal evaporation) (Fig. 3.19) or an electron beam (e-beam evaporation) in a vacuum system.) (Fig.3.20) and sputtering (in which an ion beam liberates atoms from a target, allowing them to move through the intervening space and deposit on the desired substrate) are commonly used PVD process, for instance for the deposition of aluminium or gold conductors.

Two physical principles are used in PVD: evaporation and sputtering. In evaporation, the source is placed in a small container with tapered walls, called the crucible, and is heated up to a temperature where evaporation occurs. Various techniques are utilized to reach the high temperatures needed, including the induction of high currents with coils wound around the crucible and the bombardment of the material surface with an electron beam (e-beam evaporators). This process is mainly used to deposit metals, although dielectrics can also be evaporated. In a typical system the crucible is located at the bottom of a vacuum chamber whereas the wafers are placed lining the dome-shaped ceiling of the chamber).

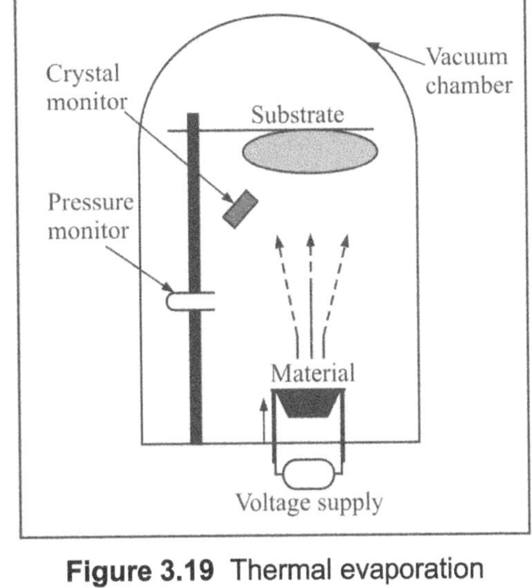

Figure 3.19 Thermal evaporation deposition system

- Principles
- Vaporization of solid material
- Re-condensing of the vapor on a cool substrate
- Heating of material: passage of a large current through a filament container made of W, a, or Mo
- Vacuum
- Mean free path for molecules in air at 25 °C is ~ 45 and 4500 cm at pressures of 10^{-4} and 10^{-6}
- Fig. 3.19 presents thermal Evaporation Deposition System Torr. lower than 10-5 Torr for a straight line path.

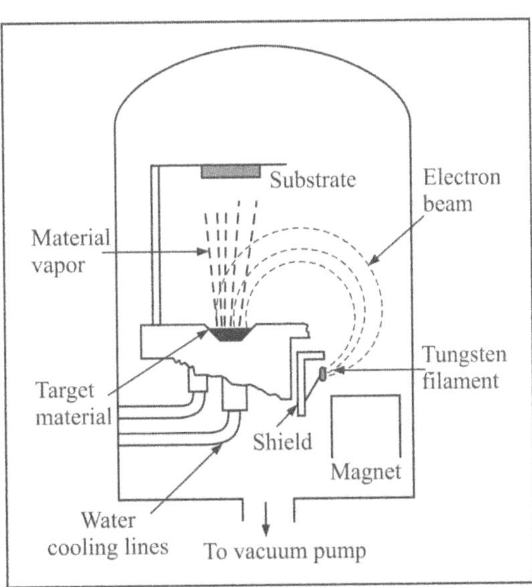

Figure 3.20 Schematic representation of an e-beam deposition system

Principles

- Physical vapor deposition
- High DC current applied to W filament that generates excited electrons
- Beam sweep controlled by magnets
- Vacuum
- Roughing pump (10^{-4} Torr)
- Cryogenic pump (10^{-7} Torr

Ion implantation allows more precise control of the dose (the total amount of impurities introduced per area unit) and the impurity profile (the concentration versus depth). In ion implantation the impurities are ionized and accelerated towards the semiconductor surface. The penetration of impurities into the material follows a Gaussian distribution. After implantation, an annealing process is needed to activate the impurities and repair the damage in the crystal structure produced by ion collisions. A drive-in process to redistribute the impurities, done in a standard furnace such as those used for oxidation or diffusion, may be required as well.

Some micro fabrication techniques utilize these effects to pattern the deposited layer. One way to improve the step coverage is by rotating and/or heating the wafers during deposition.

3.2.2.1.3.2 Sputtering

In sputtering a target of the material to be deposited is bombarded with high energy inert ions (usually argon) (Fig. 3.21). The outcome of the bombardment is that individual atoms or clusters are removed from the surface and ejected towards the wafer. The physical nature of this process allows its use with virtually any existing material. Examples of interesting materials for micro-fabrication that are frequently sputtered include metals, dielectrics, alloys (such as shape memory alloys), and all kinds of compounds (for example, piezoelectric lead zirconate titanate (PZT)). The inert ions bombarding the target are produced in direct-current (DC) or RF plasma. In a simple parallel-plate system the top electrode is the target and the wafers are placed horizontally on top of the bottom electrode. In spite of its lower deposition rate, step coverage in sputtering is much better than in evaporation. However, the films obtained with this deposition process are non conformal.

Principles

- High voltage DC or RF glow discharge (Ar atoms) impact cathode surface
- Cathode source ejects or sputters atoms from surface
- Substrate mounted on negatively charged plate
- Vacuum
- Roughing pump (10^{-3} Torr)

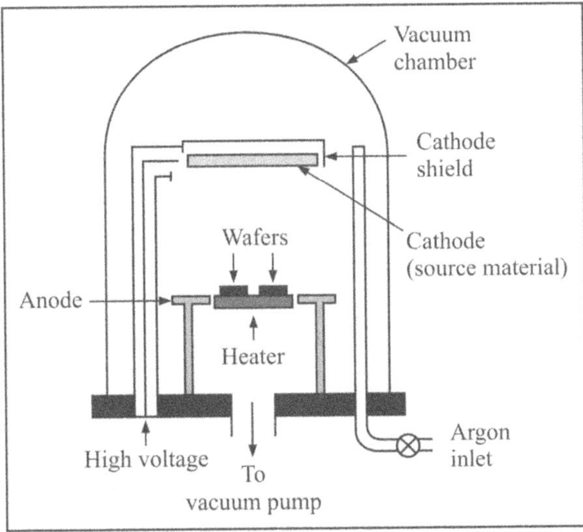

Figure 3.21 Schematic of sputtering system

Both evaporation and sputtering systems are often able to deposit more than one material simultaneously or sequentially. This capability is very useful to obtain alloys and multilayer films (*e.g.*, multilayer magnetic recording heads are sputtered). For certain low-reactivity metals such as Au and Pt the previous deposition of a thin layer of another metal is needed to improve adhesion. Ti and Cr are two frequently used adhesion promoters. Stress in evaporated or sputtered layers is typically tensile. The deposition rates are much higher than for most CVD techniques. However, due to stress accumulation and cracking, thickness beyond 2 μm is rarely deposited with these processes.

3.2.2.1.3.3 Thin-Film Deposition and Doping

Most fabricated micro/nanostructures contain materials other than that of the substrate, which are obtained by various deposition techniques or by modification of the substrate. Following is a list of a few typical applications for the deposited and/or doped materials used in micro/nanofabrication, which gives an idea of the required properties:

- Mechanical structure
- Electrical isolation
- Electrical connection
- Sensing or actuating
- Mask for etching and doping
- Support or mold during deposition of other materials (sacrificial materials)
- Passivation

Most of the deposited thin films have properties different from those of their corresponding bulk forms (for example, metals shows higher resistivities in thin-film form).

In addition, the techniques utilized to deposit these materials have a great impact on their final properties. For instance, internal stress (compressive or tensile) in a film is strongly process dependent. Excessive stress may crack or detach the film from the substrate and therefore must be minimized, although it may also be useful for certain applications. Adhesion is another important issue that needs to be taken into account when depositing thin films. In some cases such as the deposition of noble metals (*e.g.*, gold) an intermediate layer (chromium or titanium) may be needed to improve adhesion. Finally, step coverage and conformality are two properties that can also influence the choice of one or another deposition technique. Fig.3.22 (a-d) illustrates these concepts.

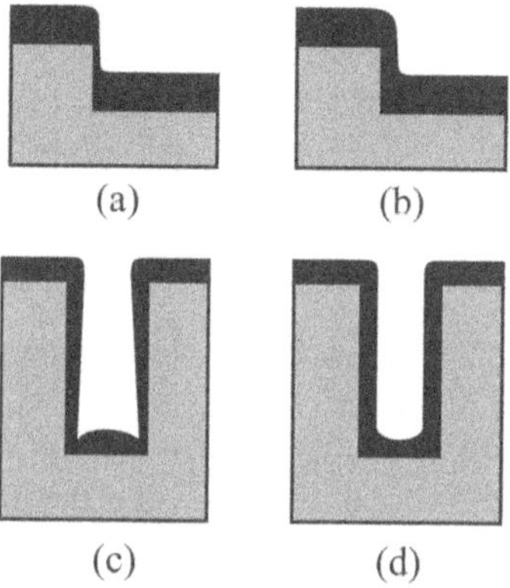

Figure 3.22 (a-d) Step coverage and conformality: (a) poor step coverage, (b) good step coverage, (c) nonconformal layer, and (d) conformal layer

Various thin film deposition techniques in use are summarized in Table 3.1

Table 3.1 Summary of Thin film deposition Techniques

	Method	Deposition rate	Deposition Temperature	Step Coverage	Typical materials
Physical	Sputtering	Up to 100 nm/min	RT	Fair	Most metals, metallic compounds
Physical	Evaporation	Typically 1-10 nm/min	RT	Poor	Low melting point metals
Physical	ALD	~1 nm/min	80 °C to 500 °C	Excellent	Oxides, Nitrides
Chemical	Thermal oxidation	1mm/min (dry) 6 nm/min (wet)	850 °C to 1050 °C	Good	Silicon dioxide
Chemical	Thermal CVD	10-100 nm/min 5 □m/min (epi)	300 °C to 900 °C	Good	Oxides, Nitrides, W, Poly
Chemical	PBCVD	few hundred nm/min	100 °C to 400 °C	Fair	Oxides, Nitrides amorphous 9i

3.2.2.1.3.4 Oxidation

Thermal oxidation is a special type of thin film deposition method which can only be done on top of silicon. In thermal oxidation, a high temperature, together with either water vapor (wet oxidation) or molecular oxygen (dry oxidation) is used to oxidize the silicon (Fig.3.23) Silicon is consumed in the process so that 46% of oxide thickness will lie below the original surface and 54% above it. Oxidation of silicon is a process used to obtain a thin film of SiO_2 with excellent quality (very low density of defects) and thickness homogeneity. Although it is not properly a deposition, the result is the same; *i.e.*, a thin layer of a new material covering the surface is produced. The oxidation process is typically carried out at temperatures in the range of 900-1200 °C in the presence of O_2 (dry oxidation) or H_2O (wet oxidation). The reactions for oxide formation are

$$Si(solid) + O_2\,(gas) \Rightarrow SiO_2\,(solid)$$

and

$$Si(solid) + 2H_2O\,(steam) \Rightarrow SiO_2\,(solid) + 2H_2\,(gas)\cdot$$

Although the rate of oxide growth is higher for wet oxidation, this is achieved at the expense of lower oxide introduced into an electric resistance furnace such as that represented in Fig. 3.23.

Tens of wafers can be processed in a single batch in such equipment. By strictly controlling the timing, temperature, and gas flow entering the quartz tube the desired thickness can be achieved with high accuracy. Thicknesses ranging from a few tens of Angstroms to 2 μm can be obtained in reasonable times quality (density).

Since silicon atoms from the substrate participate in the reaction, the substrate is consumed as the oxide grows (≈ 44% of the total thickness lies above the line of the original silicon surface). The oxidation of silicon also occurs at room temperature, however a layer of about 20 Å (native oxide) is enough to passivate the surface and prevent further oxidation. To grow thicker oxides, wafers are despite the good quality of the SiO_2 obtained by silicon oxidation (also called thermal oxide), the use of this process is often limited to the early stages of the fabrication, since some of the materials added during the formation of structures may not withstand the high temperatures. The contamination of the furnace, when the substrates have been previously in contact with certain etchants such as KOH or when materials such as metals have been deposited, also poses limitations in most cases.

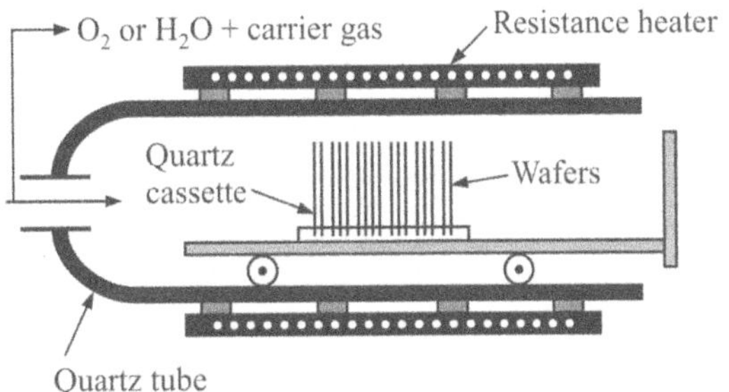

Figure 3.23 Schematic of oxidation furnace

3.2.2.1.3.5 Doping

The introduction of certain impurities in a semiconductor can change its electrical, chemical, and even mechanical properties. Typical impurities or dopants used in silicon include boron (to form p-type regions) and phosphorous or arsenic (to form n-type regions). Doping is the main process used in the microelectronic industry to fabricate major components such as diodes and transistors. In micro/nanofabrication technologies doping has additional applications such as the formation of piezo resistors for mechanical transducers or the creation of etch stop layers. Two different techniques are used to introduce impurities into a semiconductor substrate: diffusion and ion implantation.

Diffusion is the process which became dominant in the initial years following the invention of the integrated circuit to form n- and p-type regions in the silicon (Figure 3.24)The diffusion of impurities into silicon occurs only at high temperature (above 800 °C). Furnaces used to carry out this process are similar to those used for oxidation. Dopants are introduced in the furnace gaseous atmosphere from liquid or solid sources. Figure 3.24 illustrates the process of creating an n-type region by diffusion of phosphor from the surface into a p-type substrate. A masking material is previously deposited and patterned on the surface to define the areas to be doped. However, because diffusion is an isotropic process, the doped area will also extend underneath the mask. In microfabrication, diffusion is mainly used for the formation of very highly boron-doped regions (p^{++}), which are usually used as an etch stop in bulk micromachining.

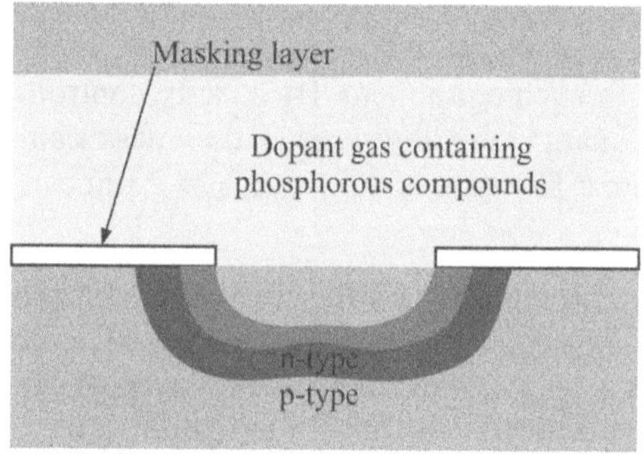

Figure 3.24 Formation of an n-type region on a p-type silicon substrate by diffusion of phosphorous

3.2.2.1.3.6 Pulsed Laser and Atomic Layer Deposition

Pulsed laser and atomic layer deposition techniques have attracted a considerable amount of attention recently. These two techniques offer several unique advantages compared with other thin-film deposition methods that are particularly useful for next-generation nanoscale device fabrication. Pulsed laser deposition (PLD) is a simple technique that uses an intense (1 GW within 25 ns) UV laser (*e.g.*, a KrF excimer) to ablate a target material (20). Plasma is subsequently formed from the target and is deposited on the substrate. Multitarget systems with Auger and reflection high-energy electron diffraction (RHEED) spectroscopes are commercially available. Fig. 3.25 shows a typical PLD deposition setup. The main advantages of the PLD are its simplicity and ability to deposit complex materials with preserved stoichiometry (so-called stoichiometry transfer). In addition, fine control over film thickness is also possible by controlling the number of pulses. The stoichiometry transfer property of the PLD allows many complex targets such as ferroelectrics, superconductors, and magneto strictives to be deposited. Other deposited materials include oxides, carbides, polymers, and metallic systems (*e.g.*, FeNdB).

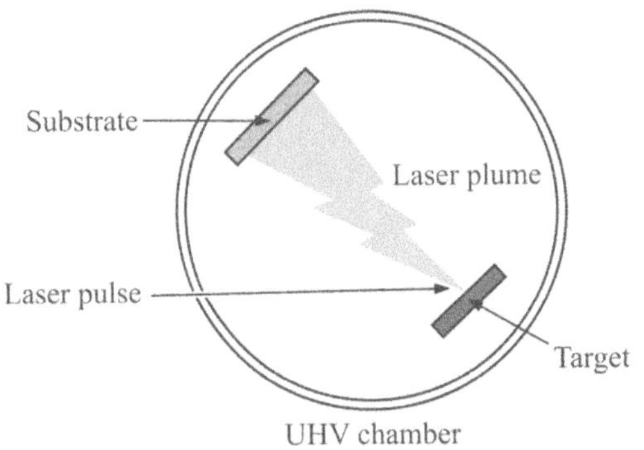

Figure 3.25 A typical PLD deposition setup

3.2.2.1.4 Atomic Layer Deposition (ALD)

ALD is a gas-phase self-limiting deposition method capable of depositing atomic layer thin films with excellent large area uniformity and conformality (21). It enables simple and accurate control over film composition and thickness at the atomic layer level (typical growth rates of a few Å/cycle). Although most of the attention recently has been directed towards depositing high-k dielectric materials (Al_2O_3, and HfO_2) for next-generation complementary metal oxide semiconductor (CMOS) electronics, other materials can also be deposited. These include transition metals (Cu, Co, Fe, and Ni), metal oxides, sulfides, nitrides, and fluorides. Atomic level control over film thickness and composition are also attractive features for applications in MEMS such as conformal three dimensional (3-D) packaging and air-gap structures.

ALD is a modification of the CVD process and is based on two or more vapor phase reactants that are introduced into the deposition chamber in a sequential manner. One growth cycle consists of four steps (Fig.3.26). First, a precursor vapor is introduced into the chamber, resulting in the deposition of a self-limiting monolayer on the surface of the substrate. Then, the extra unreacted vapor is pumped out and a vapor dose of a second reactant is introduced. This reacts with the precursor on the surface in a self-limiting fashion. Finally, the extra unreacted vapor is pumped out and the cycle is repeated.

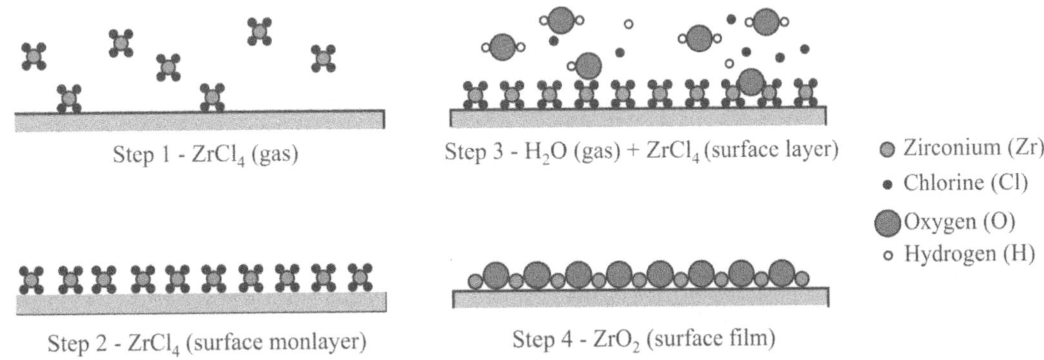

Figure 3.26 Four main steps of growth cycle of monolayers of ALD

There are four main types of ALD reactors

(i) Closed system chambers

(ii) Open system chambers

(iii) Semi-closed system chambers

(iv) Semi-open system chambers

The most commonly used closed system type chambers is shown in Fig. 3.27. In this the reaction chamber walls are designed to effect the transport of the precursors.

Applications ALD

(a) High-K dielectrics for CMOS,

(b) Semiconductor memory (DRAM)

(c) Cu interconnect barrier,

(d) Deposition in porous structures

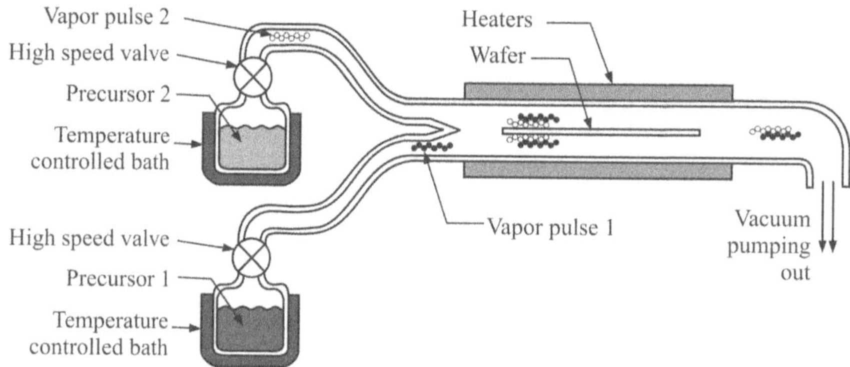

Figure 3.27 Closed system type atomic layer deposition chamber (technology backgrounder: atomic layer deposition," IC knowledge LLC, 24 April 06. <www.icknowledge.com/misc_technology/Atomic%20Layer%20Deposition%20Briefing.>

Advantages

(a) Stoichiometric films with large area uniformity and 3D conformality.

(b) Precise thickness control.

(c) Low temperature deposition possible.

(d) Gentle deposition process for sensitive substrates.

Disadvantages

(a) Deposition Rate slower than CVD

(b) Number of different material that can be deposited is fair compared to MBE

3.2.2.1.5 Electro Deposition

Electroplating

Electroplating (or electro deposition) is a process typically used to obtain thick (tens of μm) metal structures. Essential in plating is the need to guarantee a constant supply of the metal to be plated to the wafer. Fig. 3.28 shows the schematic of experimental set-up for direct current electro deposition and pulse reverse electro deposition

Recommended current densities for electro deposition processes are on the order of 5-100 mA/cm². As can be deduced from the process mechanism, the surface to be electroplated has to be electrically conductive, and preferably of the same material as the deposited one if good adhesion is desired. In order to electro deposit metals on top of an insulator (the most frequent case) a thin film of the same metal, called the seed layer, is previously deposited on the surface. Masking of the seed layer with a resist permits selective electroplating of the patterned areas. Fig. 3.29 illustrates a typical sequence of the steps required to obtain isolated metal structures.

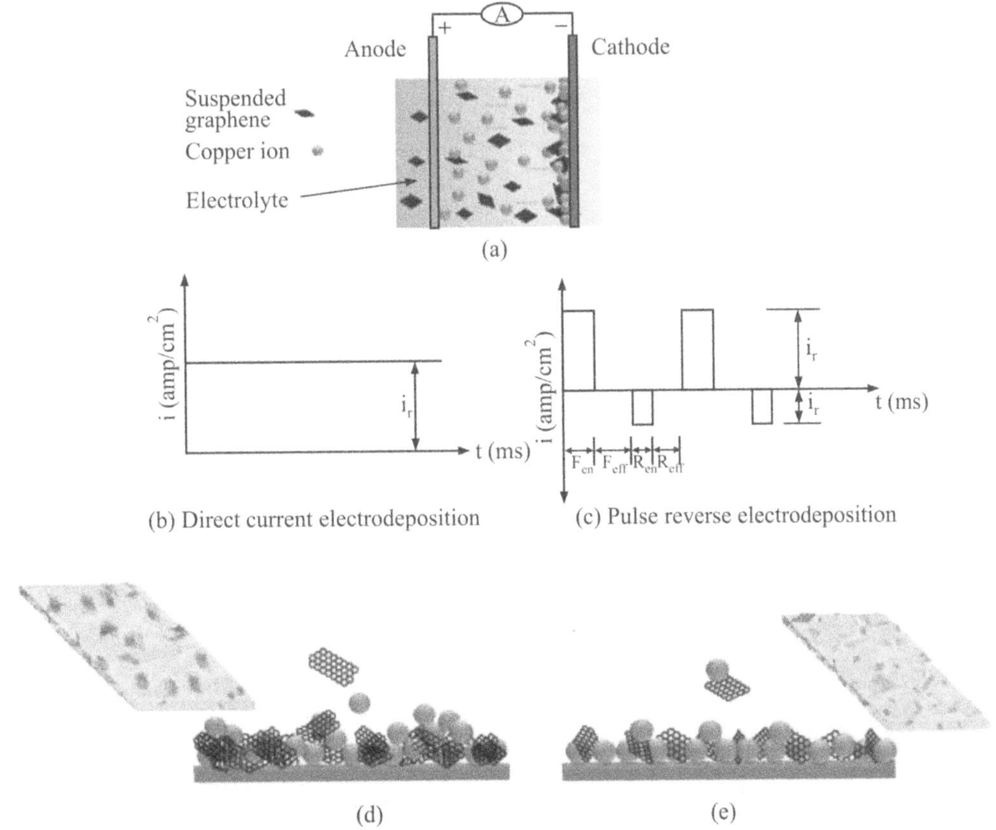

Figure 3.28 Schematic of direct current electro deposition and reverse pulse electro deposition

Depending on needed dimension control and demands on step coverage, plating in box or plating in field combined with wet chemical etching is used to structure the layers. The first uses a two step lithography process to define the structure. In the first step a box is created. After the plating a second lithography process is used to place a lid on that box. The unwanted metal is then etched away, leaving precisely defined structures protected by photoresist. Plating in field covers the whole wafer with metal. A one step lithography process then defines the patterns to be created and the unwanted material is etched away. There is a fundamental limitation of plating: plating in box gives submicron accuracy in lateral dimensions, but as the structure is moulded in a resist it results in nearly straight walls of the structure. Those walls lead to step coverage problems when an insulating layer is deposited on top of it. Wet chemical etching, more special tapered etching gives more gentle slopes, but the dimension control is the weak point, being in the range of the thickness of the layer.

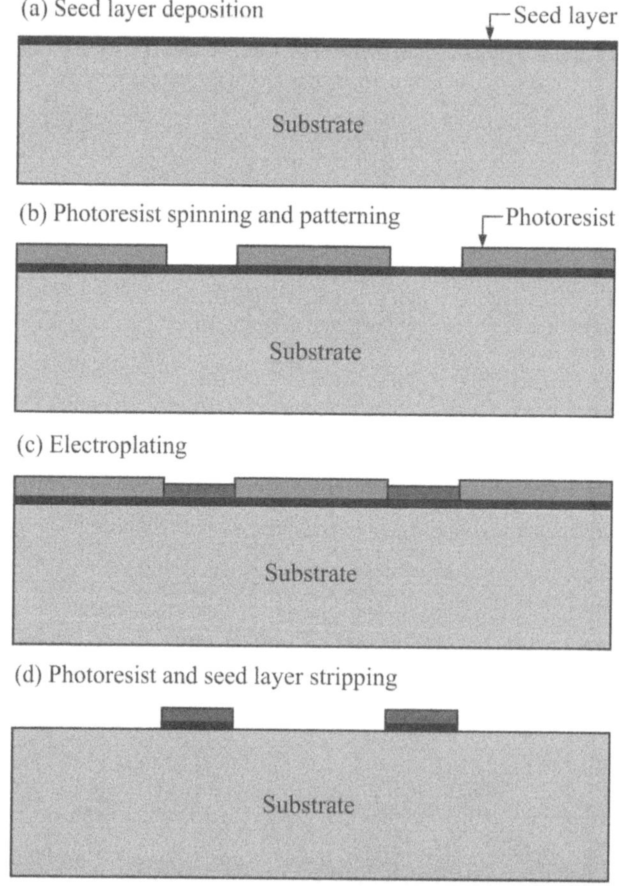

Figure 3.29 (a–d) Formation of isolated metal structures by electroplating through a mask: (a) seed layer deposition, (b) photoresist spinning and patterning, (c) electroplating, and (d) photoresist and seed layer stripping

3.2.2.1.6 Quantum Structure Nanofabrication - Using Strain-induced Self-assembly

A more recent technique for fabricating quantum wires and dots involves strain-induced self-assembly (22) (Fig. 3.30).

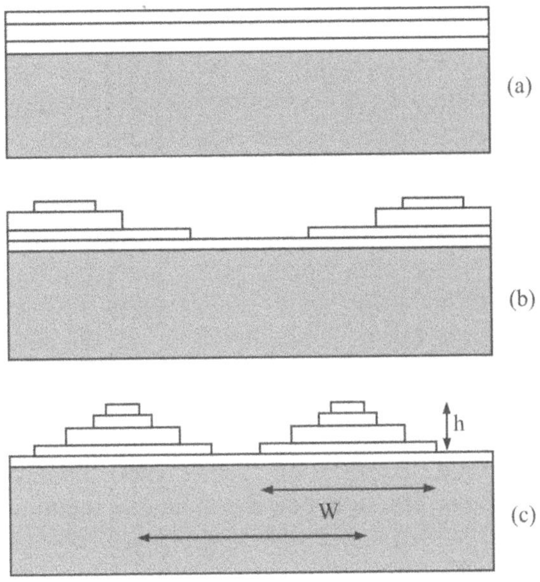

Figure 3.30 (a–c) Stranski–krastanow growth mode, (a) 2-D wetting layer, (b) growth front roughening and breakup, and (c) coherent 3-D self-assembly

The term self-assembly represents a process where a strained two-dimensional (2-D) system reduces its energy by a transition into a 3-D morphology.

The most commonly used material combination for this technique is the In_x Ga_{1-x} As/GaAs system, which offers a large lattice mismatch (7.2% between InAs and GaAs) (23,24) although recently Ge dots on Si sub-strate have also attracted considerable attention (25). This method relies on lattice mismatch between an epitaxially grown layer and its substrate to form an array of quantum dots or wires

Fig. 3.30 shows a schematic of the strain-induced self-assembly process. When the lattice constant of the substrate and the epitaxial layer differ considerably, only the first few deposited mono-layers crystallize, in the form of an epitaxial strained layer in which the lattice constants are equal. When a critical thickness is exceeded, a significant strain that occurs in the layer leads to the breakdown of this ordered structure and to the spontaneous formation of randomly distributed islets of regular shape and similar size (usually < 30 nm in diameter). This mode of growth is usually referred to as the Stranski–Krastanow mode. The quantum dot size, separation, and height depend on the deposition parameters (i.e., total deposited material, growth rate, and temperature) and material combinations. As can be seen, this is a very convenient method to grow perfect crystalline nanostructures over a large area without any lithography and etching. One major drawback of this technique is the randomness of the quantum dot distribution. It should be mentioned that this technique can also be used to fabricate quantum wires by strain relaxation bunching at step edges to or in contact with a surface with sub-nanometer accuracy. Piezoelectric positioners are typically used in order to achieve such accuracy. High resolution images can be acquired by raster scanning the tips over a surface while simultaneously monitoring the interaction of the tip with the surface.

3.2.2.1.7 Physical and Chemical Self-assembly

The central theme behind the self-assembly process is spontaneous (physical) or chemical aggregation of colloidal nanoparticles (28). Spontaneous self-assembly exploits the tendency of mono dispersed nano or submicro colloidal spheres to organize into a face-centered cubic (fcc) lattice. The force driving this process is the desire of the system to achieve a thermodynamically stable state (minimum free energy). In addition to spontaneous thermal self-assembly, gravitational, convective, and electro hydrodynamic forces can also be used to induce aggregation into complex 3-D structures. Chemical self-assembly requires the attachment of a single-molecular organic layer (self-assembled monolayer or SAM) to the colloidal particles (organic or inorganic) and subsequent self-assembly of these components into a complex structures using molecular recognition and binding.

3.2.2.1.7.1 Physical Self-assembly

This is an entropy-driven method that relies on spontaneous organization of colloidal particles into a relatively stable structure through non covalent interactions; for example, colloidal polystyrene spheres can be assembled into a 3-D structure on a substrate which is held vertically in the colloidal solution (Fig.3.31) (26-28).

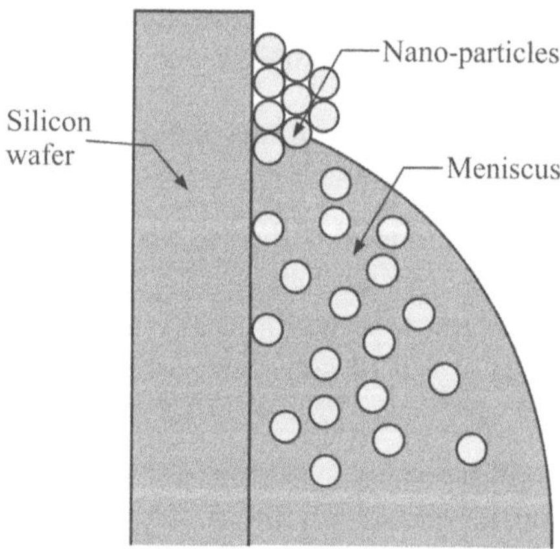

Figure 3.31 Colloidal particle self assembly on to solid substrates upon drying in vertical position

Upon the evaporation of the solvent, the spheres aggregate into a hexagonal close-packed (hcp) structure. The interstitial pore size and density are determined by the polymer sphere size. The polymer spheres can be etched into smaller sizes after forming the hcp arrays, thereby altering the template pore separations. This technique can fabricate large patterned areas in a quick, simple, and cost-effective way. A classic example is the natural assembly of on-chip silicon photonic-bandgap crystals (29) which are capable of reflecting the light arriving in any direction in a certain wavelength range (30). In this method, a thin layer of silica colloidal spheres is assembled on a silicon substrate. This is achieved by placing a silicon wafer vertically in a vial containing an ethanolic suspension of silica spheres. A temperature gradient across the vial aids the flow of silica spheres.

One can also deposit colloidal particles onto a patterned substrate (template-assisted self-assembly, TASA) (31, 32). This method is based on the principle that, when an aqueous dispersion of colloidal particles is allowed to dewet from a solid surface which is already patterned, the colloidal particles are trapped by the recessed regions and assemble into aggregates with shapes and sizes determined by the geometric confinement provided by the template. The patterned arrays of templates can be fabricated using conventional contact-mode photolithography which gives control over the shape and dimensions of the templates, thereby allowing the assembly of complex structures from colloidal particles. The cross-sectional view of a fluidic cell used in TASA is shown in Fig. 3.32. The fluidic cell has two parallel glass substrates to confine the aqueous dispersion of the colloidal particles. The surface of the bottom substrate is patterned with a 2-D array of templates. When the aqueous dispersion is allowed to dewet slowly across the cell, the capillary force exerted on the liquid pushes the colloidal spheres across the surface of the bottom substrate until they are physically trapped by the templates. If the concentration of the colloidal dispersion is high enough, the template will be filled by the maximum number of colloidal particles determined by the geometrical confinement. This method can be used to fabricate a variety of polygonal and polyhedral aggregates which are difficult to generate (33).

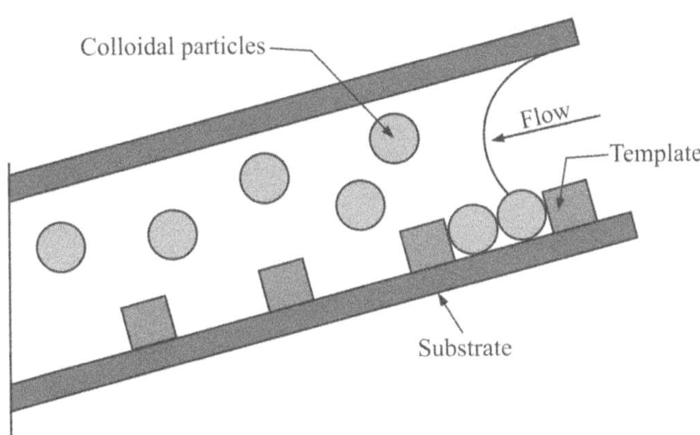

Figure 3.32 A cross-sectional view of the fluidic cell used for template-assisted self-assembly

If the size and geometry of the colloidal particles is precisely controlled, it is in theory possible to also control the spatial distribution of the colloids.

For example, mono-dispersed spherical particles can be close-packed into hexagonally arranged monolayers, i.e., the conformation of highest density for this geometry. Nonetheless, a number of parameters including colloid concentration, solvent evaporation rate, wetting characteristics of the substrate, and competition with multiple-layer formation influence the resulting array. It has been reported that the achievable resolution offered by this technique, when the colloidal monolayer is utilized as a resist, can be as low as 5 nm in all three dimensions.

These colloid monolayers can be used as a protective barrier against etching and, consequently, transfer their semi-random distribution pattern to the substrate material. More specifically, these etch masks would result in the removal of the substrate material located below the colloid interstices. The same analogy can be made for applications in selective film deposition. Depending on the application, the colloidal particles can be either removed or left in place.

3.2.2.1.7.2 Chemical Self-assembly

Organic and supramolecular SAMs play a critical role in colloidal particle self-assembly. SAMs are robust organic molecules which are chemically adsorbed onto solid substrates. Most often they have a hydrophilic (polar) head which can be bonded to various solid surfaces and a long hydrophobic (nonpolar) tail which extends outward. SAMs are formed by the immersion of a substrate in a dilute solution of the molecule in an organic solvent or water (liquid phase) or by exposure to an atmosphere containing such a molecule (gas phase). The resulting film is a dense organization of molecules arranged to expose the end group. The durability of the SAM is highly dependent on the effectiveness of the anchoring to the surface of the substrate. SAMs have been widely studied because the end group can be functionalized to form precisely arranged molecular arrays for various applications ranging from simple, ultrathin insulators and lubricants to complex biological sensors. Chemical self-assembly uses organic or supramolecular SAMs as the binding and recognition sites for fabricating complex 3-D structures from colloidal nanoparticles. The most commonly used organic monolayers include:

- Organosilicon compounds on glass and oxidized silicon
- Alkanethiols, dialkyl disulfides, and dialkyl sulfides on gold
- Fatty acids on alumina and other metal oxides
- DNA

Octadecyltrichlorosilane (OTS) is the most common organosilane used in the formation of SAMs, mainly because of the fact that it is simple, readily available, and forms good, dense layers (34). Alkyltrichlorosi lane monolayers can be prepared on clean silicon wafers whose surface is SiO_2 (with almost 5×10^{14} SiOH groups/cm^2). Since the silicon chloride bond is susceptible to hydrolysis, a limited amount of water has to be present in the system in order to obtain good quality monolayers. Monolayers made of methyl-and vinyl-terminated alkylsilanes are autophobic to the hydrocarbon solution and hence emerge uniformly

3.2.2.1.7.3 Molecular Self-assembly

Molecular self assembly is an alternative to lithographic techniques for the fabrication of features and structures at the nanometer scale.

Nature uses self assembly in infinitely subtler ways: The whole natural world is self assembled. Spontaneous organization of molecules into stable well defined aggregates (nano meter length scale. Molecules can be transported to surfaces through liquids to form self assembled nanolayers (SAMs).

It is based on thermodynamically favored interactions of molecules such as peptides, proteins, and DNA, and other organic or inorganic molecules (35) are spontaneously brought together to energetically stable conformations favored by non covalent forces including hydrophobic, vander Waals, and electrostatic interactions, as well as hydrogen bonding.

This technique has a number of advantages including the ability to fabricate three-dimensional structures and the potential for molecular control of the material (36). Molecular self assembly is able to control pattern formation to the sub-nanometer scale. Fig. 3.33 depicts the formation of poly thipene nano wires formed by self assembly.

Self assembly of monolayers of amphiphilic peptide β-hairpins was shown to result in organized structures containing 2.5 nm hairpins spaced by less than 0.3 nm (35).

Layer by layer assembly, one of the self assembly methods, is based on the consecutive deposition of multiple thin polyion films from solution utilizing the electrostatic attraction that develops between oppositely charged molecules as a driving force for assembly (36, 37).

Time for adsorption of polyions onto surfaces ranges from minutes to hours depending on factors such as polyion type and concentration.

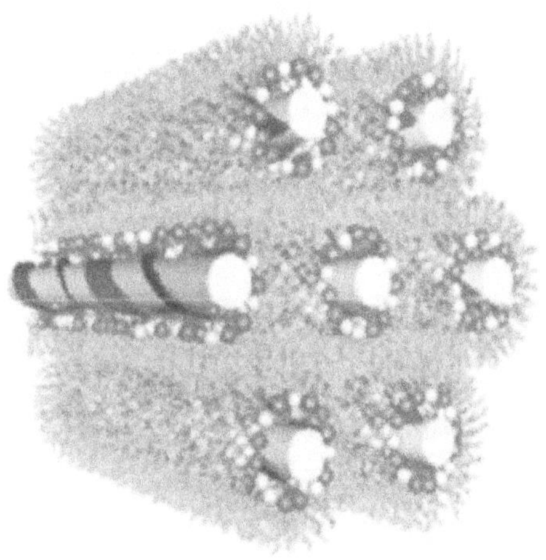

Figure 3.33 Poly thiopene wires formed by self assembly

Intermediate washes between sequential depositions are performed to avoid contamination from previous solutions. This technique has been widely accepted because it permits the use of numerous materials, including biomolecules such as proteins (37) and DNA (38) provides great control over the film structure, thickness, and function. In addition, because film formation occurs by adsorption of polyanions or polycations from solution, the number of possible morphologies and sizes are only limited by those of the substrate to be coated.

Self-assembly is by no means limited to molecules or the nano domain and can be carried out on just about any scale, making it a powerful bottom-up assembly and manufacturing method (multiscale ordering).

Another attractive feature of this technique relates to the possibility of combining self-assembly properties of organic molecules with the electronic, magnetic, and photonic properties of inorganic components.

Template manufacturing is another bottom up technique which utilizes material deposition (electroplating, CVD, etc.,) into nano templates in order to fabricate nanostructures.

Due to the simplicity and flexibility of electrochemistry for plating and surface finishing of a broad range of materials, its principle has recently been widely used for electrochemical fabrication of various metallic nano-structures based on various templates. For example, electrochemical deposition has been used to deposit large arrays of nanostructures in nano porous templates, such as porous alumina. This template-based deposition typically provides metal nano wires as small as 25 nm in diameter and a few micrometers in length (39).

3.2.2.2 Scanning Probe Techniques

3.2.2.2.1 Scanning Probe Lithography

The invention of scanning probe microscopy in the 1980s revolutionized atomic-scale imaging and spectroscopy. In particular scanning tunneling and atomic force microscopes (STM and AFM) have found widespread applications in physics, chemistry, material science, and biology. The possibility of atomic-scale manipulation, lithography, and nanomachining using such probes was considered from the beginning and has matured considerably over the past decade.

Scanning probe microscopy (i.e., STM, AFM, etc.) uses small (<50 nm) tips to image surfaces with atomic resolution. This ability suggests opportunities for their use in generating nanostructures and nanodevices. In this form they are referred to as "Scanning Probe Lithography" (SPL), which uses the tip of an AFM to selectively *remove* certain areas on a surface and "Dip-pen nanolithography" (DPN), which, similarly, uses the AFM tip to *deposit* material on a surface with nanometre resolution.

Scanning probe microscopy (SPM) systems (Fig. 3.34) are capable of controlling the movement of an atomically sharp tip in close proximity.

Surface scanning the tip in the x –y-plane while recording the tunnel current permits the mapping of the surface topography with resolution on the atomic scale.

In a more common mode of operation the amplified current signal is connected to the z-axis piezoelectric positioner through a feedback loop so that the current and therefore the distance are kept constant throughout the scanning. In this configuration the picture of the surface topography is obtained by recording the vertical position of the tip at each x-y-position.

The STM system only works for conductive surfaces because of the need to establish a tunneling current. The atomic force microscopy was developed as an alternative for imaging either conducting or non conducting surfaces. In AFM the tip is attached to a flexible cantilever and is brought into contact with the surface (Fig. 3.34) (b). The force between the tip and the surface is detected by sensing the cantilever deflection. A topographic image of the surface is obtained by plotting the deflection as a function of the x-y-position. In a more common mode of operation a feedback loop is used to maintain a constant deflection while the topographic information is obtained from the cantilever vertical displacement. Some scanning probe systems use a combination of the AFM and the STM modes, i.e., the tip is mounted on a cantilever with electrical connection so that both the surface forces and tunneling currents can be controlled or monitored. STM systems can be operated in ultrahigh vacuum (UHV STM) or in air, whereas

AFM systems are typically operated in air. When a scanning probe system is operated in air, water adsorbed onto the sample surface accumulates underneath the tip, forming a meniscus between the tip and the surface. This water meniscus plays an important role in some of the scanning probe techniques described below.

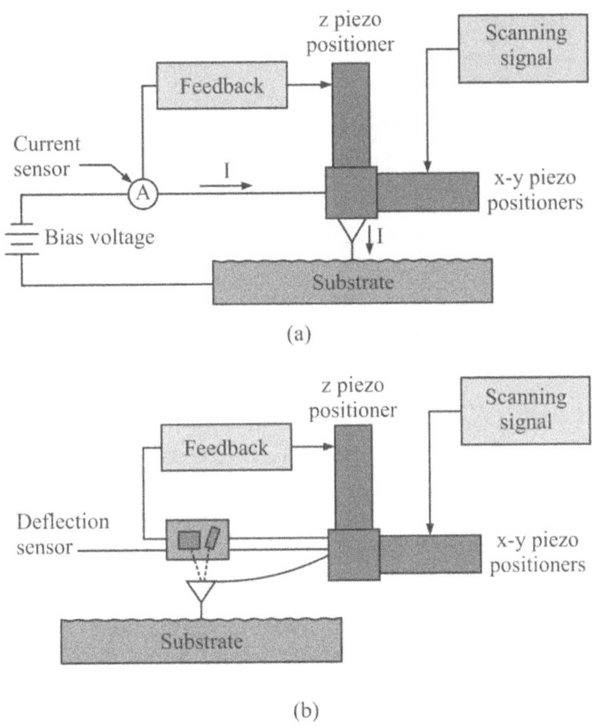

Figure 3.34 (a, b) Scanning probe systems: (a) STM and (b) AFM

3.2.2.2.2 Scanning-Probe-Induced Oxidation

Nanometer-scale local oxidation of various materials can be achieved using scanning probes operating in air and biased at a sufficiently high voltage. A tip bias of –2 to –10 V is normally used, with a writing speed of 0.1=100 μm/s in ambient humidity of 20-40%. It is believed that the water meniscus formed at the contact point serves as an electrolyte such that the biased tip anodically oxidizes a small region of the surface (40). The most common application of this principle is the oxidation of hydrogen-passivated silicon. A dip in HF solution is typically used to passivate the silicon surfaces with hydrogen atoms. Patterns of oxide written on a silicon surface can be used as a mask for wet or dry etching. Patterns with 10 nm line width have been successfully transferred to a silicon substrate in this fashion (41). Various metals have also been locally anodized using this approach, such as aluminum or titanium (42). An interesting variation of this process is anodization of deposited amorphous silicon (43). Amorphous silicon can be deposited at low temperature on top of many materials (Figure 3.35). The deposited silicon layer can be patterned and used as, for example, the gate of a 0.1 μm CMOS transistor, or it can be used as a mask to pattern an underlying film. The major drawback of this technique is poor reproducibility due to tip wear during the anodization. However, using AFM in noncontact mode has proved to overcome this problem (44).

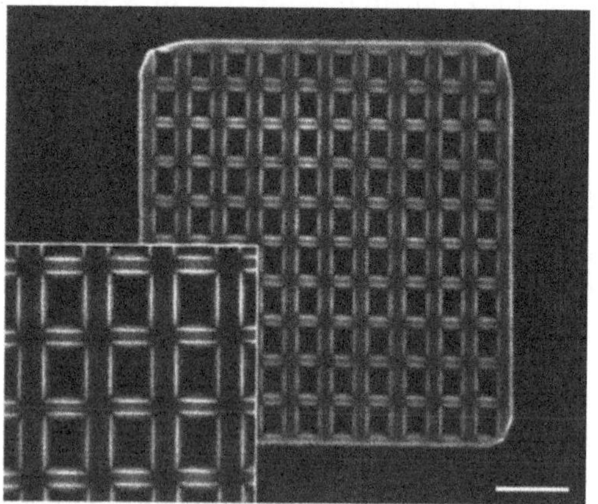

Figure 3.35 SEM image of an inverted truncated pyramid array fabricated on a silicon SOI wafer by SPM oxidation and subsequent etch in TMAH (pitch is 500 nm) (43)

3.2.2.2.3 Probe Resist Exposure and Lithography

Electrons emitted from a biased SPM tip can be used to expose a resist in the same way e-beam lithography that does (Fig. 3.36) (44). Various systems have been used for this lithographic technique; these include constant-current STM, noncontact AFM, and AFM with constant tip–resist force and constant current. The systems using AFM cantilevers have the advantage that they can perform imaging and alignment tasks without exposing the resist. Resists well characterized for e-beam lithography (*e.g.*, PMMA or SAL601) have been used with scanning probe lithography to achieve reliable sub-100 nm lithography. The wafers are cleaned and the native oxide (in the case of silicon or poly) is removed with a HF dip. Subsequently 35-100 nm thick resist is spin-coated on top of the surface. The exposure is done by moving the SPM tip over the surface while applying a bias voltage sufficiently high to produce emission of electrons from the tip (a few tens of volts). Development of the resist is performed in standard solutions following the exposure. Features below 50 nm in width have been achieved with this procedure.

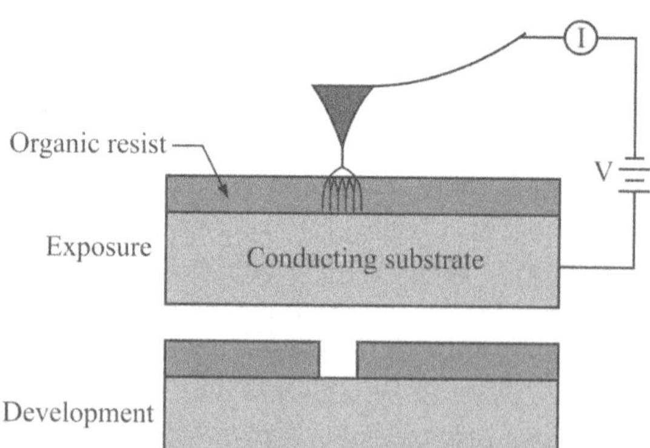

Figure 3.36 Scanning probe lithography with organic resist

3.2.2.2.4 Dip-Pen Nanolithography

In dip-pen nanolithography (DPN) the tip of an AFM operated in air is inked with a chemical of interest and brought into contact with a surface. The ink molecules flow from the tip onto the surface as with a fountain pen. The water meniscus that naturally forms between the tip and the surface enables the diffusion and transport of the molecules, as shown in Fig. 3.37.

In this technique high precision actuators move atoms from place to place. Micro tips emboss or imprint materials. Electron (or ion) beams are directly moved over a surface.

Chemical reactors create conditions for special growth. Biological agents sometimes used to help processes. Materials are harvested for integration.

Inking can be done by dipping the tip into a solution containing a low concentration of the molecules followed by a drying step (e.g., blow-drying with compressed difluoroethane). Line widths down to 12 nm with spatial resolution of 5 nm have been demonstrated with this technique. Species patterned with DPN include conducting polymers, gold, dendrimers, DNA, organic dyes, antibodies, and alkanethiols. Alkanethiols have been also used as an organic monolayer mask to etch a gold layer and subsequently etch the exposed silicon substrate. One can also use a heated AFM cantilever to control the deposition of a solid organic ink.

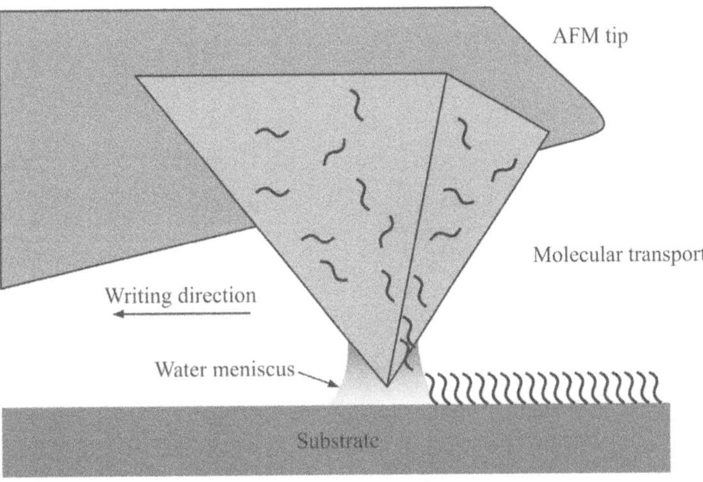

Figure 3.37 Schematic representation of the working principle of dip pen nanolithography

This technique was recently reported by Sheehan *et. al.*, in which 100 nm lines of octadecylphosphonic acid (melting point 100 °C) were written using a heated AFM probe (46).

3.2.2.2.5 Other Scanning Probe Nanofabrication Techniques

A great variety of nanofabrication techniques using scanning probe systems have been demonstrated. Some of these are proof of concept demonstrations and their utility as viable and repeatable fabrication processes has yet to be evaluated. For example, a substrate can be mechanically machined using a STM/AFM tip acting as a plow or engraving tool (47). This can be used to create structures directly in the substrate, although it is more commonly used to pattern resist for a subsequently etch, lift-off or electrodeposition step. Mechanical nanomachining with SPM probes can be facilitated by heating the tip above the glass-transition temperature of a polymeric substrate material. This approach has been applied to SPM-based high-density data storage in polycarbonate substrates (48,49).

3.2.2.2.6 CMP (Chemical Mechanical Polishing)

Chemical Mechanical Polishing (CMP) was first developed by IBM to enable lithography on multi-layer stacks. Due to the differences in surface height focus problems arose in next lithography steps, in this way making further miniaturisation impossible. At this moment CMP is worldwide accepted as a cleanroom process that is inevitable in the production of integrated circuits. Originally developed for the planarisation of SiO_2 surfaces only, the CMP spectrum has broadened to a wide variety of processes including metal CMP like copper damascene.

In a CMP process wafers to be polished are moving in a sweeping motion over a polishing pad while applying some pressure and the backside of the wafer (backpressure).

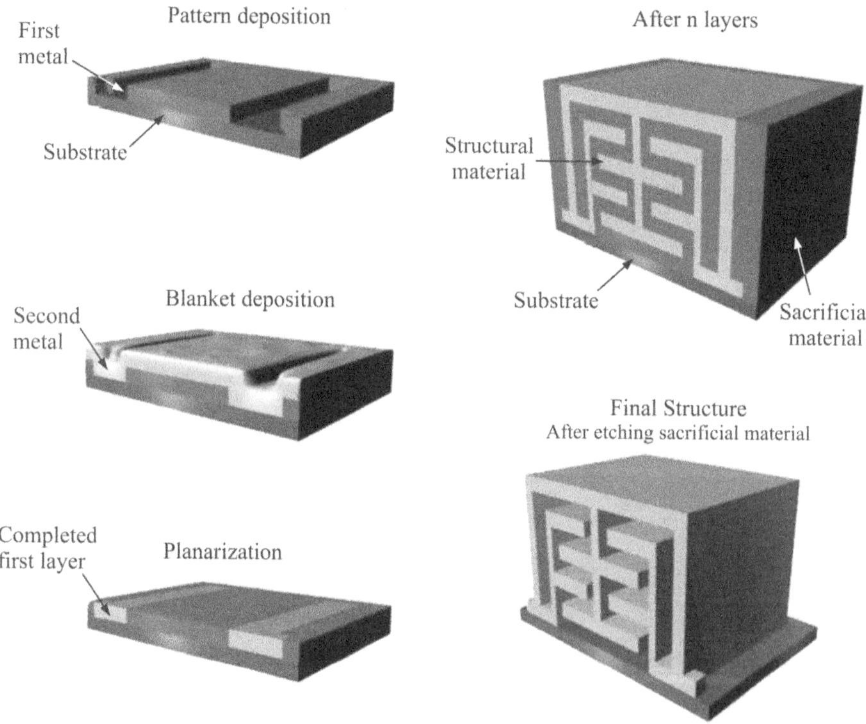

Figure 3.38 CMP/electrodeposition process (Courtesy Microfabrica)

The combination of CMP and electrodeposition is an especially interesting way to construct 3-D objects with basically planar technologies. Microfabrica uses an additive process based on multi-layer selective electrodeposition of metals, following the process sequence: patterned layer deposition, blanket layer deposition and planarisation (Fig. 3.38).

References

1. Parker, N. W. et al. (2000). High-throughput NGL electron-beam direct-write lithography system. *Proc. SPIE* **3997**: 713.doi:10.1117/12.390042.
2. L. F. Thompson, C. G. Willson and M. J. Boeden (1983). Introduction Microlithography American Chemical Society, Washington, D. C.
3. P. Raj-Choudary (1997). Hand Book of Micro Lithography, Micro Machimning and Micro Fabrication, Vol.1, Micro Lithography (SPIE) IEEE, Bellingham, Washington/London.
4. L. Ming, C Bao-quin, Y Lian Chun, R.X Quixa (2001). The sub micron fabrication technology,Proc.6th Fransisco, Vol 1 452-455.
5. Chen Y and Pepin A. (2001). Nanofabrication: conventional and nonconventional, methods. Electrophoresis, 22: 187-207.
6. Chen Y, Rousseaux F, Haghiri-Gosnet A, et al. (1996). Proximity X-ray lithography as a quick replication technique in nanofabrication: recent progress and perspectives. Microelectronic Engineering, 30: 191-4.
7. Li N, Tourovskaia A and Folch A. (2003). Biology on a chip: micro fabrication for studying the behavior of cultured cells. Crit Rev Biomed Eng, 31: 423-88.
8. Xia Y, Whitesides GM. (1998). Soft lithography. Ann Rev Mat Sci, 28: 153-84.
9. Voldman J, Gray ML and Schmidt MA. (1999). Microfabrication in biology and medicine. Ann Rev Biomed Eng, 1: 401-25.
10. Tan W, Desai TA. (2003). Microfluidic patterning of cells in extracellular matrix biopolymers: effects of channel size, cell type, and matrix biopolymers: effects of channel size, cell type, and matrix composition on pattern integrity. Tissue Eng, 9: 255-67.
11. Andersson H, van den Berg A. (2004). Microfabrication and microfluidics for tissue engineering: state of the art and future opportunities. Lab on a Chip, 4: 98-103.
12. Burmeister F, Badowsky W, Braun T, et al. (1999). Colloid monolayer lithography-a flexible approach for nanostructuring of surfaces. Applied Surface Science, 144-155: 461-6.
13. Curtis A and Wilkinson C. (2001). Nanotechniques and approaches in biotechnology. Trends Biotechnol, 19: 97-101.
14. Schaffer E. Thurn-Albrecht T, Russell TP.et.al (2000). Electrically Induced Structure formation and pattern transfer, Nature, 403, 874-7.
15. Liu T, Burger C, Chu B, (2003). Nanofabrication in polymer matrices, Prog. Colloidal Polymer. Sci 28: 5-26.
16. Thurn-Albrecht T, Schotter J, Kästle GA, et al. (2000). Ultrahigh-Density nanowire arrays grown in self-assembled diblock copolymer templates. Science, 290: 2126-9.
17. J. Vossen (1976). Thin Film Process (Academic, NewYork).
18. Xi a Y, Rogers J.A, Paul K.E and Whitesides G.M. (1999). Unconventional methods for fabricating and patterning nanostructures. Chem. Rev. 99: 1823–1848. doi:10.1021/cr980002q [PubMed]
19. Wlkinson C.D.W, Curtis A.S.G and Crossan J (1998). Nanofabrication in cellular engineering. J. Vac. Sci. Technol. B. 16: 3132–3136. doi:10.1116/1.590451.

20. H. UKerbs, M. Stamer, J. FauPel, E. Suske, T. Scharf, C. Fusche, N. Sebit, H. Khijewisky, D. Nilekie, E. Panchenko and M. Buback (2003). Pulsed Laser Deposition(PLD, A versatile Thin Film Technique; Adv Solid, State Physics, 43, 505-517.

21. M. Leskela and M. Ritala (2003). Atomic Layer deposition chemistry: Recent Developments and future challenges, Angew. Chem. Int. Ed, 42, 5548-5554.

22. F. Guffarth, R. Hertz, A. Schliwa, O. Stier, N. N. Ledentsov, A.R Kovsh, V.M . Ustinov D. Bimberg, M. Grundmann and N.N. Ledenstov (1999). Quantum Dot Heterostructures, Wiley, Newyork.

23. B. C. Lee, S. D, Lin, C.P Lee, H. M lee, J.C Will and K.W Sun (2002). Selective growth of single in as quantum dots using strain engineering, Appl. Phys. Letters 80, 326-328.

24. K. Burner (2002). Si/Ge nanostructures, Rep. Prog. Phys .65, 27-72.

25. Plass, J.A. Last, N.C Barrtelt and G. L. Kellogg (2001). Self assembeled domain patterns, Nature, 412,875.

26. Y. A. Viasov, x-z. Bo, Sturm, D.J. Norris (2001). On Chip natural self assembly of silicon photonic band gap crystals, Nature 414, 289-293.

27. C. Gigault, K. Dainoki-Veress and J. R. Dutcner (2001). Changes in the morphology of self assembled polysterene microsphere monolayers produced by annealing; J. Colloid interface Sci 243,143-155.

28. J. C. Huteen and C.R Martin (1997). A general template based method for the preparation of nano materials. J.Mater.Chem.7, 1075-1087.

29. J.D Joannopoulos. P.R Villeneuve, S. Fan: Photonic Crystals: Putting a new twist on light, Nature 386, 143-149(1997).

30. D Clark, R. Ferrington, J. Tien, K. E. Paul and G.M Whitesides (2002). Template Directed Self assembly of10-um sized hexagonal plates, J. am. Chem Soc, 124, 5419-5426.

31. S.A, D.T Mitchell and C.R Martin (1999). Using template synthesized micro and nanowires as building blocks for self–assembly of supramolecular architectures, Chem. Matr, 11, 1183-1185.

32. Y. Yin. Y. Liu, B. Gates and Y. Xia (2001). Template assisted self assembly: A practical route to complex aggregates of mono dispersed colloids with well defined sizes shapes and structures. J. Am. Chem. Soc, 123, 8718-8729.

33. S.R Wasserman, Y.T. Tao and G.M. Whitesides (1989). Structure and reactivity of alkyl silioxane monolayers formed by reaction of alkyl trichlorosilanes on silicon substrates, Langmuir, 5, 1074-1087.

34. C.P Tripp and M.L. Hair (1992). An infrared study of the reaction of octadecyltrichlorosilane with silica Langmuir, 8, 1120-1126.

35. Rajagopal K and Schneider J. (2004). Self-assembling peptides and proteins for nanotechnological applications. Curr Opin Struct Biol, 14: 480-6.

36. Decher G. (1997). Fuzzy nanoassemblies: toward layered polymeric multi-composites. Science, 277: 1232-7.

37. Lvov YM, Lu Z, Schenkman JB, et al. (1998). Direct electrochemistry of myoglobin and cyto-chrome in alternate layer-by-layer films with DNA and other polyions. J Am Chem Soc, 120: 4073-80.

38. Taton TA, Mucic RC, Mirkin CA, et al. 2000. The DNA-mediated formation of supramolecular mono- and multilayered nanoparticle structures. J Am Chem Soc, 122: 6305-6.

39. G. A. Gleves, Z. T. M. Murakami, M. J. Kranz, J. A. Haber (2006). Multigram synthesis of copper nano wires using ac electro deposition into porous aluminium oxide templates,J.Mater.Chem.16, 3075-3083.

40. M. Calleja, J. Anguita, R. Garcia, K. Birkelund, F Perez-Murano, J. A. Dagata (1999). Nanometer-scale oxidation of silicon surfaces by dyanamic force microscopy; reproducibility, kinetics and nano fabrication, Nanotechnology 10, 34-38.

41. E. S. Snow, P. M. Campbell and P. R. Perkins (1997). Nano Fabrication with proximal probes. Proc. IEEE 85: 601- 611.

42. H. Sugumura, I. Uchida, N. Kitamura and H. Masunara (1993). Tip Induced anodization of titanium surfaces by scanning tunneling microscopy; a humidity effect on nano lithography, Appl. Phys. Letters, 63, 1288-1290.

43. Kramer, J. Jorritsma, H. Birk and C. Schonnberger (1995). Nanometer lithography on Silicon and Hydrogenated amorphous silicon with low energy electrons, J. Vac. Sci Techno. B 13, 805-811.

44. F.S.S Chien, W. F. Hsieh, S. Gwo, A. E. Vladar and J. A. Dagata (2002). Silicon nanotubes fabricated by scanning-probe oxidation and tetra-methyl ammonium hydroxide etching, J. App. Phy. 91, 10044-10050.

45. H. Son, K. W. Guarini and C. F. Quate (2001). Scanning Probe Lithography, Kluwer, Academic Boston.

46. C. A. Mirkin (2001). Dip-pen nanolitography, automated fabrication of custom multi component, sub-100-nano meter surface architectures, MRS Bulletin 26, 535-538.

47. P. E. shehan, L. J Whitman, W. P. king and B. A. Nelson (2004). Nanoscale deposition of solid inks via thermal dip pen nanolithography; Appl. Phys. Letters 85, 1589-1591.

48. G. M. Whitesides and B. Grzybowski (2002). Self assembly at all scales, Sciences 295, 2418-2421.

49. P. Kazamaier and N. Chopra (2000). Bridging size scales with self assembling supramolecular materials; MRS Bulletin, 25, 30-35.

4

Micro Electro Mechanical Systems (MEMS) and Nano Electro Mechanical Systems (NEMS)

4.1 Micro Electro Mechanical Systems (MEMS) and Nano Electro Mechanical Systems (NEMS)

4.1.1 Introduction

Electro mechanical systems (EMS) generally consist of two integrated components regardless of scale. The mechanical parts include moveable structures such as mirrors, beams, cantilevers and gears (1, 2, 3), which respond to applied forces by deflection or vibration. The electronic elements, such as small motors and integrated circuits, act as transducers to transform mechanical motion into optical or electrical signals and vice-versa. Scaling these down, there are two distinct levels of EMS: Micro Electro Mechanical Systems (MEMS) and Nano Electro Mechanical Systems (NEMS).

Micro-Electro-Mechanical Systems, or MEMS, is a technology that in its most general form can be defined as miniaturized mechanical and electro-mechanical elements (i.e., devices and structures) that are made using the techniques of micro-fabrication. The critical physical dimensions of MEMS devices can vary from well below one micron on the lower end of the dimensional spectrum, all the way to several millimeters.

The term used to define MEMS varies in different parts of the world. In the United States they are predominantly called MEMS, while in some other parts of the world they are called "Microsystems Technology" or "micro-machined devices".

MEMS devices can vary from relatively simple structures having no moving elements, to extremely complex electromechanical systems with multiple moving elements under the control of integrated microelectronics. The one main criterion of MEMS is that there are at least some elements having some sort of mechanical functionality whether or not these elements can move. MEMS have been studied for decades and are now finding increasing application in industrial and commercial sectors (4). While MEMS technology is technically outside the nanotechnology domain, many researchers consider it as an integral part of the nanotechnology field (5).

Nano Electro Mechanical Systems (NEMS): These represent devices of dimensions in the nanometer range, leading to low mass, high mechanical resonance frequencies, potentially large quantum mechanical effects such as zero point motion, and a high surface-to-volume ratio useful for surface-based sensing mechanisms. Uses include accelerometers, or detectors of chemical substances in the air. Indeed, it is expected that the functionalities shown by MEMS could also be performed by NEMS (1, 2). However NEMS is in the earlier stages of research and development, generally at the experimental & laboratory level. The application prospects of NEMS are somewhat at beginning stage compared with the more mature MEMS. At present there are only few NEMS devices which are rapidly expanding and NEMS facing a number of technical hurdles before application is realized.

4.1.2 Functional Elements of MEMS

While the functional elements of MEMS are miniaturized structures, sensors, actuators, and microelectronics, the most notable (and perhaps most interesting) elements are the microsensors and microactuators (Figure 4.1). Microsensors and microactuators are appropriately categorized as "transducers", which are defined as devices that convert energy from one form to another. In the case of microsensors, the device typically converts a measured mechanical signal into an electrical signal.

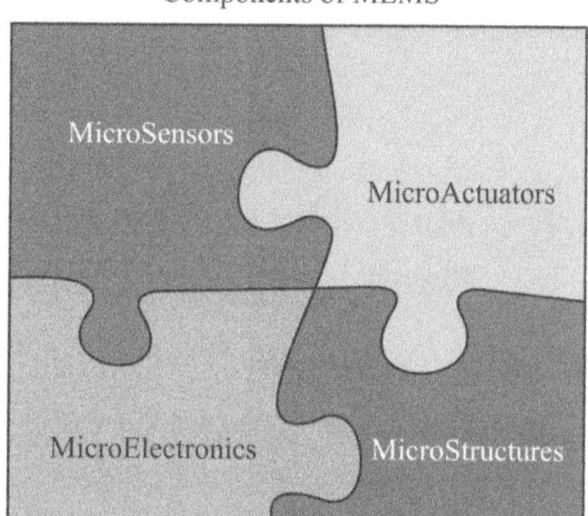

Figure 4.1 MEM components

MEMS are also referred to as Micro System Technology (MST) as MEMS is a process technology used to create these tiny mechanical devices or systems, and as a result, it is a subset of MST. There are considerable overlaps between fields in terms of their integrating technology and their applications and hence it is extremely difficult to categorize MEMS devices in terms of sensing domain and/or their subset of MST. The real difference between MEMS and MST is that MEMS tends to use semiconductor processes to create a mechanical part. In contrast, the deposition of a material on silicon for example, does not constitute a MEMS product but is an application of MST.

Micromachining relates to the creation of small parts such as lenses, gearwheels etc., with mechanical functionality. Micro-opto-electro-mechanical systems (MOEMS) are also a subset of MST and, together with MEMS, form the specialized technology fields using combinations of miniaturized optics, electronics and mechanics. Such microsystems incorporate the use of microelectronics batch processing techniques for their design and fabrication. The class of MEMS devices operating as actuators or sensors for biochemical processes and instrumentation is called BioMEMS.

While the device electronics are fabricated using `computer chip' IC technology, the micromechanical components are fabricated by sophisticated manipulations of silicon and other substrates using micromachining processes. Processes such as bulk and surface micromachining as well as high-aspect-ratio micromachining (HARM) selectively remove parts of the silicon or add structural layers to form the mechanical and electromechanical components.

4.1.3 Inter Dependence of MEMS and Nano Technologies

While integrated circuits are designed to exploit the electrical properties of silicon, MEMS takes advantage of other nano material properties like optical, mechanical etc. Although MEMS and Nanotechnology are sometimes cited as separate and distinct technologies, in reality the distinction between the two is not so clear cut. In fact, these two technologies are highly dependent on one another. The well-known scanning tunneling-tip microscope (STM) which is used to detect individual atoms and molecules on the nanometer scale is a MEMS device. Similarly the atomic force microscope (AFM) which is used to manipulate the placement and position of individual atoms and molecules on the surface of a substrate is a MEMS device as well. In fact, a variety of MEMS technologies are required in order to interface with the nano-scale domain. Likewise, many MEMS technologies are becoming dependent on nanotechnologies for successful new products. For example, the crash airbag accelerometers that are manufactured using MEMS technology can have their long-term reliability degraded due to dynamic in-use stiction effects between the proof mass and the substrate. A nanotechnology called Self-Assembled Monolayers (SAM) coatings are now routinely used to treat the surfaces of the moving MEMS elements so as to prevent stiction effects from occurring over the product's life.

Many experts have concluded that MEMS and nanotechnology are two different labels for what is essentially a technology encompassing highly miniaturized things that cannot be seen with the human eye. Similar broad definition exists in the integrated circuits domain which is frequently referred to as microelectronics technology even though state-of-the-art IC technologies typically have devices with dimensions of tens of nanometers.

Whether or not MEMS and nanotechnology are one in the same, it is quite clear that there are overwhelming mutual dependencies between these two technologies that will only increase in time. Perhaps what is most important are the common benefits afforded by these technologies, including: increased information capabilities; miniaturization of systems; new materials resulting from new science at miniature dimensional scales; and increased functionality and autonomy for systems.

4.1.4 Overview and History of Development

From a very early vision in the early 1950's, MEMS has gradually made its way out of research laboratories and into everyday products. In the mid-1990's, MEMS components began appearing in numerous commercial products and applications including accelerometers used to control airbag deployment in vehicles, pressure sensors for medical applications, and inkjet printer heads. Later also other products were developed like: oscillators, displays, lab on chip and microphones.

MEMS has several distinct advantages as a manufacturing technology.

In the first place, the interdisciplinary nature of MEMS technology and its micromachining techniques, as well as its diversity of applications has resulted in an unprecedented range of devices and synergies across previously unrelated fields (for example biology and microelectronics).

Secondly, MEMS with its batch fabrication techniques enables components and devices to be manufactured with increased performance and reliability, combined with the obvious advantages of reduced physical size, volume, weight and cost.

Thirdly, MEMS provides the basis for the manufacture of products that cannot be made by other methods.

These factors make MEMS as pervasive technology as integrated circuit microchips.

However, three points makes it very different: MEMS products tend to be application specific, giving rise to a wide range of very different products.

Secondly, the number of MEMS products will be always less than that for semiconductor ICs. A good example is the inkjet printer. The four inkjet nozzles are operated using printed circuit boards with tens of other silicon devices.

Thirdly, contrary to IC manufacturing, there is no such thing as a standard building component like the transistor. This leads to a more diverse technology base with more development and engineering work and therefore to a more expensive and more difficult to maintain technology.

History

The history of MEMS is useful to illustrate its diversity, challenges and applications. The following chronology summarizes (Table 4.1) some of the key MEMS milestones (1-5).

Table 4.1 Chronology of MEMS development events

1950s	1958	Silicon strain gauges become commercially available.
	1959	"There's plenty of room at the bottom" - Richard Feynman gives a keynote presentation at California Institute of Technology. He issues a public challenge by offering $1000 to the first person to create an electrical motor smaller than 1/64th of an inch.
1960s	1961	First silicon pressure sensor demonstrated.
	1967	Invention of surface micromachining. Westinghouse creates the Resonant Gate Field Effect
		Transistor. Description of use of sacrificial material to free micromechanical devices from the silicon substrate.
1970s	1970	First silicon accelerometer demonstrated.
	1979	First micromachined inkjet nozzle.
1980s	Early 1980s	First experiments in surface micromachined silicon.
	1982	Disposable blood pressure transducer.
	1982	"Silicon as a Mechanical Material". Instrumental paper to entice the scientific community - reference for material properties and etching data for silicon.
	1982	LIGA Process demonstrated.
	Late 1980s	Micromachining leverages microelectronics industry and widespread experimentation and documentation increases public interest.
	1988	First MEMS conference.

Table 4.1 *Contd...*

1990s		Novel methods of micromachining developed with an aim of improving sensors.
	1992	MCNC starts the Multi-User MEMS Process (MUMPS) sponsored by Defense Advanced Research Projects Agency (DARPA).
	1992	First micro machined hinge and introduction of the Bosch DRIE process.
	1993	First surface micromachined accelerometer sold (Analog Devices, ADXL50).
	1994	Deep Reactive Ion Etching is patented.
	1995	BioMEMS rapidly develops.
2000s		Massive industrialisation and commercialisation.
	2001	Triaxis accelerometers appear on the market.
	2002	First nanoimprinting tools announced.
	2003	MEMS microphones for volume applications introduced.
	2003	Discera start sampling MEMS oscillators.
	2004	TIs DLP chip sales rose to nearly $900 million.
	2005	Analog Devices shipped its two hundred millionth MEMS-based inertial sensor.
	2006	Packaged triaxis accelerometers smaller then 10 mm^3 are becoming available.
	2006	Dual axis MEMS gyros appear on the market.
	2006-2015	Perpetual releases vibration energy harvester. A number of innovative micro/nano devices for bio-medical applications at commercial scale.

4.1.5 MEMS - Applications

4.1.5.1 Technical attributes

MEMS is fast becoming a mature field with prototypes already in the marketplace, generating $3.4 billion in sales globally in the year 2000 and expect to be $31.5 billion by 2020 (1). The advantages provided by MEMS include low power, low mass, low cost and high functionality 6). With the integration of sensors, actuators and computer circuitry, MEMS can sense and measure environmental changes caused by pressure, light, motion, electromagnetic signal and chemicals, and make decisions and act on the information received (4). MEMS have been used for automotive pressure sensors for pollution control, disposable blood pressure sensors, and accelerometers for car airbags (4).

MEMS devices are largely manufactured by photolithographic micro-machining techniques (similar to those used for making integrated circuits) to integrate mechanical and electronic parts on the surface of a silicon chip (4). Progress in the fabrication of optical MEMS by self-assembly techniques (techniques where micro-components are assembled into a desired pattern autonomously) has been reported (7).

4.1.5.2 Generic MEMS Applications

- *Medical diagnostics and chemical analysis sensors*: Absorption of molecules on a micro beam can alter the mass of the beam and therefore the beams vibrational frequency. MEMS sense the existence of some chemicals by detecting the changes in vibrational frequency of micro-beams (8). MEMS sensors might consume only 20% of the operating power of current devices and be integrated into soldiers masks or equipment (6).

- *Electronic display control devices*: Electrically controlled micro-mirrors reflect light directly onto display screens to illuminate either pixels or pictures elements (8). This provides for high resolution, brightness, and large screen display. It might also be possible to employ such devices for small head-mounted monocular displays (7).
- *High-density data storage systems*: Two applications of MEMS are Scanning Tunnelling Microscopes (STM) and AFM (3). Originally, STM and AFM were designed to see and manipulate individual atoms on an objects surface using a needle-like probe. With an array of such probes applied to the surface of an appropriate storing medium, binary data can be stored by creating a dent for 1 and no dent for 9. As the size of a dent is at atomic scale, the storing density could be very high for example a postage stamp-size memory card with several gigabytes storing capacity is practicable.
- *Aerodynamic control*: MEMS-made up of smart skins (consisting of sensors and movable tiny plates) could be applied to the wings of aircraft. Sensors could then monitor airflow near the wing, and adjust positions accordingly. This would produce an enhanced aerodynamic-control, which would enable an aircraft to turn more quickly and withstand turbulence (3).

4.1.5.3 Defense Applications of MEMS

- MEMS radio-frequency switches are used augment existing military surveillance, communication and radar-ranging satellites;
- Construction of a smart fusing and arming device in anti-torpedo weapons;
- Improved ruggedness and survivability of land equipment in the development of shock-resistant instrumentation for munitions or projectiles;
- Embedded tyre-pressure sensors for armoured vehicles to monitor tyres for signs of deterioration and possible blowouts;
- MEMS-based sensors and transmitters to enhance the information exchange between commanders and soldiers by providing individual geo-location and physical status data;
- MEMS sensors to be employed for protection against biological and chemical weapons by providing fast detection, diagnostics, and potentially automated methods of treatment.

4.1.5.4 Industrial Applications

High volume MEMS can be found in a diversity of applications across multiple markets Table 4.2. In the context of emerging technologies, MEMS products were initially centered around technology-product paradigms rather than product-market paradigms. Consequently, MEMS devices have found numerous applications across a diversity of industries. For example, MEMS accelerometer initially found a launching application in airbag sensors, from this they were further developed to serve other applications in the automotive sector such as Electronic Stability Program (ESP) and rollover detection. Due to technological advancements, production costs went down and the sensors became affordable for consumer applications. Parallel to this, accelerometers were introduced in high end applications such as medical pace makers.

Table 4.2 MEMS in various application areas
(*Source:* Nexus Market Analysis for MEMS and Micro Systems III (2005-2009))

	Consumer electronics	**Medical**	**Communi cations**	**Aerospace, defence**	**Automotive**	**Other**
Inertial sensors	Pedometer, game control, image stabilisatio in video camera's hard disk protection	Motion tracking pacemaker		Missile guidance navigation, laser range finder	Airbags, vehicles dynamic control, navigation systems, active suspension, roll detection	Seismic, exploration, robotics, machine vibration
Optical devices	Microdisplay s, autofocus lenses	Micro spectrometers for patient self testing/ monitoring	Variable optical attenuators ,optical switching, tuneable filters			Micro-mirrors in bar code readers and projectors
RF MEMS			Tuneable capacitors and reasonators for mobile phones, switches in base stations	Switches and tuneable capacitors for radar and communicati ons		Switches, relays resonators, e-fuses
Microphes	Mobile phones, note books camcorder	Hearing aids			Hands free calling	
Micro-fluidic	**Inkjet heads**	Lab on Chio for diagnostics (Point of Care) and analytics, respirators, insulin pump, micro needles				Micro dispending, gas and liquid chromato-graphy, mass spectromet ers, micro cooling micro reactors

Table 4.2 ***Contd...***

	Consumer electronics	Medical	Communi cations	Aerospace, defence	Automotive	Other
Pressure sensors	**Hiking altimeters, scuba gear**	Blood pressure, kidney dialysis, breathing		Flight control system, cabin pressure, hydraulic systems	Manifold Air Pressure, Tire Pressure Manage-ment Systems	
Flow sensors		Test long function			Air intake of engine, air quality in cabin	
IR sensors	Cabin temperature control, crash prevention, anti fog systems, seat occupancy, tire and break monitoring		Thermo-meter, diagnostics	Security monitoring		Home climate, microwave control, white goods, industrial HVAC (Humidity, Ventilation, Air Condi-tioning)
Others	Fingerprint sensors for authent-ication			Fingerprint sensors for authent-ication		Probecards for electrical testing

4.1.5.5 Some Commercially Established MEMS Applications

4.1.5.5.1 Automotive airbag sensor

Automotive airbag sensors were one of the first commercial devices using MEMS. They are in widespread use today in the form of a single chip containing a smart sensor, or accelerometer, which measures the rapid deceleration of a vehicle on hitting an object. The deceleration is transmitted by a change in voltage. An electronic control unit subsequently sends a signal to trigger and explosively fill the protective airbag.

The airbag sensor has been pivotal to the success of MEMS and micromachining technology. With over a hundred million devices in production and sold in the last 15 years.

4.1.5.5.2 MEMS for Ink Jet Printers

MEMS has enabled more and more heating elements and piezoelectric crystals to be incorporated into a printer head. Early printers had 12 nozzles with resolutions of up to 92 dpi. Today they can have up to 600 nozzles which can all fire a droplet simultaneously enabling 1200 dpi at high speed. Epson, Lexmark, Hewlett-Packard, Olivetti, and Canon all use a form of these MEMS in their inkjet printers. Although the inkjet market for paper printing is under pressure from alternatives such as laser-printing, new applications are emerging, such as industrial printing of electronic circuits, promising a flexible low-cost way of producing electronic products. Another new application is for printing of biochemical materials and printing on soft substrates.

4.1.5.5.3 MEMS Bio Medical Applications

The experience gained from the early MEMS applications has resulted in an enabling technology for new biomedical applications (e.g. Lab on Chip) and optical or wireless communications, also referred to as respectively Micro Opto Electro Mechanical Systems (MOEMS), and Radio Frequency MEMS (RF MEMS).

4.1.5.5.3.1 Lab-On-Chip

Biochips are biological microchips that host reactions between DNA, proteins, chemical and biological reagents on glass, silicon or plastic plates in order to extract information.

The main advantages of Lab on Chip components include ease-of-use, speed of analysis, low sample and reagent consumption and high reproducibility due to standardisation and automation. As is clear from the term Lab on Chip, the goal is to place the entire process of a laboratory onto a single chip. Microfluidic systems therefore typically contain silicon micromachined pumps, flow sensors, microchannels, microreactors and chemical sensors. (Figure 4.2). They enable fast and easy manipulation and analysis of small volumes of liquids. The ability to receive test results in a few hours or even minutes, rather than a week or so, will make a vast difference in diagnosis and treatment, but also in the patients well being. There are three main application areas for Lab on Chip: medical diagnostics, clinical diagnostics and life science research.

Lab on chip concepts have been developed and are being developed for many applications such as leukaemia, breast cancer, dipolar disorder and a several infectious diseases .

From a unit shipment standpoint, Point of Care diagnostics is the largest (potential) segment, although in terms of turnover, the segment of life science research is larger, thanks to higher chip prices. Price of components here can run into several hundreds of dollars, while for POC diagnostics a few dollars or less is acceptable at most. Chips for specialized clinical diagnostics tests operate somewhere in the middle.

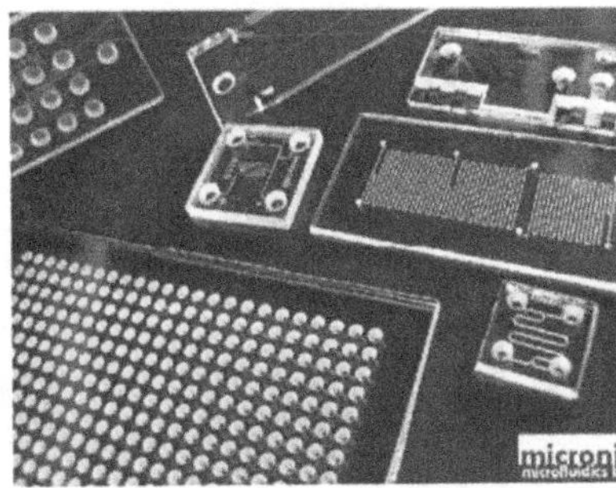

Figure 4.2(a) Microfluidic components
(Courtesy: Micronit Microfluidics Technologies)

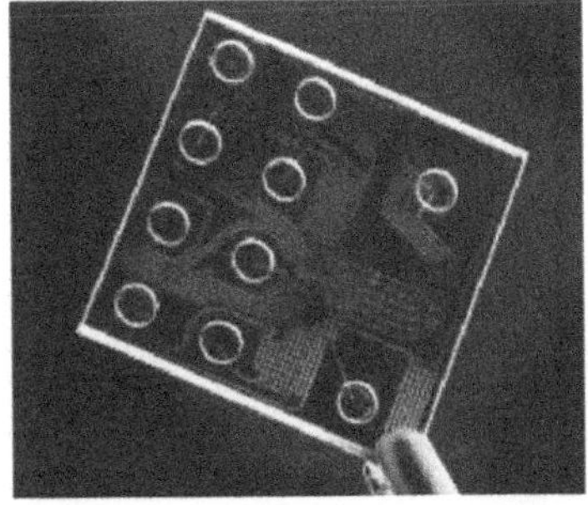

Figure 4.2(b) Lab on Chip
(Courtesy: Agilent Technologies)

4.1.5.5.3.2 Micro Opto Electro Mechanical Systems (MOEMS)

Optical communications has emerged as a practical means to address the network scaling issues created by the tremendous growth in data traffic caused, in part, due to the rapid rise of the use of internet. The most significant MOEMS based products include, variable optical attenuators, optical switching units and tuneable filters. Their small size, low cost, low power consumption, mechanical durability, high accuracy, high switching density as well as the low cost batch processing of these MEMS-based devices make them a perfect solution to the problems of the control and switching of optical signals in telephone networks.

4.1.5.5.3.3 Radio Frequency MEMS (RF MEMS)

RF MEMS constitute one of the fastest growing areas in commercial MEMS technology. RF MEMS are designed specifically for use within the electronics in mobile phones and other wireless communication applications such as radar, global positioning satellite systems (GPS) and steerable antennae. The main advantages gained from such devices include higher isolation and therefore less power loss and an ability to be integrated with other electronics. MEMS has enabled the performance, reliability and function of these devices to be enhanced while driving down their size and cost at the same time.

4.1.5.5.4 Microphones

MEMS based microphones were developed in the eighties, but commercial success was difficult to achieve due to it being a replacement product. The new devices had to compete with very well entrenched traditional products such as the electret condenser microphone which is inexpensive ($0.50 or less) and shows a good performance in terms of frequency response and sensitivity. However, as MEMS microphones have the advantage of better temperature resistance, they are therefore better suited for a highly automated assembly line; in contrast to the electret condenser microphone, which have to be soldered by hand. Other advantages include the option of integration with electronics and the ability to fabricate arrays into one chip. The main application areas for such devices include digital cameras, mobile phones, electronic note books/pads, hands free sets of mobile phones in cars. Clearly hearing aids is also an important application area.

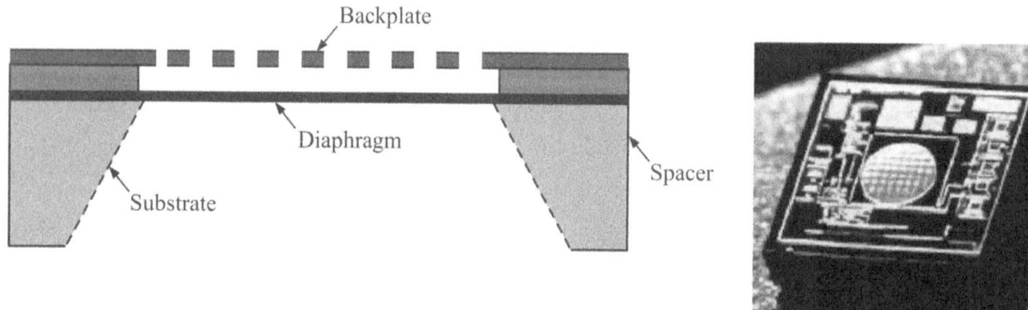

Figure 4.3 (a) Schematic cross section of a MEMS Microphone

Figure 4.3 (b) MEMS microphone (*Courtesy:* Akustica)

The MEMS microphones are fabricated on the surface of a silicon wafer (Figure 4.3). A membrane layer is deposit on a silicon wafer on which a spacer layer is deposited and used to separate this diaphragm from the back plate. By a combination of bulk and surface micromachining the membrane is released, so it can vibrate freely with incoming sound. The changing capacitance of the charged capacitor formed by the back plate and diaphragm transforms the sound signal into an electrical signal.

4.1.6 MEMS Markets

4.1.6.1 Automotive

Automotive has been the first mass market for MEMS products and was for a long time the main driving force for the MEMS industry. There are currently over 100 sensors in each modern, high end, car of which about 30 + MEMS products, mainly accelerometers, gyros, inclinometers, pressure- and flow sensors (engine management: air intake, oil and coolant pressure, particle and NO_2 emission). The increasing complexity of the cars, due to demands on safety, driver and passenger comfort and environmental restrictions will cause this market to grow for the coming years. Expected growth areas are: IR sensors for air quality, accelerometers for motor maintenance, microscanners for displays, energy scavengers for tire pressure management etc. Due to increasing compatibility with IC production technologies, the number of suppliers is increasing and prices are expected to decrease over the coming years.

4.1.6.2 IT Peripherals

The major products, in terms of microsystems, within the IT peripherals market are read/write heads and inkjet print heads. Both product categories are under pressure due to alternative technologies offered: respectively by solid state memories and laser printing. New MEMS applications in this field include microphones, accelerometers and RF MEMS products.

4.1.6.3 Telecommunication

The optical telecom market is expected to grow steadily over the coming years and MOEMS will undoubtedly play an important role in this growth. There are currently many MOEMS based concepts and technologies which are being proposed and tested, but adoption by the industry is slow due to their reluctance to introduce unproven technology. The surplus of technical possibilities and the competition from wireless communication is making this market a difficult one to operate in. The wireless market in general is becoming an interesting sector with many new functionalities on offer by RF MEMS components. Currently, this market is also growing slowly, although MEMS resonators in particular are increasingly replacing conventional quartz resonators.

4.1.6.4 Consumer Electronics and Life Style Products

Consumer Electronics is currently the most interesting area for suppliers of accelerometers, microphones and other MEMS products. The increasing number of functionalities within smaller and smaller devices is behind the drive to decrease the size of electronic components and subsystems. Apples I-Phone and Nintendos Wii console are interesting examples; both use accelerometers for image stabilisation and gaming control. Other interesting opportunities include: microphones and zoom lenses in mobile phones and oscillators in watches. High end mobile phones also employ inertial sensors such as accelerometers and gyroscopes for applications such as scrolling, character recognition, gaming and image stabilization.

4.1.6.5 Medical and Life Science Applications

According to Frost Sullivan there is a significant shift in the present healthcare model at a global level from the simple therapeutic to a preventive and predictive model on an individual basis. One of the enablers behind this will be microfluidic based Point of Care (PoC) instruments and other Lab on Chip (LoC) devices. The result will be more effective, personalized, safe and cost-effective therapy, better diagnosis and treatment; and, most importantly, increased patient satisfaction.

Microfluidics and LoC technologies offer advantages such as increased sensitivity, mobility, and efficiency in assays as well as helping to multiply the number of tests performed per day in laboratories. The advent of microfluidics will promote time and cost savings. This will be the strongest driver for the uptake of these technologies.

The adoption of microfluidics will also significantly reduce the time and cost of producing large amounts of biological material, which helps drug developing companies to reduce initial expenses of drug development. To this end, most of the diagnostic companies have included microfluidics within their development roadmaps, regarding it a key enabling technology. If microfluidics can create products with a sufficiently low cost of processing and the companies have the patience to overcome the regulation hurdles, such devices will take their place amongst the group of large volume MEMS products. Other applications of MEMS in the medical fields include: accelerometers in pace makers, microphones in hearing aids, micro-needles and micro pumps for medicine delivery.

4.1.6.6 Household Applications

Initially MEMS devices were considered to be too expensive for this market. Of late, as a spin off from the development activities for consumer electronics, MEMS sensors became of increasing interest for this market sector. Pressure and temperature sensors are the most abundantly found MEMS devices for household applications, and accelerometers for washing machines are regarded as a growth area.

4.2 Micro Fabrication Techniques for MEMS

4.2.1 Micro Fabrication Techniques

MEMS fabrications techniques mainly evolved from the Planar process technology in micro fabrication of semiconductor devices which are mostly based on semiconductor technologies like photolithography, sputtering, evaporating, Low Pressure Chemical Vapour Deposition (LPCVD), Plasma Enhanced Chemical Vapour Deposition (PECVD), wet and dry etching, and chemical mechanical planarization which are discussed in detail in Chapter 3. To these technologies some specific MEMS technologies are added like bulk and surface machining; wafer bonding; backside alignment, Deep Reactive Ion Etching (DRIE) etc.

The general process flow for planar processing of MEMS is depicted in Figure 4.4.

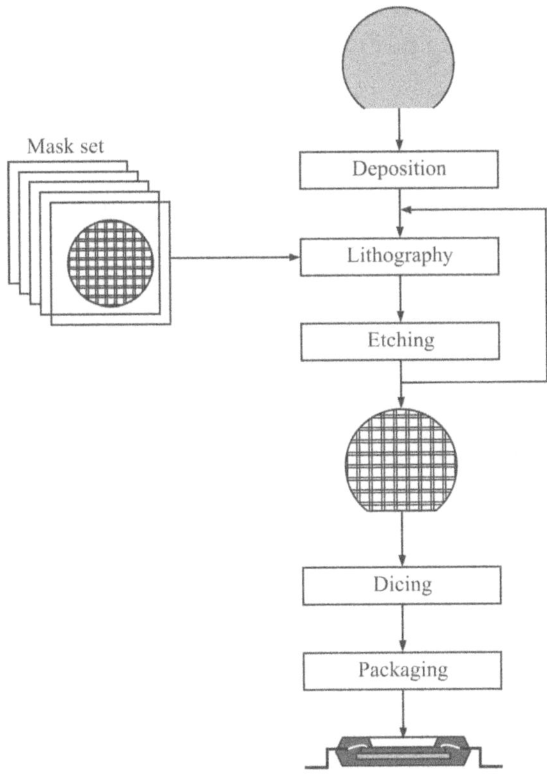

Figure 4.4 General process flow for planar processing of MEMS

4.2.2 MEMS Manufacturing Technologies

4.2.2.1 Silicon Micro Machining for MEMS

Silicon micromachining has been a key factor for the vast progress of MEMS. Silicon micromachining refers to fashioning microscopic mechanical parts out of a silicon substrate or on a silicon substrate. Silicon micromachining comprises two technologies: bulk micromachining, in which structures are etched into silicon substrate and surface micromachining, in which the micromechanical layers are formed from layers and films deposited on the surface.

The major high volume MST/MEMS products (e.g. microphones) now entering the market are produced on 200 mm wafers. Thinner wafers have many benefits in MEMS for instance to reduce the cost of wafer through etch while cost of processing as thinner wafers are more difficult to handle. Therefore selecting the right thickness of the substrate is not just a case of establishing which thickness is the best for the product design; it also should take into account which thickness is easily available and the process is less expensive.

Bulk micromachining and surface micromachining are the two major micromachining processes of silicon, and silicon wafer bonding is usually necessary for silicon microfabrication.

LIGA and 3D microfabrications have been used for high aspect ratio and 3D-microstructure fabrication for MEMS.

4.2.2.2 Bulk Micromachining of Silicon

The term *bulk micromachining* comes from the fact that this type of micromachining is used to realize micromechanical structures within the bulk of a single-crystal silicon wafer by selectively removing ('etching') wafer material.

The microstructures fabricated using bulk micromachining may cover the thickness range from submicron to full wafer thickness (200-500 μm), and the lateral size range from submicron to the lateral dimensions of a full wafer. Bulk micromachining technique allows to selectively remove significant amounts of silicon from a substrate to form membranes on one side of a wafer, a variety of trenches, holes, or other structures (Figure 4.5).

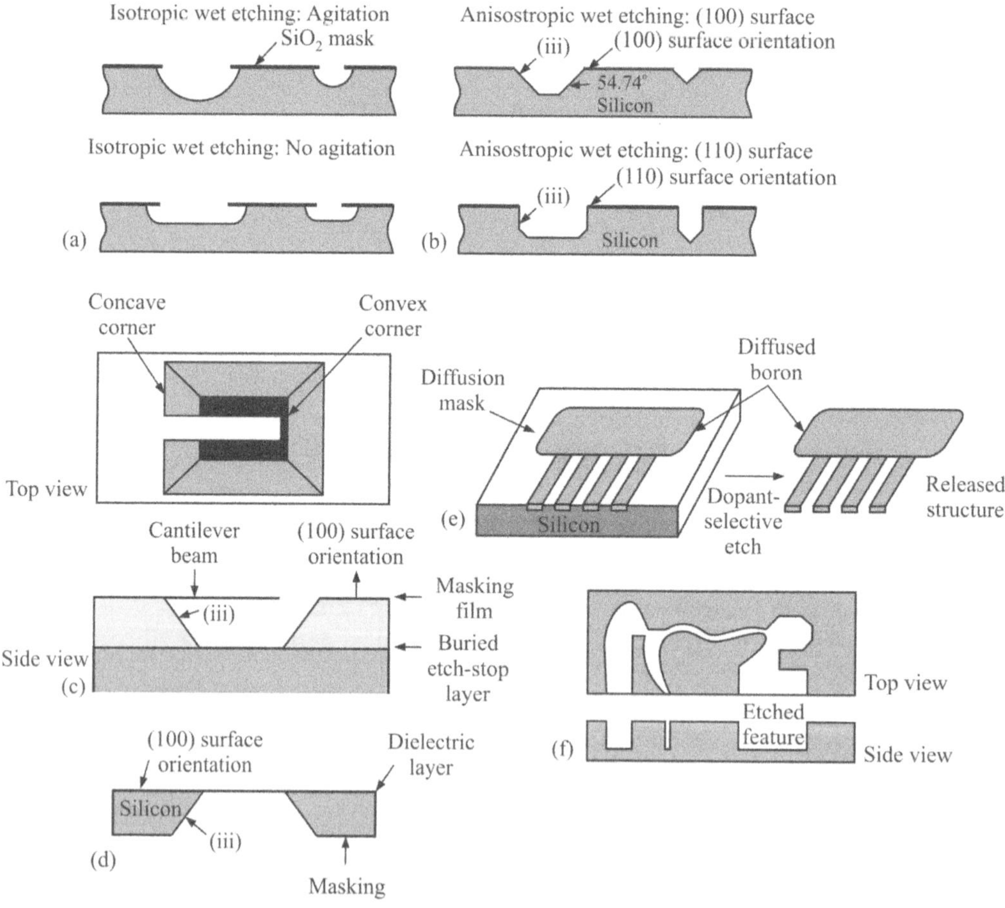

Figure 4.5 Bulk silicon micromachining. (a) Isotropical etching, (b) Anisotropic etching,(c) Anisotropic etching with buried etch-stop layer,(d) Dielectric membrane released by back-side bulk etching, (e) Dopant-dependant wet etching, (f) Anisotropic dry etching (9)

Bulk micromachining technique can be divided into wet etching and dry etching of silicon according to the phase of etchants. Liquid etchants, almost exclusively relying on aqueous chemicals, are referred to as *wet etching*. Vapor and plasma etchants are referred to as *dry etching*.

4.2.2.3 Surface Micromachining of Silicon

Surface micromachining does not shape the bulk silicon but instead builds structures on the surface of the silicon by depositing thin films of 'sacrificial layers' and 'structural layers' and by removing eventually the sacrificial layers to release the mechanical structures (Figure 4.6). The dimensions of these surface micro machined structures can be several orders of magnitude smaller than bulk-micro machined structures. The prime advantage of surface-micro machined structures is their easy integration with IC components, because the wafer is also the working area for IC elements. It should be noted that as miniaturization is immensely increased by surface micromachining, the small mass structure involved may be insufficient for a number of mechanical sensing and actuation applications.

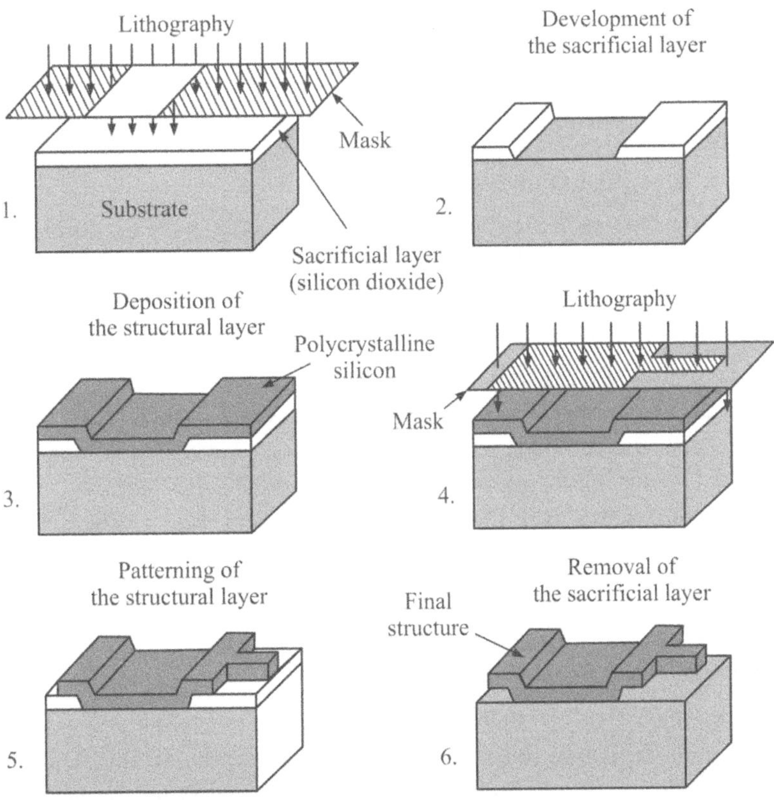

Figure 4.6 Processing steps of typical surface micromachining (9)

Surface micromachining requires a compatible set of structural materials, sacrificial materials, and chemical etchants. The structural materials must possess the physical and chemical properties that are suitable for the desired application. In addition, they must have satisfactory mechanical properties, for e.g., high yield and fracture stresses, minimal creep and fatigue, and good wear resistance. The sacrificial materials must have good mechanical properties to avoid device failure during fabrication. These properties include good adhesion and low-residual stresses to eliminate device failure by delamination and/or cracking. The etchants to remove the sacrificial materials must have excellent etch selectivity and they must be able to etch off the sacrificial materials without affecting the structural ones. In addition, the etchants must have proper viscosity and surface tension characteristics. The common IC compatible materials used in surface micromachining are

1. Poly-Si/Silicon dioxide; low-pressure chemical vapor deposition (LPCVD)-deposited poly-Si as the structural material and LPCVD deposited oxide as the sacrificial material. The oxide is readily dissolved in HF solution without the poly-Si being affected. Together with this material system, silicon nitride is often used for electrical insulation.
2. Polyimide/aluminum; in this case polyimide is the structural material and aluminum is the sacrificial material. Acid-based etchants are used to dissolve the aluminum sacrificial layer.
3. Silicon Nitride/poly-Si; Silicon Nitride is used as the structural material, whereas poly-Si is the sacrificial material. For this material system, silicon anisotropic etchants such as KOH and EDP are used to dissolve poly-Si.
4. Tungsten/Silicon Dioxide; chemical vapor deposition (CVD) deposited tungsten is used as the structural material with oxide as the sacrificial material. HF solution is used to remove the sacrificial oxide. Other IC compatible materials such as silicon carbide, diamond like carbon, zinc oxide, gold, etc., are also used.

Silicon microstructures fabricated by surface micromachining are usually planar structures (or two-dimensional). Other techniques involving the use of thin film structural materials released by the removal of an underlying sacrificial layer have helped to extend conventional surface micromachining into the third dimension. By connecting polysilicon plates to the substrate and to each other with hinges, 3D micromechanical structures can be assembled after release Another approach to 3D structures used the conformal deposition of polysilicon and sacrificial oxide films to fill deep trenches previously etched in the silicon substrate.

4.2.2.4 Wafer Bonding for MEMS

Silicon micromachining has limitations in forming complex 3D microstructures in a monolithic format; multichip structures are then proposed for advanced MEMS, in which wafer-to-wafer bonding is critical in the formation (10). The wafer bonding for MEMS can be carried out by three methods: anodic bonding, intermediate-layer bonding-assisted bonding, and direct bonding. More details about wafer bonding are discussed in Chapter 3.

4.2.2.5 Three-Dimensional Microstructures using Surface Micromachining

Three-dimensional surface microstructures can be fabricated using surface micromachining. The fabrication of hinges for the vertical assembly of MEMS was a major advance towards achieving 3-D microstructures (11). Optical microsystems have greatly benefited from surface-micromachined 3-D structures. These microstructures are used as passive or active components (micromirror, Fresnel lens, optical cavity, etc.) on a silicon optical bench (silicon microphotonics). An example is a Fresnel lens that has been surface micro-machined in polysilicon and then erected using hinge structures and locked in place using micromachined tabs, thus liberating the structure from the horizontal plane of the wafer (12, 13). Various microactuators (e.g., comb drive, and vibromotors) have been used.

4.2.2.6 High Aspect Ratio (HAR) Silicon Micromachining (14-26)

The bulk and surface micromachining technologies discussed above fulfill the requirements of a large group of applications. However certain applications, require the fabrication of high-aspect-ratio structures that is not possible with the afore mentioned technologies. For HAR micro machining (14) three technologies, are in use i.e., LIGA, HEXSIL, and HARPSS, (15) capable of producing structures with vertical dimensions much larger than their lateral dimensions by means of x-ray lithography (LIGA) and DRIE etching (HEXSIL and HARPSS).

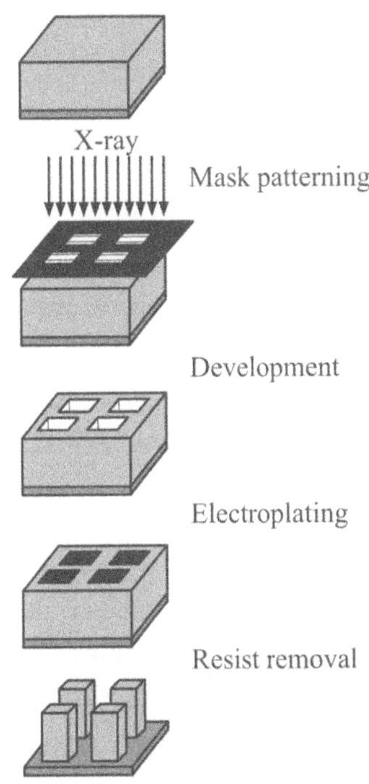

Figure 4.7 LIGA-Process Flow

4.2.2.6.1 LIGA

LIGA process is one of those microfabrications. LIGA is a German acronym for Lithographie, Galvanoformung, Abformung (lithography, galvanoforming, molding). It was developed by the research Center Karlsruhe in the early 1980s in Germany using X-ray lithography for mask exposure, alvanoforming to form the metallic parts, and molding to produce microparts with plastic, metal, ceramics, or their combinations. The general process flow for LIGA is given in Figure 4.7

In LIGA process with standard UV photolithography and photoresist the maximum thickness achievable is on the order of a few tens of microns and the resulting metal structures show tapered walls. LIGA is a technology based on the same plating-through-mask idea but can be used to fabricate metal structures of thickness ranging from a few microns to a few millimeters with almost vertical side-walls. This is achieved using x-ray lithography and special photoresists. Due to their short wavelength, X-rays are able to penetrate through a thick photoresist layer with no scattering and define features with lateral dimensions down to 0.2 μm (aspect ratio > 100 : 1).

With the LIGA process, microstructure's height can be up to hundreds of microns to millimeter scale, while the lateral resolution is kept at submicron because of the advanced X-ray lithography. Various materials can be incorporated into LIGA process allowing electric, magnetic, piezoelectric, optic, and insulating properties of sensors and actuators with a high-aspect ratio, which are not possible to make with the silicon-based processes. Besides, by combining the sacrificial layer technique and LIGA process, advanced MEMS with moveable microstructures can be built. However, the high production cost of LIGA process because of the fact that it is not easy to access X-ray source limits the application of LIGA. Another disadvantage of LIGA process relies on the fact that structures fabricated using LIGA are not truly three dimensional, because the third dimension is always in a straight feature. As we know that complex thick 3D structures are necessary for some advanced MEMS, it means that other 3D-microfabrication processes need to be developed for MEMS.

4.2.2.6.2 HEXSIL

The second method for fabricating high-aspect-ratio structures, which is based on a template replication technology, is hexagonal honeycomb polysilicon (HEXSIL) (14). Figure 4.8 shows a simplified process flow. A high-aspect-ratio template is first formed.

Next, a sacrificial multilayer is deposited to allow the final release of the structures. The multilayer is composed of one or more PSG nonconformal layers to provide fast etch release (≈ 20 μm/min in 49% HF) alternated with conformal layers of either oxide or nitride to provide enough thickness for proper release of the structures. The total thickness of the sacrificial layer has to be larger than the shrinkage or elongation of the structures caused.

4.2.2.6.3 HARPSS

The high-aspect-ratio combined poly- and single-crystal silicon (HARPSS) technology is another technique capable of producing high-aspect-ratio electrically isolated polycrystalline and single-crystal silicon microstructures with capacitive air gaps ranging from sub micrometer to tens of micrometers (15). The structures, tens to hundreds of micrometers thick, are defined by trenches etched with DRIE and filled with oxide and poly layers. The release of the microstructures is done at the end by means of a directional silicon etch followed by an isotropic etch. The small vertical gaps and thick structures possible with this technology find application in the fabrication of a variety of MEMS devices, particularly inertial sensors (16) and RF beam resonators (17). Figure 4.9 shows the process flow in a cross section of a single crystal silicon beam resonator.

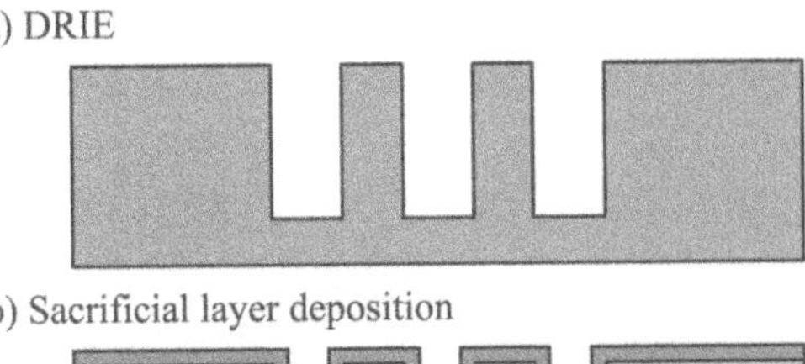

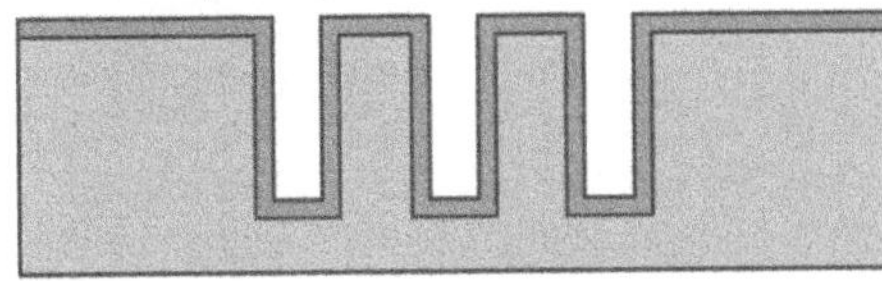

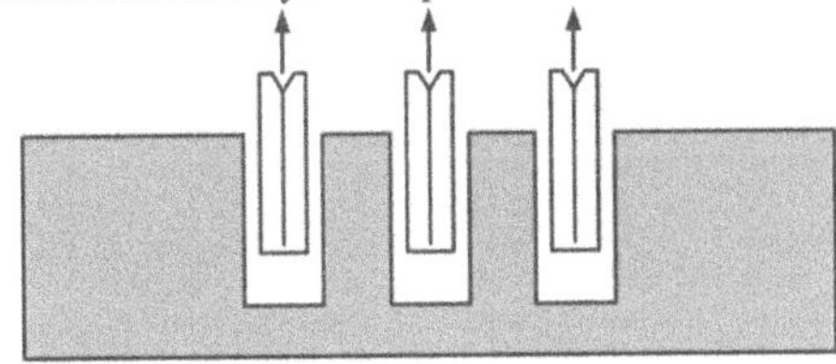

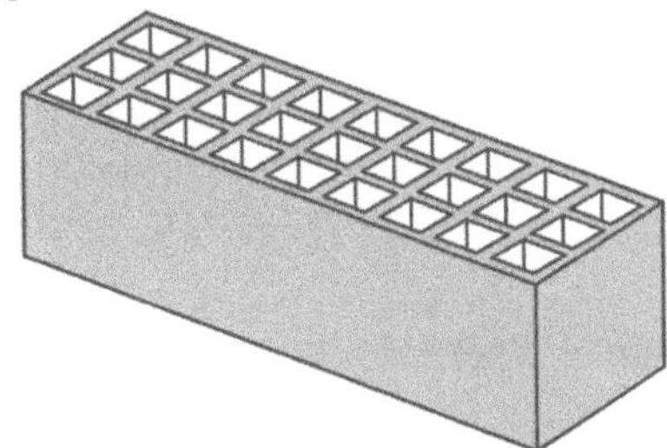

Figure 4.8 (a–f) HEXSIL process flow: (a) DRIE, (b) sacrificial layer deposition, (c) structural material deposition and trench filling, (d) etch structural layer from the surface, (e) etch sacrificial layer and pull the structure out, and (f) example of a HEXSIL-fabricated structure in a silicon substrate using DRIE.

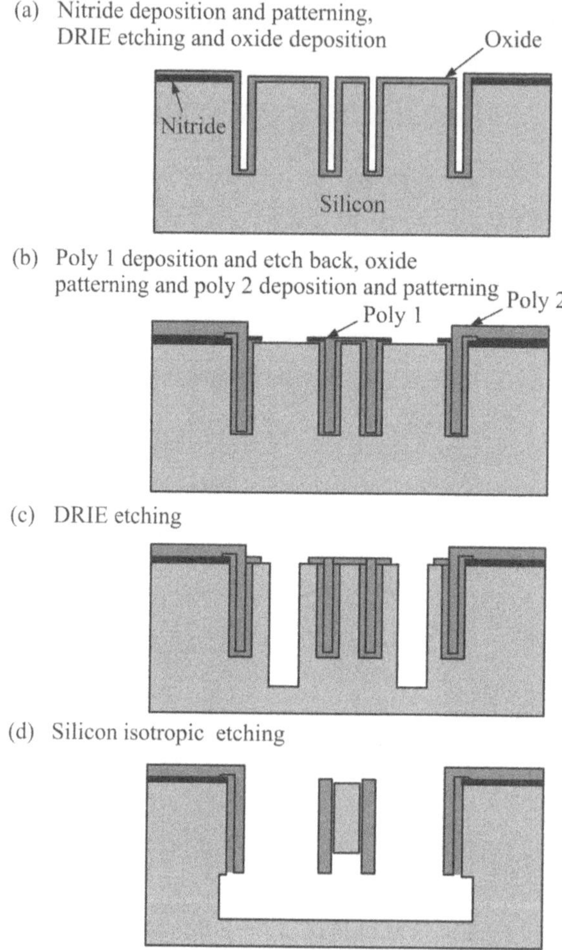

Figure 4.9(a-d) HARPSS process flow: (a) nitride deposition and patterning, DRIE etching and oxide deposition, (b) poly 1 deposition and etch back, oxide patterning, and poly 2 deposition and patterning, (c) DRIE etching, and (d) silicon isotropic etching.

4.3 Industrial Process Control

The reliability, scalability, sensitivity, and cost-effective solutions offered by micro-electro-mechanical systems (MEMS) technology all have the potential to provide viable sensor solutions for industrial automation. The trend toward MEMS-enabled miniaturization is certainly strengthened by the increasing interest in miniaturized, remotely powered, autonomous sensor systems. The small size of MEMS devices makes them especially suitable in volume constrained applications; also, the low energy consumption of these devices are essential for wireless monitoring. The ability to integration is a prerequisite for multi-sensor network applications. With regards to cost, industrial automation is less price-sensitive than the higher-volume markets such as that of consumer electronics, but much more demanding in application support and offering total solutions. The diversity is also higher. Applications of interest include: corrosion control, pressure and flow sensing, gas sensors, accelerometers for vibration control and energy harvesting.

4.4 Challenges and Future Development

4.4.1 Challenges

4. 4.1.1 Quality and Standards

The MEMS industry is still in developing stage when compared to the more established industries of microelectronics, semiconductors amongst others. In addition, the multi-disciplinary nature of Micro and Nano Technologies (MNT) is specific to the characteristics of this novel technology. Furthermore, the technologies and products are developed to address a multi-market scenario. These characteristics, particular to MNT, make it very difficult for standards to be set and followed within this industry. This applies equally to product performance standards and to fabrication standards. The resultant scarcity of clear standards allows operators to produce solutions (and products) which generally do not conform to wider system requirements but dictate their own, and to slow integration, interaction and maintenance. Clearly, organizations with influence and/or market monopoly tend to enforce their preferred solutions, which may not be optimum for all users/buyers. Equally problematic are the small ventures, which attempt to penetrate protected markets and an established way-of-doing-things.

The lack of fabrication standards is also a major issue. Processes used for MNT may not fall within the realm of the well-established ones (particularly in the case of silicon processing). In addition, as MNT products are unlikely to require full-time utilization of the production equipment, such products will, generally, be consigned to a secondary priority relative to the main product of a fabrication line. Also, in order to meet specific product requirements or to satisfy certain customers, fabrication standards may not be followed, resulting in inconsistencies in quality. This situation is difficult to control, particularly whilst MNT markets remain uncertain.

4.4.1.2 Hurdles for Technology and Commercialization

The big challenge in MEMS 10 years ago was commercialization. Although technologically promising, MEMS is difficult to sell: technology was not proven, there was no reliable infrastructure behind it and only a few people understood the potential and were able to link the technology with a specific application. Earlier the biggest issue was seen to be the industrialization of the technology and more specifically the packaging and integration. After spending much time to bring the idea to the market, developing core (chip based) technology and applications simultaneously, the package was often neglected. Laboratory prototypes were difficult to translate to high-volume and low-cost production concepts. Both issues have been improved considerably over the years: many MEMS devices are on the market now and both specific and generic MEMS packages are increasingly becoming available, especially for high volume sensors for automotive and consumer electronics.

4.4.1.3 Miniaturization Issues

In order to successfully manufacture MEMS devices, basic physics and operating principles need to be fully understood and appreciated at both a macro and a micro level. In essence, scaling is often not straightforward. Sometimes little advantages in terms of performance, size/weight, reliability and cost can be gained with a MEMS device. Increased surface area (S)-to-volume (V) ratios at micro scales have both considerable advantages and disadvantages.

Micro-level issues

(i) Friction is greater than inertia. Capillary, electrostatic and atomic forces as well as friction can be significant.

(ii) Heat dissipation improves with smaller dimensions, contrary to heat storage; consequently thermal transport properties could be a problem or, conversely, a great benefit.

(iii) Fluidic or mass transport properties are extremely important. Tiny flow channels are prone to blockages but can conversely regulate fluid movement.

(iv) Material properties (Young's modulus, Poisson's ratio, grain structure) and mechanical behaviour (residual stress, wear and fatigue etc.) may be size dependent.

(v) Integration with on-chip circuitry is complex and device/domain specific.

Miniature device packaging and testing is not straightforward. Certain MEMS sensors require environmental access as well as protection from other external influences. Likewise, many devices are fragile or sensitive to external influences.

(vi) Testing is not rapid and is expensive in comparison with conventional IC devices.

Cost for the successful commercialization of a MEMS device, needs to leverage the IC batch fabrication resource and be mass-produced. Hence mass-market drivers must be identified to generate the high volume production.

Miniaturization and cost are strongly related, especially for products made with wafer scale technologies; Basically, the smaller the device the more devices one can place on a wafer, the lower the processing cost per die (assuming the processing cost per wafer remains the same). This principle led to a continuous drive towards the miniaturization of the MEMS products, as is for instance seen at accelerometers.

4.4.2 The Future of MEMS

4.4.2.1 Future MEMS on Market Trends

The future of MEMS is integrally linked to market trends in general and driven by the increasing demand to monitor and control our environment and the equipment and instruments we use in our daily lives. This demand does, undoubtedly, lead to the need for more sensors in cars, more sensors in industrial equipment and installation and, more sensors for our ambient intelligence. In order to avoid the need for a multitude of wires, such sensors must be self sustaining and able to communicate. As a result, not only more sensors are needed, but also small energy generating modules and wireless transmission components. Clearly, the increased numbers of devices will drive size reduction which in turn will enable higher levels of integration.

Studies also show that there is an increasing trend to spend on health care, due to the growing numbers of elderly people in developed countries and a growing medical and pharmaceutical infrastructure in developing countries. There is also a trend to transfer diagnostics from the central analytical laboratory to a surgeries or clinics that are smaller and closer to the patients. This fuels a growing demand for small and affordable devices for diagnostics, especially for point of care diagnostics.

In addition, national security is of increased importance related to the growing fear from terrorist attacks and outbreaks of infectious human or animal diseases. This drives a need for small multi parameter instruments to test air, water, blood etc for microbiological threats.

Simultaneously there are some general technical trends which will influence the MEMS industry. In the first place, the long awaited, ending of the Moore's law will lead to an increasing attention in the IC industry for other ways to improve the performance of ICs. Then so called Moore than Moore trend will add new technologies to the rather mono cultural landscape of semiconductors to enable extra functionalities, to exploring further miniaturization by using die stacking etc.

There is also a tendency to explore more flexible and more affordable production technologies. This will be driven by the production research into typical low cost, large surface area devices like solar cell, displays, wearable electronics and disposable diagnostics devices. (Re) introduction of printing technologies is one of the possibilities, low cost substrates like plastics are another.

Last but not least, a plethora of products will come from the large amount of nanotechnology research investments; in many cases MEMS will act as an interface between the nano and human size world.

The following are some special MST/MEMS products expected to dominate in future markets

4.4.2.2 Small, Self Sustaining, Wireless Sensor Networks

As discussed above, the most important market trends are those reliant on merging functionalities in products such as cameras, music players, computing, mobile phones, and portable consumer electronics in general, as well as the merging of MST/MEMS technologies with high volume electronics. In this context, some examples of MST/MEMS products being introduced into portable electronics products include silicon based microphones and liquid lenses for autofocus and zoom options. Another important trend is the increasing need for memory capacity - embedded as well as stand-alone.

One general technology trend is especially worth mentioning: the increasing use of sensors and other small electronic devices in and around our direct environment. We will be surrounded by more electronics and will be interacting more with such intelligence. This creates a need for smaller and cheaper electronic products, which are energy efficient and self sustaining as well as able to communicate via wireless with other electronic devices.

The big advantage of wireless sensor systems is that by removing wires and batteries, there is the potential for embedding sensors in previously inaccessible locations; it also becomes feasible to use a large numbers of sensors because there is no longer the need for regular replacement or recharging of batteries.

4.4.2.3 Medical diagnostic devices

A shift from clinical to point of care diagnostics, accompanied by a slow but significant drift from medical diagnostics to more lifestyle oriented diagnostics: allergy, drugs, cholesterol, alcohol, but also diagnostics in sport. This together will make this market more consumer-oriented.

The last eight years there have been more applications for registration of diagnostic devices than two decades before. Therefore perhaps the fastest growing market for MEMS with many interesting applications for Point of Care monitoring, of which glucose level monitoring will represent the largest single segment. Cheap disposable sensors are regarded as the sweet spot in diagnostics.

But there are also many niches like: MEMS based retinal implants, robot surgery devices and diagnostic pills.

4.4.2.4 DNA Chips

DNA chips have been commercialized by many companies, and a range of equipment is available for DNA chip sample processing, data analysis etc. Future lab-on-a-chip technology may include implantable "pharmacy-on-a-chip" devices precisely and timely releasing drugs into a body, without needing needles or injections. The delivery of insulin is one such application, as is the delivery of hormones, chemotherapy drugs and pain killers. First generation devices are being developed which release their medication upon signals from an outside source, wired through the skin. Proposed second generation devices may be wireless and third generation MEMS chips could interact with MEMS sensors embedded in the body to respond to the body's own internal signals.

4.4.2.5 Micro fluidic devices

Furthermore, there are many uses expected for micro fluidics in all kind of analytical equipment such as mass spectrometry and chromatography. A micro fluidics platform integrates sample definition, injector, detectors and diagnostics leading to fast analysis, increased ease of use, handheld and remote operation. They will not only be used for home defense, but also for environmental measurements and industrial process control.

There is also a rising interest in the chemical industry in using micro fluidics. The main reasons being:

For small quantities chemists tend to use up batch process, but often continuous processing is more easy to control, while the residence time distribution (the time the material stays in the reactor) and the heat transfer are much better controlled.

Also scaling up is much easier. While the process conditions in chemistry are so dependent on the scale of the reactor, different processes are used in the lab and in the industrial plant. Micro fluidic scaling up is not so much a matter of moving from a small reactor to a big one, it is a matter of increasing the number of reactors, without changing the process.

Furthermore it can increase the catalytic efficiency due to its inherent larger amount of available surface.

As a consequence, compared to traditional installations, micro fluidic based installations tend to be smaller, more energy efficient, safer, higher yielding and easier to expand. One approach to achieve this is to offer complete standardized systems ("backbones") facilitating the easy construction of miniaturized chemical installations. However, it must be stated that efficient use of micro reactors is only possible using special chemical processes.

4.4.2.6 Cantilevers/nanowires

Cantilevers are like miniature diving boards but measuring only a few hundreds of micrometers long and a few tens of micrometers wide. For sensor applications, the cantilevers are coated with an application specific layer and placed in a sensor device. A liquid or gas is passed through them. When a molecule or microbe fits to the coating it is locked on the surface. This reaction can be measured by a change in momentum of the cantilever or a change in resistance. The device can be designed to search for specific materials.

For example, if the organism to be detected is *E.coli*, the cantilever can be coated in antibodies specific to *E. coli* cells. An array of cantilevers each coated with a specific material can be used to detect different molecules or organisms simultaneously.

Cantilevers and more especially nanowires are extremely sensitive and can measure deflections of just 1 nanometer, so are able to detect the presence of very small quantities of contamination. The applications for this new technology are abundant, for instance IBM proposes to use cantilevers to detect DNA. Also bacteria and even parts of bacteria can be identified, making the sensors ideal for testing the quality of water and food samples.

In addition to conventional aircraft, application of MEMS technologies could enable the development of micro, unmanned aerial vehicles (UAVs). These small flight vehicles would perform as aerial robots to transport loads to remote sites/hazardous locations or as reconnaissance devices.

4.5 Nano Electro Mechanical Systems (NEMS)

NEMS integrate electrical and mechanical functionality on the nanoscale which is a next miniaturization step after MEMS. In NEMS transistor-like nano electronics will be integrated with mechanical actuators, pumps, or motors, and may thereby form physical, biological, and chemical sensors.

4.5.1 Generic Applications of NEMS

Due to their scale relative to atomic and molecular structures (and in addition to those applications described for MEMS), NEMS are likely to prove to be useful for sensing and measurement.

Some potential NEMS-specific applications are (22-25)

- Measurements of force or displacement in fundamental science and metrology;
- Magnetic resonance force microscopy to image individual molecules with atomic-scale resolution;
- Fast logic gates, switches or mechanical computers;
- Non-invasive medical diagnostics;
- Highly functional mass/force sensors;
- Displays; and
- Ultrahigh-density data storage.

A key application of NEMS is atomic force microscope tips. The increased sensitivity achieved by NEMS leads to smaller and more efficient sensors to detect stresses, vibrations, forces at the atomic level, and chemical signals (24). AFM tips and other detection at the nanoscale rely heavily on NEMS. If implementation of better scanning devices becomes available, all of nanoscience could benefit from AFM tips.

NEMS can be used in areas where previously it was impossible due to size considerations. One example of the potential military application of NEMS technology is the concept of surveillance dust, in expensive, ultra-miniature NEMS- or MEMS-based sensors to monitor sensitive areas on the battlefield (24). To be more adoptable, NEMS components should be cheap for the concept to be plausible, considering that the number of NEMS particles in a dust must be huge. However, it has been suggested that given that the manufacturing process for these are similar to those used in making integrated circuits, cheap NEMS should be possible which will be a major advantage of NEMS technology (25).

Table 4.3 provides possible nano-device applications (26-28).

Table 4.3 Potential nano-device applications and indicators for realization

Outstanding Properties	Potential applications	Key indicators	Level of maturity
Light, cheap, low power-consumption, high functionality	*Sensors, high-resolution display, high density storage, aerodynamic control*	Efficient mass-manufacturing techniques to scale up laboratory-demonstrated applications would be desirable	Matured/ Emerging
As above, but with higher sensitivity and lower power-consumption.	*As above, but especially for tools in sensing and measurements at the molecular scale*	Invention of effective methods to enhance stability of NEMS. Development of novel techniques of efficient transduction in NEMS.	Laboratory demonstrated
Self-replicating programmable robot-like nanomachines	*Build or destroy anything atom by atom*	Many fundamental barriers exist (see the main text)	Highly speculative.

4.5.2 Technical Hurdles for NEMS

It has been noted that, technically, there are three significant obstacles that need to be overcome in order to realise the full potential of NEMS.

(i) *Instability*: Having high-sensitivity means that NEMS have a propensity for instability, as they are designed to be responsive to atoms and molecules. In cases where gaseous atoms or molecules are present, natural fluctuations may constrain the application of NEMS.

(ii) *Interaction between nano*- and macro-world: Interaction between the nano- and macro-worlds is not trivial. Transducers in NEMS act as messengers for transferring information from nanoscale up to the macroscale and feedback and control back down. As such, they need to be sufficiently sensitive to resolve displacement of mechanical parts in the picometre to femtometre (10^{-12}m ñ 10^{-15} m) range, across a very large bandwidth (up to a few gigahertz frequencies). To further complicate the situation, most of the transduction techniques used in MEMS are not applicable at the nanoscale. Therefore, building transducers for NEMS that satisfy these two requirements is a significant challenge.

(iii) *Novel physical properties*: A challenge at the fundamental level of science is to deepen understanding of the physics of surface atoms. As devices shrink further and further, surfaces will play a more and more important role in dominating physical properties. Understanding the roles played by surface atoms continues to be a major challenge.

4.5.3 Nano Fabrication Techniques for NEMS

The integration of the knowledge gained from micro- and nano-fabrication can lead to design principles for nanodevices that can detect substances, analyze their environment, and perform tasks such as the release of a specific molecule. These vehicles will combine responsive polymers, nanoparticles, nucleotides, to develop NEMS systems. Both micro- and nanofabrication offer a number of possibilities for the study of chemical, biological, and physical processes at the cellular and molecular scale, and for the design of synthetic devices capable of interacting with biological systems at these levels.

Some of the advantages of nanofabricated devices include

(a) Ability to control the features to the nanometer scale for reproducible mass production of structures and devices,

(b) Ability to miniaturize already-existing systems for the study of cellular or molecular processes,

(c) The capacity of including electronics within structural devices through the use of the well-developed semiconductor techniques, and

(d) The high throughput possible with some of the micro- and nanofabrication methods.

4.5.3.1 Methods of Fabrication of NEMS

These are discussed in detail in chapter 3. There are two main approaches used for NEMS fabrication: (i)Top down approach and (ii) Bottom-up approach.

4.5.3.1.1 Top down approach

The top-down approach uses the traditional micro fabrication methods, i.e., optical and electron beam lithography, to manufacture devices. While being limited by the resolution of these methods, it allows a large degree of control over the resulting structures. Typically, devices are fabricated from metallic thin films or etched semiconductor layers.

4.5.3.1.2 Bottom-up approach

In this process chemical properties of molecules will be exploited to cause single-molecule components to (a) self-organize or self-assemble into some useful conformation, or (b) rely on positional assembly. These approaches utilize the concepts of molecular self-assembly and/or molecular recognition. This allows fabrication of much smaller structures, albeit often at the cost of limited control of the fabrication process.

A combination of these approaches may also be used, in which nanoscale molecules are integrated into a top-down framework.

4.5.3.1.3 NEMS-Materials-carbon nano tubes, graphene etc.

Carbon nanotubes, diamond and graphene are some of the frequently used materials as they possess useful properties which meet the needs of NEMS. The mechanical properties of carbon (such as large Young's modulus) are fundamental to the stability of NEMS while the metallic and semiconductor conductivities of carbon based materials allow them to function as transistors. Both graphene and diamond exhibit high Young's modulus, low density, low friction, excessively low mechanical dissipation, (29) and large surface area. The low friction of CNTs, allow practically frictionless bearings and has thus been a huge motivation towards practical applications of CNTs as constitutive elements in NEMS, such as nanomotors, switches, and high-frequency oscillators (30-31). Carbon nanotubes and graphene's physical strength allows carbon based materials to meet higher stress demands, when common materials would normally fail and thus further support their use as a major materials in NEMS technological development (32).

Along with the mechanical benefits of carbon based materials, the electrical properties of carbon nanotubes and graphene allow it to be used in many electrical components of NEMS. Nanotransistors have been developed for both carbon nanotubes (31) as well as graphene (32). Transistors are one of the basic building blocks for all electronic devices, so by effectively developing usable transistors, carbon nanotubes and graphene are both very crucial to NEMS.

4.5.3.1.4 Challenges in NEMS Fabrication

Despite all of the useful properties of carbon nanotubes and graphene for NEMS technology, both of these products face several limitations to their implementation.

One of the main problems is carbon's response to real life environments. Carbon nanotubes exhibit a large change in electronic properties when exposed to oxygen (33). Similarly, other changes to the electronic and mechanical attributes of carbon based materials must fully be explored before their implementation, especially because of their high surface area which can easily react with surrounding environments.

Carbon Nanotubes will have varying conductivities, being either metallic or semiconducting depending on their helicity when processed (34). Because of this, very special treatment must be given to the nanotubes during processing, in order to assure that all of the nanotubes have appropriate conductivities. Graphene also has very complicated electric conductivity properties compared to traditional semiconductors as it lacks an energy band gap and essentially changes all the rules for how electrons move through a graphene based device (35). This means that traditional constructions of electronic devices will likely not work and completely new architectures must be designed for these new electronic devices.

4.5.3.1.5 Hurdles in Commercialization of NEMS

Key hurdles currently preventing the commercial application of many NEMS devices include low-yields and high device quality variability. Before NEMS devices can actually be implemented, reasonable integrations of carbon based products must be created. A recent step in that direction has been demonstrated for diamond, achieving a processing level comparable to that of silicon (36). The focus is currently shifting from experimental work towards practical applications and device structures that will implement and profit from of such novel devices (37). The next challenge to overcome involves understanding all of the properties of these carbon-based tools, and using the properties to make efficient and durable NEMS with low failure rates (38).

4.5.3.1.6 Future NEMS

NEMS devices, if implemented into everyday technologies, could further reduce the size of modern devices and allow for better performing sensors. Carbon-based materials have served as prime materials for NEMS use, because of their exceptional mechanical and electrical properties. Once NEMS interactions with outside environments are integrated with effective designs, they will likely become useful products to everyday technologies.

Using Computer simulations/modeling of continuum mechanics and molecular dynamics (MD), important behaviors of NEMS devices can be predicted via without experiments (39-41). These simulations greatly help engineers to efficiently analyze the stability of NEMS devices without resorting to ultra-fine meshes and time-intensive simulations (40). Simulations have other advantages as well: they do not require the time and expertise associated with fabricating NEMS devices; they can effectively predict the interrelated roles of various electromechanical effects; and parametric studies can be conducted fairly readily as compared with experimental approaches. Computer simulations greatly help in designing and Fabricating optimized NEMS with desired characteristics at low cost. Global Market of NEMS projected to reach $108.88 million by 2022 (42).

References

1. Wilson, M., et al (2002). Nanotechnology: basic science and emerging technologies: Chapman and Hall.
2. Griffiths and David J. (2004). Introduction to Quantum Mechanics (2nd ed.). Prentice Hall. ISBN 0-13-805326-X.

3. Gabriel, K.J. (1995). Engineering Microscopic Machines. Scientific American; September: 118-121.
4. Roukes, M. (2001). Nanoelectromechanical systems face the future. Physics World; February: 25-31.
5. Ouellette, J. (1996). MEMS: Mega promise for micro devices. The Industrial Physicist; September: 9-12.
6. Cornell, B. and L. Mar. Nanotechnology - the technology of the 21st century: the economic impact of emerging nanometer scale technologies www.dest.gov.au/science/pmseic/documents /Nanotech.pdf.
7. Callahan, S.L. (2000). Nanotechnology in a New Era of Strategic Competition. Joint Forces Quarterly; Autumn: 20-26.
8. Srinivasan, U., et al. (2002). MEMS: some Self-Assembly Required. Optics & Photonics News; November: 20-24.
9. Kovacs, G T A, Maluf, N I and Petersen, K E, (1998). Bulk micromachining of silicon, Proc. IEEE, 86(8), 1536 -1551.
10. Schmidt. M A. (1998). Wafer-to-wafer bonding for microstructure formation, Proc. IEEE, 86 (8), 1575-1585.
11. Fujita, H, (1998). Microactuators and micromachines, Proc. IEEE, 86(8), 1721-1732.
12. K.S.J Pister, M.W. Judy, S.R Burgett, R.S Fearing (1991). Micro fabricated hinges; 1 mm vertical features with surface micromachining, Proc, 6th Int Conference, Solid state sens. Actuators, Sanfrancisco 647-6450.
13. L.Y. Lin, S.S. Lee, M.C. Wu, K.S.J. Pister (1995). Micromachined integrated optics for free space inter connection, Proc. IEEE Micro Electro Mech. Syst. Workshop. Amesterdam 77-82.
14. C. G. Kller and R. T. Howe (1995). HEXSIL Dimorpns for vertical actuation, Dig. Tech. Pap. 8th Intl Conf. Solid State Sens. Actuators Euroscens. IX Stockholm DD. 99-102.
15. F. Ayazi and K. Najan (1998). High aspect ratio combined poly and single-crystal silicon (HARPSS) MEMS technology, J. Microelectromech. Syst. 9, 288-294.
16. F. Ayazi. K. Njan; A Harpss (2001). Polysilicon Vibrating ring gyroscope, J. Microelectromech. Syst, 10, 169-179.
17. Y.S. No and F. Ayazi (2001). The HARPSS process for fabrication of nano-precission silicon electromechanical resonators, Pro1st IEEE conf. Nanotechnol Maui 489-494.
18. R.S. Muller and K.Y Lau (1998). Surface-micromachined microoptical elements and systems, Proc. IEEE 86(8), 1705-1720.
19. Fatikow S and Rembold U (1997). Microsystems Technology and Micro-robotics, Springer, New York.
20. Ball, P. (1997). Made to Measure: new materials for the 21st century. Princeton: Princeton University Press.
21. Schmidt M.A. (1998). Wafer to Wafer Bonding for Micro Structure Formation of Proc. IEEE 86(8), 1575-1585.
22. Fujita H. (1998). Micro Actuators and Micro Machines. Proc. IEEE 86(8), 1721-1732.
23. Roukes, M (2001). Plenty of Room Indeed. Scientific American; September: 48.

24. Roukes, M (2001). Nanoelectromechanical systems face the future. Physics World; February: 25-31.

25. Craighead, H.G. (2000). Nanoelectromechanical Systems. Science; 290: 1532-1535.

26. Kaku, M. Visions (1998). How science will revolutionize the 21st Century and Beyond. Oxford: Oxford University Press.

27. Callahan, S.L. (2000). Nanotechnology in a New Era of Strategic Competition. Joint Forces Quarterly; Autumn: 20-26.

28. Bower, J.L. and C.M. Christensen. (1995). Disruptive Technologies: Catching the Wave. Harvard Business Review; January-February: 43-53.

29. Saalfeld, F.E. and J.F. Petrik (2001). Disruptive Technologies - A concept for moving innovative Military Technologies Rapidly to Warfighters. Armed Forces Journal International; May: 48-52.

30. Y. Tao, J. M. Boss, B. A. Moores and C. L. Degen (2012). Single-Crystal Diamond Nanomechanical Resonators with Quality Factors exceeding one Million. Ar Xiv, 134

31. Kis, A. and Zettl, A. (2008). Nanomechanics of carbon nanotubes. Philosophical Transactions of the Royal Society A 366(1870): 1591–1611. Bibcode: 2008 RSPTA. 366.1591K. doi:10.1098/ rsta.2007.2174.

32. Hermann S, Ecke R, Schulz S and Gessner T (2008). Controlling the formation of nanoparticles for definite growth of carbon nanotubes for interconnect applications. Microelectronic Engineering 85(10): 1979-1983. doi:10.1016/j.mee. 2008.06.019.

33. Dekker. Cees, Tans. Sander J., Verschueren and Alwin R. M. (1998). Room-temperature transistor based on a single carbon nanotube. Nature 393 (6680): 49-52. Bibcode: Natur. 393, 49T. doi: 10.1038/29954

34. Westervelt, R. M. (2008). Applied Physics: Graphene Nanoelectronics. Science 320 (5874): 324-325. doi:10.1126/science.1156936. PMID 18420920

35. Collins, PG, Bradley, K, Ishigami, M and Zettl, A (2000). Extreme oxygen sensitivity of electronic properties of carbon nanotubes. Science 287 (5459): 1801-4. Bibcode: 2000 Sci... 287. 1801C. doi: 10.1126/science. 287. 5459. 1801. PMID 10710305

36. Ebbesen, T. W, Lezec, H. J, Hiura, H, Bennett, J. W, Ghaemi, H. F and Thio, T (1996). Electrical conductivity of individual carbon nanotubes. Nature 382 (6586): 54-56. Bibcode: 1996 Natur. 382... 54E. doi:10.1038/382054a0

37. Y. Tao and C. L. Degen (2013). Facile Fabrication of Single-Crystal-Diamond Nanostructures with Ultra High Aspect Ratio. Advanced Materials.

38. Loh. O, Wei. X, Ke. C, Sullivan. J and Espinosa, HD (2011). Robust carbon-nanotube-based nano-electromechanical devices: Understanding and eliminating prevalent failure modes using alternative electrode materials. Small (Weinheim an der Bergstrasse, Germany) 7 (1): 79-86. doi:10.1002/smll.201001166.PMID 21104780

39. Dequesnes, Marc, Tang, Zhi and Aluru, N. R. (2004). Static and Dynamic Analysis of Carbon Nanotube-Based Switches. Journal of Engineering Materials and Technology 126 (3): 230. doi:10.1115/1.1751180

40. Ke, Changhong, Espinosa and Horacio D. (2005). Numerical Analysis of Nanotube-Based NEMS Devices—Part I: Electrostatic Charge Distribution on Multiwalled Nanotubes. Journal of Applied Mechanics 72 (5): 721. Bibcode: 2005JAM.... 72..721K.doi:10.1115/1.1985434.

41. Ke, Changhong, Espinosa, Horacio D. and Pugno, Nicola (2005). Numerical Analysis of Nanotube Based NEMS Devices—Part II: Role of Finite Kinematics, Stretching and Charge Concentrations. Journal of Applied Mechanics 72 (5): 726. Bibcode: 2005 JAM.... 72.. 726K. doi: 10. 1115/1. 1985435.

42. Global Market NEMS Projection (2007).

5

Quality Control, Standards and Metrology for Nano-Technology

5.1 Quality Control in Nano-Technology

5.1.1 Introduction

The continuous miniaturization in manufacturing technologies now allows fabrication of nano-sized samples as well as nanoscale precision and nanoscale features. This progress is achieved by a variety of techniques such as precision engineering, beam sputtering, and photolithography as well as molecular manipulation. Nano-technology products are now found in pharmaceutical industry, microelectronics, and in precision engineering. Nano-technology is not only a simple continuation of micro technology but also marks the ultimate end of materials science as the dimensions where materials properties stop and molecular properties start. One can also say that nanotechnology is where molecular features and materials meet.

The special characteristics, particular to Micro Nano Technologies (MNT) make it very difficult for standards to be set and followed within this industry. This applies equally to product performance standards and to fabrication standards. The resultant scarcity of clear standards allows operators to produce solutions (and their integration, interaction and maintenance) with varying standards. Clearly, organizations with influence and/or market monopoly tend to enforce their preferred solutions, which may not be optimum for all users/buyers.

There are a variety of industrially important nanostructures like nanocomposites, nanowires, nanopowders, nanotubes, fullerenes, nanofibers, nanocages, nanocrystallites, nanoneedles, nanofoams, nanomeshes, nanoparticles, nanopillars, thin films, nanorods, nanofabrics, quantumdots etc., which need special protocols for quality control. The most common way to classify nano structures is by their dimensions.

5.1.2 Classification of Nanostructures according to Dimensionality

Nanostructures are divided into 3 classes:

1. **Nanoscale in one dimension (D):** Thin films layers and surfaces.
2. **Nanoscale in two dimensions (2D):** Carbon nanotubes, inorganic nanotubes, nanowires, biopolymers.

3. **Nanoscale in three dimensions (3D):** Nanoparticles, fullerenes, dendrimers, quantum dots.

According to Ojkowski and Fecht classification (1) nanomaterials can be classified for quality control as given below:

(a) ***(0-D) Nanoparticles:*** Crystals made of metals (Quantum dots), with sizes below 100 nm, free standing or suspended in a fluid or embedded in another material fall in this category. The characteristic dimension important for special applications depends on the physical phenomenon important for that application

(b) ***(1-D) Nanowires, nanorods, nanotubes:*** Wires with thickness below 100 nm, free standing or suspended in a fluid or embedded in another material belong to this category

(c) ***(2-D) Coatings, thin-film-multilayers:*** This category includes coatings made of nanocrystalline metals, or nanocomposites. The structural analysis of the coating will reveal grains (single crystals) of metal of diameter in the nanometre scale, joined by grain boundaries. Another example of a 2-D nanomaterial is the surface of a non-nanocrystalline metal, where due to special treatment (for instance shot-peening) a nanostructured layer is formed. In this category such a structure is considered as a 2D layer. When a surface layer is approximately one nanometre thick single crystalline film of metal on a single crystal surface, a sandwich of such structures, or an array of metal particles on the surface, is not included in this category. Such a structure has been considered to belong to the category of nanostructures (2). Thus it is necessary to distinguish between nanostructured layers, self-assembled patterns of nano-wires, nano-crystallites, single standing thin films and coating made of nanocrystals.

(d) ***(3-D) Bulk; powders; other nanostructures:*** This category comprises of pieces of metal that can be handled as individual items of material, but are composed of nano crystals. The dimensions of the material are in all directions significantly larger than the grain size. The difference compared to thick nano-coatings, where the dimension is larger than the grain size in all dimensions is that a 3-D nanostructured metal is assumed to be a free standing material, while a coating cannot be handled as an item separate from the substrate.

Nanocrystalline metals can be produced by compacting single crystals of size in the range of nanometres, with clearly defined high angle grain boundaries between them.

***Powders*:** These consist of nanocrystals produced, for instance, by mechanical milling techniques. The powder particle size is much larger than the size of the individual nanocrystallites as it is composed of, for instance of nanocrystals of size 50 nm compared to 20 Pm particles. It can, in part, have an amorphous structure. Formally speaking this is a 3-D structure, but this is opposed to the common understanding that a powder cannot possess a 3-D structure. Therefore such materials have been classified in a separate category – 'Powders'.

***Other Nanostructures*:** These are structures made of various materials, or consist of one material, with some aspects of the structure measured at the nanoscale. Examples include filters with nanopores, foams with nanopores, single or multilayer films on the surfaces of materials with a thickness of a few nm, self-assembled nanoparticles on the surface of material, etc. For nanostructures the material itself does not matter as much as its form.

In the case of foams and membranes the voids are nanometric, not the material itself, hence it is a case of a nanostructure, and not a nanomaterial. Nanometric films can hardly be called a material either. They exist only jointly with the substrate. Together they form a nanostructure, but cannot be described as a nanomaterial. The key to the properties results from behaviour taking place at the interface between the materials.

In the case of coatings these are a material covered with another nanomaterial. One can imagine an infinite variety of combinations of nanopowders, nanowires, intermixed or embedded in bulk materials or films, membranes etc. Their categorization in terms of materials is not always possible. The proper terminology categorization for such structures needs to be developed yet.

In view of these challenges Nano technology will only become a coherent field of endeavour through the confluence of follows important technological streams: (a) New and improved control of the size and manipulation of nanoscale building blocks (b) New and improved characterization (spatial resolution, chemical sensitivity etc) of materials at the nanoscale (c) New and improved understanding of the relationship between nanostructure and properties and how these can be engineered and (d) better safety and consumer acceptance.

In order to correlate properties with nanostructure, a quantitative characterization of nanostructures in materials (10-100 nm) is needed. Some methods and procedures need to be unified to achieve comparable and reliable results. Standard protocols need to be prescribed regarding how to measure such parameters important in the fabrication and application of nanoparticles and nano crystalline or nanostructured materials such as:

- density of nano-sized particles
- particle size distribution or grain size distribution
- particle shape
- complex shapes
- what is really measured
- how to use diffraction methods to determine shape, size and structure of nanoparticles
- how to describe the characteristic dimension of nanostructured metals
- difficulties with assessment of nanoporosity in sintered materials.

5.2 Standards for Nano Technology

5.2.1 Need for Nano Standards and Current Challenges

Precise control of dimensions of objects is the key issue of quality control in nanotechnology and the science of nano-objects. The dimensions of these objects are below 100 nm, and the precision requested frequently is of the order of 0.1 nm. This requires new nano standards and methods of measurement to be developed as those applied for conventional materials in many cases cannot be simply applied to nanostructures and which lead to the development of nano standards metrology. Further special protocols for nanostructures and nano materials need to be developed.

Nano technology standards are needed

(a) To support commercialisation and market development
(b) To provide a basis for procurement – technical/quality/environmental management
(c) To support appropriate legislation/regulation.

Standards should be

(a) Informative defining what MUST be done in e.g., a specific test method, or *INFORMATIVE*, providing information only.
(b) Voluntary unless called in a contract or regulation.
(c) Provide a means of validated quantification

However currently the following challenges exist

- No internationally agreed terminology/definitions for nanotechnology(ies).
- No internationally agreed protocols for toxicity testing of nanoparticles.
- No standardized protocols for evaluating environmental impact of nanoparticles.
- Existing "methods of test" may not be suitable for nanoscale devices and nanoscale dimensions.
- Measurement techniques and instruments need to be developed and/or standardized.
- New calibration procedures and certified reference materials are needed for validation of test. Certified reference materials are needed for validation of test instruments at the nanoscale.
- Multifunction nanotechnology systems and devices will need complex measurements.

5.2.2 Current Status

Currently Nanotechnology relies on standards through at least three concepts:

1. Documentary standards define agreed-upon terminology or standard language for a field of science, engineering, or technology; they are agreed-upon means for conducting measurements; agreed-upon performance characteristics of instruments or commercial products; and particularly, they are documented agreements on means to facilitate trade and commerce.
2. Standards often refer to standard reference materials, materials that are certified by a national standards laboratory to have specified characteristics traceable to an international system of the fundamental system of physical units of measurement.
3. Standards generally refer to the fundamental physical realization of the units of measurement defined in the International System of Units (SI). Like metrology standards: length, mass, time, quantity of matter – primary and secondary standards.

The SI's base units of standards are widely accepted in science and technology and set measurement standards agreed to through the Convention of the Meter, a diplomatic treaty between fifty-four nations. The International Bureau of Weights and Measures (BIPM), located in Sèvres near Paris, France, has the task of ensuring world-wide uniformity of measurements and their traceability to the SI basic units.

There are seven base quantities: **length, mass, time, electric current, thermodynamic temperature, amount of substance,** and **luminous intensity**. Although these seven quantities are by convention regarded as independent, their respective base units—the **meter, kilogram, second, ampere, kelvin, mole,** and **candela**—are in a number of instances interdependent. In nanotechnology all the base units have relevance to the instruments used and the measurements performed by researchers in academia, government, and industry. These base units also form the fundamental foundation for commerce and business. In particular, the unit of length at the nanoscale, the nanometer, is derived from the base unit meter by subdividing the meter by a factor of one billion. Another unit derived from the base units that is of particular use in nanotechnology is the unit of force, the Newton. The force exerted by the cantilevers used in atomic force microscopes is typically specified in terms of nano Newtons or one-billionth of a Newton (Newton is force about equal to the weight of an apple).

The SI globally accepted nanotechnology standards are vital to continued progress in the field's research and development, and for safe, secure, and responsible commercialization of nanotechnology in the years ahead. Standards are important to businesses, consumers, and researchers.

Around the world, there are numerous nanotechnology standards-setting groups. Some of the leading standards setting organizations and their relevant nanotechnology committees are:

- International Standardization Organization (ISO) Technical Committee (TC) 229 on Nanotechnologies
- ASTM (formerly known as the American Society for Testing and Materials) International's Committee E56 (Nanotechnology)
- International Electrotechnical Commission Technical Committee 113 (Nanotechnology Standardization for Electrical and Electronics Products and Systems)
- Institute of Electrical and Electronics Engineers' Nanotechnology Council.

These groups develop voluntary standards. Standards that are the best formulated, with the strongest basis in science, are most likely to be adopted by the global community. U.S. leadership and participation in the international standards-setting process allows the United States to help shape the strategic and technical direction of nanotechnology development everywhere.

U.S. Federal research related to measurement within science and technology is led by the National Institute of Standards and Technology (NIST). The standards community is actively involved in nanotechnology standardization, including ASTM International's Committee E56 on Nanotechnology and the American National Standards Institute (ANSI)-accredited U.S. Technical Advisory Group (TAG) to ISO TC 229, Nanotechnologies. ANSI now hosts a Nanotechnology Standards Database. ANSI administers the U.S. TAG to ISO TC 229, which is responsible for formulating positions and proposals on behalf of the United States with regard to ISO standardization activities related to nanotechnology. The United States holds the leadership of the ISO/TC 229 Working Group on health, safety, and environment (Fig. 5.1).

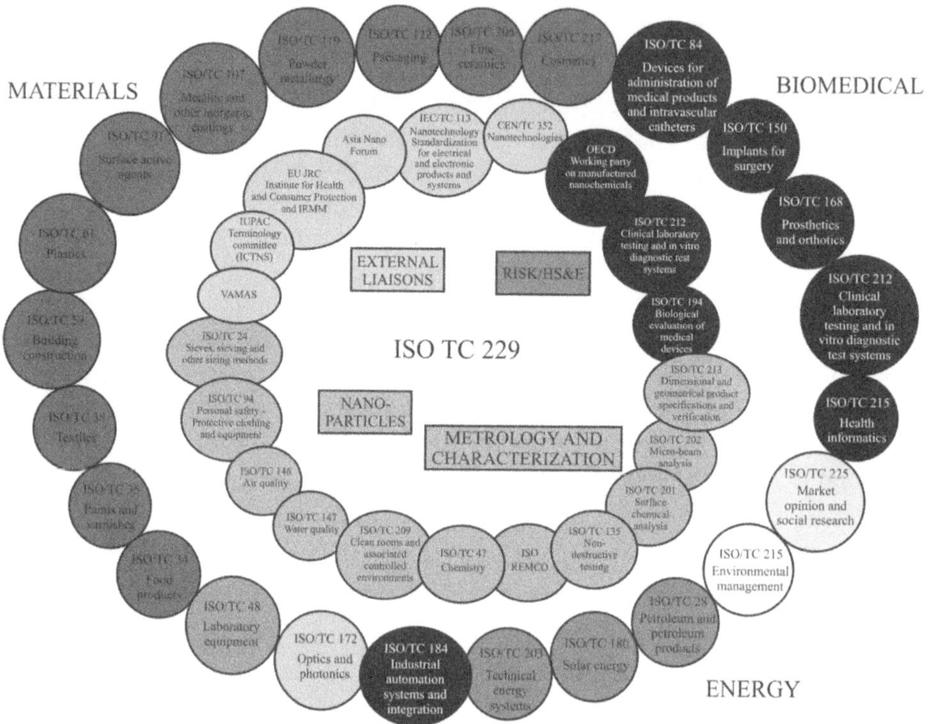

Figure 5.1 Current potential interactions for ISO and TC229

Indian Standards for Nano-Technologies

Bureau of Indian Standards: Indian Standard Nanotechnologies -- Materials specifications -- Guidance on specifying nano-objects: ICS 07.030

This draft Indian Standard which is identical with ISO/TS 12805:2011 'Nanotechnologies-Materials specifications - Guidance on specifying nano-objects' issued by the International Organization for Standardization (ISO).

The committee has now decided to adopt this standard under dual numbering system and make it align with ISO/TS 12805. The text of ISO Standard has been approved as suitable for publication as an Indian Standard without deviations. Certain terminology and conventions are, however, not identical to those used in Indian Standards. Attention is particularly drawn to the following:

(a) Wherever the words 'International Standard' appear, referring to this standard, they should be read as 'Indian Standard'

(b) Comma (,) has been used as a decimal marker while in Indian Standards the current practice is to use a point (.) as the decimal marker.

In this adopted standard, reference appears to certain International Standards for which Indian Standards also exist. The corresponding Indian Standards which are to be substituted in their respective places are listed below along with their degree equivalence for the editions indicated: International Standard Corresponding Indian Standard.

Degree of Equivalence: ISO/TS 27687 Nanotechnologies --Terminology and definitions for nanoobjects - Nanoparticle, nano fibre and nano plate.

Doc.: MTD 33 (5115) Identical ISO/TS 80004-1 Nanotechnologies --Vocabulary -- Part 1: Core terms, Doc.: MTD 33 (5111) Identical.

5.2.3 Calibration

It is difficult to provide samples using different precision instruments that can be calibrated at nanoscale. Reference or Calibration standards are important for repeatability to be ensured. But there are no international standards for calibration and the calibration artefacts provided by the company along with their equipment is only good for calibrating that particular equipment. Hence it is difficult to select a universal calibration artefact using which we can achieve repeatability at nanoscale. At nanoscale while calibrating care needs to be taken for influence of external factors like vibration, noise, motion and internal factors like the interaction between the artefact and the equipment which can measure are difficult to assess.

5.2.4 Traceability

In metrology at macro scale achieving traceability is quite easy and artefacts like scales, laser interferometers, step gauges, and straight edges are used. At nanoscale a crystalline highly oriented pyrolytic graphite (HOPG) mica or silicon surface is considered suitable to be used as calibration artefact for achieving traceability (3). But it is not always possible to ensure traceability. Like what is a straight edge at nanoscale and even if take the same standard as that for macroscale there is no way to calibrate it accurately at nanoscale. This so because the requisite internationally or nationally accepted reference standards are not always there. Also the measurement equipment required to ensure traceability has not been developed and there is a need for establishing nanoscale standards. Also there is a need to establish some kind of uncertainty estimation model. Traceability is one of the fundamental requirements for manufacturing and assembly of products when multiple producers are there.

5.2.5 Tolerance

It is the permissible limit or limits of variation in dimensions, properties, or conditions without significantly affecting functioning of equipment or a process. Tolerances are specified to allow reasonable leeway for imperfections and inherent variability without compromising performance. In nanotechnology the systems have dimensions in the range of nanometers. Defining tolerances at nanoscale with suitable calibration standards for traceability is difficult for different nano manufacturing methods. There are various integration techniques developed in the semiconductor industry that are used in nano manufacturing.

Integration techniques:

- In Hetero integration direct fabrication of nanosystems from compound substrates is done. Geometric tolerances are required to achieve the functionality of the assembly.
- In Hybrid integration nanocomponents are placed or assembled on a substrate fabricating functioning nanosystems. In this technique, the most important control parameter is the positional accuracy of the components on the substrate.
- In Monolithic integration all the fabrication process steps are integrated on a single substrate and hence no mating of components or assembly is required. The advantage of this technique is that the geometric measurements are no longer of primary importance for achieving functionality of nanosystem or control of the fabrication process.

5.3 Nano Metrology

5.3.1 Introduction

Nano metrology deals with the science of measurement at the nanoscale level and plays an important role in- order to produce nanomaterials and devices with a high degree of accuracy and reliability in nano manufacturing. In order to apply practical metrology to the field of nano technology *i.e,* to make measurement in the nanometer range traceable to the SI units of length and angle, practical measurement standards must be constructed. A challenge in this field is to develop or create new measurement techniques and standards to meet the needs of next-generation advanced manufacturing, which will rely on nanometer scale materials and technologies. The needs for measurement and characterization of new sample structures and characteristics far exceed the capabilities of current measurement science. Anticipated advances in emerging nanotechnology industries will require revolutionary metrology with higher resolution and accuracy than has previously been envisioned (4).

Nano metrology requirements are different for the industry and research institutions. Nano metrology for research has progressed faster than that for industry mainly because implementing nanometrology for industry is difficult. In research oriented nano metrology resolution is important whereas in industrial nano metrology accuracy is given precedence over resolution. Further due to economic reasons it is important to have low costs and fast in industrial nano metrology while it is not important for research nano metrology.

Various techniques for measuring at nanoscale have been developed in the last few decades. Most of them based on some physical phenomena observed on particle interactions or forces at nanoscale. Some of the most commonly used techniques are Atomic Force Microscopy (AFM), X-Ray Diffraction, Scanning Electron Microscopy (SEM), Transmission Electron Microscopy (TEM), High Resolution Transmission Electron Microscopy, and Field Emission Scanning Electron Microscopy. These are a wide range of analytical techniques which may be used for nano materials characterization depending on the type of information needed. For high resolution surface investigations, two commonly used techniques are Scanning Electron Microscopy (SEM) and Atomic Force Microscopy (AFM). Each of these techniques resolves surface structure down to the nanometer scale. However, the image formation mechanisms are quite different, resulting in different types of information about the surface structure.

5.3.2 History of SEM & AFM Development

The first SEM was constructed in 1938 by von Ardenne by rastering the electron beam of a Transmission Electron Microscope (TEM) to essentially form a Scanning Transmission Electron Microscope (STEM) (5,6). In 1942, Zworkin et. al. developed the first SEM for bulk samples. This configuration contains many of the basic principles of today's SEMs (6,7). Cambridge Scientific Instruments produced the first commercial instrument in 1965. A number of improvements have occurred since this time, resulting in an increase in resolution from 50 nm in 1942 to ~ 0.7 nm today. Besides the development of morphological imaging, the SEM has been developed to detect signals which are used to determine compositional information, such as X-rays, backscattered electrons, cathode luminescence, Auger electrons, and specimen current.

5.3.3 Atomic Force Microscope (AFM)

The development of the AFM was preceded by the development of the Scanning Tunneling Micro-scope (STM) in 1981 at IBM Zurich Research Laboratory by Binnig and Rohrer (7,8). Its ability to view the atomic lattice of a sample surface earned the inventors the Nobel Prize in Physics in 1986. Although the STM provides sub-angstrom resolution in all three dimensions, it is limited to conductive and semi-conductive samples. To image insulators as well as conductors, the Atomic Force Microscope (AFM) was developed in 1986(9), and the first commercial AFMs were produced in 1989 by Digital Instruments.

AFM consists of scanning a sharp tip on the end of a flexible cantilever across a sample surface while maintaining a small, constant force (Fig 5.2).

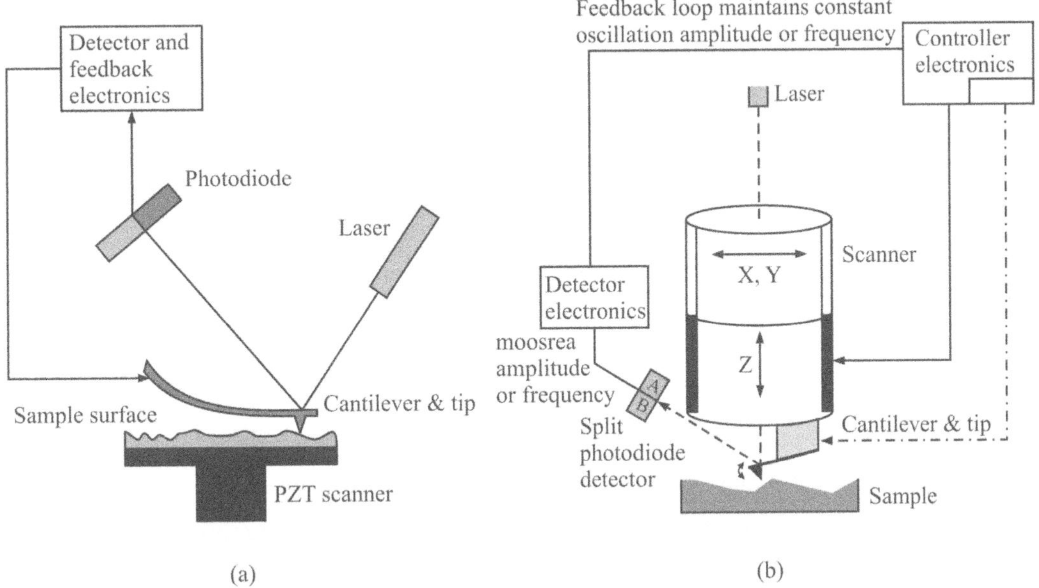

Figure 5.2 (a) Schematic of AFM showing cantilever tip as a probe, (b) Schematic of primary components of Atomic Force Microscope (AFM)

AFM provides three-dimensional surface topography at nanometer lateral and sub-angstrom vertical resolution on insulators and conductors (Fig. 5.3). From this beginning, the field of Scanning Probe Microscopy (SPM) was born which consists of a family of techniques that involves scanning a sharp tip across the sample surface while monitoring the tip-sample interaction to form a high resolution image.

Figure 5.3 Tapping mode AFM image of 1.4 Å monoatomic steps on epitaxial silicon deposited on (100) Si. 1 μm scan

5.3.4 Scanning Electron Microscopy (SEM)

Scanning electron microscopy has an ultimate lateral resolution of a nanometer (depending on the type of electron gun employed), making it a useful alternative to light microscopy for detailed examination of material having micro and nano structures. SEM makes use of a tightly focused, energetic electron beam, which is swept across the surface of the sample in a raster-scan pattern (this is similar to how the cathode ray tube in the TV works). Secondary electrons are emitted from the sample and detected, which forms the image.

The following are the main components of SEM:

- Electron Source ("Gun"),
- Electron Lenses,
- Sample Stage,
- Detectors for all signals of interest,
- Display / Data output devices,

Infrastructure Requirements: Power supply, Vacuum system, Cooling system, Vibration-free floor, Room free of ambient magnetic and electric fields.

SEMs always have at least one detector (usually a secondary electron detector), and most have additional detectors. The specific capabilities of a particular instrument are critically dependent on which detectors it accommodates. Fig. 5.4 shows the schematic of SEM.

The operation of the SEM consists of applying a voltage between a conductive sample and filament, resulting in electron emission from the filament to the sample. This occurs in a vacuum environment ranging from 10^{-4} to 10^{-10} Torr. The electrons are guided to the sample by a series of electromagnetic lenses in the electron column. The resolution and depth of field of the image are determined by the beam current and the final spot size, which are adjusted with one or more condenser lenses and the final, probe-forming objective lenses. The lenses are also used to shape the beam to minimize the effects of spherical aberration, chromatic aberration, diffraction, and astigmatism. Fig. 5.5 shows photograph of a typical SEM instrument, showing the electron column, sample chamber, EDS detector, electronics console, and visual display monitor.

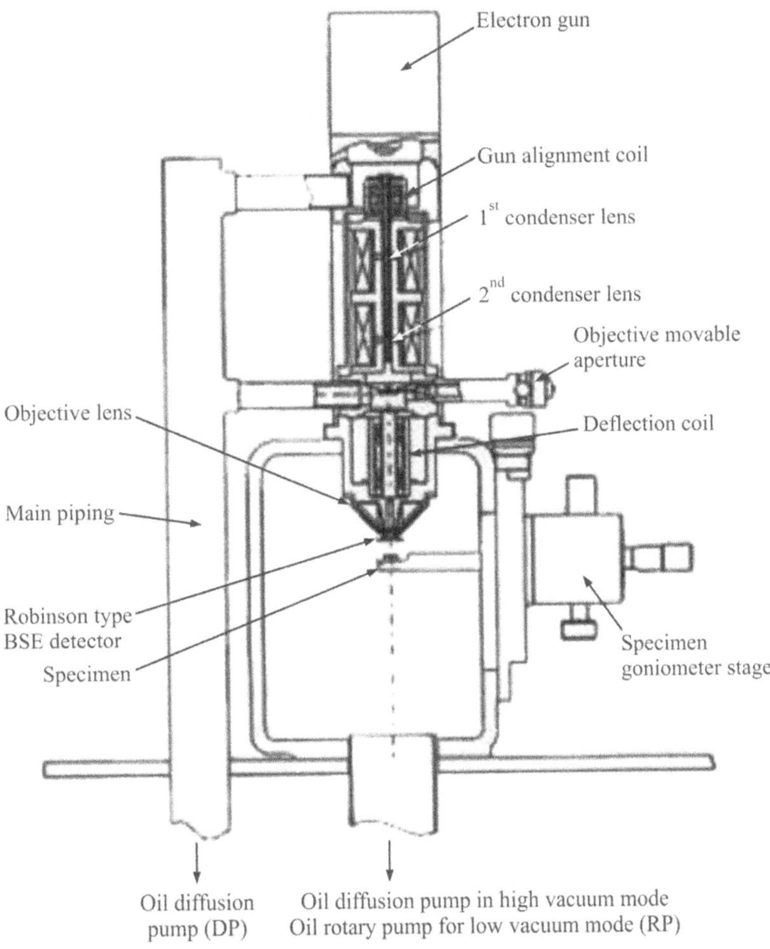

Figure 5.4 Schematic of primary components of Scanning Electron Microscope (SEM)

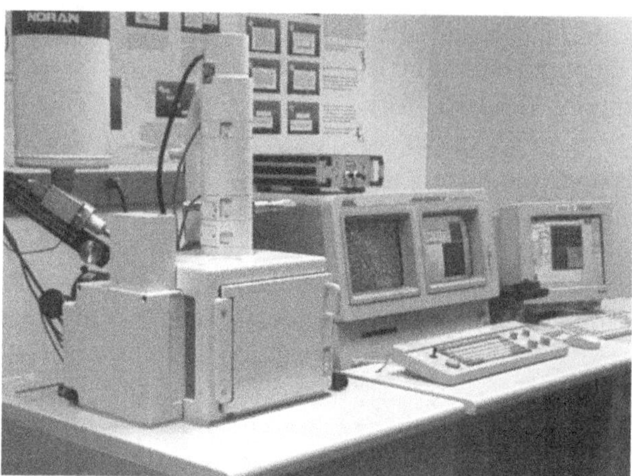

Figure 5.5 Photograph of SEM instrument

The electrons interact with the sample within a few nanometers to several microns of the surface, depending on beam parameters and sample type. Electrons are emitted from the sample primarily as either backscattered electrons or secondary electrons. Secondary electrons are the most common signal used for investigations of surface morphology. They are produced as a result of interactions between the beam electrons and weakly bound electrons in the conduction band of the sample.

Some energy from the beam electrons is transferred to the conduction band electrons in the sample, providing enough energy for their escape from the sample surface as secondary electrons. Secondary electrons are low energy electrons (< 50 eV), so only those formed within the first few nanometers of the sample surface have enough energy to escape and be detected. High energy beam electrons which are scattered back out of the sample (backscattered electrons) can also form secondary electrons when they leave the surface. Since these electrons travel farther into the sample than the secondary electrons, they can emerge from the sample at a much larger distance away from the impact of the incident beam which makes their spatial distribution larger. Once these electrons escape from the sample surface, they are typically detected by an Everhart-Thornley scintillator-photomultiplier detector. The SEM image formed is the result of the intensity of the secondary electron emission from the sample at each x,y data point during the rastering of the electron beam across the surface.

5.3.5 Comparison of AFM and SEM Techniques

There are a number of different ways to compare and contrast these two techniques with respect to each other. Although investigations that use both SEM and AFM to characterize a material are common, there are just a few studies which present complementary nature of the techniques (9-16). These techniques can be compared with respect to 3 factors:

1. Surface Structure,
2. Composition, and
3. Environment.

The comparisons are discussed below for typical equipment configurations and operating procedures.

5.3.5.1 Surface Structure

Although both SEM and AFM are similar in lateral resolution, there are situations in which one technique can provide a more complete representation of the sample surface, depending on the information desired. One principle difference is in how the two techniques process vertical changes in topography. Atomically smooth surfaces can occur either naturally, such as on mineral surfaces, or by processing, such as polishing and epitaxial growth on semiconductor, data storage, and optical surfaces. A Tapping Mode AFM image of an epitaxial silicon surface is shown in Fig. 5.6. Unlike SEM, the AFM can measure in all three dimensions (x, y, and z) with a single scan. Since the AFM has a vertical resolution of < 0.5 Å, it can resolve the 1.4 Å monoatomic silicon steps on the surface as well as calculate an RMS roughness of 0.7 Å (17). On a smooth sample the SEM has difficulty resolving these features due to the subtle variations in height.

5.3.5.2 Thin Films

On most thin films, the SEM and AFM produce a similar representation of the sample surface. A common application of surface investigations of thin films consists of determining changes in morphology with variations of deposition parameters, such as temperature, pressure, time, etc. Fig. 5.6 shows SEM and AFM images of a polysilicon thin film at approximately the same lateral magnification. The two images show similar surface structure, however, they differ in the other types of information that can be acquired on this sample.

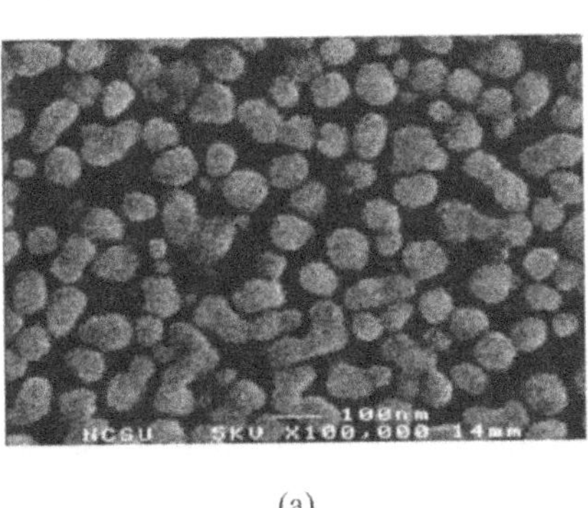

(a)

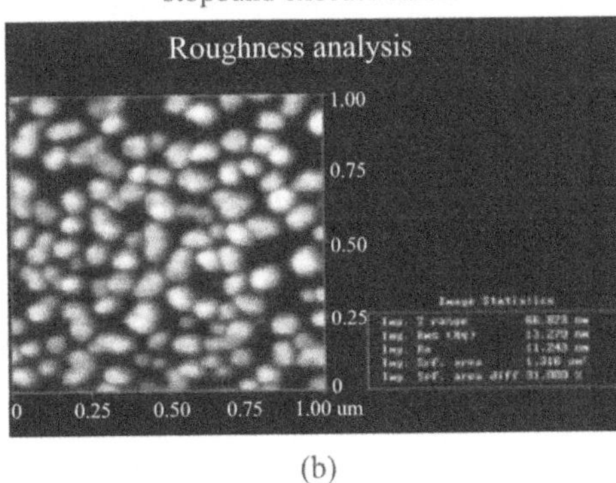

(b)

Figure 5.6 (a) SEM image of rugged polysilicon thin film. 100,000x, Bar = 0.1 μm; (b) Tapping Mode AFM image of the same with roughness measurement 1 μm scan

The three-dimensional nature of the AFM can be used to calculate changes in roughness and surface area variations due to differences in deposition parameters. For the SEM, a large area view of the variations in surface structure can be acquired all at once (such as several mm's), whereas a 100 μm × 100 μm area is typically the largest area viewed by an AFM. These images are an example of "rugged" polysilicon films which are used as capacitors in memory devices. By making these films rough, the surface area is increased which make it possible to hold more charge without increasing the lateral dimensions of the capacitors on the chips. By adjusting the deposition parameters and using the AFM to analyze the surface area of the films, the deposition parameters needed to produce a film with the maximum surface area can be determined (18).

5.3.5.3 High Aspect Ratio Structures

Semiconductor processing commonly requires measurements of high aspect ratio structures such as trenches and via holes. In a SEM, these structures are typically measured in cross section by cleaving the wafer and imaging the sample on end to obtain the dimensions of the structure. In contrast, the AFM image of a trench or via is made by scanning over the sample surface. The ability of the AFM to measure these structures nondestructively makes it possible for the wafer to be returned to the production line after the measurement is acquired. To image some higher aspect ratio structures, the proper tip shape is needed for the AFM to scan narrow openings and steep sidewalls. Although the SEM measurement is destructive to the sample, the ability to image the undercuts of these lines is a useful application that AFMs are not typically designed to perform.

5.3.5.4 Rough Surfaces

One of the key advantages of the SEM with respect to other types of microscopy is its large depth of field. This ability makes it possible to image very rough surfaces with millimeters of vertical information within a single image. A SEM image of non-woven polyethylene oxide fibers can be seen in Fig. 5.7a. The depth of field and small beam size makes it possible to image the fibers far below the top layer. This ability also makes it possible to measure very rough surfaces over larger lateral areas as well. Although the AFM can measure vertical surface variations below 0.5 Å, its ability to measure a tall structure comes from how far the scanner can move vertically. Standard scanners typically have 5 to 6 of μm vertical range, however, in some configurations the vertical range approaches 10 μm larger. For scanning areas that have heights of greater than 5 to 10 of μm variation, the SEM would be better suited for the analysis.

Another example of a complex three dimensional surface structure which shows how the SEM and AFM can complement each other can be seen in Fig. 5.7b. The convoluted three dimensional Y_2O_3 oxide crystal shown growing out of a relatively flat Y_2O_3 thin film on a Si substrate is easily imaged in the SEM. Although the AFM would have problems imaging the obtuse angels and enclosed areas of this surface, the roughness of the Y_2O_3 film can be measured whereas in the SEM image the surface roughness is not evident. Therefore, the two techniques together give a more complete "picture" of the sample.

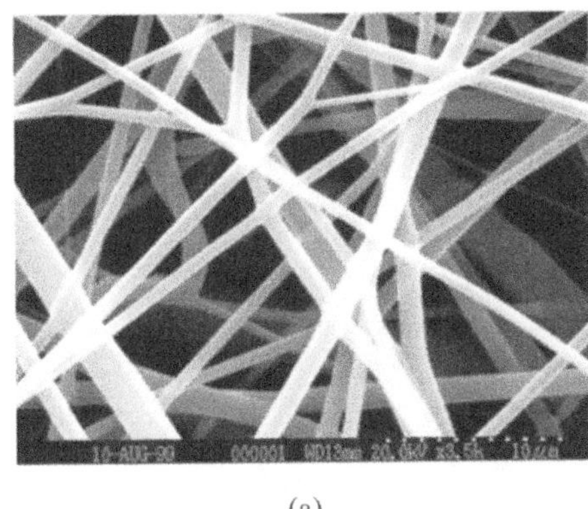

(a)

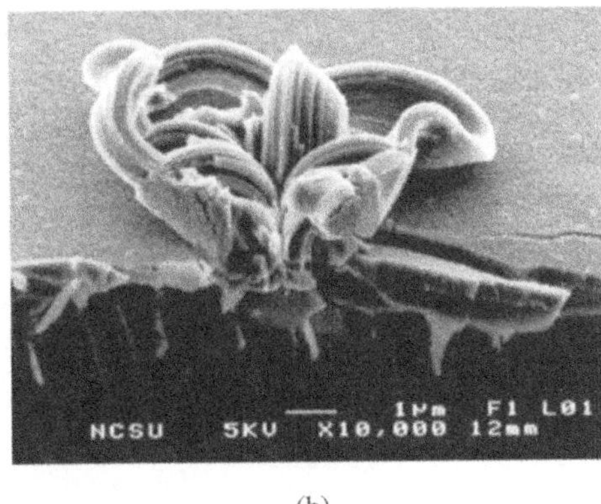

(b)

Figure 5.7 (a) SEM image of a non-woven textile sample of polyethylene oxide fibers. The large depth of field of the SEM makes it possible to image fibers which are 10's of µm's below the upper layer of fibers. Bar = 10 µm; (b) SEM image of Y_2O_3 crystal. Bar = 1 µm

5.3.5.5 Composition

Both SEM and SPM provide compositional information through a variety of techniques. SEM is the only one of the two techniques which provides elemental analysis, however, both SEM and AFM when associated with other techniques can provide compositional information through analyzing materials and physical properties of the sample.

AFM/SPM: Although an AFM does not provide elemental analysis, it can supply compositional information by differentiating materials based on physical properties, such as stiffness, elasticity, compliance, friction, adhesion, magnetic and electrostatic fields, carrier concentration, temperature distribution, spreading resistance, and conductivity. Many of these techniques consist of looking simultaneously at another signal while performing standard AFM imaging. One of the most common techniques for mapping differences in materials properties is Phase Imaging. Phase Imaging is conducted during Tapping Mode AFM operation by monitoring the phase lag between the oscillating drive signal used to drive the cantilever and the oscillating detection signal from the photodiode detector. This signal will indicate differences in visco elasticity and/or adhesion across the imaged area. This technique is commonly applied to mapping the distribution of polymers in a heterogeneous system, or mapping the distribution of filler, such as silica or carbon black, in a polymer matrix. An example of Phase Imaging on a polyethylene film is shown in Fig. 5.8. Other ways to get similar information are by Force Modulation AFM, which maps differences in elasticity across the sample surface, and Lateral Force Microscopy (LFM), which maps differences in friction across the sample surface.

There are also techniques that can be used to investigate long range forces across the imaged area. Magnetic Force Microscopy (MFM) and Electric Force Microscopy (EFM) map the magnetic and electrostatic field gradients, respectively, which extend from the sample surface. These techniques are performed by using either a magnetic or conductive probe to map the attractive and repulsive forces between the tip and the sample. MFM is commonly used to detect the domain structure of magnetic bits written on magnetic media, to evaluate the performance of magnetic heads, and to investigate the magnetic structure of experimental materials.

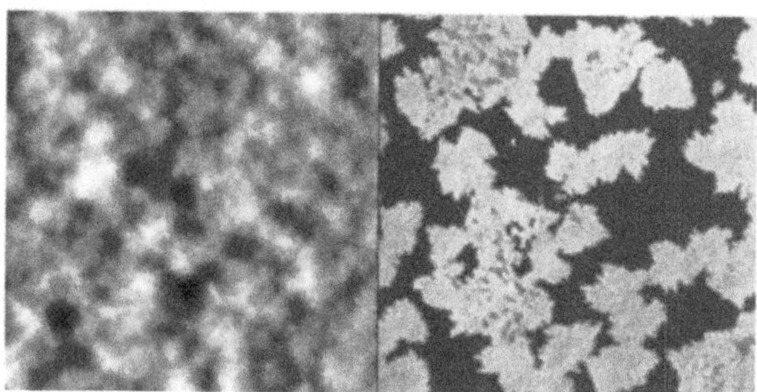

Figure 5.8 Phase image of two components which are used to form a polyethylene (PE) film. The phase image (right) clearly shows the distribution of the two polymers due to differences in stiffness which is not evident from the topographic image (left). 2 μm scan

5.3.5.6 Environment

One of the primary differences between these two types of microscopy is the environment in which they are performed, i.e., SEM is conducted in a vacuum environment, and AFM is conducted in an ambient or fluid environment. There are several issues which make environment an important issue. First, there is a frequent need in fields such as biology and biomaterials to study hydrated samples. These two techniques compensate for this need by different means: an environmental chamber for a SEM, and a fluid cell for the AFM. Further, the SEM is required to work in a vacuum environment due to the nature of the technique which brings up the issues of vacuum compatibility of the sample, the conductivity of the surface, and vacuum maintenance. To image poorly conductive surfaces without sample charging may require conductive coatings or staining, which may alter or obscure the features of interest; or it may require low voltage operation, or an environmental chamber, which may sacrifice resolution.

For SEM, hydrated samples are addressed by placing a specimen in an environmental chamber with either an electron transparent window or a small aperture for the beam to enter the chamber. The chamber is typically flushed with an inert gas saturated with water vapor. Common applications are to either investigate hydrated surfaces to preserve their surface structure when hydrated, or to reduce charging on insulating samples.

One of the primary attractions to the AFM is its ability to image insulating surfaces at high resolution in fluid. Imaging samples in a hydrated state with an AFM is commonly performed by enclosing the sample and probe in a fluid environment, as shown in Fig. 5.9.

Since AFM is not based on conductivity, the image and scanning mechanism is not disturbed by the presence of the fluid. Common applications for AFM investigations in fluid are in the biological sciences, biomaterials, crystal growth, force interaction studies, and for investigating processes *in situ*. With the appropriate accessories, AFM can also be used in varied gaseous environments and at elevated temperature. The latter is particularly important for research and development of polymers.

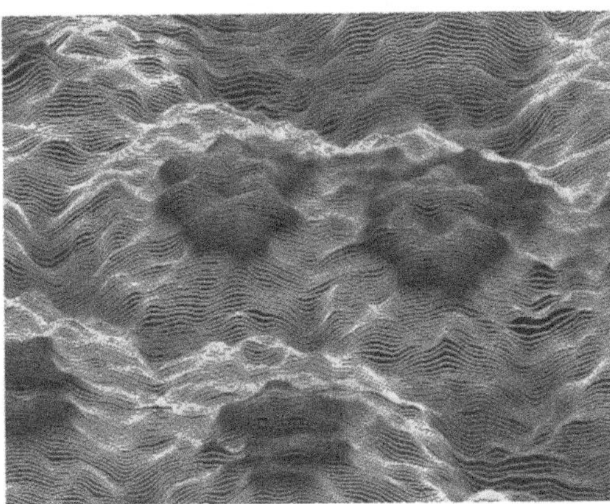

Figure 5.9 Fluid cell for an AFM which allows imaging in an enclosed, liquid environment

5.3.5.7 Special Applications

Although SEM and AFM appear very different, they actually share a number of similarities. Both techniques raster a probe across the surface to detect some interaction with the surface to form an image. Both have a lateral resolution which is similar in scale (although under certain conditions AFM is superior). And both techniques have image artifacts that the operator is trained to identify. The SEM has had a much longer time to mature as a technique and to develop an understanding of how to identify and avoid artifacts, but the rapid adoption and implementation of AFM has resulted in a similar understanding of artifacts.

However, one should be wary of combined systems in which an AFM is placed inside the SEM chamber. One of the true advantages of the AFM is its ability to perform high resolution measurements outside of a vacuum environment. Placing it inside a vacuum environment reduces its flexibility and increases its operating time. Combined AFM/SEM systems often have reduced capabilities and typically compromise the performance of both instruments. SEM and AFM are complementary techniques that provide a more complete representation of a surface when used together than if each were the only technique available. These techniques overlap in their capabilities to provide nanometer scale lateral information. However, they deviate in the fact that the AFM can provide measurements in all three dimensions, including height information with a vertical resolution of < 0.5 Å, whereas the SEM has the ability to image very rough samples due to its large depth of field and large lateral field of view.

The SEM can provide elemental analysis using X-ray detection, whereas the AFM can provide compositional information based on physical properties. The fact that the two techniques operate in different environments can be a strength when used together since the AFM does not encounter vacuum issues (difficult sample preparation, sample modification, etc.) and may image samples in an enclosed fluid or other environment. The vacuum environment of the SEM makes it possible to conduct a number of techniques that require vacuum, such as X-ray analysis. By having both techniques side-by-side in an analytical facility, the overall scope of analytical capabilities will be broadened, adding to the flexibility of the facility.

5.3.6 Scanning Tunneling Microscope (STM)

Scanning Tunneling Microscopy (STM) is another instrument commonly used in nano technology. It is used to measure 3-D topology of the specimen. The STM is based on the concept of quantum tunneling. When a conducting tip is brought very near to the surface to be examined, a bias (voltage difference) applied between the two can allow electrons to tunnel through the vacuum between them. Measurements are made by monitoring the current as the tip's position scans across the surface, which can then be used to display an image (Fig. 5.10).

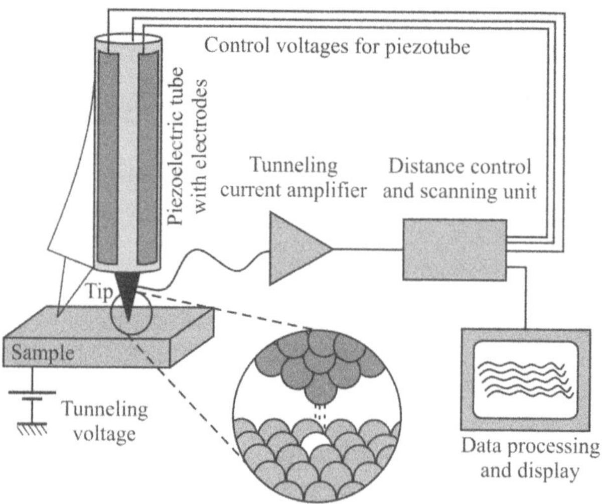

Figure 5.10 Schematic of Scanning Tunneling Microscope (STM)

5.3.7 Transmission Electron Microscopy (TEM)

Transmission electron microscopy (TEM) provides the visualization of the internal structure of specimens by the use of transmitted electrons. Because electrons have a much smaller wavelength than that of visible light, microscopy-using electrons offers a much greater (50,000x) resolving power than light microscopy. Because electrons have a very limited range transiting matter, TEM samples must be very thin and they must be placed in a vacuum inside the instrument. Fig. 5.11 shows the photograph of TEM Hitachi H-7100 which can be used to study morphology and structure of the carbon nanotubes, nanofibers and vapor growth carbon fibers.

The above instruments which produce realistic pictures of the surface are excellent measuring tools for Research and Industrial applications of nanotechnology. The requirement in industrial nano metrology is for higher accuracy than resolution as compared to research nano metrology.

The following is the list of measurement techniques used in the characterization of nano materials for specific characters.

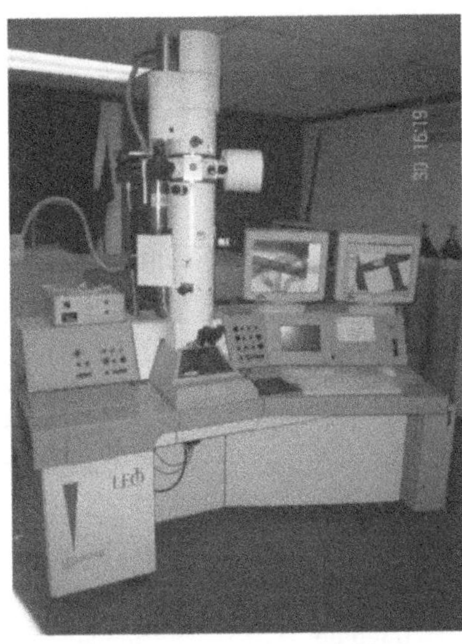

Figure 5.11 Photograph of TEM hitachi H-7100

List of other techniques used for characterization of nano materials

1. X-ray absorption Spectroscopy, 2. X-Ray Diffraction, 3. Small Angle X-Ray Scattering, 4. Transmission Electron Microscopy 5. Capacitance Spectroscopy, 6. Polarization Spectroscopy, 7. Auger Electron Spectroscopy, 8. Raman Spectroscopy, 9. Small Angle Neutron Scattering, 10. Cyclic Voltammetry, 11. Linear Sweep Voltammetry, 12. Nuclear Magnetic Resonance, 13. Mössbauer Spectroscopy, 14. Fouier Transform Infrared Spectroscopy, 15. Photoluminescence Spectroscopy, 16. Electroluminescence Spectroscopy, 17. Differential Scanning Calorimetry, 18. Secondary Ion Mass Spectrometry, 19. Cathodoluminescence Spectroscopy, 20. Electron Energy Loss Spectroscopy 21. Energy Dispersive X-Ray Spectroscopy, 22. Four point probe and I-V technique, 23. X-Ray Photoelectron Spectroscopy, 24. Scanning Near-field Optical Microscopy, 25. Single-molecule Spectroscopy, 26. Neutron Diffraction.

References

1. Daniszewska, W. Ojkowski, H. Fecht, K.J. Kurzydlowski, U. Narkiewicz, G. A. Salishchev, M. J. Zehetbauer, M. Kulczyk, M. Chmielecka and D. Kuzienko (2005). Metallic Nano-Materials and Nanostructures: Development of Technology Roadmap, Solid State Phenomena, 114 (2000) 345. E-MRS Fall Meeting, Warsaw, September.
2. http://www.nanotec.org.uk
3. R. V. Lapshin (1998). Automatic lateral calibration of tunneling microscope scanners (PDF). Review of Scientific Instruments (USA: AIP) 69(9): 3268–3276. doi:10.1063/1.1149091. ISSN 0034-6748.
4. Van Ardenne, M. (1938) Z. Phys. 109, 407.

5. Goldstein J. I., Newbury D.E., Echlin P., Joy D.C., Fiori C. and Lifshin E. (1981). Scanning Electron Microscopy and X-ray Microanalysis, Plenum Publishing Corp., New York, 3.
6. Zworykin V.K., Hillier J. and Snyder R.L. (1942). ASTM Bulletin 117, 15.
7. Binning G, Roher H, Gerber C, Weibel E (1982), Surface Studies by Scanning Tunneling Microscopy, Phys. Rev. Lett. 49, 57.
8. Binnig G B, Quate C F, and Gerber Ch. (1986). Atomic Force Microscope, Phys, Rev. Lett., 12, 930.
9. Russell, P.E. and Troughton, E.B. (1999). Comparative Study of Field Emission-Scanning Electron Microscopy and Atomic Force Microscopy to Access Self-Assembled Monolayer Coverage on Any Type of Substrate, Microscopy and Microanalysis 5, 413.
10. Castle J.E. and Zhdan P.A. (1997). Characterization of Surface Topography by SEM and SFM: Problems and Solutions, J. Phys. D: App Phys, 30, 722.
11. Lemoine P., Lamberton R.W. and Ogwu, A.A. (1999). Complementary Analysis Techniques for the Morphological Study of Ultrathin Amorphous Carbon Films, J. App. Phys., 86, 6564.
12. Van Helleputte H.R.J.R., Haddeman T.B.J., Verheijen M.J. and Baalbergen J-J. (1995). Comparative Study of 3D Measurement Techniques (SPM, SEM, TEM) for Submicron Structures, Microelectronic Engineering, 27, 547.
13. Nessler, R. (1999). Scanning Microscopy Technologies: Scanning Electron Microscopy and Scanning Probe Microscopy, Scanning 21, 137.
14. Fischer A., Jentoft F.C., Weinberg G., Schlogl R., Niesen T.P., Bill J., Aldinger F., De Guire, M.R. and Rühle, M. (1999). Characterization of Thin Films Containing Zirconium, Oxygen, and Sulfer by Scanning Electron and Atomic Force Microscopy, J. Mat. Res. 14, 3725.
15. Campbell A.L, Bunning T.J., Stone M.O., Church D. and Grace M.S. (1999). Surface Ultrastructure of Pit Organ, Spectacle, and Non Pit Organ Epidermis of Infrared Imaging Boid Snakes: A Scanning Probe and Scanning Electron Microscopy Study, J. Struct. Biol. 126, 105.
16. Reynolds G.W. and Taylor J.W. (1994). Correlation of Atomic Force Microscopy Sidewall Roughness Measurements with Scanning Electron Microscopy Line-Edge Roughness Measurements on Chemically Amplified Resists Exposed by X-ray Lithography, J. Vac. Sci. Technol. B 17 (1999), 2723. Meeting of the ECS, Vol. 94-1, Pennington, NJ, 461.
17. Strausser Y.E., Doris B., Diebold A.C. and Huff H.R. (1994). Measurement of Silicon Surface Microroughness with AFM, Extended Abstracts of the 185th Meeting of the ECS, Vol. 94-1, Pennington, NJ, 461.

6

Applications of Polymers/Organics, Liquid Crystals, Nano-Composites and Ferro-Fluids in Nano Technology

6.1 Applications of Polymers/Organics in Nano Technology

Polymer nanoscience deals with the study and application of nanoscience to polymer-nanoparticle matrices which have at least one dimension of less than 100 nm. The transition from micro- to nano-particles causes change in its physical as well as chemical properties which is called nano effect.

Nano effect: Besides the particles shape, surface or morphology, dispersion and aggregation in a matrix significantly contribute to the properties and performance of a composite.

How nanoparticles affect the properties of nanocomposite materials is of concern to engineers and scientists who are interested in the applications of nanocomposites. The nanomaterials or nanocomposites and the theories related to them are based on a concept of "critical scale," or nanoscale. For a critical scale of nanoparticles, the nanocomposite properties produce a superior property compared with the pure counterpart matrix if the nanoparticles are below a critical size.

Increase in the ratio of the surface area to volume, and the size of the particle are the two major factors responsible for nano effect. The increase in surface area-to-volume ratio, which increases as the particles get smaller, leads to an increasing dominance of the behavior of atoms on the surface area of particle over that of those interior of the particle. This affects the properties of the particles when they are reacting with other particles. Because of the higher surface area of the nano-particles, the interaction with the other particles within the mixture is more and this increases the strength, heat resistance, etc., and many factors do change for the mixture. An example of a nanopolymer is silicon nanospheres which show quite different characteristics when their size is 40–100 nm and they are much harder than silicon, their hardness being between that of sapphire and diamond.

Thus when a material structure has one dimension less than the critical scale of 100 nm, can appear totally different the behaviour of its material properties. Based on such phenomena, scientists determined the abnormal phenomena to be produced on the scale level between single-atom molecules and clusters of many thousands of molecules.

The application of Polymers/Oganics in nanotechnology are presently being pursued using two approaches: top-down nanotechnology and bottom-up nanotechnology (Figure 6.1). Technologies that are used in top-down nanotechnology are mostly lithographic techniques such as (extreme) UV lithography, nanoimprint lithography, e-beam lithography, soft lithography and scanning probe lithography. The largest challenges for these techniques lie in enhancing the resolution and making these technologies cheaper and faster. In bottom-up nanotechnology, (self-) assembly of molecules or colloids is used to create nanostructures or nanopatterns. The most important concern in bottom-up nanotechnology is the control over the spatial position of the molecules or nanoparticles. For the fabrication of devices, a combination of both techniques is required to provide an interface with the technical environment.

In both approaches, polymers play an important role, because they combine a number of favourable features, including flexibility, processability, low cost, size in the nanometer range, diverse functionalities and microphase separation.

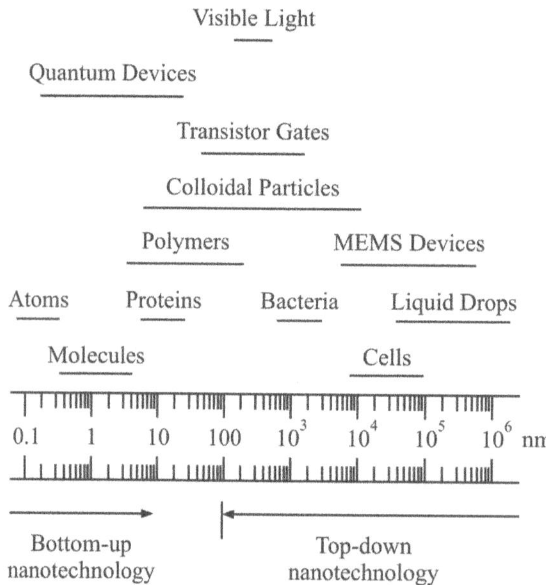

Figure 6.1 Overview of relevant sizes in nanotechnology
Reproduced with modifications from Geissler et al. (1)

6.1.1 Polymer in Top-Down Nanotechnology

Many lithographic techniques that are presently in use in the top-down approach to fabricate nanostructures have originally been developed for precision engineering or microtechnology. UV lithography and imprint lithography are examples of existing techniques where technological advancements have lead to higher resolutions from micrometer up to nanometer sizes. In UV lithography, polymers are used as resists. Exposure of the polymer resist to UV radiation through a mask induces reactions in the polymer that make it either more soluble (positive resist) or less soluble (negative resist). After development, a pattern of resist remains that can be used as a mask for etching of the substrate, or for the fabrication of metal patterns via a lift-off procedure.

In nanoimprint lithography (NIL), a pattern is imprinted in a polymer resist layer. This imprinting can either be done above the Tg of the polymer (thermal NIL) (2) or at room temperature in a liquid prepolymer that is subsequently cured via UV radiation (mold-assisted transport) (3). Nano imprint lithographic tools are commercially available that can imprint structures as small as 20 nm, but in literature already imprinting of structures of 6 nm has been reported (4, 5).

A new technique that was developed for top-down nanotechnology is scanning probe lithography (SPL), in which an AFM tip is used to pattern surfaces (6, 7) for instance by indentation of polymer films for high-density data storage. The advantage of this technique is that high resolutions can be obtained (8, 9). However, the serial nature of this technique leads to low throughput and high costs. Higher throughputs can be achieved in SPL by the development of arrays of thousands of tips, which has been successfully demonstrated for imprinting polymer with the development of the 'millipede' (10-11).

Above mentioned techniques have utilized polymers either as a resist layer or as a substrate. However, Whitesides et al., (12) investigated the use of elastomeric stamps as the structuring device for patterning of substrates. This has led to the development of many new techniques for top-down nanotechnology collected under the name Soft Lithography (12). The technical details of various soft lithography are discussed in detail in Chapter 3.

The first soft lithographic technique that was developed by Kumar et al. (13) showed that a patterned elastomeric stamp could be used to print patterns of self-assembled monolayer (SAM) on a metal surface. The patterned SAM was subsequently applied as a mask to selectively etch the metal, and many other soft lithographic techniques were developed, such as replica molding (REM) (14), microtransfer molding (TM) (15), micromolding in capillaries (MIMIC) (16), solvent-assisted micromolding (SAMIM) (17), capillary force lithography (CFL) (18,19) and edge spreading lithography (ESL) (20).

The largest advantages of soft lithography techniques are their low cost, since no sophisticated machines are needed, the possibility to pattern curved objects, which makes the technique suitable for low-cost roll-to-roll manufacturing techniques, and the nearly unlimited amount of patternable materials and substrates. Not only can polymers be used as the stamp material, but also as the substrate or as an ink (21, 22) in soft lithography techniques. The resolution that can be achieved with soft lithography is increasing, and sub-50 nm structures have been obtained with composite polymer due to the development of stiffer Poly dimethyl siloxane PDMS (Figure 6.2(a)) (23, 24), while 5-nm-wide structures were obtained using REM with supported carbon nanotubes as the mold (Figure 6.2(b)) (25).

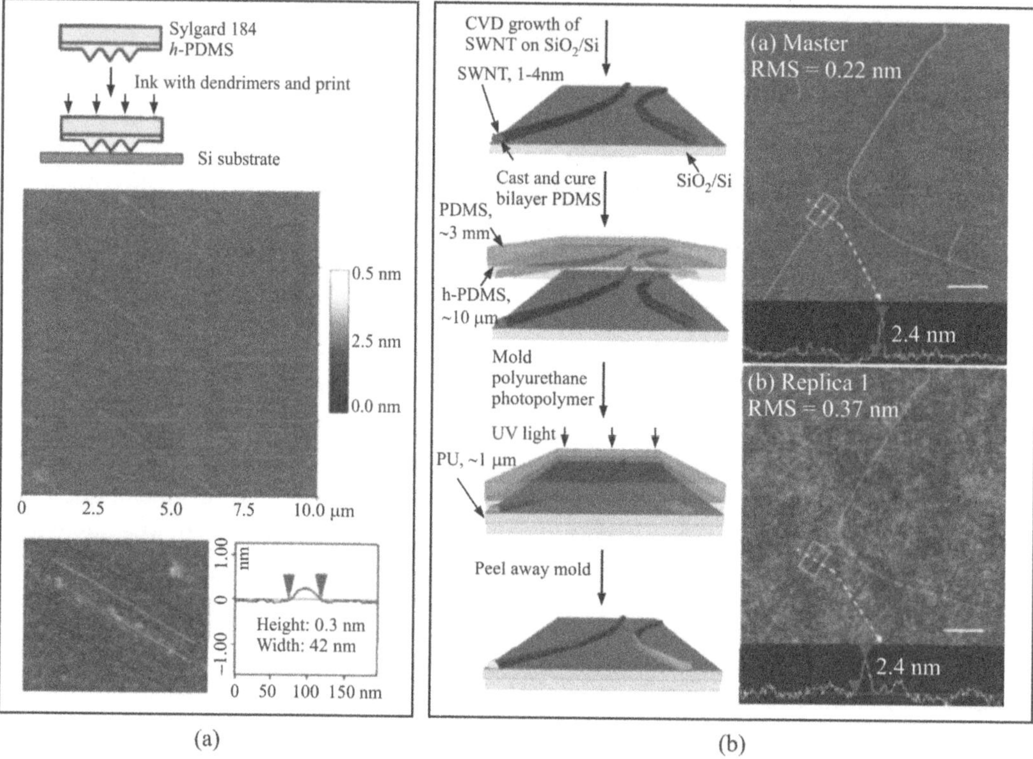

Figure 6.2 State-of-the-art soft lithography: (a) Using h-PDMS stamps, 42-nm-thin lines of dendrimer were microcontact printed on silicon wafer. Reproduced from Li et al. (24) (b) Replication molding of supported carbon nanotubes yielded poly (urethane) replicas with feature sizes smaller than 5 nm. Reproduced from Hua et al. (25)

6.1.2 Polymer in Bottom-Up Nano-Technology

Bottom up nano Engineering is the aim of many scientist and engineers ever since the idea of controlling materials down to nano scale came to reality. Essentially it is the art of fabricating quantative amounts of similar nanoshaped aggregates from designed molecules with specific properties. Whereas, molecular engineering is a long lasting success story, the possibility of nano shaped aggregates via controlled bottom up growth has emerged only within last ten years.

6.1.3 Application of Organic Polymer based Nanofibers

Nanofibers as one alternative to nano aggregates has attracted attention in the recent past mainly in their form as inorganic semiconducting nano wires such as silica nanowires for light guiding, nano wire sensors and nano lasers.

However organic polymer based nano fibers (e.g., Polyethylene oxide nano fibers) produced by electro spinning (Figure 6.3) which can be used to produce semiconducting nano wires promise to revolutionise the submicron integrated photonics due to their small sizes and extra ordinary morphological an optoelectronics properties.

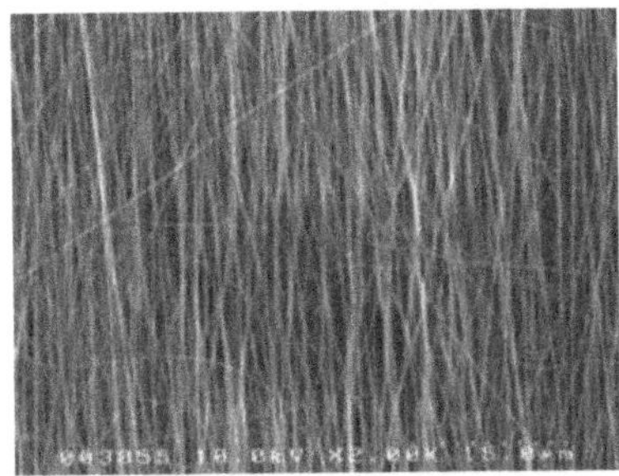

Figure 6.3 SEM image of polyethylene oxide nano fibers made by electro spinning

A more direct molecular nano technology is expected to start with organic molecules as building blocks whose oligomeric properties are easy to tune and to control compared to polymers. The expected hurdles here are the controlled growth of similar nano aggregates of predefined shapes and their transfer on more complicated target substrates. The huge advantage however is the tremendously increased design flexibility, enhanced integration into devices and potentially considerable improved performance comparable to performance from liquid crystal based flat screens to organic light emitting device based screens.

Self assembly of rod like organic oligomers on a muscovite mica substrate upon evaoporation in high vaccuum leads to well shaped mutually aligned nano aggregates. Upon irradiation with UV light the distinct nature of these aggregates becomes obviously emission intense polarized blue light. The morphology of nano structures strongly depends on the functionlization of organic oligomers. Not only morphology but also optical properties can be tailored. By choice of an appropriate functionalization the structures act as nano scaled frequency doubler.

Figure 6.4 shows four fluorescence microscope images of diverse organic nano aggregates. Typical dimensions are several 10 nanometers in height and few hundred nano meters in width and length starting from few ten micrometers to several hundred micro meters. Flexibility is a great advantage for organic molecules compared to inorganic compounds. The developments and extrapolation of these organic oligomers is crucial for future nano technology.

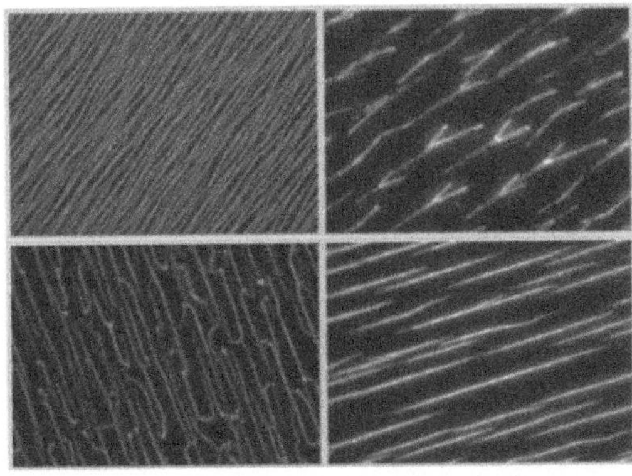

Figure 6.4 Fluorescence microscope images of different nano organics

Among organic molecules oligomers built up of π-conjugated system have recently seen a tremendous development due to their interesting optical electrical and opto electrical properties. The use of organic rather than inorganic compounds to fabricate nano structures usually has the advantage of higher luminisence efficiency at the same material density, more flexible spectroscopic properties and general easier and cheaper processing. Especially rod like properties of oligo thiopenes, perylenes, pentacenes and oligo-p-phenylenes has been extensively studied in this context. Among those the latter ones and especially the p-hexaphynelene has been found to form very interesting needle like parallel oriented nano aggregates upon vapor deposition by a sophisticated high vacuum surface growth process. Owing to their typical height and width on a nanometer scale and length up to one millimeter, the fibers bridge the gap between microscopic and macroscopic dimensions. These nano fibers show high quantum yields being about 30% of anisotropic blue luminescence which makes them promising candidates for new building blocks for opto electronic devices like organic light emitting diodes (OLEDS). Waveguides as well as ultraviolet pumped organic nano lasers are other potential applications. In addition electrical properties have been investigated in view of potential applications of nano–fibers in organic field effect transistors (OFETs) or electroluminescent nano lasers.

Since the para-hexaphenelene has proven to be versatile building block for generating long parallel aligned organic nano fibers by self assembly with outstanding optical properties an important step in exploiting full potential of organic nano aggregates is to functionalise the molecular building blocks. Such piramordial functionalisation is advantageous compared to modifying existing nano aggregates. Although the aggregates can be modified by adjusting growth conditions like substrate temperature, deposition rate, or varying or modifying growth substrate before deposition of the organic material which is possible only in narrow limits.

But chemical functionalization on molecular basis widens the scope of possibilities for modification. This not only allows tailoring and improvement of morphology and optical properties but also create new properties. It is challenging to explore the prospects and limits of this kind of nano engineering in which chemically functionalized oligomers undergo similar self assembly process and allow creation of quantative amounts of crystalline nano fibers with tailored morphologies and optical, electrical, mechanical and even some new properties.

6.1.4 Supra Molecular Polymers

A supramolecular polymer is a polymer whose monomer repeat units are held together by non covalent bonds. Non-covalent forces that hold supramolecular polymers together include coordination, π-π interactions, and hydrogen bonding. One system that has been demonstrated uses quadruple hydrogen bonds to form supramolecular polymers. Functionalization of polymers with the quadruple hydrogen bonding unit from the Meijer

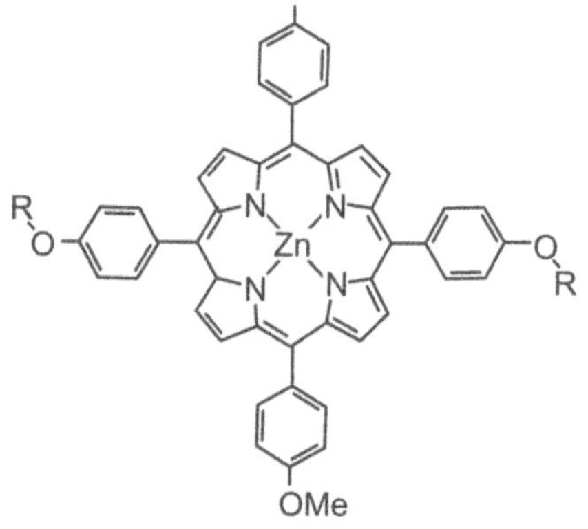

Figure 6.5 Porphyrin dye structure

group introduces reversible cross-links and a virtual increase in the polymer's molecular weight. In general, polymers with a higher molecular weight possess better material properties (Figure 6.5).

Further when they are combined with inorganic particles they form into well-defined supramolecular aggregates. Control of this assembly process is the main objective in bottom-up nanotechnology. Supramolecular polymers and block copolymers are promising candidates for bottom-up nanotechnology, since their morphology can be tuned by changes in molecular design or assembly conditions. Examples of applications of polymers in a few subfields of bottom-up nanotechnology will be briefly discussed in the following paragraphs.

In the field of nanoelectronics, the self-assembly of π-conjugated molecules via hydrogen bonding is used to form fiber-like supramolecular polymer stacks, which can potentially be used as conductive wires in organic opto-electronic devices (26-29). Unfortunately, the current flow through these molecular wires is not yet reliable. The reliability of the molecular junction was successfully improved by growing a self-assembled monolayer of thiols between the top and bottom electrode, using a conductive polymer layer between the SAM and the top electrode to prevent electrical shorts (30).

6.1.5 Application of Polymers in Nano Medicine

Nanomedicine is the medical application of nanotechnology. Examples of applications that are currently under investigation include the development of drug delivery systems (31) and targeting nanoparticles for imaging, and fabrication of implants or scaffolds for tissue engineering. Polymers are good candidates for application in all these areas, since there are many biocompatible and biodegradable polymers available, and the mechanical, chemical and surface properties of polymers can easily be modified, as well as their size.

Functionalization of polymer scaffolds with peptide sequences that promote cell adhesion can for instance be done via supramolecular interactions (32-33). An advantage of using supramolecular interactions for modification of polymers is that a modular approach can be used and polymers with different functionalities can be obtained simply via mixing. The strength of the interaction of the functional group with the polymer can be tuned, thereby achieving controlled release of the functional group (i.e., drugs or peptide sequences).

Supramolecular interactions can be based on ionic interactions (34-36), coordinative bonds (36-42) or hydrogen bonds (43-53). One hydrogen bonding array that has been investigated extensively in literature is the bisurea motif, a strong and self-complementary motif that is based on the formation of bifurcated hydrogen bonds between urea groups (54). It has been used in organo gelators (55), hydrogelators (56), DNA-based coatings (57), as a patterning tool in self-assembled monolayers (58) and in polymer micelles (59). Bisurea motifs have also been used as the hard block segment in thermoplastic elastomers (TPEs) (52, 60-67). The directional hydrogen bonding between the urea groups leads to the self-assembly of the hard block segments into these polymers are very suitable for the fabrication of tissue engineering scaffolds. Furthermore, the bisurea nanofibers can be used for the functionalization of the polymer, since it has been shown that a peptide molecule containing a bisurea molecule was bound tightly to a bisurea TPE with an identical bisurea (Figure 6.6(a)) (67) and Figure 6.6(b) AFM image (64).

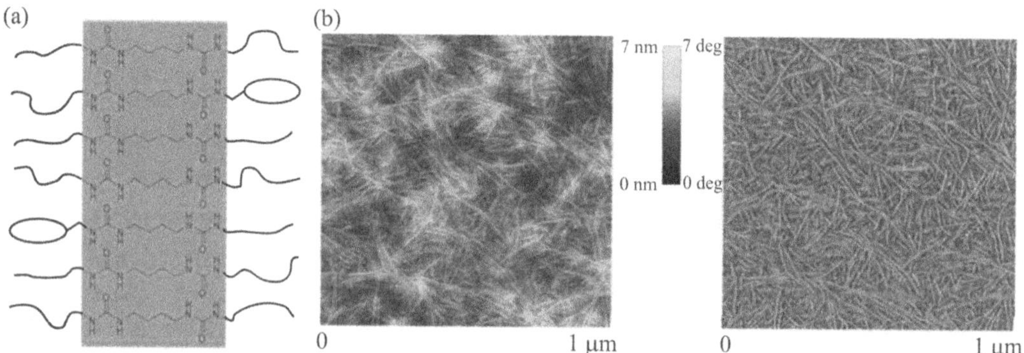

Figure 6.6 (a) Hydrogen bonding between bisurea hard block units in a bisurea-based thermoplastic elastomer, and incorporation of functional groups. (b) AFM height and phase image of nanofibrous morphology of a bisurea-based thermoplastic elastomer. Reproduced from Versteegen et al. (64)

Control over the position of functional groups has been investigated by using self-sorting in multi-functionalized polymers (68-73) or by using block copolymers. Block copolymers are useful in nanotechnology because they spontaneously form nanostructures due to microphase separation (74-77). Their morphology can be tuned between spherical, cylindrical, interconnected network or lamellar or by varying the sizes of the blocks. If the two blocks have a different etch-resistance, the block copolymer morphology can be directly transferred into a silicon wafer via etching (78). Ordering and alignment of the block copolymer microdomains can be achieved by combining bottom-up assembly of block copolymers with top-down fabrication of patterned substrates, since the block copolymers will align to the edges of the patterns (79-80).

The differences in chemical composition of the two blocks can be used for the controlled functionalization of one of the two blocks (81, 82). Also the selective binding of nanoparticles to one of the two blocks has been achieved (83).

A nice example of a combination of top-down and bottom-up approaches in patterning and functionalizing a polymer surface was demonstrated by Xu et al. (Figure 6.7) (84). First, photolithography was used to create a pattern of thymine-functionalized poly styrene (Thy- PS) and poly vinyl-*N*-methylpyridinium (PVMP). Subsequently, molecular recognition was used to selectively modify this pattern with diamino pyridine- functionalized PS via hydrogen.

The principle of molecular recognition via supramolecular interactions is not only useful for the functionalization of polymers. Molecular recognition inside nanoporous polycarbonate membranes can be used to selectively filter dyes and proteins (85), which is of interest for the development of biosensors.

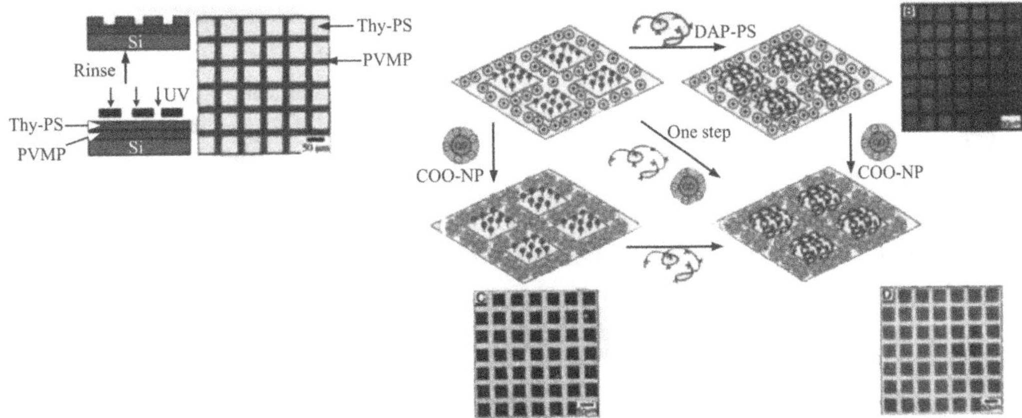

Figure 6.7 Combination of top-down photolithography and bottom-up self-assembly to create patterns of polymers and nanoparticles. Reproduced from Xu et al. (96)

6.1.6 Challenges for Application of Polymers in Nanotechnology

With top-down and bottom-up techniques meeting at the nanoscale, many possibilities and challenges arise for nanotechnology. Regarding the use of polymers in nanotechnology, each specific field has its own challenges. For UV lithography, the current photoresists are not suitable for extreme UV, since their absorption coefficients at 14.5 nm are too high. New polymers have to be found that have low absorption, high sensitivity, high etch resistance and the ability to form high resolution images with low line-edge roughness (86).

In soft lithography, the use of commercially available Polydimethy Siloxane (PDMS) as the material for elastomeric stamps is not ideal, since this PDMS is very soft and hydrophobic. Alternative stamp materials are currently under investigation; some of these new stamp materials exhibit a higher modulus, thereby preventing sagging and pairing of the stamp and improving the resolution that can be reached with microcontact printing (87,88), others are more hydrophilic than PDMS and can be used for printing polar inks or proteins (87, 89-92). Another problem in microcontact printing is the diffusion of ink over the substrate. Solutions are being sought in two directions, namely limiting the diffusion of the ink by using a higher molecular weight ink (93, 94), such as a polymer, or preventing the use of an ink by functionalization of the stamp surface with reactive or catalytic groups (95-97).

For the use of polymeric substrates in plastic electronics, adhesion between the substrate and the metallic wires is crucial for the reliability of these devices. However, promising substrates like poly ethylene terephthalate (PET) and poly ethylene naphthalene (PEN) are not very adhesive. Currently, adhesive layers are employed prior to metal sputtering and the films are subsequently patterned using UV lithography and etching techniques. Preferably, the administration of the adhesive layer to these films, patterning and metallization should occur in a roll-to-roll process. Perhaps soft lithographic techniques such as microcontact printing can meet these requirements.

In bottom-up nanotechnology, the spatial control of molecules is the largest challenge. Exact positioning of molecules or atoms on surfaces is possible using manipulation with an STM tip (98). However, this is a low-throughput technique that is far from being commercialized. Self-assembly and phase-separation of polymers are much faster and cheaper techniques to obtain control over the location of functional groups. With the discovery of more and more complex block copolymer morphologies and the development of more accurate self-sorting systems, realization of spatial control over functional groups is within reach. This control can be expanded to large areas with complex patterns via the combination of top-down and bottom-up technologies, where lithographic techniques are used to create templates for assembly.

6.1.7 Fuel Cell Applications

Fuel cell applications involve polymers in the proton exchange membrane, binder for the electrodes and matrix for bipolar plates. The electrodes typically comprise carbon black particles (0.5– 1.0 mm) with Pt catalyst particles of 2–5 nm and a polymeric binder (usually Nafion). Platinum nanoparticles deposited onto single-walled carbon nanotubes with Nafion as a binder were shown to give improved performance over the conventional carbon black-based electrodes (100). Nanoparticle incorporation in the proton exchange membrane has been noted in numerous publications to improve mechanical properties as well as to enhance proton conductivity in direct methanol fuel cells.

6.1.8 Application of Liquid Crystals and Flexible Flat Panel Display

6.1.8.1 Fundamentals of Liquid Crystal Displays

6.1.8.1.1 Liquid Crystals

The term liquid crystal (LC) is used to describe a substance in a state between liquid and solid but which exhibits the properties of both. Thus a liquid crystal is a fourth state of matter: it has properties between those of a conventional liquid and those of a solid crystal. Like a liquid, it flows, and like a crystal, it can display long-range molecular order (Figure 6.8). Molecules in liquid crystals tend to arrange themselves until they all point in the same specific direction. This arrangement of molecules enables the medium to flow as a liquid. Depending on the temperature and particular nature of a substance, liquid crystals can exist in one of several distinct phases.

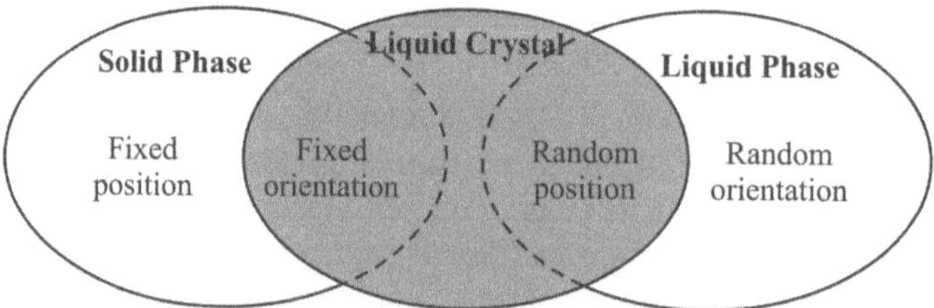

Figure 6.8 The liquid crystal phase is in between a solid and a liquid phase (Image credit: L. Filipponi, iNANO, Aarhus University, Creative Commons Attribution Share (Alike 3.0)

In terms of classifications, LCs (together with polymers and colloids), are often classified as "soft matter" (Figure 6.9) and treated under the branch of physical chemistry of condensed matter.

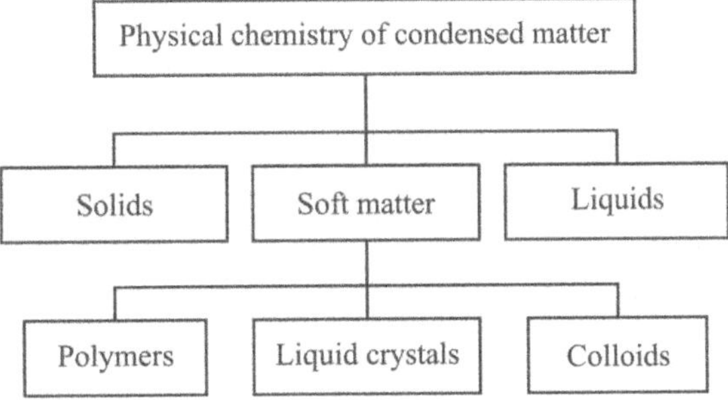

Figure 6.9 Classification of liquid crystals with reference to different forms of matter

Liquid-crystalline systems change their molecular and supermolecular organisation drastically due to very small external perturbations: The molecules in liquid crystal displays for instance are re-oriented by relatively weak electrical fields. If one dissolves a small amount of chiral molecules in an achiral liquid-crystalline host phase, this results in remarkable macroscopic chirality effects, ranging from helical superstructures to the appearance of ferroelectricity.

6.1.8.1.2 Liquid Crystal Self-assembly

Liquid crystals are partly ordered materials, somewhere between their solid and liquid phases. This means that LCs combine the fluidity of ordinary liquids with the interesting electrical and optical properties of crystalline solids. The molecules of liquid crystals are often shaped like rods or plates or some other forms that encourage them to align collectively along a certain direction (Figure 6.10).

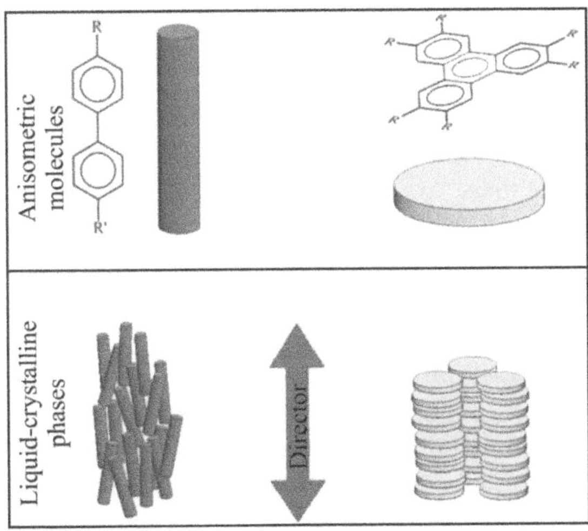

Figure 6.10 Example of the self-organisation of anisometric (i.e., with asymmetrical parts) molecules in liquid-crystalline phases

On the left: rod-like molecules form a nematic liquid, in which the longitudinal axes of the molecules are aligned parallel to a common preferred direction ("director"). On the right: disc-like (discotic) molecules arrange to molecule-stacks (columns), in which the longitudinal axes are also aligned parallel to the director. As a result of their orientational order, liquid crystals exhibit anisotropic physical properties, just like crystals.

Liquid crystals are temperature sensitive since they turn into solid if it is too cold and into liquid if it is too hot. This phenomenon can, for instance, be observed on laptop screens when it is very hot or very cold.

6.1.8.1.3 Liquid Crystals Phases and their Properties

A liquid crystal is formed by the self-assembly of molecules in ordered structures, or phases (Figure 6.11). An external perturbation, such as a change in temperature or magnetic field, even very small, can induce the LC to assume a different phase. Different phases can be distinguished by their different optical properties

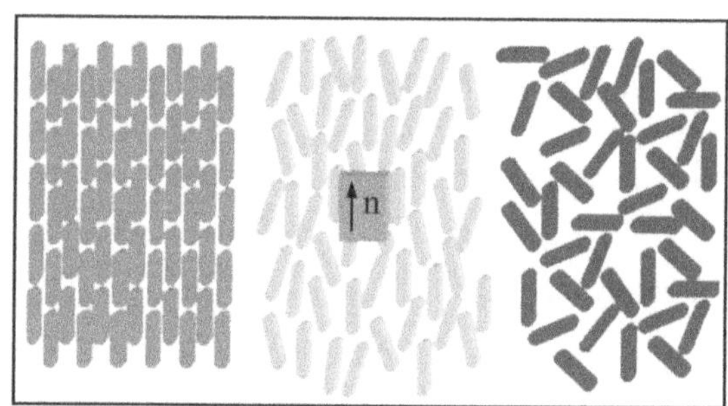

Figure 6.11 Schematic representation of molecules in a solid (left, molecules are well organised), in a liquid crystal (centre molecules have a long range distance order and in a liquid (right) molecules are not ordered. (Image credit: Copyright IPSE Educational Resources, University of Wisconsin Madison)

Liquid Crystals are divided into following groups

(i) ***Thermotropic LC:*** They consist of organic molecules, typically having coupled double bonds, and exhibit a phase transition as temperature is changed (**Figure 6.12 (a)**);

(ii) ***Lyotropic LC:*** They consist of organic molecules, typically amphiphilic (water-loving) and exhibit a phase transition as a function of both temperature and concentration of the LC molecules in a solvent (typically water) (**Figure 6.12 (b)**);

(iii) ***Metallotropic LC:*** They are composed of both organic and inorganic molecules, and their LC transition depends not only on temperature and concentration, but also on the organic-inorganic composition ratio.

(iv) ***Lyotropic liquid*:** crystalline phases are abundant in living systems, such as biological membranes, cell membranes, many proteins (like the protein solution extruded by a spider to generate silk), as well as tobacco mosaic virus. Soap is another well known material which is in fact a lyotropic LC. The thickness of the soap bubbles determines the colours of light they reflect.

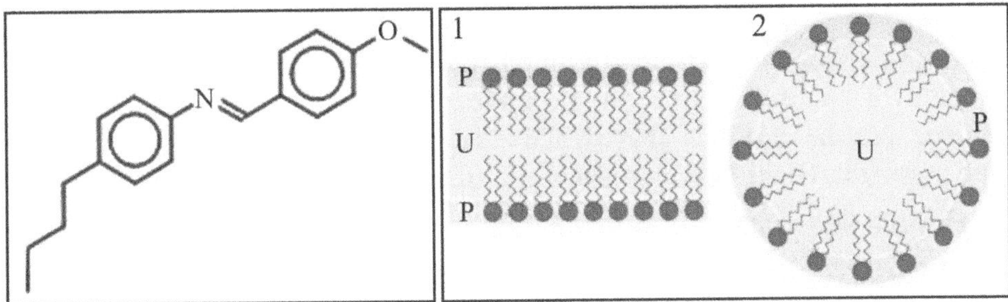

Figure 6.12 (a) Chemical structure of N- (4-Methoxybenzylidene)-4-butylaniline (MBBA); (b) Structure of lyotropic liquid crystal: 1 is a bilayer and 2 is a micelle. The red heads of surfactant molecules are in contact with water, whereas the tails are immersed in oil (blue). (Image credit: Wiki commons, Creative Commons Attribution Share Alike 3.0)

6.1.8.2 Liquid Crystals Applications

The order of liquid crystals can be manipulated with mechanical, magnetic or electric forces. What is interesting is that this change of order can be obtained with very small variations of these forces. The properties of liquid crystals are useful in many applications. The colour of some liquid crystals depends on the orientation of its molecules, so any perturbation that disturbs this orientation (e.g., a difference in magnetic or electric field; a difference in temperature; or the presence of certain chemicals) can be detected with a colour change.

One important feature of liquid crystals is the fact that an electrical current affects them. A particular sort of nematic liquid crystal, called twisted nematics (TN), is naturally twisted. Applying an electric current to these liquid crystals will untwist them to varying degrees, depending on the current's voltage. LCDs use these liquid crystals because they react predictably to electric current in such a way as to control the passage of light (99).

The working of a simple LCD is shown in Figure 6.13. It has a mirror (**A**) in back, which makes it reflective. There is a piece of glass (**B**) with a polarizing film on the bottom side, and a common electrode plane (**C**) made of indiumtin oxide on top. A common electrode plane covers the entire area of the LCD. Above that is the layer of liquid crystal substance (**D**). Next comes another piece of glass (**E**) with an electrode in the shape of the rectangle on the bottom and, on top, another polarizing film (**F**), at a right angle to the first one.

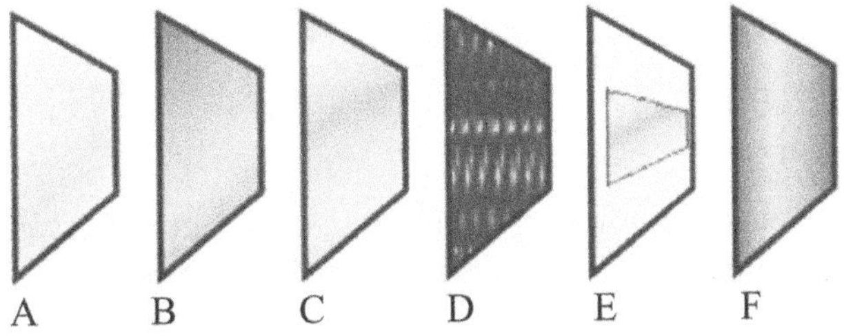

Figure 6.13 Schematic of working principle of LCD

The electrode is hooked up to a power source like a battery. When there is no current, light entering through the front of the LCD will simply hit the mirror and bounce right back out. But when the battery supplies current to the electrodes, the liquid crystals between the common-plane electrode and the electrode shaped like a rectangle untwist and block the light in that region from passing through. That makes the LCD show the rectangle as a black area.

6.1.8.2.1 Flexible Substrates

The primary flexible substrate candidates are plastics and thin glass. Plastic substrates are inexpensive, roll-to roll processable and can be laminated to multi-layers, but they also impose limitations with respect to thermal processing and barrier performance. Companies are developing coatings for these substrates as well as new plastic substrates to compensate for these constraints. Thin glass substrates exhibit better thermal stability and have higher visual transparency than plastics, but cannot fully bend and are not compatible with roll to roll processing.

The use of thin metal substrates is a complementary approach to the glass and plastic displays. Flexible metallic substrates provide excellent barrier properties, thermal and dimensional stability over a broad temperature range. In addition, they offer potential integration with backplane technology for active-matrix displays.

6.1.8.3 Flexible Display Technology

Display technology has progressed enormously in the last decade. Until a few years ago, there were bulky televisions with cathode ray tube (CTR) technology, and mobile phones with black and white displays that could only show text. Nowadays, LCD televisions are becoming the norm and mobile phones that can show photos and movies are the norm, many even with touch displays. This progress has been enabled by intense research in the field, driven by a billion-dollar industry and the constant need for more functional devices combining image and video quality, low power consumption and low cost. Thickness and flexibility are also becoming an important requirement. Here, some of the latest advances in display technology and the impact that nanotechnologies have had in their development are reviewed. **Organic light-emitting diodes** (OLEDs), **quantum dot light-emitting diodes** (QD-LEDs) and **electronic paper** (e-paper) are described in detail. Another area of development is **field emission display** (FED), which, in a sense, is the progression of the old cathode ray technology to the nanoscale. Field emission display uses an element, such as carbon nanotubes, as a source of electrons that strike a coloured phosphor. This technology is not commercial yet and remains at prototype level.

6.1.8.4 Encapsulation

Since flexible displays utilize polymer materials, a barrier layer is essential in protecting and enclosing the functional materials and layers from oxygen and water degradation. Since organic materials tend to oxidize and hydrolyze, oxygen and water permeation through a flexible substrate is of particular importance flexible electronics. Although single-layer barrier layers do provide the packaged materials with some protection, it appears that multiple layers are necessary for organic light emitting diode applications for long-term stability.

6.1.8.5 Organic and Inorganic Conducting Layers

Indium tin oxide is the typical conducting layer used in display technology because of its excellent sheet resistance and optical clarity. However, the process temperature required for ITO on glass is incompatible with plastic substrates. Therefore lower temperature processes have to be developed for ITO in order for it to be considered for flexible display applications. When ITO is deposited on a polymeric substrate, it can crack under tensile strain and cause catastrophic failure.

Conducting polymers are also being considered for flexible display applications. Although their sheet resistance and optical properties are not as attractive as ITO, they do have exceptional mechanical properties and low process temperatures. As ITO and conducting polymer technology compete for the conducting substrate solution, there is a new conducting substrate technology based on nanotechnology. Flexible and transparent electrodes have been formed from carbon nanotube dispersions in the combination with wet coating processes and printing techniques.

6.1.8.6 Electro-optic Materials

The various types of electro-optic materials for flexible display fall into three categories – emissive, reflective, and transmissive. For emissive applications, small molecules and polymers are being used for Organic Light Emitting Diodes (OLED) applications. In order to have a truly low power display, a reflection mode of operation will have to be implemented on flexible substrates. Polymer-dispersed liquid crystals, encapsulated electrophoretics, gyricon, and bichromic ball composites all operate in the reflective mode. For electronic book and paper applications, an efficient reflective mode display is crucial to eliminate the power consuming backlight.

6.1.8.7 Flexible LCD and OLED Displays

Current displays rely mainly on two approaches: liquid crystal displays (LCD) and plasma display panels (PDP). These two technologies have basically replaced the old cathode tube technology which is being phased out. Plasma display panels require much more energy than liquid crystal displays to operate and are currently not as successful as LCDs, especially at a time when energy consumption is a major concern. LCD technology, on the other hand, is becoming extremely common due to a considerable price drop in the last few years.

Liquid crystal displays offer a number of advantages compared to the old cathode tube technology: first of all, displays are much thinner and less bulky due to their inherent structure and assembly, and they provide a much better image quality. However, **LCD**s have two main problems connected to inherent composition: first, LCD displays can be **difficult to see at oblique angles**, as laptop computer users are well aware; second, they require backlight, which consumes power. The need for backlight explains why, in an LCD, black looks more like deep grey than true black.

Flat panel displays generally consist of four layers: a back substrate providing mechanical strength; a plane of switches on the substrate that addresses each pixel; a light-controlling layer; and a front panel that holds the top electrodes, encapsulates the light-control layer and offers support (100). In liquid crystal displays, the substrate is usually glass coated with amorphous silicon or organic conductor, in which the pixel-switches (TFTs) are patterned. Trapped between these electronics and the front glass is the liquid crystal material, which acts as a light-controlling layer. The spacing between the two of glass pieces must be carefully controlled to make light-control layer work. Making this sandwich of materials flexible requires finding a set of technologies that can combine to create a matrix of individually addressable pixels that will flex. Since the rigidity of a device increases with the cube of its thickness, reducing the thickness of the glass substrate is an obvious step to take. One method to accomplish this is to etching the glass after the display is complete. Some researchers also tried to replacing the substrate with a flexible plastic, but producing reliable amorphous silicon electronics on a flexible substrate is very difficult using conventional lithographic patterning techniques. In addition, as the display is flexed to different radiuses, maintaining a fixed electrode gap is extremely demanding.

One flexible LCD very close to commercialization is the cholesteric LCD from Kent Displays Inc. This display utilizes a liquid crystal material originally derived from animal cholesterol, hence the name cholesteric. This LCD will be a full-color screen, 160 mm across the diagonal, which is slightly larger than Pocket PC screens. In addition, each color pixel in the display consists of a red, a blue, and a green cell stacked on top of each other, instead of side by side as in today's full-color laptop LCDs. As a result, the cholesteric LCD's resolution is far superior than that of current laptop displays.

6.1.8.7.1 OLED Display

Organic light emitting diodes (OLED) display is another promising technology for flexible flat panel displays offering a number of advantages over LCDs (101). Flexible OLEDs are very lightweight and durable. Their use in devices such as cell phones and PDAs can reduce breakage. Potentially, OLEDs can be embedded in fabrics to create "smart" clothing.

6.1.8.7.2 How OLEDs Work

In this technology, the display is formed of an emissive and conductive layer sandwiched between an anode and a cathode. A voltage is applied across the OLED so that the anode is positive with respect to the cathode. This creates a flow of electrons from the anode to the cathode. Therefore, electrons are removed from the conductive layer and added to the emissive layer. The conductive layer becomes positively charged and the emissive layer negatively charged. This causes electron holes to appear at the boundary between the two layers. Electrostatic forces induce electrons and holes to get closer and they recombine, emitting energy as a photon. Emission is in the visible region and this is what generates the colours in the display. Molecules used in OLEDs as the name suggests are organic molecules such as organo-metallic chelates and conjugated dendrimers (two types of macromolecules that can reach the nanoscale). The molecules can be directly evaporated or printed on the polymer substrate, which is a notable advantage, and can lead to very complex multilayer structures. The molecules are deposited in rows and columns onto the flat carrier (simply by printing) and the resulting matrix of pixels can emit different colours. Therefore, the colour generation in an OLED is fundamentally different from that in an LCD, which is powered by a fluorescent lamp (backlight) colour-filtered to produce red, green and blue pixels. Thus, when an LCD screen displays full white colour, two thirds of the light is absorbed by the filters. This explains some fundamental advantages of OLEDs over LCD.

Thus all OLEDs will have four basic components: substrate, anode, organic layers, and cathode. Flexible substrate materials are usually plastic, thin glass or metal foils. The anode is a transparent layer of metal of low work function which serves to remove electrons when a current flows through the device. The cathode is a metal layer of high work function which injects electrons when a current flows through the device. In between the cathode and the anode are the organic layer(s) where transport and recombination of the electrons and holes occur. Depending on the device, the OLED could have one, two or multiple organic layers. (103,104). Figure 6.14 shows the structure of a bilayer device. Finally a top cover glass is used to encapsulate the device.

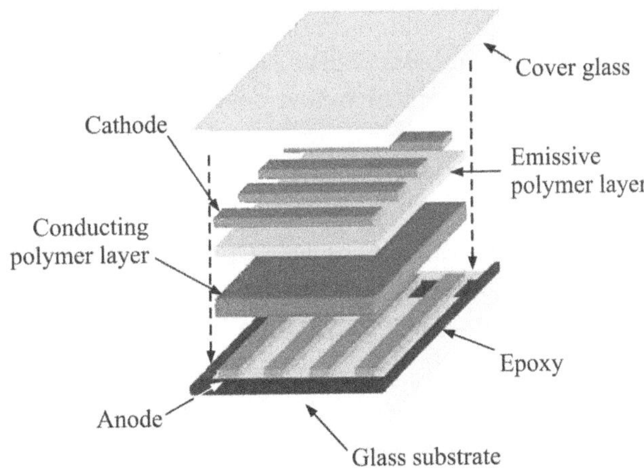

Figure 6.14 Bilayer OLED consists of two organic layers: emissive and conducting (102)

OLED pixels directly emit light, they do not require backlight, thus consuming less power than LCDs. When in the OFF mode, the OLED elements produce no light and consume no power.

- OLEDs have a much better picture quality because they naturally achieve a higher contrast ratio. For a viewer, the most noticeable effect is that the black areas in an OLED appear true black.
- OLEDs can be much thinner and lighter than LCD panels. In addition, OLED can be printed onto any suitable substrate using an inkjet printer so they can, theoretically, be incorporated into flexible substrates, which opens up numerous new display possibilities (roll-up displays or displays included in fabrics).

Unlike LCDs, OLED panels can be seen well under sunlight and at different angles.

LED technology is currently used in small displays (OLEDs are used in numerous colour mobile phones) but it is not yet established in television panels due to some challenges presented by the OLED technology. Many of these challenges can be addressed with the use of nanomaterials and nanotechnologies.

6.1.8.7.3 Issues in OLED Production in Mass Scale

There are some issues to be addressed in industrial production of OLED displays. First, the organic molecules used in the OLED have an inherent tendency to degrade in time. Increasing the lifetime of OLEDs is a research priority. This is particularly true for the colour blue (e.g. the lifetime of a blue OLED is 14 000 hours, whereas a typical LCD panel has a lifetime of 60 000 hours). Much research is being dedicated to the synthesis and testing of new organic molecules that can increase the lifetime of OLEDs. The problem of OLED lifetime is particularly important in those electronics that require a long lifetime (e.g. televisions). In devices that are used intermittently (mobile phones etc.), lifetime is less important. This explains why OLED technology is already fairly well diffused in mobile devices.

A second problem is in the nature of the top cathode layer of the OLED, which needs to be a transparent electrode. Currently, ITO (indium tin oxide) is used since it offers a combination of conductivity and transparency. The problem is the availability of ITO, which is declining, and in the way it is processed, which is by chemical vapour deposition (CVD). This deposition method requires very high temperatures which limit the type of substrate that can be used. ITO is slightly opaque and this affects image quality.

A very active field of research is the search for suitable nanomaterials to replace ITO in OLED cathodes. One possibility is carbon nanotubes. Currently, there are numerous methods to fabricate CNTs (alternatives to CVD) such as spray coating and roll-to-roll printing. Furthermore, numerous companies are emerging that produce CNTs and the price is dropping.

Another issue in OLEDs is that the organic molecules contained within them are very sensitive to water and other substances. Therefore, the coating and packing of the OLED components is of fundamental importance. One of the problems is that current packaging materials are either too brittle (which is a problem in flexible screens) or require too high a temperature during the production process, which would destroy the device. Currently, alternating layers of organic and inorganic materials are used. This gives some degree of flexibility and blocks the entry of foreign substances (although not totally). Nanomaterials could help solve this challenge: for example, a method was recently developed whereby nanoparticles are used to fill the pores in these organic and inorganic layers and this markedly improved the packaging of these devices.

6.1.8.7.4 How Active-matrix OLEDs Work

In industrial production to make displays out of OLEDs, a technology called active-matrix display is utilized (Figure 6.15). Active matrix OLED (AMOLED) display consists of OLED pixels that have been deposited or integrated onto a TFT array to form a matrix of pixels that illuminate light upon electrical activation. The active-matrix TFT backplane acts as an array of switches that control the amount of current flowing through each OLED pixel. The TFT array continuously controls the current that flows to the pixels, signaling to each pixel how brightly to shine. Typically, this continuous current flow is controlled by at least two TFTs at each pixel, one to start and stop the charging of a storage capacitor and the second to provide a voltage source at the level needed to create a constant current to the pixel.

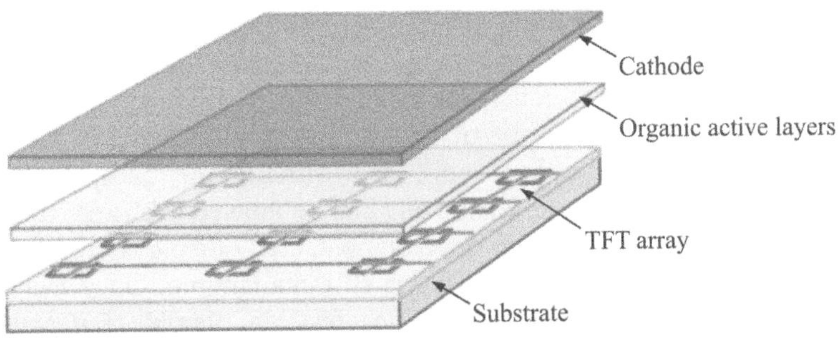

Figure 6.15 Active matrix OLED configuration (102)

6.1.8.7.5 Advantages and Disadvantages

The radically different manufacturing process of OLEDs renders it far superior to traditional flat panel displays. Roll to roll processing of OLEDs will significantly lower mass production cost. Since no backlight is needed in OLED displays, flexibility can be achieved much easier than in LCDs. In addition, OLEDs offer bright colorful images with a wide viewing angle and low power.

The greatest challenge in flexible OLED display is encapsulation and packaging. To protect the OLED from water and oxygen degradation, OLEDs are conventionally sealed with a glass lid using an ultraviolet-cured epoxy resin (Figure 6.16). Flexible OLED packaging, however, is much more challenging. The standard sandwich construction that works well for glass-based displays is insufficient flexible OLED displays where the ability to conform or flex the display is key. A variety of methods to provide flexible encapsulation is currently being developed.

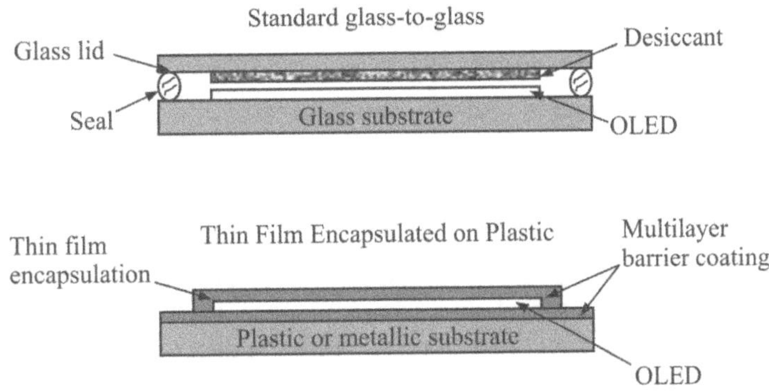

Figure 6.16 Standard OLED encapsulation vs. encapsulation for flexible OLEDs

Even when the technological challenges are met, there still is a piece missing from the flexible displays puzzle, the production equipment. Currently there is no infrastructure to produce plastic displays in any volume. It will be a few years before there are sufficient roll-to-roll lines to produce displays that will significantly increase the market share. In addition, OLED's commercialization is restrained by key patents held by Kodak and other firms. It is expected that OLED display technology become widespread once the patents had expired.

As the components and manufacturing processes of flexible electronics mature, the concept of flexible flat panel display will eventually become a reality. Flexible displays offer tremendous advantages over conventional flat panel displays, like light weight, durability, low power consumption, portability etc. In particular, OLED displays offer bright sharp images at wide viewing angles and bright light, but are difficult to encapsulate. Gyricon and electrophoretic displays have thin form factors and can be viewed at a great range of angels, but their high resolution displays still require development. LCD displays are already mature, but making the sandwich-structured device flexible is still challenging. Once technical difficulties are overcome and roll to roll processing becomes feasible, flexible flat panel displays will widely commercialize and enter all our lives.

6.1.8.7.6 Other Application of OLEDs

Organic light-emitting diode (OLED) technology can be used for two main applications: displays and light sources. The possibility of using OLEDs instead of traditional light bulbs is very attractive in terms of energy consumption (conventional incandescent lights lose most of their energy as heat). OLED technology is thus being considered for use in lights and signs. The fact that OLEDs can be fabricated on flexible substrates opens the door to OLEDs in textiles, labels and other flexible materials.

Electronic paper (103) (e-paper): e-paper is a display technology that aims to mimick the appearance of ordinary ink on paper. There are various technologies to create e-paper, among which electrophoretic paper is the most established. Unlike conventional flat panel displays, electrophoretic displays rely on reflected ambient light, rather than the emission of light from the display, and can retain text or images without constant refreshing, thereby requiring much less power. This also means that they can be used under sunlight without the image fading. e-paper is also very light and flexible. E-paper has numerous commercial applications, such as e-books (i.e., e-readers that can display the digital version of a book), e-papers, electronic labels, general signage, timetables at bus stations, etc.

6.2 Nano – Composites

6.2.1 Nano Composite Materials

Nanocomposite materials are solid ones with multiple phases, which is a combination of two or more materials with different physical and chemical properties. In composite materials, one phase is usually continuous and called the matrix, while the other phase is a reinforcing material called the dispersed phase. In self-assembled systems, the ordered structure is the dispersed phase, while the template film can be the continuous phase.

Types of Nanocomposite Materials

1. ***Organic-inorganic Nanocomposites*:** In organic-inorganic composites, nanocomposites refer to composites in which the inorganic phase has nanoscale morphology such as particles, fibers and tubes. There is an additional type of organic-inorganic nanocomposite, where either the organic phase or the inorganic phase has nanoscale morphology in an organic-inorganic composite. For example, in emulsion systems of W/O or O/W, clay nanoparticles and silica are often used to stabilize the emulsions,

which then form an organic-inorganic nanocomposite. In designing a system of thiol / FeOOH, heating the mixture would lead to the formation of thiol/FeO or thiol/FeO 2334 nanocomposites.

2. ***Polymer-Inorganic Nanocomposites***: Generally, it refers to a nanocomposite in which organic polymer forms a composite with inorganic nanoparticles, e.g., BiI3–nylon nanocomposite and montmorillonite–nylon (polyester) nanocomposites.

The Classification of Nanocomposites:

1. Classification I. This classification is based on the nanomaterials dimensional (D) morphology.
   (a) 0-D nanopowder or nanoparticle;
   (b) 1-D nanowire, nanowisp, nanosilk, nanotube or nanocrystalline whisker;
   (c) 2-D nanomaterial with morphology of layer, lamellar or belt structure;
   (d) 3-D nanomaterial with morphology of cylinder, solid or block structure.
2. Classification II. This classification is based on polymer morphology, shape or state. In the nanocomposites, in which the inorganic particle is a 0-D nanoparticles.
   (a) 1-D nanocomposite when the polymer in the composite is in solution;
   (b) 2-D nanocomposite when the polymer in the composite is in film or membrane shape;
   (c) 3-D nanocomposite when the polymer in the composite is in a powder state or morphology;
   (d) 0-D nanocomposite formed with two different nanopowders.
3. Classification III. This classification is based on a nanophase of tube- or non-tube like material in the nanocomposites.
   (a) Tube-like, e.g., single-wall carbon nanotubes, TiO Tube, or AlO tube. 223
   (b) Non-tube-like, e.g., layer of clay, lamellar, silk, belt or band-like shape.

Sometimes, the classifications are mixed to describe nanocomposites, e.g., carbon nanotubes or nanocrystalline whiskers with polymer powder construct a 1 and 3 type nanocomposite.

Nanolayers of clay with polymer powder form 2 and 3 type of nanocomposites, etc.

6.2.2 Clay Composites

The clay mineral (also called a phyllosilicate) is usually of a layered type and a fraction of hydrous, magnesium, or aluminum silicates (104-105). Hectorite, saponite, and montmorillonite are the most commonly used smectite type layered silicates for the preparation of nanocomposites. Montmorillonite (MMT) has the widest acceptability for use in polymers because of their high surface area, and surface reactivity (106-107). It is a hydrous aluminosilicate clay mineral with a 2:1 expanding layered crystal structure, with aluminum octahedron sandwiched between two layers of silicon tetrahedron. Each layered sheet is approximately 1 nm thick (10 Å), the lateral dimensions of these layers may vary from 30 nm to several microns or larger, depending on the particular layered silicate. The aspect ratio is about 10–1000 and the surface area is in the range of 750 m^2/g.

Graphite has a similar geometry (layered structure) with nanoclay, therefore a clay-polymer reinforcement concept is applicable (108). Graphite flakes have been known as host materials for intercalated compounds. By applying rapid heating, some of the graphite-intercalated compounds (GICs) expand and a significant increase in volume takes place. Many literature citations identify the expanded graphite flakes with polymer systems for lightweight and conductive nano composites.

6.2.3 Polymer Composites

Polymer matrix based nano-composites have generated a significant amount of attention in the recent years. This area emerged with the recognition that ex-foliated clays could yield significant mechanical property advantages as a modification of polymeric systems (109-111). The achieved results were at least initially viewed as unexpected ("nano-effect") offering improved properties over that expected from continuum mechanics predictions. More recent results have, however, indicated that while the property profile is interesting, the clay-based nano-composites often obey continuum mechanics predictions. There are situations where nano-composites can exhibit properties not expected with larger scale particulate reinforcements.

With inorganic particle and nanoparticle inclusions, nucleation of crystallization can occur. At the nano -dimension scale, the nanoparticle can substitute for the absence of primary nuclei thus competing with the confined crystallization. At higher nanoparticle content, the increased viscosity (decreased chain diffusion rate) can lead to decreased crystallization kinetics. Thus, the crystallization process is complex and influenced by several competing factors. Nucleation of crystallization (at low levels of addition) evidenced by the onset temperature of crystallization (T_c) and crystallization half-time has been observed in various nanocomposites (poly-(3-caprolactone)–nanoclay polyamide 66–nanoclay (110-111), polylactide–nanoclay (112), polyamide 6–nanoclay (113) polyamide 66–multi-walled carbon nanotube (116), polyester–nanoclay (114), poly(butylene terephthalate)–nanoclay (115), polypropylene–nano-clay (sepiolite) (116), polypropylene/multi-walled carbon nanotube (117).

In polymer-layered silicate (PLS) nanocomposites, stacking of the layers leads to a regular van der waals gap between the layers called the interlayer or gallery. Isomeric substitution (for example tetrahedral Si^{4+} by Al^{3+} or octahedral Al^{3+} by Mg^{2+} or Fe^{2+}) within the layers generates negative charges that are counterbalanced by alkali and alkaline earth cations (typically Na^+ or Ca^{2+}) situated inside the galleries (118). This type of layered silicate is characterised by a moderate surface charge known as the cation exchange capacity (CEC). To render layered silicates miscible with other polymer matrices, it is required to convert the normally hydrophilic silicate surface to an organophilic one, which can be carried out by ion-exchange reactions with cationic surfactants (119).

In 1990 a Toyota research group produced and characterized the first polymer/layered silicate nanocomposite (PLSN) material using Nylon-6. The unprecedented improvements that were achieved in a number of material properties showed tremendous promise for layered silicate material as a filler. In the years since, a flood of research has been carried out on nanocomposites using different polymer matrices, different layered silicates and multiple forms of compatibilizer. The layered silicates employed in this technology are most commonly phyllosilicates. Phyllosilicates make up a large family of naturally occurring clays whose basic layer structure is comprised of individual sheets of tetrahedral, oxygen rich layers and octagonal layers with silicon atoms at their core. In case of the most commonly used layered silicate in this family (montmorillonite), Mg^{2+} and Al^{3+} cations occupy the spaces at the centre of the tetrahedral sheets. Each silicate layer has a thickness of roughly a single nanometer and individual layers in their natural state are spaced approximately one nanometer apart. Although there are a wide variety of silicate types, all layered silicates share the physical trait of high aspect ratio with length dimensions varying from 30 nm to several microns. The high aspect ratio and nanometer scale dimensions give layered silicates a large surface area for interaction between silicate surface and polymer matrix. This large surface area is the primary difference between nanosilicate and traditional filler materials. Nanosilicate combined with polymer in only small quantities has been shown to dramatically increase the Young's modulus, flexural modulus and impact strength of a polymer matrix. Increases in thermal degradation temperature, decreases in thermal deformation and improvements in barrier properties from neat polymer materials have all been recorded.

Multiple types of polyamides, polyolefin, polyesters and polycarbonates have been combined with layered silicates modified to obtain maximum compatibility with the polymer matrix. It is well known that obtaining successful layered silicate nanocomposites requires excellent compatibility between the polymer matrix and the nanosilicate. A number of silicate surface modifications as well as a variety of compatibilizing compounds have been used for this purpose.

6.2.3.1 Nano Polymer Composites Mechanical Properties – Reinforcement

Polymer composites comprising nanoparticles (including nanofibers where the fiber diameter is in the nanodimension range) are often investigated where reinforcement of the polymer matrix is achieved. While the reinforcement aspects are a major part of the nanocomposite investigations reported in the literature, many other variants and property enhancements are under active study and in some cases commercialization. The advantages of nanoscale particle incorporation can lead to a myriad of application.

A common reason for adding fillers to polymers is to increase the modulus or stiffness via reinforcement mechanisms described by theories for composites (120). Properly dispersed and aligned clay platelets have proven to be very effective for increasing stiffness. This is illustrated in Figure 6.17 by comparing the increase in the tensile modulus, E, of injection molded composites based on nylon 6, relative to the modulus of the neat polyamide matrix, E_m, when the filler is an organoclay versus glass fibers.

In this example, increasing the modulus by a factor of two relative to that of neat nylon 6 requires approximately three times more mass of glass fibers than that of montmorillonite, MMT, platelets. Thus, the nanocomposite has a weight advantage over the conventional glass fiber composite. Furthermore, if the platelets are aligned in the plane of the sample, the same reinforcement should be seen in all directions within the plane, whereas fibers reinforce only along a single axis in the direction of their alignment. In addition, the surface finish of the nanocomposite is much better than that of the glass fiber composite owing to nanometer size of the clay platelets versus the 10-15 m diameter of the glass fibers.

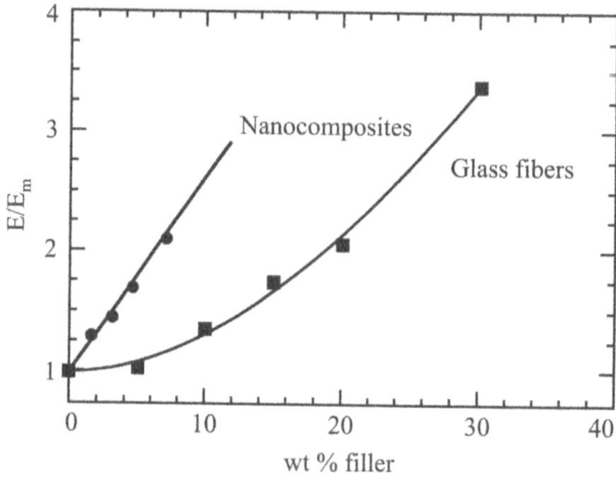

Figure 6.17 Comparison of modulus reinforcement (relative to matrix polymer) increases for nanocomposites based on MMT versus glass fiber (aspect ratio w20) for a nylon 6 matrix (123)

Theoretical predictions have shown that the modulus for well aligned platelets can be three times that for well-aligned fibres, especially as the aspect ratio of clay layers increases (121). Other studies however, suggest that the modulus increase is not entirely due to the load-carrying ability of the platelets, but is caused by the volume of polymer constrained by the platelets (121). This suggests that to optimize the increase in modulus, the degree of dispersion must be optimised to maximise the degree of matrix/filler interaction. Work on polymer platelets nanocomposites in which adding malefic anhydride (MA) to the matrix changed the degree of filler dispersion supports. This provides that despite the plasticising effect of MA, the modulus improves due to enhanced dispersion of the clay (122). Lan and Pinnavaia (123) found that, 'as the degree of exfoliation increased by changing the length of alkylammonium intercalating chain, the modulus and strength improved'.

As the polymer intercalates and swells, the layers and the area of interaction between the polymer and the filler increases and the modulus increases significantly. Figure 6.18 represents three main types of composites for layered silicate materials.

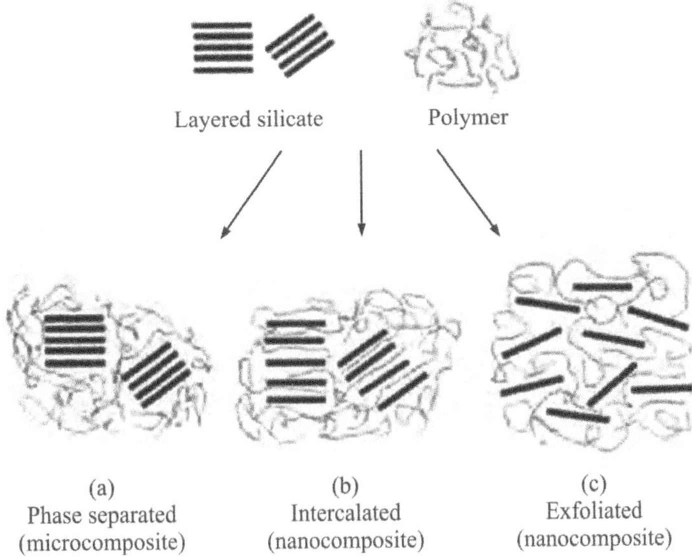

Figure 6.18 Scheme of three main types of layered silicates in polymer matrix

The polymer/clay interaction plays a significant role in controlling mechanical behavior is also evident from the fact that improvement in properties tends to be higher above the glass transition temperature than below it. However, proper dispersion is critical for achieving this (124-125). A significant improvement in flexural modulus (1.5 GPa to 2.1 GPa) mega Pascal's (KN (mm)) and tensile strength 29 (mega pascals (N/mm^2)) MPa to 40 MPa) and impact strength (18 J/m to 24 J/m) of PP/clay nanocomposites is reported on addition of about 15 per cent PP-g-MAH to PP-organoclay system. Xu, et al. (126) have reported similar improvements in tensile strength and impact strength of the composites with a nanoclay addition of 10 per cent to 15 per cent. Nanocomposites are optically transparent and/or optically active.

6.2.3.2 Polymer Nano-composites Biomedical applications

The applicability of polymer nanotechnology and nanocomposites to emerging biomedical/ biotechnological applications is a rapidly emerging area of development. One area of intense research involves electrospinning for producing bio-absorbable nanofiber scaffolds for tissue engineering applications. This might be construed as a nano-composite as the resultant scaffold allows for cell growth yielding a unique composite system. Another area also involving nanofibers is the utilization of electrically conducting nanofibers based on conjugated polymers for regeneration of nerve growth in a biological living system.

Polymer nanocomposites based on hydroxyapatite ($Ca_{10}(PO_4)_6(OH)_2$) have been investigated for bone repair and implantation (127-128). Hydroxyapatite, a major constituent of hard tissue, exhibits undesirable mechanical properties if directly employed thus polymer-based matrix composites are desired. Biodegradation of the matrix is also desired to allow infiltration of new bone growth at the repair site. Often natural polymers (poly-saccharides, polypeptides, collagen, chitosan) or synthetic bio-degradable polymers are employed as the matrix in these studies.

Collagen derived gelatin (129) and poly-2-hydroxyethyl-methacrylate/poly (3-caprolactone) (130) nanocomposites based on hydroxyapatite are examples of systems studied for bone repair systems.

Electrospinning of biodegradable polymer solutions is a popular method to produce nanofiber scaffolds for tissue engineering applications. Poly (L-lactic acid)/exfoliated montmorillonite clay/salt solutions were electrospun followed by salt leaching/gas foaming (130). The resultant scaffold structure contained both nano-and micro-sized pores offering a combination of cell growth and blood vessel invasion micro-dimensions along with nanodimensions for nutrient and metabolic waste transport.

Polymer matrix nanocomposites have been proposed for drug delivery/release applications. The addition of nanoparticles can provide an impediment to drug release allowing slower and more controlled release, and reduced swelling (131) and improved mechanical integrity (132) of hydrogel-based nanocomposites.

6.2.3.3 Commercial Applications of Polymer-based Nano-composites

An area where nanocomposites could achieve a dramatic commercial prominence is in advanced composites. Carbon fiber reinforced composites have a limit on the achievable properties (particularly in cross-ply composites) due to the low modulus and strength of the matrix phase. As discussed earlier Modification of the matrix phase with carbon nanotubes at the lower scale of dimensions and carbon nanofibers at a higher dimensional scale would allow for significant improvement by Young's module and strength.

6.2.4 Carbon Nanotube-Reinforced Composites

In micrometer-size carbon tubes, the hexagonal symmetry of the carbon atoms in planar graphene sheets will be distorted, because the lattice is curved and must match along the edges (with dangling bonds) to make perfect cylinders. This leads to a helical arrangement of carbon atoms in the nanotube shells. Depending on the helicity and dimensions of the tubes, the electronic structure changes considerably (133, 134). Hence, although graphite is a semi-metal, carbon nanotubes (CNTs) can be either metallic or semiconducting. Nanotubes are closed by fullerene-like end caps that contain topological defects (pentagons in a hexagonal lattice). The electronic character of the ends of these tubes differs from the cylindrical parts of the tubes and is more metallic due to the presence of defects in these regions (135). The discovery of nanotubes has complemented the excitement and activities associated with fullerenes (136). Although fullerenes have fascinating physical properties, their relevance in the nanocomposite field is limited.

The properties of CNTs are unique compared to other graphite fibres. Their structure remains distinctly different from that of traditional carbon fibres, which have been used industrially for several decades (e.g., as reinforcements in tennis rackets, airplane body parts and batteries). Nanotubes represent the ideal, most perfect, ordered carbon fiber, the structure of which is entirely known at the atomic level. Table 6.1 shows measured and theoretical properties of both SWNT and MWNT.

Table 6.1 Theoretical and experimentally measured properties of CNT.

Property	Nanotubes	Graphite
Lattice structure	(Cylindrical) hexagonal lattice helicity Nanotubes: ropes, tubes arranged in triangular lattice with lattice parameters of a = 1.7 nm, tube – tube distance = 0.315.	Planar hexagonal, plane-to-plane distance c = 0.335 nm
Specific gravity	0.8 – 1.8 g cc^{-1} (theorectical)	2.26 g cc^{-1}
Elastic modulus	~1 TPa for SWNT ~0.3 – 1 TPa for MWNT	1 Tpa (in-plane)
Strength	50 – 500 GPa for SWNT, 10–60 GPa for MWNT	
Resistivity	~50 – 50 micro-ohm-cm	50 (in-plane)
Thermal conductivity	3000 W m^{-1} K^{-1} (theorectical)	3000 W m^{-1} K^{-1} (in-plane) 6 W m^{-1} K^{-1} (c axis)
Thermal expansion	Negligible (theoretical)	-1×10^{-6} K^{-1} (in plane) 29×10^{-6} K^{-1} (c axis)
Oxidation in air	700 °C	450-650 °C

Monotonic increase of indentation (Vickers hardness) by up to 3.5 times on loading up to 2 per cent SWNTs and a doubling of thermal conductivity with 1 percent SWNTs was reported (140). Also, 1 percent MWNT loading in polystyrene increases the modulus and breaking stress by up to 42 and 25 per cent, respectively (139). Similarly, the strength of 1 per cent PS/MWNTs is found to increase by 25 per cent (13-16 MPa) and strain at yield by 10 per cent (139). One of the major benefits expected from incorporating CNTs in polymers is an increase in both the electrical and thermal properties of polymers. The electrical characteristic of interest in polymers is percolation threshold. For CNTs/ epoxy system a very low percolation threshold of below 1 per cent is reported. Likewise, the electrical conductivity of CNTs/PMMA composites is reported to increase by about nine times with the addition of 5-8 per cent CNTs. While the polystyrene is insulating, the films doped with MWNTs are conducting (conductivity 7.1×10^{-2} $Ohm^{-1}cm^{-1}$). Electrical conductivity (140) of 0.5 percent MWNT/PFA is 1.3×10^{-2} $Ohm^{-1}cm^{-1}$.

6.2.5 Nanofiber-reinforced Composites

Carbon nanofibers (CNF) are a unique form of vapour-grown carbon fibres that fill the gap in physical properties between conventional carbon fibres (5–10 µm) and carbon nanotubes (1–10 nm). The reduced diameter of nanofiber provides a larger surface area with surface functionalities in the fiber (141). Typically CNF are not concentric cylinders; the length of the fibre can be varied from about 100 µm to several cm, and the diameter is of the order of 100-200 nm with an average aspect ratio greater than 100. The most common structure of CNF is the truncated cones, but there are wide ranges of morphologies (cone, stacked coins, etc). CNF have the morphology where these are hollow at the centre (much like a MWNT) and have a larger diameter than MWNT but the individual layers are not arranged in concentric tubes.

Applications: Tandon and Ran (142) enhanced the thermos mechanical properties of conventional aerospace carbon fibre-reinforced (IM7) composites using carbon nanofiber. They manufactured IM7/CNF matrix unidirectional laminate aerospace structures using the filament-winding technique. Glasgow and Tibbetts (143) oxidised the surface of carbon nanofiber to improve the tensile behaviour in PP composites. Lafdi and coworkers (144) showed improved flexural strength and modulus in epoxy-based composites with oxidised nanofibres. Finegan and Tibbetts (145) incorporated CNF in a PP with improved strength and stiffness. Thermal transport across bonded radiator panels is important where thermally-conductive adhesives play an important role. Electrically-conductive bonded joints are needed in spacecraft to eliminate the build up of static charge on the structure due to the impingement of charged particles. Gibson, et al (146). modified the epoxy-based adhesives formulated with silver-coated and uncoated vapour-grown carbon nanofibers for several aerospace applications such as electrical conductivity, thermal transport, and mechanical properties. But they concluded that it does not help to remove the waste heat (146). Lao, et al (147) used CNF, clay platelets, and silica nanoparticles to find the relationship between the flammability and mechanical properties of nylon-11. They achieved a combination of enhanced mechanical and flammability properties in clay platelets and CNF. In their analysis, clay-based nanocomposites showed better flammability, while the CNF-based nanocomposites showed better mechanical properties.

6.2.6 Metal Matrix Nano-Composites

The inadequacy of metals and alloys in providing both strength and stiffness to a structure has led to the development of metal matrix composites (MMCs), whereupon the strength and ductility is provided by the metal matrix and the strength and/or stiffness is provided by the reinforcement that is either a ceramic or high stiffness metal based particulate or fibre.

Metal matrix nano composites including (Al, Fe, Au, Ag, etc.), Metal oxides (ZnO, Al_2O_3, $CaCO_3$, TiO_2, etc) can be designed to possess qualities such as low coefficient of thermal expansion and high thermal conductivity which make them suitable for use in electronic packaging applications. Metal matrix composites today are extensively used in automobile and aerospace applications. (148-153).

Carbon nanotube-metal matrix composites (CNT-MMC) are an emerging class of new materials that are being developed to take advantage of the high tensile strength and electrical conductivity of carbon nanotube materials. Critical to the realization of CNT-MMC possessing optimal properties in these areas are the development of synthetic techniques that are

(a) economically producible,

(b) provide for a homogeneous dispersion of nanotubes in the metallic matrix, and

(c) lead to strong interfacial adhesion between the metallic matrix and the carbon nanotubes. Since the last decade, a number of investigations have been carried out using CNT as reinforcement in different materials, namely polymer, ceramic and metals.

6.2.6.1 Processing Techniques

Carbon nanotube reinforced metal matrix (MM-CNT) composites are prepared through a variety of processing techniques. Figure 6.19 shows the various processes that have been adopted for synthesis of CNT-reinforced MMCs. Powder metallurgy is the most popular and widely applied technique for preparing MM-CNT composites. Electrodeposition and electroless deposition are the second most important techniques for deposition of thin coatings of MM-CNT composites as well as for deposition of metals on to CNTs. For low-melting-point metals such as Mg and bulk metallic glasses, melting and solidification is a viable route. Apart from these techniques, scattered efforts have been made on indigenous methods for preparing MM-CNT composites.

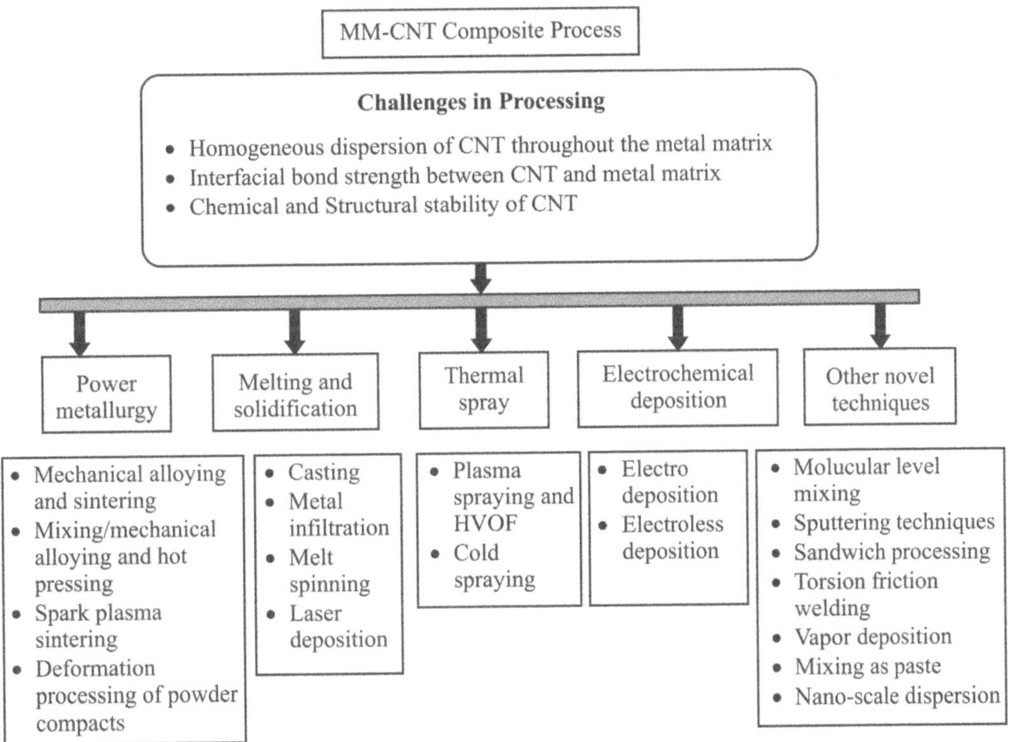

Figure 6.19 Metal matrix–carbon nanotube nano composites production processes

The selection of nanoparticles depends on the desired thermal, mechanical, and electrical properties of the nanocomposites. For example, Al nanoparticles are often selected due to their high conductivity; calcium carbonate ($CaCO_3$) particles are chosen because of their low cost and silicon carbide (SiC) nanoparticles are used because of their high hardness, corrosion resistance, and strength (154).

6.3 Defense Applications of Nano Composites

Nano composites are finding a large number of Defense applications some of which are discussed below.

6.3.1 High Performance Fibre/Fabrics

By infiltrating nanotube mats or woven fibres with a polymer, continuous sheets or films of a nanocomposite can be produced (155-156). The nanotubes will contribute to mechanical properties (strength and stiffness) of the film as well as to electrical conductivity.

Using an electro-spinning technique (157-158) pure polymer and polymer nanocomposite fibres which are highly aligned and with diameters in the range of 200 nm to 300 nm are produced. These exhibit highly improved optical and mechanical properties.

For non-woven fabrics to achieve electrical conductivity, a simple and flexible method has been reported, where the non-woven mat was exposed to a high intensity light source (e.g., from a flash tube), which resulted in immediate joining of fibres at cross-over points of contact. Using a mask, fibres can be joined in any desired pattern (158). Such materials find wide applications in defence as electrically conductive fabrics, sensors, electromagnetic shielding, microwave absorption, electrical energy storage (capacitors), actuators, and materials for micro UAVs.

6.3.2 Ballistic Protection

There are few reports of the promising application of nanocomposites in body armour. Shear thickening fluids (159-160) consist of a fluid containing a dispersion of particles and this fluid stiffens and resists deformation if sheared rapidly by an external force. Reports from the US Army Research Laboratory indicate promising results when combining inorganic nanoparticles (of silica) in polyethylene glycol. When this shear thickening fluid is impregnated into conventional Kevlar, the ability of the material to absorb energy is greatly improved. In one example, the ballistic performance (in terms of absorbed energy) was more than doubled so that four layers of Kevlar impregnated with the shear thickening fluid absorbed as much energy as would have been absorbed by 10 layers without the shear thickening fluid. This will lead to a more flexible armour with reduced weight. Such materials find applications for body/personal armour where flexibility of movement is required besides protection against blunt weapons (stones, sticks and bars) for arms and legs.

6.3.3 Microwave Absorbers

Nanocomposites as microwave absorbers are receiving much attention. Nguyen and Diaz (161) have reported a method to synthesize polypyrrole (PP) nanocomposites containing iron oxides (g and a), tin oxide, tungsten oxide and titanium dioxide. Pyrrole containing a dispersion of nanoparticle metal oxides was polymerised *in situ* and the magnetic properties reported. The electrical conductivity and dielectric losses can be tuned by varying the concentration and orientation of the nanotubes additions. Glatowski, et al (155) have been awarded a patent in this area, covering a wide range of thermoplastic and thermosetting matrices containing oriented nanotubes. Only a few weight percent of nanotubes need be added to the polymer to achieve useful properties. Efforts have been made to utilise CNTs for developing economical microwave (in the range 8 GHz to 24 GHz) absorbers (162,163). These materials have wide applications in electrical energy storage (condensers) integrated into load-carrying structures for UAVs, high strength CNTs-polymer fibres for energy absorption, electromagnetic shielding, etc.

6.3.4 Refractive Index Tuning

In many optical applications such as telecommunications and optical computing, polymer optical fibres are very attractive to adjust the refractive index of the connecting optical fibre (due to ease of mass production and low cost). This can be done by the addition of nanoparticles with various refractive indices to the polymer. Bohm et. al., (164) reported additions of nanoparticle zirconia, alumina, and silica to poly (methyl methacrylate) and they were able to adjust the refractive index over a sufficient range. Levels up to 10 Wt % loading were reported. Tuning the refractive index of surface coatings is important in signature management. Damage resistance (abrasion and scratching) of the fibre is likely to be improved by addition of ceramic nanoparticles.

6.4 Application of Ferro-Fluids in Nano Technology

6.4.1 What are Ferro-fluids?

Ferro-fluids (FF) are permanently stable colloid suspensions of ferromagnetic nano particles (usually magnetite-Fe_3O_4)) in a carrier liquid which can be polar or non polar (165-169). As a result of spontaneous magnetization ferro-fluids, they will have magnetic moment and represent an elementary magnetic dipole. Elementary dipoles influence each other. To prevent their aggregation they need to be covered by a stabilizing dispersing agent (surfactant) which prevent particle agglomeration even under a strong magnetic field gradient. Suspension stability of these FF refers to how long the suspension remains permanently homogenous i.e., how long the ferromagnetic particles do not separate from the carrier liquid and do not settle at the bottom of the container (do not sediment), neither create mutual aggregations. To reach high suspension stability, ferro fluids must contain ferromagnetic particles of the size 5 to 15 nm (1 nm = 10^{-9}m) nanoparticles.

Ferro-fluids were first invented in 1963 by NASA's Steve Papell as a liquid rocket fuel that could be drawn toward a pump inlet in a weightless environment by applying a magnetic field.

6.4.2 Forces involved in the Formation of FF

The formation of a ferro-fluid involves various types of forces that hold the components together (Fig 6.20). For example, magnitite is held together by ionic interactions. Ionic attractions between hydroxide anions and tetramethyl ammonium cations allow colloidal suspension of the magnatite to exist in solution. Without the tetramethyl ammonium hydroxide as a surfactant, the magnetite nano particles tend to cluster together. Therefore it is necessary to have the appropriate surfactant to stabilize an aqueous ferro-fluid.

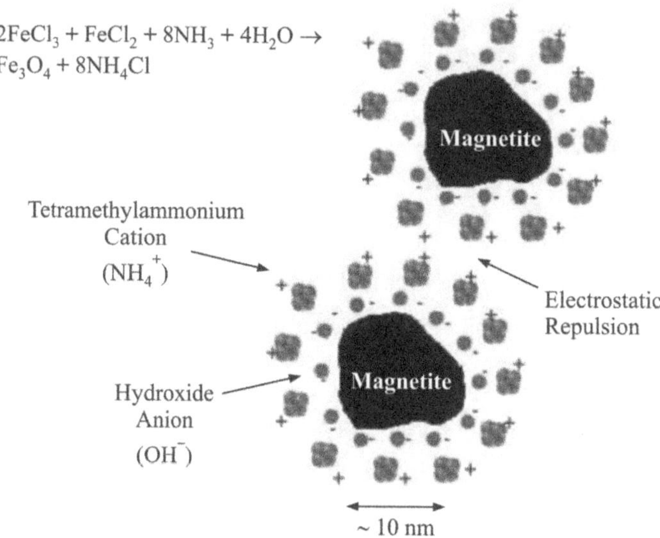

Figure 6.20 Schematic of formation of ferrofluid colloidal suspension in presence of surfactants, (Ref:Berger,P.;Adelman,N.B.Beckman,K.J,Campbell,D.J,Ellis,A.B;Lisensky.G.C. Journal of Chemical Education, 1999, 76, (943-8)

Surfactants are formed by the chains of polar molecules (e.g., fatty acid), long 1 to 2 nm. Every chain is at one end bound with a nano particle and at the other end loosely attracted by the molecules of the carrier medium (Fig. 6.21). Detergent/surfactant is thus a surface active material that prevents direct contact between nanoparticles, causes repulsive forces between them and so prevents their aggregation.

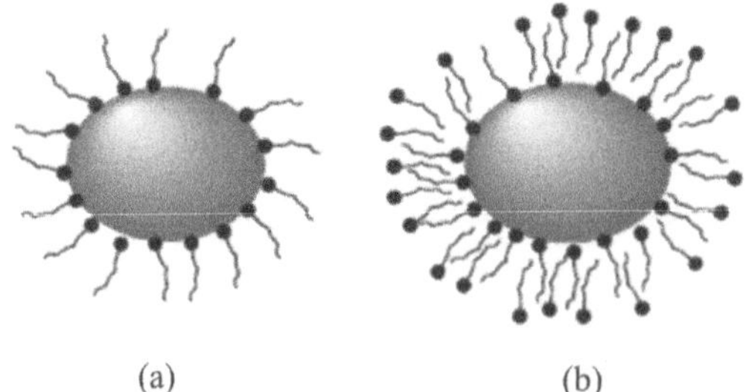

Figure 6.21 Sketch of surfacted ferrofluid grains: (a) single-layered grains; (b) double-layered grains

According to the type of coating, the FF's are classified into two main groups: surfacted (SFF) if the coating is a surfactant molecule, and ionic (IFF) if it is an electric shell.

6.4.3 Surfacted Ferro-fluids (SFF)

Surfacted ferrofluids are formed by magnetic particles (usually magnetite, Fe_3O_4) coated with surfactant agents (amphiphilic molecules, as oleic acid and aerosol sodium di-2 ethylhexylsulfosuccinate) in order to prevent their aggregation. Steric repulsion between particles acts as a physical barrier (170), that keeps grains in the solution and stabilizes the colloid. If the particles are dispersed in a nonpolar medium, as oil, one layer of surfactant is needed to form an external hydrophobic layer. The polar head of the surfactant is attached to the surface of the particles and the carbonic chain is in contact with the fluid carrier. On the other hand, if the particles are dispersed in a polar medium, as water, a double surfactation of the particles is needed to form a hydrophilic layer around them. The polar heads of surfactant molecules can be cationic, anionic or nonionic. In Fig. 6.21 the sketches of an oil-based (Fig. 6.21 a) and water-based (Fig. 6.21 b) ferrofluid grains are presented.

Surfacted ferrofluids are widely used in technological devices, being commercially available (171,172). The surfactant must be matched to the carrier type and must overcome the attractive van der waals and magnetic forces between the particles. The colloid and thermal stabilities, crucial to many applications, are greatly influenced by the choice of the surfactant.

A typical ferrofluid may contain by volume 5% magnetic solid, 10% surfactant and 85% carrier. Surfactants are added during the synthesis of ferrofluids to surround the small particles and overcome their attractive tendencies. For example, oleic acid can be added to oil-based ferrofluids. This surfactant is a long-chain hydrocarbon with a polar head that attaches itself to the surface of the magnetite particle. The long chains comprising the tails act as a repellent cushion and prevent the close approach of another particle.

For the aqueous-based synthesis of ferrofluids using magnetite, Fe_3O_4, the surfactant, tetramethylammonium hydroxide, $[N(CH_3)_4][OH]$, is used. The nanoparticles created by the synthesis are thought to be coated with hydroxide ions from the surfactant, which themselves attract a sheath of largely positive tetramethylammonium cations. This structure creates electrostatic interparticle repulsion that can overcome the van der waals forces that would otherwise cause the particles to agglomerate.

6.4.4 Ionic Ferro-fluids(IFF)

In ionic ferrofluids (173-177) nanoparticles are electrically charged to keep the colloidal system stable. Magnetic particles (usually maghemite, σ Fe_2O_3 and different ferrites, MFe_2O_4, where M = Mn, Co, Zn, Cu, Ni) are obtained through a chemical precipitation method, and an acid-alkaline reaction between particles and the bulk keeps the surface of them electrically charged. Usually, the liquid carrier is water, and the pH of the solution can vary from about 2 to 12, depending on the sign of the surface charge of the particles. Acid IFF ($pH < 7$) have positively charged particles, and alkaline IFF ($pH > 7$) have negatively charged particles. The surface charge density of the particles, typically of the order of $10\ {}^{\prime}C = cm^2$ is a function of the solution's pH. The *Point of Zero Charge* (*PZC*) is defined as the value of the pH at which the particles are not electrically charged (i.e., the ferrofluid precipitates).

In Fig. 6.22 the sketches of an acid (Fig. 6.23 a) and alkalinc (Fig. 6.22 b) ferrofluid grains are presented.

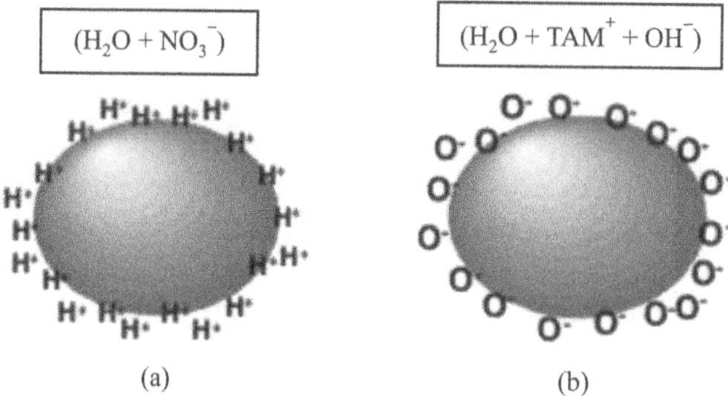

Figure 6.22 Sketch of ionic ferrofluid grains. The counter -ions and the water molecules are also indicated: (a) acid ferrofluid grain; (b) alkaline ferrofluid grain, (TAM^+ OH^-) is the tetramethylammonium hydroxide

Ionic citrated or tartrated maghemite ferrofluids have both characteristics (steric and electrostatic repulsion) to prevent aggregation of the particles. In the case of citrated ferrofluids, particles are coated with small chain molecules such as COOH-CH_2-COH-COOH-CH_2-COOH. In the presence of water, these attached molecules ionize. Besides the steric repulsion, there is also an electrostatic interaction.

In ionic ferrofluids, long-range electrostatic interactions between charged particles give rise to repulsive interactions, which guarantee colloidal stability. On the other hand, in surfacted ferrofluids, there are steric repulsion forces, of short-range nature

6.4.5 Magnetic Behavior of Ferro-fluids

If the ferrofluid is not in the magnetic field, magnetic moments of nanoparticles are randomly oriented and the liquid is non-magnetic, (Fig. 6.23).

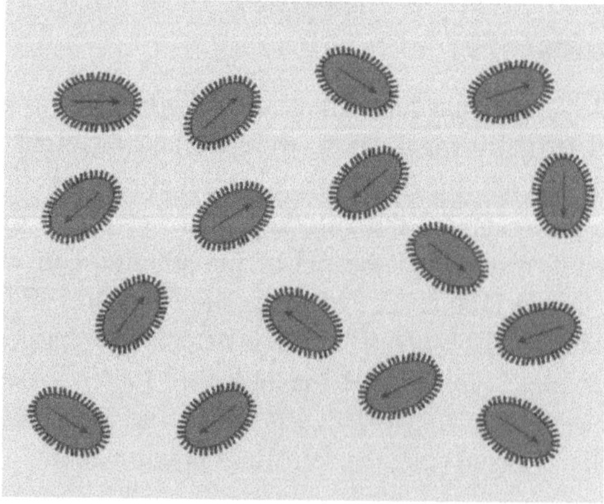

Figure 6.23 Ferrofluid without the influence of outer magnetic field: magnetic moment of nanoparticles has random distribution

If the ferrofluid is in the magnetic field, the nanoparticles are polarized and turn in the direction of the magnetic field and make chains lying in the directions of the lines of force. This process leads to considerable changes of physical chemical qualities of ferro-fluids which impart great technical importance on them. Crystals of ferromagnetic materials change their shape when they are placed in a magnetic field. This phenomenon is called magnetostriction resulting different other physical effects (178-182) which can be exploited for various technological applications like actuators, sensors etc., which are of great importance.

Thus when a magnetic field is applied to a ferrofluid, the magnetic moments of the particles orient along the field lines almost instantly. The magnetization of the ferrofluid responds immediately to the changes in the applied magnetic field and when the applied field is removed, the moments randomize quickly. This phenomenon is called **magnetostriction**.

In a gradient field the whole fluid responds as a homogeneous magnetic liquid which moves to the region of highest flux. This means that ferrofluids can be precisely positioned and controlled by an external magnetic field. The forces holding the magnetic fluid in place are proportional to the gradient of the external field and the magnetization value of the fluid which indicates that the retention force of a ferrofluid can be adjusted by changing either the magnetization of the fluid or the magnetic field in the region.

For example when a strong magnet is placed near the ferrofluid, spikes are observed (Figure 6.24.).The spikes arise from the tendency of the particles to line up along the magnetic field lines to lower their energy. Surface tension of the fluid, however, limits the extent to which the particles can align themselves with the field.

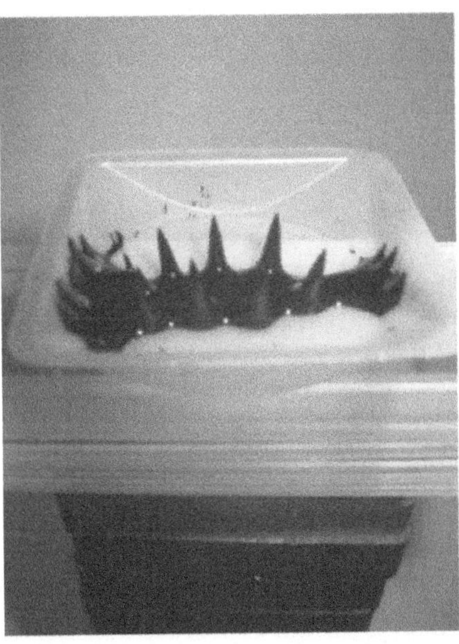

Figure 6.24 Spiking effect of an aqueous based ferrofluid in the presence of a magnet underneath

6.4.5.1 Mechanic-elastic Qualities (Viscosity) of Ferro-fluids under Magnetic Field

In the absence of magnetic field the nano particles move in carrier liquid by thermal (Brown) motion. However due to the magnetic viscosity phenomenon ferrofluid remains liquid even in a strong magnetic field, it means its particles do not sediment and do not aggregate in a strong magnetic field. Unless exposed to the magnetic field, it is isotropic, but in the magnetic field it becomes strongly anisotropic. The dependence of the magnetic inductance (B) on the intensity of the magnetic field (H) has in ferrofluids a course similar to that of solid ferromagnetics: with growing H increases B and asymptomatically approaches the state of saturation. In the linear magnetic field as a result of losses (hysteresis and eddy currents), during remagnetization of nanoparticles these are heated and the carrier liquid is heated as well, which leads to decreasing of its viscosity.

6.4.6 Ferromagnetic Materials

The most common materials for ferromagnetic nanoparticles are magnetite iron II and iron III oxide (Fe_3O_4), maghemite(Fe_2O_2), cobalt (Co), iron (Fe) or iron nitride Fe_xN). The carrier liquid can be water, various oils, usually syntactic on hydrocarbon base, glycol and their compounds. Typical magnetic liquid contains (in volume): 5% ferromagnets, 10 % detergent and 85 % carrier liquid. Its relative permeability $\mu_r \approx 5$ drops with temperature and at Curie temperature gets the value $\mu_r = 1$, saturation magnetization is about 1,3 T and working temperature is from −125 to 200 °C. With higher temperature and temperature changes chemical deterioration of detergent chains occur on the surface of nanoparticles, which leads to destabilization of ferrofluid. Ferrofluids durability is from 8 to 10 years. High quality ferrofluids with long durability are more expensive.

6.4.6.1 Synthesis of Ferro-fluids

Ferrofluids do not exist in nature; they are developed synthetically. Older production technologies were based on long term magnetic crushing of magnetite or ferrite particles in ball mills in detergent solution. The process of grinding lasted from 500 to 1000 hours and after finishing the centrifugal separation of bigger particles followed. At present faster and more effective ways are used, based on various chemical processes leading to precipitation of nano-particles from solutions of ferrous salts. The obtained product must be purified by separating removing bigger particles by centrifugation or by sedimentation caused by gravity or non-homogenous magnetic field.

The formation of a ferrofluid involves various types of forces that hold the components together (Figure 6.20). For example, magnitite is held together by ionic interactions. Ionic attractions between hydroxide anions and tetramethyl ammonium cations allow colloidal suspension of the magnatite in solution. Without the tetramethyl ammonium hydroxide as a surfactant, the magnetite nano particles tend to cluster together. Therefore it is necessary to have the appropriate surfactant to stabilize an aqueous ferrofluid.

Magnetite can be produced by mixing Fe(II) and Fe(III) salts together in a basic solution. The particles must remain small and separated from one another in order to remain suspended in the liquid medium. Magnetic and van der waals interactions must be overcome to prevent the particles from agglomerating into larger particles. Thermal motion of magnetite particles that are smaller than ~100 angstroms in size is sufficient to prevent agglomeration due to magnetic interactions.

There are essentially two methods to prepare these nano particles, by size reduction (183) and chemical precipitation (184). In size reduction, magnetic powder of micron size is mixed with a solvent and a dispersant in a ball mill in order to grind for a period of several weeks. Chemical precipitation is probably the most used method to prepare magnetic nanoparticles nowadays. Different procedures have been developed to achieve this goal. In general, these procedures start with a mixture of $FeCl_2$ and $FeCl_3$ and water. Co-precipitation occurs with the addition of ammonium hydroxide, and then the system is subjected to different procedures to peptization, magnetic separation, filtration and finally dilution.

6.3.7 Ferro-fluids Applications

Most applications of magnetic fluids (Ferro Fluids FF) are based on their following properties:

1. Ferro fluids (FF) go to where the magnetic field is strongest and stays there
2. FF absorb electromagnetic energy at convenient frequencies and heats up
3. FFs physical properties may change with the application of a magnetic field

These properties make the Ferro fluids useful for many technological, biological and medical purposes, as well as a help in materials science and engineering research.

Ferrotec Corporation (formerly Ferrofluidics) has led the development of Ferrofluidic® technology since 1968 and has worked closely with many companies as their new product teams incorporate ferrofluids in next-generation products.

6.4.7.1 Application of Magnetostriction Properties of Ferro-fluids

Generally, magnetostriction is the change in shape of materials under the influence of an external magnetic field. The magnetostrictive effect was first described in the 19th century (1842) by an English physicist James Joule. He observed that a sample of ferromagnetic material, i.e., iron, changes its length in the presence of a magnetic field. Joule actually observed a material with negative magnetostriction, but since that time materials with positive magnetostriction have been discovered. The causes of magnetostriction are similar for both types of material. This change in length is the result of the rotation of small magnetic domains. This rotation and re-orientation causes internal strains in the material structure. The strains in the structure lead to the stretching, in the case of positive magnetostriction, of the material in the direction of the magnetic field. During this stretching process the cross-section is reduced in a way that the volume is kept nearly constant. The size of the volume change is so small that it can be neglected under normal operating conditions. Applying a stronger field leads to stronger and more definite re-orientation of more and more domains in the direction of magnetic field. When all the magnetic domains have become aligned with the magnetic field the saturation point has been achieved. Fig. 6.25 shows the idealized behaviour of length change versus applied magnetic field.

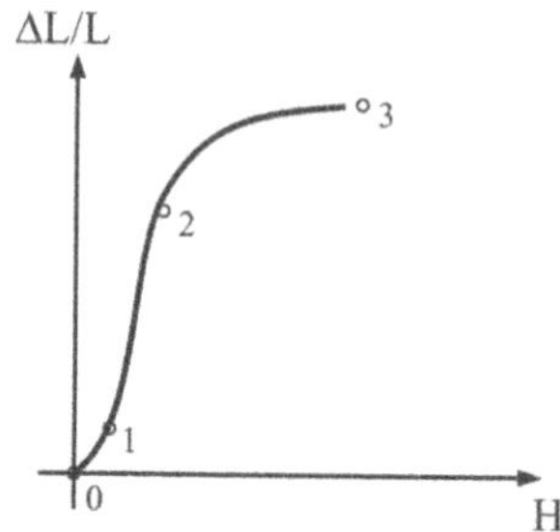

Figure 6.25 Strain versus magnetic field

The physical background to the re-orientation of magnetic domains is depicted (simplified schematically) in Fig. 6.26. In the region between 0 and 1, where the applied magnetic field is small, the magnetic domains show almost no common orientation pattern. Depending on how the material was formed there may be a small amount of a common orientation pattern, which would show itself as a permanent magnet bias. The resulting strain depends very much on how homogeneous is the base structure of the magnetostrictive material and the material formulation. In the region 1-2 ideally there should be an almost linear relationship between strain and magnetic field. Because the relationship is a simple one, it is easier to predict the behaviour of the material and so most devices are designed to operate in this region. Beyond point 2, the relationship becomes non-linear again as a result of the fact that most of the magnetic domains have become aligned with the magnetic field direction. At point 3 there is a saturation effect, which prevents further strain increase.

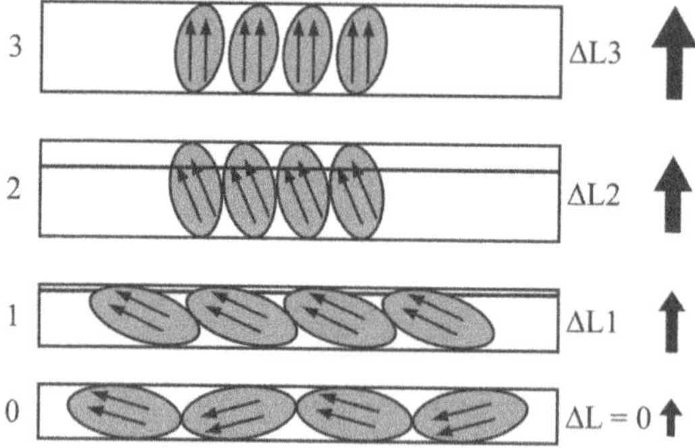

Figure 6.26 Strain versus magnetic field, schematically

The behaviour of the magnetostrictive materials in various applications is complex, because the changing conditions during operation cause changes in material properties. A full understanding of the complexity will enable engineers to use the potential advantages of magnetostrictive materials and to optimize an actuator based on "giant" magnetostrictive materials. Fig. 6.27 shows the idealized behaviour of length change in response to applied magnetic field. When a magnetic field is established in the opposite direction, the field is understood to be negative, but the negative field produces the same elongation in the magnetostrictive material, as a positive field would. The shape of the curve is reminiscent of a butterfly and so the curves are referred as butterfly curves.

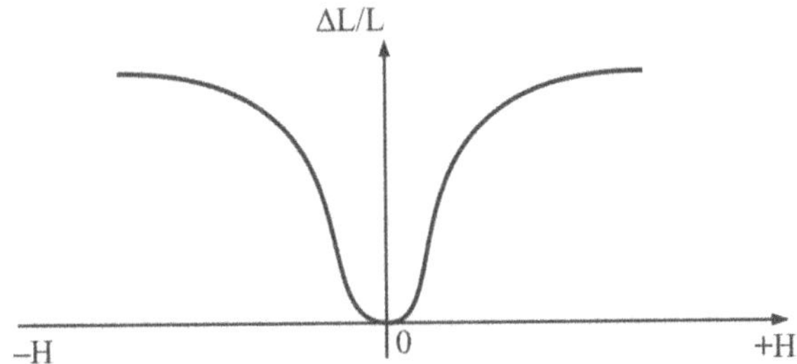

Figure 6.27 Strain versus symmetric magnetic field

6.4.7.1.1 Magnetostriction Effects

Magnetostriction is, in general, a reversible exchange of energy between the mechanical form and the magnetic form. The ability to convert an amount of energy from one form into another allows the use of magnetostrictive materials in actuator and sensor applications. Fig. 6.28 shows various physical effects which are related to the magnetostrictive effect.

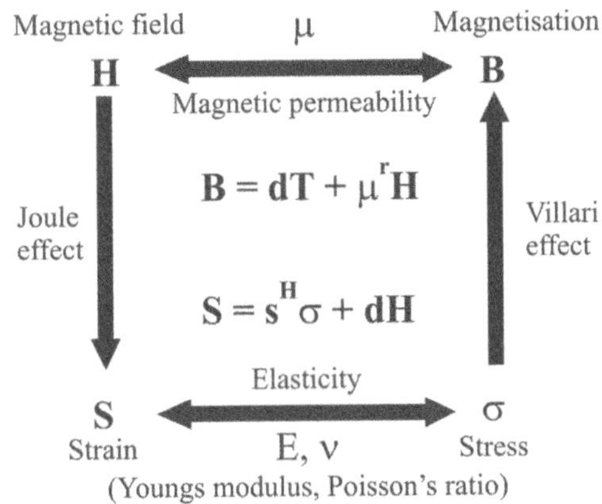

Figure 6.28 Magnetostrictive effects (180)

6.4.7.1.2 Joule Effect

The most understood effect which is related to magnetostriction is the Joule effect. This is the expansion, positive magnetostriction, or contraction, negative magnetostriction, of a ferromagnetic rod in relation to a longitudinal magnetic field. This effect is mainly used in magnetostrictive actuators. Magnetostriction is a reversible material feature. In the absence of the magnetic field, the sample shape returns to its original dimensions. The increase in length (longitudinal strain) or the contraction of diameter (lateral strain) is roughly proportional to the applied magnetic field and this can be used for various purposes in an actuator mechanism.

6.4.7.1.3 Villari Effect for Sensors

Another widely utilized effect related to magnetostriction is **the Villari Effect**. This effect is based on the fact that when a mechanical stress is imposed on a sample, there is a change in the magnetic flux density which flows through the sample as a result of the creation of a magnetic field. The change in flux density can be detected by a pickup coil and is proportional to the level of the applied stress. The **Villari effect** is reversible and is used in sensor applications.

6.4.7.1.4 E-Effect – Change of Youngs Modulus with Magnetic Field

The E-Effect is another effect related to magnetostriction which is the change of the Young's Modulus as a result of a magnetic field. The E/E in Terfenol-D (a high-performance sonar transducer) is in the range of more than 5 and can be employed in tuneable vibration and broadband sonar systems [185]. Due to the change of the Young's Modulus there is a change in the velocity of sound inside magnetostrictive materials, and this can be observed.

6.4.7.1.5 Wiedeman Effect and Matteuci Effects

Another effect related to magnetostriction is the Wiedemann Effect. The physical background to this effect is similar to that of the Joule effect, but instead of a purely tensile or compressive strain forming as a result of the magnetic field, there is a shear strain which results in a torsional displacement of the ferromagnetic sample.

The inverse Wiedemann Effect is called Matteuci Effect: Alternating current fed to a coil creates a longitudinal magnetic field in a sample, and this in turn creates a magnetic flux density in the sample. The presence of the alternating magnetic flux can be detected by another coil, a pickup coil which measures the rate of change in the magnetic flux density. Twisting the ferromagnetic sample induces a change in the magnetization of the sample, which results in a change in the rate of change of the magnetic flux density. By detecting the magnetization change using the pickup coil, the change in shear stress can be evaluated and as a result the magnitude of the applied torque can be calculated. The Matteuci effect is modified by introducing a permanent magnetic bias in the ferromagnetic sample and this is used in sensor applications.

6.4.8 Current and Future Ferro-fluid Technologies

Currently Ferrofluid technologies are well established and capable of solving a wide variety of technical problems. There are many successful applications of this engineering material and there is immense future potential.

In many applications, ferrofluid is an active component that contributes towards the enhanced performance of the device. These devices are either mechanical (e.g., seals, bearings and dampers) or electromechanical (e.g., loudspeakers, stepper motors and sensors) in nature. In other cases, ferrofluid is employed simply as a material for nondestructive testing of other components such as magnetic tapes, stainless steels and turbine blades. When correctly applied, ferrofluid can produce dramatic improvements in a products' performance; or achieve a level of performance unattainable by any other technology or product. The performance and operating life of a product that uses ferrofluid can be significantly affected by the characteristics of the ferrofluid. From ferrofluids with low evaporation rate or vapor pressure to ferrofluids with viscosity-optimized products, the characteristics of ferrofluid can dramatically shape the capabilities of the end product.

The ferrofluid is held in place by permanent magnets and forms a tight seal, reducing friction relative to that produced in a typical mechanical seal. These rotating shaft seals are found in rotating anode x-ray generators and in vacuum chambers used in the semiconductor industry. Ferrofluids are also used in high-speed computer disk drives to eliminate harmful dust particles or other impurities that can cause the data-reading heads to crash into the disks.

Another application of ferrofluids is in improving the performance of loudspeakers. In a loudspeaker electric energy is sent through a coil located in the center of a permanent magnet. Bathing the electric coil in ferrofluid dampens unwanted resonances and provides a way to dissipate heat from the excess energy supplied to the coil. This leads to better overall sound quality.

6.4.8.1 Ferrohydrodynamic Damper

Dampers used in machinery engineering dissipate kinetic energy of vibrating mechanism and so damp mechanic shocks and consequent vibrations. Conventional hydraulic dampers have constant damping, only in special cases their damping can be managed by regulation of fluid flow with a throttle valve. Ferrohydraulic dampers provide a more elegant solution (Figure 6.29). They are filled with ferrofluid that is exposed to the magnetic field of the coil induced by a controlled current signal. It changes the viscosity of the ferrofluids and the dumping increases. The current signal is, according a particular strategy controlled by an on-line sensor that reads the causes of vibrations.

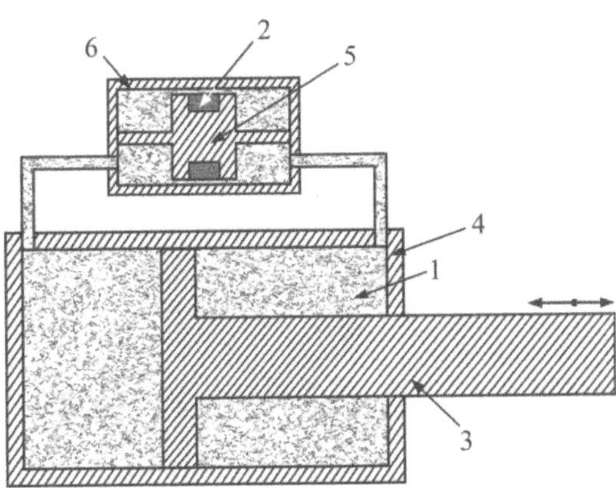

Figure 6.29 Ferrohydrodynamic Damper: 1 Ferrofluid, 2 exciting coil, 3 Plunger of damper, 4 shell, 5 - gap, 6 Throttle valve

Ferrohydrodynamic dampers were used in different equipments from fine measuring apparatuses to washing machines, lorry seats to the chassis dumper of means of transport and are very prospective. If a car moves at the speed of 72 km/h, it makes 2 cm distance in 1 ms. Conventional dampers react after 15 ms, the car then makes 30 cm before the damper reacts. The ferromagnetic damper reacts much faster, after 5 ms. The dampers reacts in 10 cm distance and the car gets over a bump without bouncing. More transmission of the vibrations to the driver cabin, limits transmission damping prevents the bouncing of wheels and thus a loss of adhesion between the tire and the surface which increases the stability of the car, mainly in curves. Magnetic dampers thus increase the safety and comfort ability of the ride, shorten braking reaction, improve the behavior of the car and extend its durability, mainly its tires.

6.4.8.2 Ferrohydrodynamic Sealing

Using ferrofluids enables perfect sealing of a rotating shaft. The way of sealing rotating shaft with ferrofluid for magnetic material shaft is as shown in Fig.6.30. In mutually separated areas there are different pressures p_1, p_2. On a permanent magnet in the shape of a short cylinder stand pole rods from magnetically soft material. Between the shafts and the pole rods there is an air gap of the size of several tenths of centimeter. In the air gap a strong magnetic field concentrates. To this area ferrofluid is pushed that is fixed by the influence of the magnetic field and has the function of sealing. For magnetic inductance in the air gap B ~ 1 T this sealing can keep the pressure difference $| p_1 - p_2| \sim 0, 2$ to 1 atm. Based on the given principle multi-layer sealing are built, that enable to increase the overpressure$| p_1 - p_2|$ to the value over 10 atm.

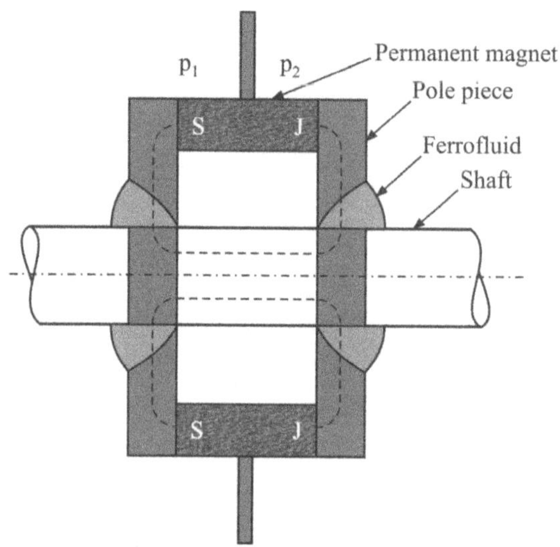

Figure 6.30 Sealing magnetic shaft using ferrofluids

In comparison to conventional (mechanical) sealing the given principle has considerable advantages. It is simpler and (and thus cheaper) more reliable and has a lower friction momentum (ferro-fluid functions as a lubricant), high tightness, long durability (producers claim up to10 years) and can function in a wide temperature range from –100 °C to 200 °C. This way of shaft sealing is used also for identical pressures ($p_1 = p_2$), as effective dust proof sealing e.g., to protect bearing operating in dusty, chemically aggressive, toxic or biologically active environment.

6.4.8.3 Speakers with Ferro-fluid

In common electro dynamic speakers in the magnetic circuit with permanent magnet there is an air gap, in which a coil vibrates fed by acoustic signal and connected with the membrane, (Figure 6.31) Unlike the conventional solution, around the coil in speakers with ferrofluid there is no air, but ferrofluid. By the influence of the strong magnetic field of permanent magnet the ferrofluid is permanently kept in the air gap. The acoustic performance is limited by the acceptable current load of the coil. As the thermal conductivity of ferrofluid is ten times bigger than thermal conductivity of air, it enables to increase current density in the coil and thus the acoustic performance of the speaker.

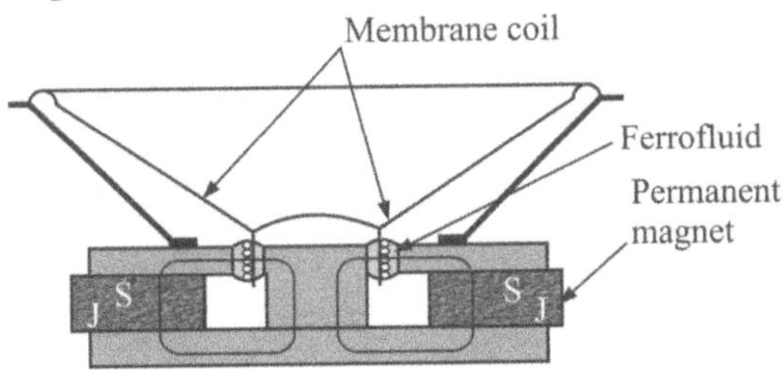

Figure 6.31 Speaker with ferrofliud in the air gap

6.4.8.4 Heat transfer

An external magnetic field imposed on a ferrofluid with varying susceptibility (e.g., because of a temperature gradient) results in a nonuniform magnetic body force, which leads to a form of heat transfer called thermomagnetic convection. This form of heat transfer can be useful when conventional convection heat transfer is inadequate; e.g., in miniature microscale devices or under reduced gravity conditions. Ferrofluids of suitable composition can exhibit extremely large enhancement in thermal conductivity (k; ~300% of the base fluid thermal conductivity). The large enhancement in k is due to the efficient transport of heat through percolating nanoparticle paths. Special magnetic nanofluids with tunable thermal conductivity to viscosity ratio can be used as multifunctional 'smart materials' that can remove heat and also arrest vibrations (damper). Such fluids may find applications in microfluidic devices and microelectromechanical systems (MEMS).

Electrical machines with ferrofluids: For power transformers ferrofluid is used as a cooling medium. Unlike transformer oil it has a higher thermal conductivity, while its electric strength is basically the same. For rotating electrical machines ferrofluid is applied in the air gap between rotor and stator. It makes cooling better, but first of all it lowers the reluctance of the magnetic circuit of the machine and thus the magnetizing current of the machine. On the other hand, it increases the hydraulic reluctance of the rotation. This requires usage of ferrofluid with low viscosity and high permeability. The mentioned way is limited to slowly running machines. Experiments show that for rotating machines to 1000 turns/min the advantages prevail related with usage of ferrofluids.

High magnetization ferrofluids are of interest as they produce volumetric efficiencies of magnetic circuit designs leading to lightweight and lower cost products. They can also be used to reduce reluctance of magnetic circuits and fringing field thus increasing useful flux density in the air gap. The domain magnetization of magnetite ultimately limits the maximum magnetization value that can be realized in a ferrofluid.

Biologists such as Michael Walker, Carol Diebel, and Cordula Haugh of the Experimental Biology Group, have also been studying how ferrofluids aid in the migratory senses of animals. They have studied how trout respond to the magnetic field of the earth. Apparently a few small branches of the trigeminal nerve enter the nose of the trout. There the nerves seem to connect with some unusual cells that do not appear to be involved in the trout's sense of smell. The cells contain crystals of a mineral that have yet to be identified. It is suspected that the crystals are magnitite. When the magnetic field is steady, spikes or action potentials are produced. If the magnetic field changes so do the spikes. This signals when the migratory process should take place. Finally, ferrofluids may be used in future biomedical applications. Researchers are attempting to design ferrofluids that can carry medication to specific locations in the body through the use of applied magnetic fields. Ferrofluids may also be used as contrast agents for magnetic resonance imaging (MRI).

Various suggestions have been made concerning the use of ferro fluids for medical applications. Ferrofluids injected into blood stream can be directed to sensitive areas with magnetic fields. The blood supply can be blocked and the artery cut during surgery without using clamps. In this case there will not be any damage to the artery as is often the case when clamps are used. Carnial aneurism which are ruptures of the arteries in the region of brain may be treated without surgery using ferrofluids. In this case ferrofluid is held by magnetic fields so as to isolate the rupture from the blood stream. Natural healing then takes place and scar tissue forms at the rupture which eventually disappears.

6.4.8.5 Ferro-fluids for Treatment of Tumors and Medical Applications

Magnetic Drug Targeting is a promising approach for the loco regional treatment of tumors. An indispensable pre-requisite is the availability of appropriate Ferrofluids as carrier for the chemotherapeutic drugs. Commercially available, biocompatible Ferrofluids of Chemicell GmbH Berlin have been successfully applied for the treatment of VX-2 squamous cell carcinoma New Zealand White rabbits (186-187).

Due to their rise in temperature in an alternating magnetic field, ferrofluids can be used in the fight against cancer. For example, it is directly injected into the cancerous cells and then exposed to an alternating magnetic field from the outside. This causes overheating (hyperthermia) which kills off the cancerous cells.

Iron oxide nanoparticles have been investigated for various applications including drug delivery, magnetic resonance imaging contrast enhancement, immunoassay and cellular therapy. These investigations often employ magnetite (Fe_3O_4) dispersed in a polymeric microsphere or microcapsule involving biodegradable and/or natural polymer. Poly (L-lysine) microspheres containing magnetic nanoparticles prepared by coacervation were prepared and characterized for potential use in targeted drug delivery applications. Iron and cobalt nanoparticles encapsulated in polydimethylsiloxane have been noted for treating retinal detachment disorder.

6.4.8.6 Ferro-fluids for Water Purification

Activated carbon (AC) supported Fe_3O_4 nanoparticle-based composites (CFFs) have been found to have enhanced capacity (5-9 times) than simple activated carbon for colour removal from industrial effluents (188).

References

1. M. Geissler and Y.N. Xia. Adv. Mater. 2004, 16 (15), 1249-1269.
2. S.Y. Chou, P.R. Krauss and P.J. Renstrom. Appl. Phys. Lett. 1995, 67 (21), 3114-3116.
3. L.J. Guo. Adv. Mater. 2007, 19 (4), 495-513.
4. M.D. Austin, W. Zhang, H.X. Ge, D. Wasserman, S.A. Lyon and S.Y. Chou. Nanotechnology 2005, 16 (8), 1058-1061.
5. T. Ito and S. Okazaki. Nature 2000, 406 (6799), 1027-1031.
6. W.K. Lee and P.E. Sheehan. Scanning 2008, 30 (2), 172-183.
7. D. Wouters and U.S. Schubert. Angew. Chem., Int. Ed. 2004, 43 (19), 2480-2495.
8. H.J. Mamin and D. Rugar. Appl. Phys. Lett. 1992, 61 (8), 1003-1005.
9. R.D. Piner, J. Zhu, F. Xu, S.H. Hong and C.A. Mirkin. Science 1999, 283 (5402), 661-663
10. M. Peter, X.M. Li.; J. Huskens and D.N. Reinhoudt. J. Amm. Chem. Society, 2004, 126(37), 11684-11690.
11. P. Vettiger, M. Despont, U. Drechsler, U. Durig, W. Haberle, M.I. Lutwyche, H.E. Rothuizen, R. Stutz, R. Widmer and G.K. Binnig. IBM J. Res. Dev. 2000, 44 (3), 323-340.
12. Y.N. Xia and G.M. Whitesides. Annu. Rev. Mater. Sci. 1998, 28, 153-184.
13. A. Kumar and G.M. Whitesides. Appl. Phys. Lett. 1993, 63 (14), 2002-2004.
14. Y.N. Xia, E. Kim, X.M. Zhao, J.A. Rogers, M. Prentiss and G.M. Whitesides. Science 1996, 273 (5273), 347-349.
15. X.M. Zhao, Y.N. Xia and G.M. Whitesides. Adv. Mater. 1996, 8(10), 837-840.
16. E. Kim, Y.N. Xia and G.M. Whitesides. Nature 1995, 376(6541), 581-584.
17. E. Kim, Y.N. Xia, X.M. Zhao and G.M. Whitesides. Adv. Mater. 1997, 9(8), 651-654.
18. K.Y. Suh and H.H. Lee. Adv. Funct. Mater. 2002, 12(6-7), 405-413.
19. K.Y. Suh, Y.S. Kim and H.H. Lee Adv. Mater. 2001, 13(18), 1386-1389.
20. M. Geissler, J.M. Mc Lellan and Y.N. Xia. Nano Lett. 2005, 5(1), 31-36.
21. S.M. Miller, S.M. Troian and S. Wagner. J. Vac. Sci. Technol., B 2002, 20(6), 2320-2327.
22. D.W. Li and L.J. Guo. J. Phys. D: Appl. Phys. 2008, 41(10).
23. R.D. Bennett, A.J. Hart, A.C. Miller, P.T. Hammond, D.J. Irvine and R.E. Cohen. Langmuir 2006, 22(20), 8273-8276.
24. H. Schmid and B. Michel. Macromolecules 2000, 33(8), 3042-3049.
25. H.W. Li, B.V.O. Muir, G. Fichet and W.T.S. Huck. Langmuir 2003, 19(6), 1963-1965.
26. F. Hua, Y.G. Sun, A. Gaur, M.A. Meitl, L. Bilhaut, L. Rotkina, J.F. Wang, P. Geil, M. Shim, J.A. Rogers and A. Shim. Nano Lett. 2004, 4(12), 2467-2471.
27. S. Yagai, S. Kubota, T. Iwashima, K. Kishikawa, T. Nakanishi, T. Karatsu and A. Kitamura. Chem. Eur. J. 2008, 14(17), 5246-5257.

28. A. Schenning and E.W. Meijer Chem. Commun. 2005, (26), 3245-3258.

29. F.J.M. Hoeben, P. Jonkheijm, E.W. Meijer and A. Schenning. Chem. Rev. 2005, 105(4), 1491-1546.

30. P. Jonkheijm, F.J.M. Hoeben, R. Kleppinger, J. van Herrikhuyzen, A. Schenning and E.W. Meijer. J. Am. Chem. Soc. 2003, 125(51), 15941-15949.

31. H.B. Akkerman, P.W.M. Blom, D.M. de Leeuw and B. de Boer. Nature 2006, 441(7089), 69-72.

32. M. Goldberg, R. Langer and X.Q. Jia. J. Biomater. Sci., Polym. Ed. 2007, 18(3), 241-268.

33. P.Y.W. Dankers and E.W. Meijer. Bull. Chem. Soc. Jpn. 2007, 80(11), 2047-2073.

34. P.Y.W. Dankers, M.C. Harmsen, L.A. Brouwer, M.J.A. van Luyn and E.W. Meijer. Nat. Mater. 2005, 4(7), 568-574.

35. J. Ruokolainen, R. Makinen, M. Torkkeli, T. Makela, R. Serimaa, G. ten Brinke and O. Ikkala. Science 1998, 280(5363), 557-560.

36. O. Ikkala and G. ten Brinke. Chem. Commun. 2004, (19), 2131-2137.

37. S. Valkama, O. Lehtonen, K. Lappalainen, H. Kosonen, P. Castro, T. Repo, M. Torkkeli, R. Serimaa, G. ten Brinke, M. Leskela and O. Ikkala Macromol. Rapid Commun. 2003, 24(9), 556-560.

38. T. Vermonden, M.J. van Steenbergen, N.A.M. Besseling, A.T.M. Marcelis, W.E. Hennink, E.J.R. Sudholter and M.A.C. Stuart. J. Am. Chem. Soc. 2004, 126(48), 15802-15808.

39. J.B. Beck and S.J. Rowan. J. Am. Chem. Soc. 2003, 125(46), 13922-13923.

40. H. Hofmeier, R. Hoogenboom, M.E.L. Wouters and U.S. Schubert. J. Am. Chem. Soc. 2005, 127(9), 2913-2921.

41. J.M.J. Paulusse, J.P.J. Huijbers and R.P. Sijbesma. Macromolecules 2005, 38(15), 6290-6298.

42. J.M.J. Paulusse and R.P. Sijbesma. Angew. Chem., Int. Ed. 2004, 43(34), 4460-4462.

43. S. Schmatloch, A.M.J. van den Berg, A.S. Alexeev, H. Hofmeier and U.S. Schubert. Macromolecules 2003, 36(26), 9943-9949.

44. C. Fouquey, J.M. Lehn and A.M. Levelut. Adv. Mater. 1990, 2(5), 254-257.

45. R.P. Sijbesma, F.H. Beijer, L. Brunsveld, B.J.B. Folmer, J.H.K.K. Hirschberg, R.F.M. Lange, J.K.L. Lowe and E.W. Meijer. Science 1997, 278(5343), 1601-1604.

46. J.H.K.K. Hirschberg, F.H. Beijer, H.A. van Aert, P.C.M.M. Magusin, R.P. Sijbesma and E.W. Meijer. Macromolecules 1999, 32(8), 2696-2705.

47. b.B.J.B. Folmer, R.P. Sijbesma, R.M. Versteegen, J.A.J van der Rijt and E.w. Merijer; Adv. Mater 2000, 12(12), 874-878.

48. V. Berl, M. Schmutz, M.J. Krische, R.G. Khoury and J.M. Lehn. Chem.--Eur. J. 2002, 8(5), 1227-1244.

49. F. Lortie, S. Boileau and L. Bouteiller. Chem. Eur. J. 2003, 9(13), 3008-3014.

50. W.H. Binder, M.J. Kunz and E. Ingolic J. Polym. Sci., Part A: Polym. Chem. 2003, 42(1), 162-172.

51. C.L. Elkins, K. Viswanathan and T.E. Long. Macromolecules 2006, 39(9), 3132-3139.

52. P. Cordier, F. Tournilhac, C. Soulie-Ziakovic and L. Leibler. Nature 2008, 451(7181), 977-980.

53. O. Colombani, C. Barioz, L. Bouteiller, C. Chanéac, L. Fompérie, F. Lortie and H. Montès. Macromolecules 2005, 38(5), 1752-1759.

54. P.Y.W. Dankers, E.N.M. van Leeuwen, G.M.L. van Gemert, A.J.H. Spiering, M.C. Harmsen, L.A. Brouwer, H.M. Janssen, A.W. Bosman, M.J.A. van Luyn and E.W. Meijer. Biomaterials 2006, 27(32), 5490-5501.

55. W. Dannecker, J. Kopf and H. Rust. Crystal Structure Communications 1979, 8(2), 429-432.

56. J. van Esch, F. Schoonbeek, M. de Loos, H. Kooijman, A.L. Spek, R.M. Kellogg and B.L. Feringa Chem.--Eur. J. 1999, 5(3), 937-950.

57. M. de Loos, A. Friggeri, J. van Esch, R.M. Kellogg and B.L. Feringa. Organic & Biomolecular Chemistry 2005, 3(9), 1631-1639.

58. M.R.J. Vos, P.H.H. Bomans, F. de Haas, P.M. Frederik, J.A. Jansen, R.J.M. Nolte and N.A.J.M. Sommerdijk. J. Am. Chem. Soc. 2007, 129(39), 11894-11895.

59. S. De Feyter, M. Larsson, N. Schuurmans, B. Verkuijl, G. Zoriniants, A. Gesquière, M.M. Abdel-Mottaleb, J. van Esch, B.L. Feringa, J. van Stam and F. De Schryver Chem.--Eur. J. 2003, 9(5), 1198-1206.

60. N. Chebotareva, P.H.H. Bomans, P.M. Frederik, N.A.J.M. Sommerdijk and R.P. Sijbesma. Chem. Commun. 2005, (39), 4967-4969.

61. D. Tyagi, G.L. Wilkes, I. Yilgör and J.E. McGrath. Polym. Bull. 1982, 8(11-1), 543-550.

62. O. Colombani and L. Bouteiller. New. J. Chem. 2004, 28(11), 1373-1382.

63. S. Das, I. Yilgör, E. Yilgör, B. Inci, O. Tezgel, F.L. Beyer and G.L. Wilkes. Polymer 2007, 48(1), 290-301.

64. I. Yilgör and E. Yilgör. Polym. Rev. 2007, 47(4), 487-510.

65. R.M. Versteegen, R. Kleppinger, R.P. Sijbesma and E.W. Meijer. Macromolecules. 2006, 39(2), 772-783.

66. R.A. Koevoets, R.M. Versteegen, H. Kooijman, A.L. Spek, R.P. Sijbesma and E.W. Meijer. J. Am. Chem. Soc. 2005, 127(9), 2999-3003.

67. E. Wisse, L.E. Govaert, H.E.H. Meijer and E.W. Meijer. Macromolecules 2006, 39(21), 7425-7432.

68. E. Wisse, A.J.H. Spiering, E.N.M. van Leeuwen, R.A.E. Renken, P.Y.W. Dankers, L.A. Brouwer, M.J.A. van Luyn, M.C. Harmsen, N.A.J.M. Sommerdijk and E.W. Meijer. Biomacromolecules 2006, 7(12), 3385-3395.

69. C.R. South, C. Burd and M. Weck. Acc. Chem. Res. 2007, 40(1), 63-74.

70. W. Gerhardt, M. Crne and M. Weck. Chem. Eur. J. 2004, 10(24), 6212-6221.

71. T.B. Norsten, E. Jeoung, R.J. Thibault and V.M. Rotello. Langmuir 2003, 19(17), 7089-7093.
72. J.M. Pollino and M. Weck. Chem. Soc. Rev. 2005, 34(3), 193-207.
73. M. Weck. Polym. Int. 2007, 56(4), 453-460.
74. S.K. Yang and M. Weck. Macromolecules 2008, 41(2), 346-351.
75. C. Park, J. Yoon and E.L. Thomas. Polymer 2003, 44(22), 6725-6760.
76. Z.H. Nie and E. Kumacheva. Nat. Mater. 2008, 7(4), 277-290.
77. A. del Campo and E. Arzt. Chem. Rev. 2008, 108(3), 911-945.
78. I.W. Hamley Angew. Chem., Int. Ed. 2003, 42(15), 1692-1712.
79. Y.S. Jung and C.A. Ross. Nano Lett. 2007, 7(7), 2046-2050.
80. M. Roerdink, M.A. Hempenius, U. Gunst, H.F. Arlinghaus and G.J. Vancso Small 2007, 3(8), 1415-1423.
81. H.C. Kim, C.T. Rettner and L. Sundstrom. Nanotechnology 2008, 19(23), 235-301.
82. R. Shenar, H.Xu. B.L. Frankamp, T.E Mates, A. Sanyal. O. Uzun and V.M. Rotello, J. Am. Chem. Soc, 2005, 127(46), 1638-16324.
83. O. Ikkala and G. ten Brinke. Science, 2002, 295(5564), 2407-2409.
84. A. Haryono and W.H. Binder. Small 2006, 2(5), 600-611.
85. H. Xu, R. Hong, T.X. Lu, O. Uzun and V.M. Rotello. J. Am. Chem. Soc. 2006, 128(10), 3162-3163.
86. E.N. Savariar, K. Krishnamoorthy and S. Thayumanavan. Nat. Nanotechnol. 2008, 3(2), 112-117.
87. D. Bratton, D. Yang, J.Y. Dai and C.K. Ober. Polym. Adv. Technol. 2006, 17(2), 94-103.
88. G. Csucs, T. Kunzler, K. Feldman, F. Robin and N.D. Spencer. Langmuir 2003, 19(15), 6104-6109.
89. D. Trimbach, K. Feldman, N.D. Spencer, D.J. Broer and C.W.M. Bastiaansen. Langmuir 2003, 19(26), 10957-10961.
90. N. Coq, T. van Bommel, R.A. Hikmet, H.R. Stapert and W.U. Dittmer. Langmuir 2007, 23(9), 5154-5160.
91. N.Y. Lee, J.R. Lim, M.J. Lee, J.B. Kim, S.J. Jo, H.K. Baik and Y.S. Kim. Langmuir 2006, 22(21), 9018-9022.
92. D.C. Trimbach, M. Al-Hussein, W.H. de Jeu, M. Decré, D.J. Broer and C.W.M. Bastiaansen. Langmuir 2004, 20(11), 4738-4742.
93. D.C. Trimbach, H. Stapert, J. van Orselen, K.D. Jandt, C.W.M. Bastiaansen and D.J. Broer. Adv. Eng. Mater. 2007, 9(12), 1123-1128.
94. M. Liebau, J. Huskens and D.N. Reinhoudt. Adv. Funct. Mater. 2001, 11(2), 147-150.
95. X.M. Li, M. Péter, J. Huskens and D.N. Reinhoudt. Nano Lett. 2003, 3(10), 1449-1453.

96. S. Kim, J. Lee, S.M. Jeon, H.H. Lee, K. Char and B.H. Sohn. Macromolecules 2008, 41(10), 3401-3404.

97. P. Mukhopadhyay, P.Y. Zavalij and L. Isaacs. J. Am. Chem. Soc. 2006, 128(43), 14093-14102.

98. Gupta AK, Gupta M. Biomaterials 2005, 26, 3995–4021.

99. Luke Collins. Roll-up displays: Fact or Fiction? IEE Review February 2003.

100. Universal Display Corporation, FOLED display technology, 2001-2005.

101. G. G. Malliaras*, J. R. Salem, P. J. Brock, and C. Scott. Electrical characteristics and efficiency of single-layer organic light-emitting diodes, Phys. Rev. B 1998, 58, R13411–R13414.

102. "How OLEDS work?" http://science.howstuffworks.com/oled.htm

103. E Ink Corp. How it works? Electronic Paper Display Technology, 2004.

104. Scott, M.A., Carrado, K.A. and Dutta, P.K. Handbook of layered materials, 2004.

105. Ray S.S. and Okamoto, M. Polymer/layered silicate nanocomposites: A review from preparation to processing. Progress Poly. Sci., 2003, 28, 1539-641.

106. Shen, J.W., Chen, X.M. and Huang, W.Y. Structure and electrical properties of grafted polypropylene/graphite nanocomposites prepared by solution intercalation. J. Appl. Poly. Sci., 2003, 88, 1864-869.

107. Kojima Y, Usuki A, Kawasumi M, Okada A, Kurauchi T, Kamigaito O. J Polym Sci Part A Polym Chem 1993, 31, 983–6.

108. Okada A, Fukushim Y, Kawasumi M, Inagaki S, Usuki A, Sugiyama S, et al. US Patent 4, 739,007, 1988.

109. Kawasumi M. J Polym Sci Part A Polym Chem 2004, 42, 819.

110. Di Maio E, Iannace S, Sorrentio L and Nicolais L. Polymer 2004, 45, 8893-8900.

111. Zhang QX, Yu ZZ, Yang M, Ma J and Mai YW. J Polym Sci Part B Polym Phys 2003, 41, 2861–9.

112. Mu B, Wang Q, Wang H and Jian L. J Macromol Sci Part B Phys 2007; 46: 1093– 104.

113. Nam JY, Ray SS and Okamoto M. Macromolecules 2003, 36, 7126–31.

114. Lincoln DM, Vaia RA and Krishnamoorti R. Macromolecules 2004, 37, 4554–61.

115. Li L, Li CY, Ni C and Rong L, Hsiao B. Polymer 2007, 48, 3452–60.

116. Phang IY, Pramoda KP, Liu T and He C. Polym Int 2004, 53, 1282–9.

117. Wu D, Zhou C, Fan X, Mao D and Bian Z. Polym Polym Compos 2005, 13, 61–71.

118. Bilotti E, Fischer HR and Peijs T. J Appl Polym Sci 2008, 107, 1116–23.

119. Schmidt, D.; Shah D. and Giannelis, E.P. New advances in polymer/layered silicate nanocomposites. Curr. Opi. Solid State Mater. Sci., 2002, 6 (3), 205-12.

120. Ray, S.S. and Okamoto, M. Polymer/layered silicate nanocomposites: A review from preparation to processing. Progress Poly. Sci., 2003, 28, 1539-641.

121. Fornes TD and Paul DR. Polymer 2003, 44, 4993.

122. Kawasumi M., Hasegawa N., Kato M., Usuki A. and Okada A. Preparation and mechanical properties of polypropylene-clay hybrids. Macromolecules, 1997, 30(20), 6333-338.

123. Lan T. and Pinnavaia T.J. Clay-reinforced epoxy nanocomposites. Chem. Mater., 1994, 6(12), 2216-2219.

124. Savage S.J. Defence applications of nanocomposite materials. FOI-Swedesh Defence Research Agency, User Report No. FOI-R-1524-SE, December 2004.

125. Hasegawa N., Kawasumi M., Kato M., Usuki A. and Okada A. Preparation and mechanical properties of polypropylene-clay hybrids using a maleic anhydride-modified polypropylene oligomer. J. Appl. Polym., 1998, 67(1), 87-92.

126. Xu W.B., Ge M.L. and He P. Nonisothermal crystallization kinetics of polypropylene/montmorillonite nanocomposites. J. Polym. Sci. Pt B: Polym. Phys., 2002, 40(5), 408.

127. Hule RA and Pochan DJ. MRS Bull 2007, 32, 354–8.

128. Kim HW, Kim HE and Salih V. Biomaterials 2005, 26, 5221–30.

129. Huang J, Lin YW, Fu XW, Best SM, Brooks RA, Rushton N, et al. J Mater Sci Mater Med 2007, 18, 2151–7.

130. Lee YH, Lee JH, An IG, Kim C, Lee DS, Lee YK, et al. Biomaterials 2005, 26, 3165–72

131. Zhang Q, Zha L, Ma J, Liang B. Macromol Rapid Commun 2007, 28, 116–20.

132. Haraguchi K, Li HJ. Macromolecules 2006, 39, 1898–905.

133. Mintmire J.W., Dunlap B.I. and Carter C.T. Are fullerene tubules metallic? Phys. Rev. Lett., 1992, 68(5), 631-34.

134. Hamada N., Sawada S. and Oshiyama A. New one-dimensional conductors: Graphitic microtubules. Phys. Rev. Lett., 1992, 68(10), 1579-581.

135. P.Y.W. Dankers and E.W. Meijer Bull. Chem. Soc. Jpn. 2007, 80(11), 2047-2073.

136. Carroll D.L., Redlich P., Ajayan P.M., Charlier J.C., Blase X., De Vita, A. and Car R. Electronic structure and localized states at carbon nanotube tips. Phys. Rev. Lett., 1997, 78(14), 2811-814.

137. Kroto H.W., Heath J.R., O'Brien S.C., Curl S.C. and Smalley R.E. C60: Buck minster fullerene. Nature, 1985, 318, 162.

138. Biercuk M.J., Laguno M.C., Radosavijevic M., Hyun J.K., Johnson A.T. and Fischer J.E. Carbon nanotube composites for thermal management. Apply. Phys. Lett., 2002, 80(15), 2767-69.

139. Benoit J.M., Corraze B. and Chauvet, O. Localization, coulomb interactions, and electrical heating in single- wall carbon nanotubes/polymer composites. Phys. Rev. B, 2002, 65(24), 241405-09R.

140. Mintmire J.W., Dunlap B.I. and Carter C.T. Are fullerene tubules metallic? Phys. Rev. Lett., 1992, 68(5), 631-34.

141. Krishnan Jayaraman, Kataki M., Yanzhong Zhang, Xiumei Mo. and Ramakrishna S. Recent advances in polymer nanofibers. J. Nanosci. Nanotech., 2004, 4(1-2), 52-65.

142. Tandon G.P. and Ran Y. Influence of vapor-grown carbon nanofibers on thermomechanical properties of graphite- epoxy composites. In Proceedings of the 17th Annual Technical Conference ASC, 2002.

143. Glasgow D.G. and Tibbetts G.G. Surface treatment of carbon nanofibers for improved composite mechanical properties. In Proceedings of SAMPE 2004, Longbeach, CA, 2004.

144. Lafdi, K. Carbon foam for thermal management. In Proceedings of SAMPE 2003, Dayton, OH, 2003.

145. Finegan C., Tibbetts G.G., Glasgow D.G., Ting J.M. and Lake M.L. Surface treatments for improving the mechanical properties of carbon nanofiber/thermoplastic composites. J. Mater. Sci., 2003, 38(16) 3485-490.

146. Gibson T., Rice B., Ragland W., Silverman E.M., Peng H. and Strong K.L. Formulation and evaluation of carbon nanofiber-based conductive adhesives. In Proceedings of SAMPE 2005, Long Beach, CA, 1-5 May 2005.

147. Lao S., Ho W., Ngyuen K. and Koo J.H. Flammabiltiy, mechanical, and thermal properties of polyamide nanocomposites. In Proceedings of SAMPE 2005, Seattle, WA, 2005.

148. Wagner H. D., Lourie O., Feldmana Y. and Tenne R. Stress-induced fragmentation of multiwall carbon nanotubes in a polymer matrix. Appl. Phys. Lett., 1998, 72(2), 188-90.

149. P. Rohatgi. JOM, 1991, 43, 10–15.

150. A. Kelly. J. Mater. Sci., 2006, 41, 905–912.

151. S. Rawal. JOM, 2001, 53(4), 14–17.

152. J. S. Shelly, R. Le Claire and J. Nichols: JOM, 2001, 53(4), 18–21.

153. Hussain F., Hojjati M., Okamoto M. and Gorga R.E. Review article: Polymer-matrix nanocomposites, processing, manufacturing, and application: An overview. J. Comp. Mater., 2006, 40(17), 1511-575.

154. Baughman R.H., Zakhidov A.A. and de Heer W.A. Carbon nanotubes – The route towards applications. Science, 2002, 297, 787-92.

155. Glatkowski P., Mack P., Conroy J., Piche J.W. and Winsor P. Electromagnetic shielding comprising nanotubes. US Patent No. 6265466, 2001.

156. Huang J Kaner, RB. Flash welding of conducting polymer nanofibres. Nature Materials, 2004, 3(11), 783-86.

157. Linda, S. Schadler. Polymer-based and polymer-filled nanocomposites. In Nanocomposite science and technology, edited by P. M. Ajayan; L. S. Schadler & P.V. Braun. WILEY-VCH Verlag GmbH Co. KGaA, Weinheim, 2003, 77-153.

158. Lee, Y.S. and Wagner, N.J. Dynamic properties of shear thickening colloidal suspensions. Rheologica Acta, 2003, 42(3), 199-208.

159. Lee Y.S., Wetzel E.D., Egres Jr., R.G. and Wagner N.J. Advanced body armor utilizing shear thickening fluids. In Proceedings of the 23rd Army Science Conference 2002. Orlando, FL, 2-5 December 2002.

160. Lee, Y.S., Wetzel, E.D. and Wagner N.J. The Ballistic impact characteristics of Kevlar®Woven fabrics impregnated with a colloidal shear thickening fluid. J. Mater. Sci., 2003, 38(13), 2825-833.

161. Nguyen, T. and Diaz, A.F. Novel method for the preparation of magnetic nanoparticles in a polypyrrole powder. Advanced Materials, 1994, 6(11), 858-60.

162. Sharon, Maheshwar, Pradhan, D., Zaharia, R. and Puri, V.J. Application of carbon nanomaterial as a microwave absorber. J. Nanosci. Nanotechnol., 2005, 5(12), 2117-20.

163. Bai X.D., Zhong D.Y., Zhang G.Y., Ma X.C., Liu S., Wang E.G., Chen Y. and Shaw D.T. Hydrogen storage in carbon nitride nanobells. Appl. Phy. Lett., 2001, 79(10), 1552-554.

164. Bohm J., Haubelt J., Henzi P., Litfin K. and Hanemann T. Tuning the Refractive Index of Polymers for Polymer Waveguides using Nanoscaled Ceramics or Organic Dyes. Adv. Engg. Mater., 2004, 6(1-2), 52-57.

165. S. S. Papell. U.S. Patent no 3, 215, 572 (1965).

166. R. E. Rosensweig and R. Kaiser. NTIS Rep. No. NASW-1219 (1967).

167. R. E. Rosensweig. Ferro hydro dynamics, Cambridge Univ. Press, Cambridge, London (1985); republished by Dover. Publ. Inc., New York (1997).

168. S. W. Charles and J. Popplewell. In Ferromagnetic Material, edited by E. P. Wohfarth, North-Holland Publishing Company, Vol. 2, Amsterdam (1980).

169. http://www.ferrotec.com/usa/index.html.

170. http://www.sigma-hc.co.jp/english/index.html.

171. R. Massart. IEEE Trans. Magn. 17, 1247 (1981).

172. J. Depeyrot, G.J. da Silva, C.R. Alves, E.C. Sousa, M. Maga-lhães, A.M. Figueiredo Neto, M.H. Sousa, and F.A. Tourinho. Braz. J. Phys. 31, 390 (2001).

173. R. Massart, E. Dubois, V. Cabuil, and E. Hasmonay, J. Magn. Magn. Mater. 149, 1 (1995).

174. E. Dubois, PhD. Thesis, Université Pierre et Marie Curie,Paris 6 (1997).

175. E. Hasmonay, PhD. Thesis, Université Pierre et Marie Curie, Paris 6 (1998).

176. H. Janocha (ed.), F. Claeyssen. Adaptronics and Smart Structures, Springer Verlag, ISBN 3-540-61484-2, Germany 1999, 124-143.

177. H. Janocha (ed.), Actuators, Springer Verlag, ISBN 3-540-61564-4, Germany 2004, 277-292.

178. C. MacWilliam, Design of a Magnetostrictive Actuator, School of Engineering and Industrial Design, University of Western Sydney, March 2004, Thesis, 1-86.

179. Y. Yamamoto. Tokai University Hiratsuka, Japan; H. Eda, J. Shimizu, Department of Systems Engineering, Ibarki University, Hitachi, Japan, Application of Giant Magnetostrictive Materials to Positioning Actuators, 1999 IEEE /ASME International Conference on Advanced Mechatronics, Altanta USA 1999, Conference paper, 215-220.

180. Y. Yamamoto, Tokai University Hiratsuka, Japan; T.Makino, Hiro Matsui, Institute of Advanced Technology, Moritex Corporation, Yokohama, Japan; Micro Positioning and Actuation Devices Using Giant Magnetostriction Materials, Conference, 2000 IEEE International Conference on Robotics and Automation, San Francisco CA USA 2000, Conference paper pp. 3635-3640.

181. M.J. Dapino, Department of Mechanical Engineering, The Ohio State University 2091 Robinson Laboratory Columbus, OH 43210-1107, USA, On magnetostrictive materials and their use in smart material transducer, Structural Engineering and Mechanics Journal, USA 2002, 1-28.

182. A.B. Flatau, M. J. Dapino, Aerospace Engr. & Engr, Mechanics, Iowa State University; F. T. Calkins, Boeing Information Space and Defense Systems, High bandwidth tuneability in smart vibration absorber, 1998 SPIE Smart Structures and Materials Conference, paper #3329-19/3327-42.

183. S. S. Papell, U.S. Patent no 3, 215, 572 (1965).

184. U.S. Patent 3, 843, 540 (1974).

185. Alexiou Ch., Schmid R., Jurgons R., Bergemann Ch., Arnold W. and Parak, F.G. (2002) in Odenbach, S. (ed.), Ferrofluids, 233-251, Springer Verlag, Berlin.

186. Alexiou C., Arnold W., Klein R. J., Parak F. G., Hulin P., Bergemann C., Erhardt W., Wagenpfeil S. and Lübbe A. S., Locoregional cancer treatment with magnetic drug targeting. Cancer Research 60 (2000) pp. 6641-6648.

187. Alexiou C., Arnold W., Hulin P., Klein R. J., Renz H., Parak F. G., Bergemann C. and Lübbe A. S., Magnetic mitoxantrone nanoparticle detection by histology, X-ray and MRI after magnetic tumor targeting. J. of Magnetism and Magnetic Materials 225 (2001), 187-193.

188. Ferrofluid/Activated Carbon Composites for Water Purification and Emi Shielding Applications; S.K. Barala, M. Arora, Chandni Puri, K.K. Saini, R.K. Kotnala and P.K. Saini. Magneto hydro dynamics. 49 (2013), 3-4, 277–281.

7

Role of Surface and Colloid Chemistry in Nano-Technology

7.1 Colloids

7.1.1 Introduction

A colloidal system represents a multiphase (heterogeneous) system, in which at least one of the phases exists in the form of very small particles: typically smaller than 1 μm but still much larger than the molecules. Such particles are related to phenomena like Brownian motion, diffusion, and osmosis. The terms micro heterogeneous system and disperse system (dispersion) referring to colloidal systems are more general because they include also bi -continuous systems (in which none of the phases is split into separate particles) and systems containing larger, non-Brownian particles. The term dispersion is often used as a synonym of colloidal system.

Colloids are typically defined as having at least one linear dimension approximately between 1 nanometer (3.94×10^{-8} inches) and 1 micrometer, or micron (3.94×10^{-5} inches). A length-scale restriction for colloids is straightforward but arbitrary and misleading; there are no absolute bounds.

It is best to think of a colloid in terms of how the material behaves with respect to inertial or body forces, such as gravity and fluid flow. For example, particles that do not quickly settle out or float to the top of their dispersed phase are considered colloidal.

Study of colloids and interfaces is highly multidisciplinary in nature combining both the concepts and applications from such diverse domains as chemical engineering and manufacturing, analytical and physical chemistry, biochemistry and molecular biology, environmental science, materials science including biomaterials and advanced materials, petroleum engineering and finally, nanotechnology.

The word 'colloid' refers to particles in micron to sub-micron ranges where surface properties and interactions (rather than the bulk properties) become increasingly important with declining size or separation distance. In contrast to the 'bulk' systems, the properties and behavior of colloidal systems depends strongly on the system size and inter-particle distances. Thus, understanding of colloids and interfaces is central to even such classical phenomena and applications such as (to name very few): adhesion, particle-aggregation, wetting, detergency, oil-recovery, flotation, nucleation, bio-surfaces, chromatography, paints, composite materials; foams, emulsions, aerosols and other colloidal dispersion.

Nanotechnology being the interdisciplinary field evolved from the study of colloids and the techniques of integrated circuit fabrication. Most of the hype in nanotechnology has centered on the future creation of micro electro mechanical (MEM) devices. All endeavors in nanotechnology must deal with fundamental issues common to colloids, interparticle forces most directly. Colloids play a major role in the semiconductor industry. Silica–alumina sols polish silicon wafers that go into diode lasers, memory chips, and micro-processors. Everyone takes advantage of colloidal suspensions, especially since the human body contains so many with its cells, proteins, and DNA. Biotechnology and the biological sciences are replete with numerous colloidal systems, including cells, cell membranes, viruses, bacteria, proteins, and DNA.

Colloid science is becoming increasingly pertinent to nano technological developments as we are pushing to use many things smaller, more functional and more efficient.

7.1.2 Types of Colloids

Colloids can be classified into lyophobic and lyophyllic type.

(a) **Lyophobic colloids:** These are liquid hating colloids (Lyo means solvent and phobic means hating). When these colloids are mixed with the suitable liquid, very weak force of attraction exists between colloidal particles and liquid and system does not pass into colloidal state readily. Therefore, lyophobic sols are difficult to prepare. Special techniques are employed to prepare these sols.

In Lyophobic sol, weak forces of interaction exist between colloidal particles and liquid. Hence, lyophobic sols are less stable. Addition of even small quantities of electrolytes can cause particles to precipitate. Unlike Lyophilic colloids, the precipitations of lyophobic colloids do not regain their original state as coagulated mass cannot be dispersed into colloidal form. This shows that lyophobic sols are also irreversible in nature.

(b) **Lyophilic colloids:** These are liquid loving colloids (Lyo means solvent and philic means loving). When these colloids are mixed with the suitable liquid, high force of attraction exists between colloidal particles and liquid. This result in formation of very stable solution called lyophilic sol. These sols are formed by substances like gums, starch and proteins. Lyophilic sol can be easily prepared by directly mixing colloid with the liquid. In Lyophilic sol, forces of interaction between colloidal particles and liquid are quite strong. Hence, Lyophilic sols are very stable and do not precipitate/coagulate easily. However addition of very large quantities of electrolytes can cause particles to precipitate. If large quantity of liquid is added to precipitations or the colloidal solution is stirred properly lyophilic sols can regain their original state. This shows that lyophilic sols are also reversible in nature.

Table 7.1 List of Lyophyllic and Lyophobic colloids

Types of Colloiods	
Lyophilic colloids	**Lyophobic colloids**
Polymers	Dispersions (solid particles in liquids)
Carbohydrates	Later
Protein	Fibres
DNA	Emulsion
Syntetic polymers	Foams
Solutions of polymers: Molecules of colloidal size	Composites
Surfactant solutions	
Micelles	
Liquid crystalline phases	

7.1.3 Colloidal Surfaces and Interfaces

A colloid may be defined as a unit (particle, drop, bubble, molecule, etc.) larger than a "normal" (i.e., non polymeric) molecule (<10 nm in diameter), but smaller than what may be considered a "true" bulk phase (>10,000 nm diameter).

An interface is a region of space through which a system changes from one phase (liquid or solid) to another (solid, liquid, or gas), that change usually occurring over a distance of one to several tens of unit (atomic or molecular) diameters.

In general in liquid dispersions, the area of the interface between the disperse and continuous phases is rather large. For instance, 1 cm^3 of dispersion with particles of radius 100 nm and volume fraction 30% contains interface of area about 10 m^2. Thus colloids and interfaces represent distinct, but related "zones" that span the size range between atoms or molecules and bulk phases. This is the reason why the interfacial properties are of crucial importance for the properties and stability of colloids.

In general, the term "surface" is applied to the region between a condensed phase (liquid or solid) and a gas phase or vacuum, while "interface" is normally applied to systems involving two condensed phases.

7.1.3.1 Importance of Colloidal Surfaces and of Interfaces in Nano-technology

Surfaces and interfaces define a boundary between a material and its surrounding environment and influence interactions with that environment. The particles in a colloidal dispersion are sufficiently large for a well-defined surface of separation, an interface, to exist between the particles and the medium. The presence of an interface also leads to effects such as capillarity and adsorption.

There are several types of interfaces that are of great practical importance. These general classifications include, solid– vacuum, liquid–vacuum, solid–gas, liquid–gas, solid–liquid, liquid–liquid, and solid–solid. (Table 7.2)

Table 7.2 Important interfaces which are of industrial importance

Interface Type	Occurrence or Application
Solid-vapor	Adsorption, catalysis, contamination, gas-liquid chromatography
Solid-liquid	Cleaning and detergency, adhesion, lubrication, colloids
Liquid-vapor	Coating, wetting, foams
Liquid-liquid	Emulsions, detergency, tertiary oil recovery

For colloids to be stable the interfacial free energy at the region through which exchange of the intensive properties of the system takes place should be thermodynamically favorable. In the absence of favorable interfacial free energy which can enlarge the boundary or interface and no other force like gravity exist to separate the two phases by density or other factors, Brownian motion/distrortion forces will disturb the interface until both faces mix. In colloids stability is a relative term with reference to thermodynamics and kinetics.

7.1.3.2 Interfaces and Inter-Particle Forces

Surface area is more important than size for defining colloids. They behave differently than classic objects, or macroscopic bodies, because of their high surface-to-volume ratio, or more accurately, surface-to-mass.

The force due to gravity acts on the mass while intermolecular forces effectively act at particle surfaces. If these interparticle or surface forces are large enough to alter the effects of gravity on an object, then it has the characteristics of a colloid. For example, gold is more than nineteen times as dense as water and should sink in it. But a nanometer-sized gold sphere (a few hundred atoms) has so little mass that the forces of water molecules hitting it from all directions are enough to make it effectively buoyant.

This effect of thermal energy in a dispersing fluid is called **Brownian motion**. The colloidal gold may remain stable for many years as long as it never collides with another interface, that is, the boundary between phases. The concept behind the adage "like dissolves like" is as important for colloids as it is for solutions.

A nanometer (nm) is one-billionth of a meter (10^{-9} m, or 3.94×10^{-8} inches), approximately six water molecules end-to-end. A micron (μm) is 1,000 nanometers or one-thousandth of a millimeter (10^{-6} m, or 3.94×10^{-5} inches); the thickness of one strand of hair is typically 50-100 microns ($1.97^{-3}.93 \times 10^{-3}$ inches) in diameter.

Fog and clouds contain water droplets of smaller diameters but still above 1 micron. Red blood cells are about 7 microns (2.76×10^{-4} inches), smoke particles under a micron to 10 nanometers (3.94×10^{-5} inches to 3.94×10^{-7} inches). Very large molecules called polymers (e.g., DNA, cellulose, nylon) are also colloids and may be as large as tens of nanometers, comparable to viruses.

The Scottish botanist Robert Brown (1773–1858), who also discovered the cell nucleus, first noticed small pollen particles wiggling and moving randomly under a microscope. Higher temperatures result in larger kinetic energies in the dispersant molecules, keeping even larger particles from sedimenting or floating. Thus, temperature is also a determining parameter in the definition of a colloid. Atomic and molecular interactions determine both chemical solubility and colloidal stability. Similar molecules attract each other more than dissimilar ones.

This universal **attraction** between everything, called London dispersion, is the result of electron motion. Surfactants concentrate at interfaces by adsorption to remove lyophobic (solvent-fearing) parts from the solvent. This behavior lowers the liquid's surface tension; that is to say, it lessens the imbalance of intermolecular forces between the solvent and its surroundings. The surface tension of pure water causes it to bead up on hydrophobic surfaces, such as a water-proofed jacket or a rain-treated windshield. Liquids with lower surface tensions, such as oils and alcohols, bead up to lesser extents. Water can be made to wet hydrophobic material by adding surfactant; this is sometimes called "breaking" the surface tension.

At low concentrations, surfactants form traditional solutions. At higher concentrations, surfactant molecules may self-assemble into colloidal-scale structures in bulk solution referred to as micelles. They can be spheres, cylinders, vesicles, or bilayer sheets with nanometer to micron dimensions. Above the critical micelle concentration (CMC), micelles begin to grow by molecular association or aggregation. Hydrophobic tails escape water by sticking together in a core shielded by a shell of head groups. Surfactant solubility may sometimes be too low at room temperature for micellization; not enough molecules can dissolve to start associating.

Though thermodynamics should be favorable for stability of the colloid by keeping total system in low energy state, one can play with kinetics to slow down the colloid unstability process so that desired reactions can take place in the given time.

The factors which help in attaining favorable interfacial energy for colloidal stability and dispersion of nano particles are

(i) repulsive surface forces,

(ii) thermal motion of the particle,

(iii) hydrodynamic resistance of the medium and,

(iv) high surface elasticity of fluid particles or films,

7.1.4 Colloidal Dispersions-Characteristic Features

The characteristic feature of colloidal dispersion lies in the following

1. Particle size
2. Particle shape (and flexibility)
3. Surface charge
4. Surface free energy.

7.1.4.1 Particle Size

At the molecular level, the surface atoms have a different chemical environment, that is, fewer nearest neighbors, from that in the bulk. As a consequence, for nano sized particles these surface atoms with changed atomic and electronic structures exhibit high chemical reactivity. This property makes surfaces and interfaces a favored medium for chemical and biological processes in nature and in technological applications (1).

The particle size distribution (PSD) directly affects the bioavailability of active pharmaceutical ingredients and the safety of intravenous lipid emulsions. In the case of suspensions, colloidal dispersions with nano particles also particle size distribution is important as they also contain a range of sizes.

As the particle size is reduced the surface area increases (as $1/d^2$) accompanied by large surface area to volume ratio per given mass of the particles. This is illustrated by (Figure 7.1) talking a cube of 1 m^3 progressively cut into smaller and smaller cubes, until it is composed of 1 nm^3 cubes. If a bulk material is subdivided into an ensemble of individual nanomaterials, the total volume remains the same, but the collective surface area is greatly increased. This is shown schematically in Figure 7.1.

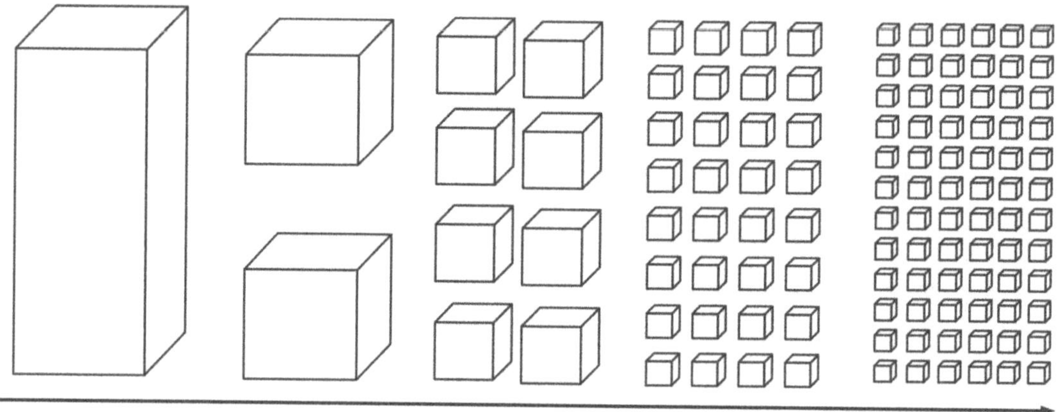

Increase of surface area

Figure 7.1 Schematic drawing showing how surface to volume increases as size is decreased (Source: L. Filipponi, iNANO, Aarhus University, Creative Commons Share Alike 3.0)

The consequence is that the surface-to-volume ratio of the material-compared to that of the parent bulk material is increased tremendously

Increase in surface area when a cube of 1 m^3 is cut into a number of cubes of 1 nm^3 from 6 m^2 to 6000 km^2 is presented in Table 7.3.

Table 7.3 Increase in surface area with decreasing particle size

Size of cube side	Number of cubes	Collective surface area
1 m	1	6 m^2
0.1 m	1000	60 m^2
0.01 m = 1 cm	10^6 = 1 million	600 m^2
0.001 m = 1 mm	10^9 = 1 billlion	6000 m^2
10^{-9} m =1 nm	10^{27}	6×10^9 = 6000 km^2

This huge increase in surface area effects not just adsorption of chemicals and other moieties onto the particle surface but also the interaction between particles and system properties, such as suspension rheology, coating and adhesion. It also allows for much faster dissolution of leading to increased bioavailability in the case of sparingly water soluble but membrane permeable molecules. As low bioavailability can lead to inefficient treatment, higher cost and risk of toxic side effects; in pharmaceutical industry there is great interest to develop reformulations based on nanotechnology. There is also a growing body of evidence that, specifically with nano particulate materials, it is the surface area and not particle size that is the defining metric that controls toxicological interaction.

The chemical groups at the material in interface determine surface properties like catalytic reactivity, electrical resistivity, adhesion, gas storage and chemical reactivity depend on the nature of the interface.

Thus, at the interfaces between the disperse phase and the dispersion medium, electrical (surface charge) effects that are normally negligible for massive solids become dominant in the description of colloidal behavior; they play an increasing part in determining the physico-chemical properties (such as surface chemistry/activity and catalysis) of the system as a whole

7.1.4.2 Particle Shape

Particle asymmetry is a factor of considerable importance in determining the overall physico-mechanical properties of colloidal systems as with shape the surface area also changes.

Shape is a function of the history of the formation of the particles, i.e., crystallization (2). Although the exact shape may be much more complex, colloidal size particles can be roughly classified as: corpuscular (spherical and ellipsoid), laminar (disc- or plate-like) or linear (rod- or needle-like).

"Globular" proteins (albumin, globulin, casein, hemoglobin) adopt a compact "random coil" configuration that approximates sphericity (3). Many active pharmacuetical ingradients exist as rod- or needle-like particles.

High molecular weight macromolecular materials (proteins, polysaccharides and synthetic polymers) usually exist in the form of long thread-like or branched-chain molecules; these materials often exhibit a considerable mechanical strength and durability not possible with corpuscular or laminar particles (4). Their shape is influenced by solution conditions (temperature, pH, salt/electrolyte concentration).

7.1.4.3 Surface Charge

All dispersed particles spontaneously acquire a surface electrical charge when brought into contact with a polar medium (i.e., water) by various charging mechanisms (5) are

1. Affinity differences of two phases for electrons
2. Ionization of surface groups
3. Differential ion adsorption from an electrolyte solution
4. Differential ion dissolution from a crystal lattice
5. Surface anisotropy
6. Isomorphous substitution.

As result of surface charges in colloids and interfaces, significant ramifications takes place in terms of stability, sensitivity to environment, electrokinetic properties, and other factors which can be exploited for wide range of applications in nanotechnology.

7.1.4.4 Surface Free Energy

Interfacial phenomena arise from the fact that atoms and molecules at interfaces, because of their special environment, often possess energies and reactivities significantly different from those of the same species in a bulk.

Nanomaterials/clusters will have a significant proportion of atoms existing at the surface (Figure 7.2). The details on surface energy are discussed more explicitly in Chapter 1.

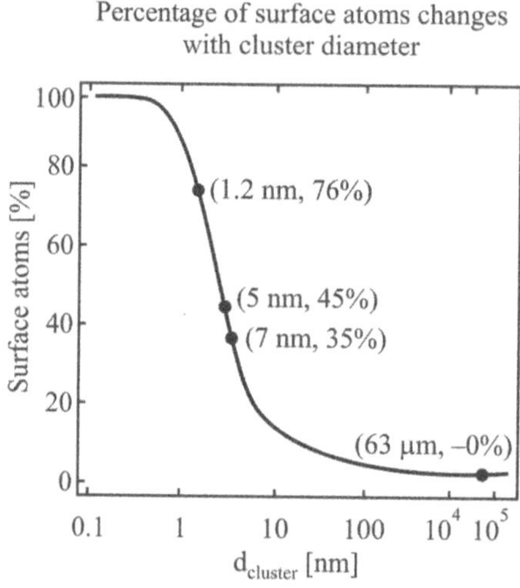

Figure 7.2 Percentage increase of surface atoms as cluster size decreases (6)

Due to large proportion of surface molecule/atoms in nano structures they posses high surface energy (Table 7.4).

For 1 g of Sodium Chloride, the original 1 g cube is successively divided into smaller cubes(7).

Table 7.4 Variation of surface energy with particle size

Side (cm)	Total surface area (cm^2)	Total edge (cm)	Surface energy (J/g)	Edge energy (J/g)
0.77	3.6	9.3	7.2×10^{-5}	2.8×10^{-12}
0.1	28	550	5.6×10^{-4}	1.7×10^{-10}
0.01	280	5.5×10^{4}	5.6×10^{-3}	1.7×10^{-8}
0.001	2.8×10^{3}	5.5×10^{6}	5.6×10^{-2}	1.7×10^{-6}
10^{-4} (1 μm)	2.8×10^{4}	5.5×10^{8}	0.56	1.7×10^{-4}
10^{-7} (1 nm)	2.8×10^{7}	5.5×10^{14}	560	170

This has a profound effect on reactions that occur at the surface such as catalysis reactions, detection reactions, and reactions that, to be initiated, require the physical adsorption of certain species at the material's surface.

Surface energy = energy required (or produced) to create a unit area of a new surface, therefore,

$$\gamma = (\partial G/ \partial A)_{\text{normal}, T, P} \quad .....(7.1)$$

G: Gibs free energy of the surface; A: Surface area of the particles

7.1.5 Mechanisms for Reduction of Overall Surface Energy

Due to the vast increase surface area, all nanostructured materials possess a huge surface energy and, thus are thermodynamically unstable or metastable. However by surface modification using various methods we can modify the surface energy to make the nano particles stable and prevent the nanomaterials from growth in size.

Reduction of surface energy involves combining individual nanostructures together to form large structures so as to reduce the overall surface area, and various methods for achieving includes:

1. Sintering individual structures merge together. Replacing solid-vapor interface by solid-solid interface
2. Agglomeration of individual nanostructures through chemical bonds and physical attraction forces at interfaces without altering the individual nanostructures.

7.1.6 Inter Molecular Forces and Colloidal Forces

7.1.6.1 Colloidal Forces

A colloidal dispersion is not a thermodynamically stable system. Colloidal particles in a dispersion medium are always subjected to Brownian motion with frequent collisions between them. Stability and other characteristics of dispersion are thus determined by the nature of the interactions between the particles during such collisions. When attractive forces dominate, the particles will aggregate and the dispersion may destabilize. When repulsive forces dominate, the system will remain in a dispersed state.

Generally, there are six important types of particle–particle interaction forces that can exist in a dispersion as summarized in Table 7.5.

Table 7.5 Summary of forces between two identical particles in a colloid

Force	Nature	Source
Electrical double layer	Repulsive	Structure of surfaces and adsorption of ions
van der Waals	Usually attractive	Universal presence
Steric interaction	Repulsive	Induced by adsorbed species
Hydration/solvation interaction	Attractive=repulsive	Hydration or solvation of adsorbent and adsorbate
Polymeric bridging	Attractive	Induced by adsorbed long-chain polymer
Hydrophobic interaction	Attractive	Intrinsic or induced by adsorption

7.1.6.2 Van Der Waals Forces

Van der Waals in 1873 (8) noted the non-ideality of gases and attributed it to the existence of molecular or atomic interactions. These interactions were correctly thought to be due to interacting dipoles but incorrectly treated statically. London (1936) (9) treated the interacting two atom system as dynamic and attributed van der Waals forces to the dispersion effect. The dispersion effect is the interaction between the instantaneous dipoles formed in the atoms by their orbiting electrons. The very rapidly changing dipole of one atom produces an oscillating electric field that acts upon the polarizability of a neighboring atom. The microscopic electric polarizability of an atom refers to its ability to respond to an external electric field by shifting its charges so as to create a dipole oriented favorably in the external field (negative end of the dipole closer to the positive pole of the external field). For an isolated atom this may occur by biasing the time-average distribution of electrons among ground and excited state to favor polar states. If excitation is easy, polarizability is high. For molecules in the gas or liquid state skeletal vibrational states and rotational states may also contribute to the polarizability. Molecules that bend easily to produce a dipolar configuration have high polarizability. In the solid state the rotational states are usually suppressed. If the atoms exist in a vacuum, the induced dipole of the neighboring atom moves in phase with the original dipole, producing an attractive atomic interaction.

Thus the Dipole or induced-dipole interactions at the atomic and molecular level are responsible for van Der Waals forces which are three type

1. **Keesom interaction:** Permanent dipole–permanent dipole interaction
2. **Debye interaction:** Permanent dipole–induced dipole interaction
3. **London (or dispersion) interaction:** Induced dipole–induced dipole interaction

These three types are collectively called van der Waals interactions which are proportional to molecular parameters that represent the polarization of molecules (polarizability, dipole moment of molecule), and have inverse sixth-power dependence on the separation between the nearest two molecules. There is a portion of interparticle repulsion at the molecular level that is also a function of interparticle separation. This force has inverse twelth-power dependence. Van der Waals interactions are always attractive at atomic and molecular levels although between colloidal objects under certain conditions, they can be repulsive.

Debye, Keesom, and London interactions for a pair of molecules can be represented as

Debye equation

$$U(x) = \frac{(\alpha_1\mu_2^2 + \alpha_2\mu_1^2)}{(4\pi\varepsilon_0)^2}x^{-6} \qquad .....(7.2)$$

Keesom equation

$$U(x) = -\frac{2}{3}\frac{\mu_1^2\mu_2^2}{(4\pi\varepsilon_0)^2 k_B T}x^{-6} \qquad .....(7.3)$$

London equation

$$U(x) = -\frac{3}{2}h\frac{\alpha_1\alpha_2}{(4\pi\varepsilon_0)^2}\frac{v_1 v_2}{(v_1 + v_2)}x^{-6} \qquad .....(7.4)$$

where x is the distance between two interacting molecules, α_1 and α_2 is the polarizability of each molecule, μ_1 and μ_2 is the dipole moment of each molecule, k_B is the Boltzmann's constant, T is temperature, h is the Plank's constant, ε_0 is the dielectric permittivity of vacuum, and v_1 and v_2 is the vibrational frequency of electron of each molecule. These three equations can be combined into one equation that can represent the net (or total) contribution to van der Waals attraction. For a simple example, that is, for a pair of identical molecules:

$$U_t(x) = -\frac{1}{(4\pi\varepsilon_0)^2}\left(2\alpha\mu^2 + \frac{2}{3}\frac{\mu^4}{k_B T} + \frac{3}{4}h\nu\alpha^2\right)x^{-6} = -\beta x^{-6} \qquad \text{.....(7.5)}$$

where the subscript t on potential means total potential and β is the van der Waals interaction parameter.

For a thermodynamics point of view, β can be well correlated with van der Waals parameters a and b in the van der Waals equation for state of gas through equations

$$a = \frac{2}{3}\frac{\pi N_A^2 \beta}{\sigma^3} \qquad \text{.....(7.6)}$$

and

$$b = \frac{2}{3}\pi\sigma^3 N_A \qquad \text{.....(7.7)}$$

where N_A is Avogadro's number and σ is the diameter of atom or molecule.

These relations provide one way to estimate van der Waals interaction parameter β between molecules from known values of a and b.

For a colloidal scale the van der Waals interactions strongly dependent on the geometry and composition of the colloidal objects involved. All molecules from one object interact with all molecules from the other object pair-wisely. The summation of these attractive forces will be the interaction force between the two colloidal objects considered.

For two spherical colloidal objects with the same size and composition, the same interactions (whether it is each of Debye, Keesom, or London interaction, or the total interaction) that can induce the association of individual molecules are responsible for the aggregation of colloidal-scale objects up to a certain length scale. However, the sixth-power dependence potential decay on the separation of two objects turns into seventh-power dependence decay at the separation in the range of 10–100 nm. This means the van der Waals interaction (attraction) between colloidal bodies decays more rapidly as the separation between them increases. The degree of this decay as a function of the distance between the colloidal objects is dependent on the geometry of the objects. The more complex geometry brings the more complicated form of expression. Hamaker constant A is a well-established parameter that can provide the relative picture and magnitude of the van der Waals interaction between colloidal objects (Hamaker, 1937) (10,11,12). It typically has the value between 10^{-19} and 10^{-20} J. For example, it can be expressed as follows for two identical interacting colloidal blocks (semi-infinite parallel plates):

$$U(x) = -\left(\frac{\rho N_A \pi}{M}\right)\beta\frac{1}{12\pi}x^{-2} = -A\frac{1}{12\pi}x^{-2} \qquad \text{.....(7.8)}$$

β is the van der Waals interaction parameter between the molecules, ρ is the density of molecules in the block, M is the molecular weight, and x is the distance between the two blocks. The intrinsic assumption made for this expression does not count on the important facts of a real situation including surface heterogeneity, screening of inside molecule by surface molecule, and interaction with a medium such as solvent.

For the colloidal objects interacting through a vacuum, the Hamaker constant should be always positive, meaning the interaction is always attractive. Typical Hamakers's constants for various colloidal objects are given in Table 7.6.

Tale 7.6 Hamakers's constants in air (vaccum) and in water all in 10^{-20} J =2.5 kT

	In air	In water
A (water)	4.4	0
A (hydrocarbons)	4-10	0.3–1
A (oxides and halides)	6-15	0.5-5
A (metals)	15-50	5-30

7.1.6.3 Electrostatic Forces and the Electrical Double Layer

Most solid surfaces in contact with water or an aqueous solution will be found to develop some type of electrical charge. The magnitude of that charge may be quite small or very large, but it will almost always exist. In macroscopic systems, the presence or absence of a surface charge may often be overlooked. However, in the microscopic world of colloids and interfaces, the presence or absence of even a small surface charge can have significant ramifications in terms of stability, sensitivity to environment, electrokinetic properties, and other factors.

As a result of the presence of electrical charges, surfaces exhibit various properties that are not present in systems having no surface charge (e.g., nonpolar solids dispersed in a nonpolar medium). Of particular interest are the so-called electrokinetic properties of colloids. As a result of their unique characteristics, charged surfaces provide a handle for manipulating a wide variety of multiphase systems that are of vital importance to life and nano technology. By the proper control of surface charges, we may be able to transform an interesting but useless system into one of great practical importance. Likewise, we may be able to take a system that is a nuisance colloid and, by changing its electrical environment, remove it completely or at least attenuate its effects.

7.1.6.3.1 Sources of Interfacial Charge

An interface may acquire an electrical charge by one or more of several mechanisms, the most common of which include

1. preferential (or differential) solution of surface ions,
2. direct ionization of surface groups,
3. substitution of surface ions,

4. specific ion adsorption, and
5. charges deriving from specific crystal structures.

While other mechanisms can be invoked, these five represent the most common and most important encountered in colloidal systems. Each type of surface charge carries with it certain characteristics that define, partially at least, the electronic nature of the resulting colloid or surface.

7.1.6.3.2 Electrostatic Theory: Coulomb's Law

The fundamental law governing interactions between charged species q_1 and q_2 in a vacuum (F_{el}) separated by a distance r, is the Coloumbs law which takes the form

$$F_{el} = \frac{q_1 q_2}{4\pi\varepsilon_0\varepsilon_r} \qquad .....(7.9)$$

where the symbols are as defined in equation 3. The work necessary to bring the two charges together from infinity to the distance r is

$$w(r) = -\int_{\infty}^{r} F_{el}\,dr = \frac{q_1 q_2}{(4\pi\varepsilon_0\varepsilon_r)} \qquad .....(7.10)$$

For charges of the same sign, w(r) will be positive and the interaction will be repulsive; if of opposite charge, it will be attractive.

Consider for a moment that a charge q_1 is isolated in space. It will produce an electric field at a point r, such that the work necessary to bring a unit electrical charge from infinity to distance r from q_1 will be equal to $q_1/(4\pi\varepsilon_0\varepsilon_r)$. That quantity of work is defined as the electrical potential at r due to the charge q_1, and is given the symbol ψ . According to Equation (7.10), the force involved in bringing a charge q_2 to within a distance r of q_1 will be simply q_2.

7.1.6.3.3 Boltzmann's Distribution and the Electrical Double Layer

While Coulomb's law is simple enough in isolation, in order to make use of it in more realistic situations, the presence of all ions in the system must be taken into consideration. In practical applications, one is concerned with solutions containing many charges (i.e., dissolved ions), particles, and surfaces that may also contain charges of the same or different sign. In order to apply Coulomb's law to solutions of electrolytes and colloids, it is necessary to employ Boltzmann's distribution law, which relates the probability of a unit (atom, molecule, ion, particle, etc.) being at a certain point with a specified free energy (or potential energy), ΔG , relative to a specified reference state. The probability is generally expressed in terms of an average unit concentration, c, at the point r relative to a concentration, c_0, at some reference distance at which the energy is taken as zero. At a temperature, T, the Boltzmann distribution is given as

$$c = c_0 \exp \frac{-\Delta G}{kT} \qquad .....(7.11)$$

When applied to the situation involving charged particles and Coulomb's law, Equation (7.9) predicts that if there exists a negative electrical potential ψ at some point in an electrolyte solution, then in the region of that point the concentration of positive charges, c_+, will be given by

$$c_+ = c_0 \exp\frac{-z^+ e\psi}{kT} \qquad \text{.....(7.12)}$$

where z^+ is the valency of the positive ion and c_0 is the concentration of the positive ion in a region where $\psi = 0$. A similar expression can be written for the negative ions in the solution

$$c_- = c_0 \exp\frac{+z^- e\psi}{kT} \qquad \text{.....(7.13)}$$

Although the solution as a whole will be electrically neutral, in the vicinity of the electrical potential there will exist an imbalance of electrical charges. Thus, for a negative, ψ there will be more positive ions in the region than negative ions. For the case where $z^- = z^+ = 1$, the excess is given by

$$c_+ - c_- = c_0\left[\exp\left(\frac{-e\psi}{kT}\right) - \exp\left(\frac{+e\psi}{kT}\right)\right] \qquad \text{.....(7.14)}$$

The region of excess charge of the opposite sign around a potential is commonly referred to as the ''ionic atmosphere'' or ''charge cloud'' associated with that potential.

7.1.6.3.4 Double-Layer Thickness: The Debye Length

Boltzmann distribution law is fundamental to the derivation of the Debye–Hückel theory of electrolyte solutions. It is applied to the determination of the ionic atmosphere around charged inter-faces. In that context, the charge cloud is more commonly referred to as the electrical double layer (EDL). The concept is illustrated schematically (Fig. 7.3) for the situation in which a particle possesses an evenly distributed charge that is just balanced by the total opposite charge, the counterions in the electrical double layer.

In reality, of course, the thermal motion of ions in solution introduces a certain degree of chaos causing the ions to be spread out in the region of the charged surface, forming a ''diffuse'' double layer in which the local ion concentration is determined by Equation (7.13). In that case, the analysis of the electronic environment near the surface is more complex and requires more detailed analysis. An additional ''reality'' is that charges occupy a finite amount of space and therefore have certain steric requirements, leading to the postulation of the so-called Stern layer. The Stern layer is a small space separating the ionic atmosphere near an interface, the diffuse double layer, from the steric ''wall'' of the charged plane just adjacent to the interface. The thickness of the Stern layer, d , is usually on the order of a few nanometers and reflects the finite size of charged groups and ions specifically associated with the surface. While the Boltzmann distribution is relatively easy to evaluate for a single point charge, the situation can become quite complex in the case of a surface having many charges.

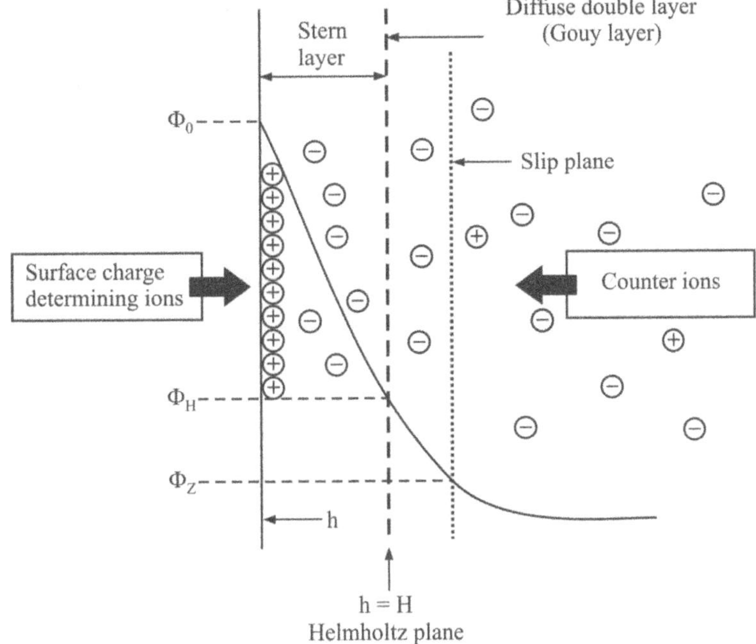

Figure 7.3 Electrical double layer structure variation of electric potential with changing inter surface distance.

From a practical (and very simplistic) point of view, it is normally assumed that the electrical potential in the solution surrounding the surface in question falls off exponentially with distance from the surface according to the Debye–Hückel approximation

$$\varepsilon = \varepsilon_0 \exp(-k_2) \quad \text{.....(7.15)}$$

where k is identified as the reciprocal of the thickness of the electrical double layer, also commonly referred to as the "Debye length." Thus at a distance of 1/k from a charged surface, the potential has fallen off by a factor of 1/e. For low surface potentials (<25 mV), the theoretical equation for the double-layer thickness, 1/k, is

$$\frac{1}{k} = \left(\frac{\varepsilon_0 \varepsilon kT}{e_1^2 \sum c_i z_i^2} \right)^{1/2} \quad \text{.....(7.16)}$$

We can see that the thickness of the electrical double layer is inversely proportional to the concentration of electrolyte in the system and to the square of the valency of the ions involved. In terms of colloidal stability, this means that the distance of separation between two particles that can be maintained under a given set of circumstances will depend on, among other things, those two factors. And their important effect gives one a handle for manipulating the characteristics and stability of many colloidal systems.

7.1.6.4 Steric and Depletion Forces

Steric forces arise from the overlap of the adsorbed layers and can be repulsive or attractive in nature depending on whether or not the outermost layers on the particles prefer to be in contact with the solvent. If the solvent power of the medium for the exposed portions of the exterior layer, for example, those of an adsorbed polymer layer, is satisfactory, they will be compatible with the medium and the suspension will remain stable (13,14).

On the other hand, if the solvent power for the adsorbed layer is minimal, there will be a tendency for the exposed portions of the adsorbed layer on one particle to interpenetrate into those of a layer on another particle and thereby promote aggregation.

Interpenetration and aggregation is possible only if the net change in Gibbs free energy due to the interpenetration of the polymer chains is negative. The Gibbs free energy change is considered in such cases to be determined essentially by the change in entropy due to the release of solvent molecules and due to the decrease in randomness of the polymer chain and by the enthalpy of desolvation of the polymer chains.

For flocculation to occur according to this mechanism, the increase in entropy due to the release of solvent molecules should outweigh the loss of entropy due to the interpenetration of polymer chains and increase in enthalpy of desolvation. It is clear that the desolvation characteristics of the adsorbed polymer species and the dependence of it on solution properties such as temperature and ionic strength will be important in determining aggregation by this mechanism. Ottewill and Walker derived an equation for the energy change due to an overlap of the adsorbed layers by using Flory's liquid lattice model for polymer solutions (13,15).

$$V_{steric}(a) = \frac{4\pi kTC_v^2}{3v_1^2\rho_2^2}(\psi_1 - \kappa_1)(\delta - a)^2(3R + 2\delta + a/2) \qquad .....(7.17)$$

where C_v is the concentration of material in the adsorbed layer

v_1 is the molecular volume of solvent molecules

ρ_2 is the density of the adsorbate

ψ_1 and κ_1 are the entropy and enthalpy parameters, respectively, of mixing proposed by Flory

δ is the adsorbed layer thickness

R is the particle radius

a is the distance separating the surfaces of two approaching particles

Steric repulsion can be explained in the following manner. When two particles with adsorbed long-chain molecules approach each other, the entropy per adsorbed molecule decreases causing desorption and a concomitant increase in the interfacial energy. This means that additional work has to be input to bring the particles together and this manifests itself as a repulsive force.

Thus Steric force is mainly induced by polymers, polyelectrolytes, or biomacromolecules that are adsorbed or grafted onto the surface of colloidal objects. It is a long-range force whose interaction can reach up to $\sim 10 \cdot R_g$ (where R_g is the radius of the gyration of the polymer chain) in aqueous solution.

Every possible interaction, including polymer–polymer, polymer–solvent, and polymer–colloid interaction, and the physical/chemical conditions of the system, including solvent (good or poor in a given polymer), temperature, and nature of charge (especially in the case of polyelectrolytes) involved can affect this force.

Polymer-induced forces can be either repulsive or attractive. depending on solvent –polymer interactions

A long-chain adsorbate with several active sites on it can induce aggregation by attaching itself to two or more particles. Polymers can provide such bridging between particles particularly under conditions where particles are not totally coated by the polymeric species.

If particles are already fully covered with polymers, bridging can take place only if there is either detachment of some portion of the polymer already on a particle to provide sites for attachment of polymer fractions adsorbed on other particles or polymer–polymer bonding itself. It has been suggested that maximum flocculation occurs when the fraction of particle surface covered by polymer molecules is close to 0.5. It is to be noted that this hypothesis has not, however, been experimentally proven.

As indicated earlier, bridging should be possible even when particles are fully covered if some detachment and reattachment of the adsorbed polymer is possible.

Adsorption of polymers is in general found to be irreversible, that is, except in the occasional case of polymers with functional groups that form covalent bonds with the surface species, only due to the cumulative effect of adsorption of a single-polymer molecule on several sites.

Detachment and reattachment of portions of such a polymer should, however, be possible and such a process should facilitate bridging. This has been shown to be so by allowing polymers to be displaced from the surface by adding competing agents into the system. Thus polyacrylamide can be desorbed from clay by adding phosphates into the solution. Absence of bridging might be interpreted to suggest the presence of strong repulsive steric interactions between dangling portions of the adsorbed polymer chains.

For steric stabilization and bridging flocculation by adsorbed polymers, there is still no satisfactory quantitative theory. This is because any theory for those processes must be couched in terms of one or the other of the theories of polymer solution thermodynamics which are at best semiquantitative. Also, to calculate the distance dependence of the steric interaction, it is necessary to predict the conformation of close-packed polymer molecules. In fact, except in two solvents, it is very difficult to predict the polymer conformation in solution (13). Despite these difficulties, some progress has been made in developing pragmatic models capable of predicting the distance dependence of the steric and bridging interactions.

7.1.6.5 Solvation and Hydration Forces

When two colloidal surfaces or objects are closer than a few nanometers, the DLVO (Derjaguin-Landau- Verwey-Overbeek) calculation, which was based on the van der Waals attractive force and the electric double-layer repulsive force in colloidal state often shows a large discrepancy from experimental results. This rather short-range interaction (compared with van derWaals and electric double-layer interactions) arises from the layering (or ordering) of solvent molecules on/near the surface of colloidal objects.

This is, in general, called solvation force (or structural force), and when water is the solvent, it is called hydration force. It often shows an oscillatory pattern as a function of the distance between two surfaces, and the pattern decreases exponentially as the distance increases.

7.1.6.5.1 Solvation Force

For an atomically smooth surface, when the separation between the two surfaces is close enough to the molecular size (diameter for the spherical molecule and size of the short axis for the rod-shaped one) of the solvent, the solvent molecules become ordered along/ between the two surfaces.

As this situation increases the molecular density in this region compared with the bulk, this region becomes entropically unfavorable, so the two surfaces experience a repulsive force. At distances for the close-packing of the molecules (also at double, triple, and so on of the molecular size of the solvent), the same situation causes a repulsive force between the two surfaces, but with the magnitude usually decreasing monotonically.

At distances in between, the molecules are disordered. This produces less molecular density between the surfaces, and hydrodynamic pressure from the other side of the surfaces causes the attractive interaction between the surfaces. This attractive force also diminishes exponentially as the distance between the surfaces increases.

Solvation force depends on the physical factors both of solvent molecule and of colloidal surface. From the solvent side, it includes shape, size, and polarity of solvent molecule. From the surface side, it includes surface properties such as hydrophobicity, hydrophilicity, surface roughness, and surface homogeneity.

7.1.6.5.2 Hydration Force

When the interaction occurs solely in water, the interaction between the colloidal objects described above differs significantly. It is always repulsive whenever the surfaces that are interacting are hydrophilic in nature. When the surfaces are hydrophobic, the hydration force becomes attractive, and it is called the hydrophobic attraction. Hydration repulsion is a short-range force that operates below ~3 nm of separation between the surfaces, as compared with the DLVO force or double-layer repulsion that operates over the range up to ~100 nm.

Hydration force seems to appear whenever the water molecules strongly bind to the hydrophilic groups on the surfaces such as hydroxyl, ammonium, phosphate, or sugar, especially in the case of biological systems. Thus, the binding nature of water molecules on the surface such as orientational degree and density, and the nature of the surface, such as softness, hardness, roughness, and homogeneity, and the density/nature of functional groups on the surfaces are important determining factors for hydration force.

Hydration force is important in many industrial processes such as froth flotation, clay processing, and clay-related nano composites preparation/processing. Also, along with hydrophobic interaction and electric double-layer repulsion, hydration force plays an important role in especially the self-assembly of many biological systems such as membranes, protein folding/unfolding, and enzymes.

7.1.6.6 Hydrophobic Effect

Hydrophobic interactions may exist naturally or be induced by the adsorbed hydrophobic species. Polar solvent molecules squeezed between two hydrophobic surfaces have reduced freedom to form structures in certain directions, since contact with the particle surfaces is essentially avoided. The hydrophobic surfaces therefore have a preference to associate with each other. It is found that hydrophobic interactions extend over a much longer range than the vander Waals force (16). Graphite and coal are a few of the solids possessing natural hydrophobicity and their aggregation is observed in polar solvents. More often, surface hydrophobicity of the solid particles can be induced by the adsorption of surfactants.

Hydrophobic forces have been experimentally first measured by Israelachvili and Pashley (17) with mica surfaces in equilibrium with cetyltrimethylammonium bromide solution. The hydrophobic force was found to decay exponentially in the range of 0-10 nm. van Oss (18) described the rate of decay with distance in terms of interaction energy as

$$\Delta G_1 = \pi R \lambda \Delta G_{l_0} \exp\left[\left(l_0 - l\right)/\lambda\right] \qquad .....(7.18)$$

where l is the correlation length of the molecules of the liquid medium; DG is the interaction energy at the distance of l_0, l_0 is the minimum equilibrium distance.

In aqueous solutions, $\lambda = 0.6$, three times of water molecule dimension, which may be attributed to the fact that the water molecules tend to occur in clusters of three or more molecules resulting from hydrogen bonding with each other.

Though the real origin of the hydrophobic force still remains to be a mystery, its effect on the dispersion stability has long been recognized. Somasundaran et al., (19) reported a clear relationship between the stability and the surfactant adsorption density as well as the chain length of the adsorbed surfactants.

When a substance that is sparingly soluble in water, such as paraffin-chain alcohols, long-chain amines, hydrocarbons, and rare gases, contacts with water. The tetrahedral bonding sites of water molecules are disrupted around the solute – water interface or at the surface of solute. Such molecules are known as hydrophobic substances. The most favorable way to reorganize energetically for water molecules is giving up one bonding site and forming a water molecular network around the solute molecules. This is called iceberg uster or iceberg formation. This is actually a loss of entropy of the system, and it is compensated for and overcome by the enthalpy term. When the solute molecules with water clusters come close together, this iceberg structure starts to break into individual water molecules. As they go back into the bulk water, they regain their original molecular motion. Now, the solute molecules that come close together obviously lose their entropy, but the overall system gains a significant amount of entropy that overcomes the entropy loss from the aggregation of solute molecules. This is why hydrophobic substances often show an unusually stronger interaction in water than in a gas state, that is, mainly van der Waals interaction.

Thus Hydrophobic effect is the loss of entropy of the system which is compensated for and overcome by enthalapy term as $G = H - T.S$ and each contribution will be changing as function of temperature and pressure. Hydrophobic effect plays a major role in self assembly of nano structures.

7.1.6.6.1 Surfactants Surface Active Materials: which can Induce Surface Hydrophobicity are known as Surfactants

Structure and Chemical Characteristics Required for Surface Activity

Surface-active materials (surfactants) possess a characteristic chemical structure that consists of:

1. Molecular components that will have little attraction for one surrounding (i.e., the solvent) phase, normally called the lyophobic group, and
2. Chemical units that have a strong attraction for that phase - the lyophilic group

Classification of Surfactants

Surfactants may be classified as emulsifiers, foaming agents, wetting agents, dispersants, or similar. Surfactants may also be generally classified according to some physical characteristic such as its degree of water or oil solubility, or its stability in harsh environments.

The four general groups of surfactants are defined as follows:

1. Anionic, with the hydrophilic group carrying a negative charge such as carboxyl ($RCOO^- M^+$), sulfonate ($RSO^- M^-$), or sulfate ($ROSO^- M^+$).
2. Cationic, with the hydrophile bearing a positive charge, as for example, the quaternary ammonium halides ($R4N^+ X^-$).
3. Nonionic, where the hydrophile has no charge but derives its water solubility from highly polar groups such as polyoxyethylene ($UOCH_2CH_2OU$), sugars or similar groups.
4. Amphoteric (and zwitterionic), in which the molecule has, or can have, a negative and a positive charge on the principal chain (as opposed to a counterion, M^+ or X) such as the sulfobetaines, $RN^+(CH_3)_2CH_2CH_2SO^-$
3

7.1.6.6.2 Surfactant Solubilizing Groups

The most common hydrophilic groups encountered in surfactants today are illustrated in Table 7.7.

Table 7.7 The most commonly encountered hydrophilic groups in commercially available surfactants

Class	General Structure
Sulfonate	$R–SO_3^- M^+$
Sulfate	$R–OSO^- M^+$
Phosphate	$R–OPO_3^- M^+$
Ammonium	$R_xH_yN+X^-$ (x = 1 – 3, y = 4 – x)
Quaternary ammonium	R_4N+X^-
Betaines	$RN^+ (CH_3)_2 CH_2 COO^-$
Sulfobetaines	$RN^+ (CH_3)_2 CH_2CH_2SO_3^-$
Polyoxyethylene (POE)	$R – OCH_2CH_2 (OCH_2CH_2)_n OH$
Polyols	Sucrose, sorbitan, glycerol, ethylene, glycol, etc
Polypeptide	R – NH – CHR – CO – NH – CHR' – CO – ... CO_2 H
Polyglycidyl	$R – (OCH_2CH [CH_2OH]CH_2)n^-$ .... – $OCH_2CH[CH_2OH]CH_2OH$

R designates some suitable hydrophobic group that imparts surface activity, M is an inorganic or organic cation, and X is an anion (halide, acetate, etc).

7.1.6.6.3 Common Surfactant Hydrophobic Groups

Most common hydrophobic group used in surfactants is the hydro-carbon radical having a total of 8–20 carbon atoms. Commercially there are two main sources for such materials that are both inexpensive enough and available in sufficient quantity to be economically feasible: biological sources such as agriculture and fishing, and the petroleum industry (which is, of course, ultimately biological). Listed below and illustrated structurally in Table 7.8 are the most important commercial sources of hydrophobic groups.

Table 7.8 The most commonly encountered hydrophobic materials used in the commercial manufacture of surfactants

CH_3 $(Ch_2)n$ Natural fatty acids (n = 12 –18)	CH_3 $(CH_2)n$ CH_3 Petroleum parafine (n = 8 – 20)
CH_3 $(CH_2)n$ CH = CH_2 Olefine	CH_3 $(CH_2)n$ CH_2–(benzene ring) Alkylbenzenes (n = 6 – 10, linear or branched

Table 7.8 *Contd...*

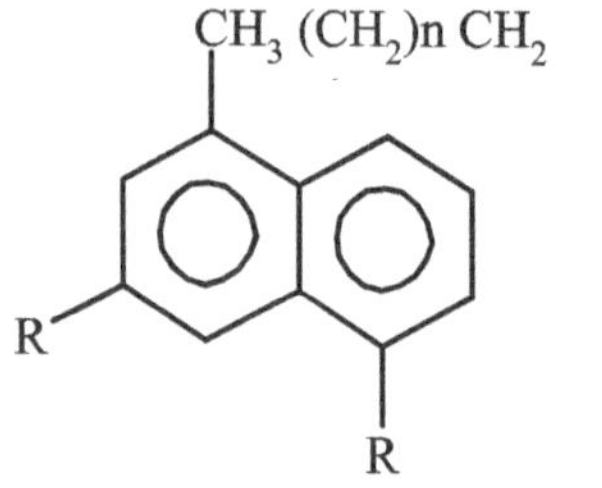 $CH_3\ (CH_2)n\ CH_2$ (naphthalene ring with substituents R, R) Alkylaromatics (n = 1-2 for water soluble = 8 or 9 for oil soluble surfactants)	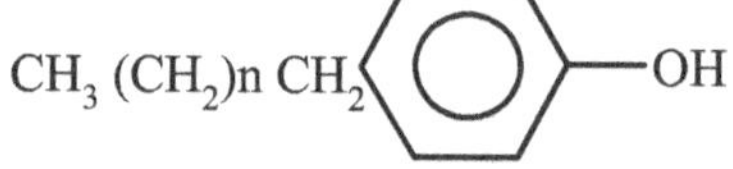 $CH_3\ (CH_2)n\ CH_2$ (benzene ring) –OH Alkylphenols (n = 6 – 10, linear or branched)
	$CF_3\ (CF_2)n\ COOH$ Fluorocarbons (n = 4 – 8, may be branched of H- terminated)
$CH_3\ (CHCH_2O\ (CH\ CH_2)n$ with X and CH_3 substituents Polyoxypropylene (n = degree of oligomerization X = oligomerization, initiator)	O $CH_3O\ (Sl\ O)n\ CH_3$ CH_3 Silicones

7.1.6.7 Capillary Condensation Forces

Phase changes can occur at the interface of solids and liquid mixtures. As mentioned earlier, a surface excess of a second species will cause a disruption of the molecular ordering and a reduction of the free energy maxima and minima arising from packing constraints.

In the case of water on the surface of mica in a nonpolar solvent due to the low contact angle of water on mica, it will form an annulus around the contact area of two such surfaces. This is analogous to capillary condensation from the vapor phase. The thermodynamically stable state for two surfaces in close proximity in a nonpolar liquid containing water is with a bridge of water connecting the two surfaces.

The water bridge should be stable when the surface separation is less than or equal to twice the Kelvin radius. The net effect of this spontaneous capillary condensation occurring at separation distances lower than about 10 Å is to provide a very strong attractive force as well as large adhesion between the surfaces. The adhesion at contact reflects the interfacial tension of the water–nonpolar liquid interface and is an order of magnitude larger than expected either from van der Waals or solvation forces (20).

7.1.6.8 Hydrogen Bond

A hydrogen bond is a strong and directional intermolecular interaction. It is directly responsible for all those unexpected but interesting and vital physical properties of water molecules. Some examples include its unusually high melting and boiling points when compared with those molecules with similar molecular weight, its density minimum at 4 °C, and the increasing of the dielectric constant when liquid water is transformed into ice, unlike most typical polar molecules.

Water is not the only molecule that can form a hydrogen bond. When highly electronegative atoms such as oxygen, nitrogen, and fluorine are covalently bonded with hydrogen, their high electronegativity strongly attracts an electron cloud of hydrogen atoms, which makes the hydrogen atom electron deficient, meaning positively polarized. This positively polarized hydrogen atom can then interact strongly with the nearby electronegative atoms. The bond length of a typical hydrogen bond is less than that of the combined van der Waals radii of the two atoms involved, but larger than that of the covalent bond of the two atoms involved. This is in line with the general order of bond strength: covalent bond (<500 kJ/mol) > hydrogen bond (10–40 kJ/mol) > van der Waals interaction (<1 kJ/mol). Water–water interaction from a hydrogen bond is about 12–20 kJ/mol.

What makes the hydrogen bond directional is its unique bonding sites. For example, even though the van der Waals interaction can sometimes make molecules align along their axis of dipole moment, it does not occur on the specific interaction site, meaning its direction of interaction is always random. On the other hand, the hydrogen bond–capable molecules always interact only through specific bonding sites that are polarized hydrogen atom and lone-pair electron. Depending on the number of these bonding sites on the molecules, different types of hydrogen-bonded structures can be formed. This includes the linear chain, ring, dimer, layered sheet, and three-dimensional network. For certain types of molecules, intramolecular hydrogen bonding can also occur (Figure 7.4).

It is especially important for biological molecules such as DNA, RNA, proteins, and membranes. For instance, it is greatly responsible for the formation and stability of the DNA double-helical structure, integrity and function of biological membranes, and cell functions such as cellular transport. Both the strength of the hydrogen bond and its directionality are critical for these events.

One of the unique features of the hydrogen bond in molecular self-assembly is that it provides the stability and the directionality to the molecular self-assembly. This causes the self-assembled structures to be formed with a rich range of morphology and important functionality.

The hydrogen bond is also a key driving force for the gelation of many organic molecules, macromolecules, and biological molecules and systems such as cells. Pollack (2001) (21)) introduces the revolutionary view on biological cells and their functions based on the gelation process in the biological systems. This concept is based on the hydrogen bond between water and biological molecules, and its cascade-like formation and destruction.

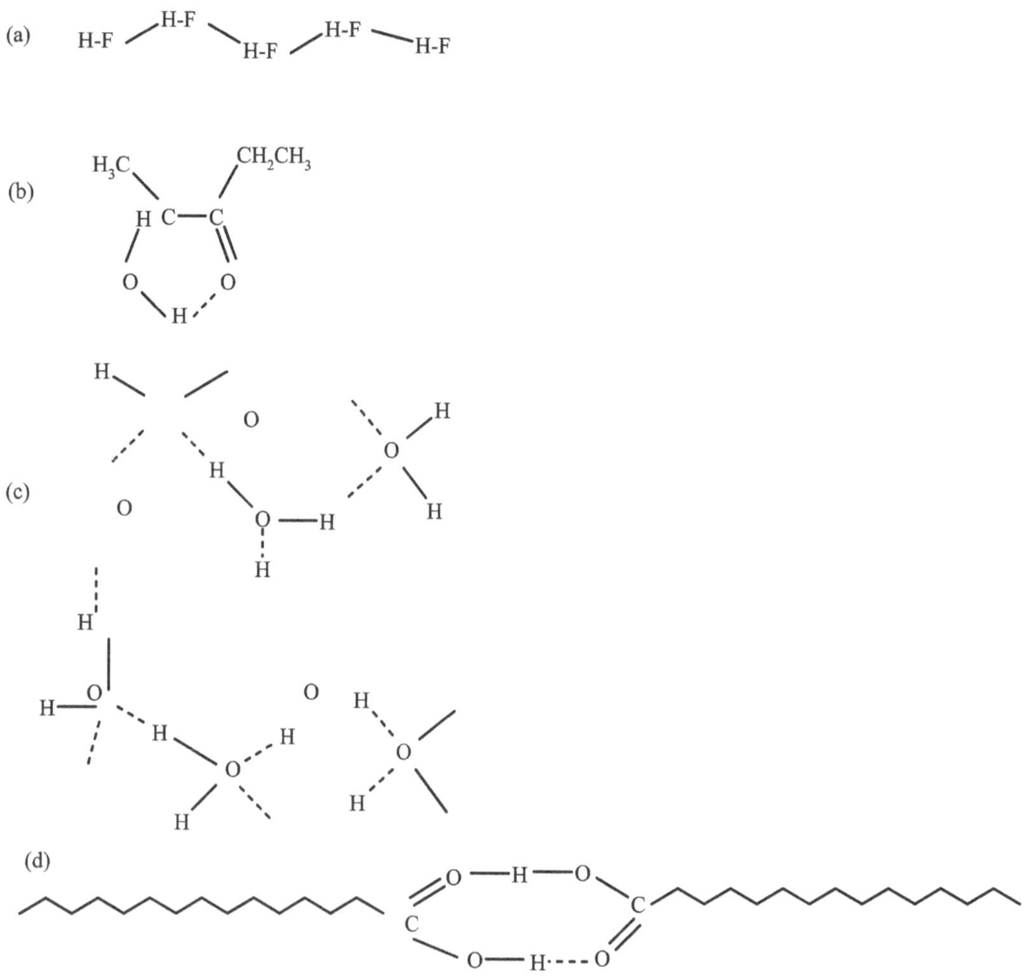

(a) H-F H-F H-F H-F H-F

(b) H_3C CH_2CH_3 H C—C O O H

(c) H O H O H O O—H H H

H H—O H H O H O H H

(d) O—H—O C C O—H···O

Figure 7.4 Formation of different types of hydrogen-bonded structures: (a) linear chain of hydrofluoric acid, (b) intramolecular ring of 2-hydroxy-3-pentanone, (c) six-membered ring network of water, and (d) dimer of pentadecanoic acid. This dimer geometry can limit the direct hydrophobic interaction between the two pentadecyl chains; rather it can provide the possibility of stacking-type directional self-assembly via hydrophobic interaction between the dimers

7.1.6.9 Other Interactions Involving Ions – Dipoles and Polarization

Some hybrid nano structures interaction mechanisms between strictly coloumbic and van der Waals attractions and still make significant contribution to overall energy of the system.

Many commonly encountered materials include ionic and molecular units that may interact by mechanisms that may be considered to be ''hybrids'' between strictly coulombic phenomena and the van der Waals interactions to be considered below.

Although weaker and perhaps more subtle, interactions involving a charged species and a neutral (but usually polar) unit can make a significant contribution to the total interaction energy of a system. Certain chemical species, including polymers, proteins, and drugs, it is apparent that they often include bonds that can impart an overall polar nature to the molecule as permanent dipoles, or they can be polarized by the effect of neighboring electric fields producing induced dipoles.

The presence of permanent or induced dipoles means that the molecules can become involved in specific interactions with charged species, other dipoles, or nonpolar molecules, and those interactions can significantly affect the physical characteristics of the system.

7.1.6.10 Balancing between Inter Molecular and Colloidal Forces for Production of Nano Structures through Self Assembly in a Colloidal State

In the production of nano structures through self assembly in a colloidal state the primary building units can be atoms, molecules, polymers, biological molecules and the primary self assembly is governed by balance between inter molecular and colloidal forces as shown in Figure 7.5.

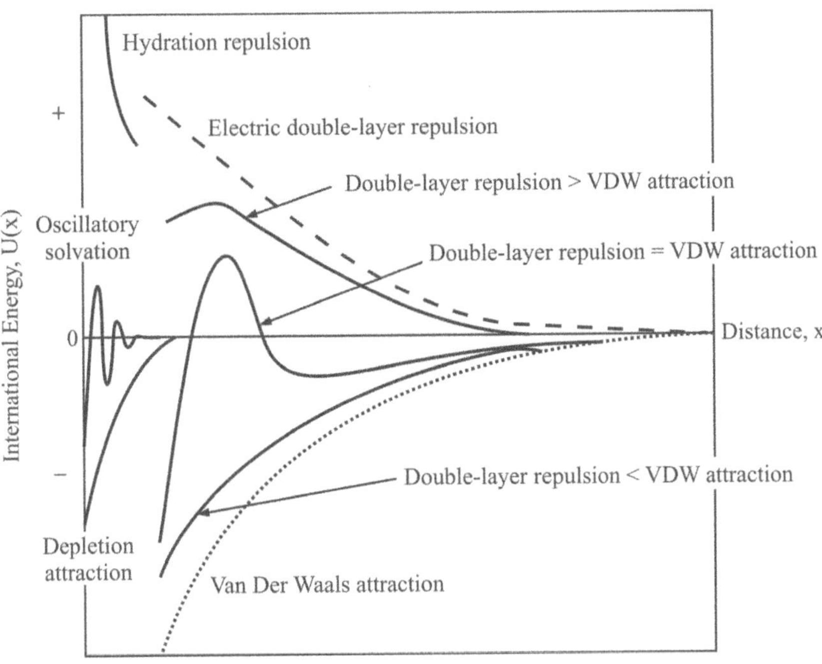

Figure 7.5 Intermolecular and colloidal potential energy

The interactions between particles that are important in determining the process feasibility of nano structure production through self assembly in a colloidal state are.

(i) London–van der Waals attractive forces (V_a),

(ii) Electrical double-layer forces (V_e) that can be attractive or repulsive in nature,

(iii) Bridging forces (V_b) that are always attractive.

(iv) Steric forces (V_{steric}) that arise from the overlap of adsorbed layers.

(v) Steric forces can be repulsive or attractive depending on the solubility properties of the adsorbed layer.

By proper balancing these interactions between the weak long range intermolecular forces and large length scale colloidal forces one can design optimum process conditions for various commercially valuable nano structures.

7.2 Liquid Phase (Colloidal Phase) Synthesis for Production of Nanostructured Materials/Nanoparticles

All materials whose critical structural features such as pore size for porous materials unit cell size for crystalline materials and the size of repeating unit in polymers are in nano meters, they are classified as nano structures. All particles regardless of their shape as long as all the three dimensions are in the range of nano meter scale they are called as nano particles.

The liquid phase production of nanometer-sized colloidal crystallites and particles with outstanding fundamental properties and applications in electronics, catalysis, and optics is a challenging task. The predominant motivation in this field arises from the unusual dependence of the properties of the colloidal materials (metals, metal oxides, or semi conductors) on particle size and particle size distribution. Apart from the intrinsic physical properties of metal-based nanoparticles, control of the stability and surface reactivity of nanosized objects is also a major advantage in the subsequent manipulation, attachment, and ordering of these nanoparticles on a given support or into self- organized colloidal nanostructures. Consequently, considerable efforts have been made in the last 10 years to control the size, stability, and morphology of inorganic nano-objects.

The general scheme followed in liquid phase synthesis of nano materials is shown Figure 7.6.

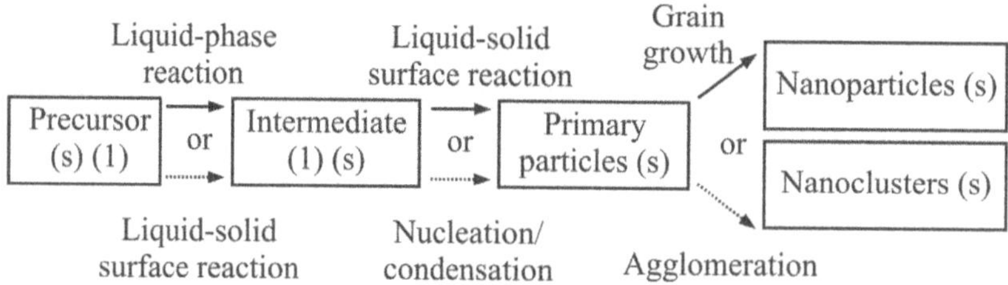

Figure 7.6 General scheme for liquid phase synthesis of nano materials

As can be seen from general scheme the mechanism of liquid phase fabrication of nano materials involve:

(i) the precursor solution typically involving a catalyst,

(ii) the nucleation and

(iii) growth stage

Particle growth has been kinetically controlled, and the resulting nano-clusters have been further protected against agglomeration by the use of terminating and stabilizing agents. Among the various strategies, nano chemistry and the host–guest inclusion approach appeared to be of particular interest in the construction of nanoscale objects under controlled conditions

7.2.1 Methods for Production of Nano Systems

Three methods are in practice in the production of nano structures in liquid phase

(a) Solvo thermal or hydrothermal synthesis

(b) Sol-Gel method

(c) Synthesis of micro porous nanomaterials in structured media (micro emulsions)

(d) Synthesis of nanoparticles by kinetic confinement-by-adsoprtion

7.2.1.1 Solvo Thermal Synthesis or Hydrothermal Method

In this process the precursors are dissolved in hot solvents (e.g., n-butyl alchohol). Normally solvents other than water will provide more facile conditions for growth of nano structures. The first stage involves preparation of a precursor solution containing a catalyst followed by nucleation and growth stage. This method being simple low cost with high yield finding attractive for various types of nano structures production.

If the solvent is water the process is called Hydrothermal Synthesis.

Figure 7.7 presents the production of TiO_2 nano crystals by Solvo Thermo process (22).

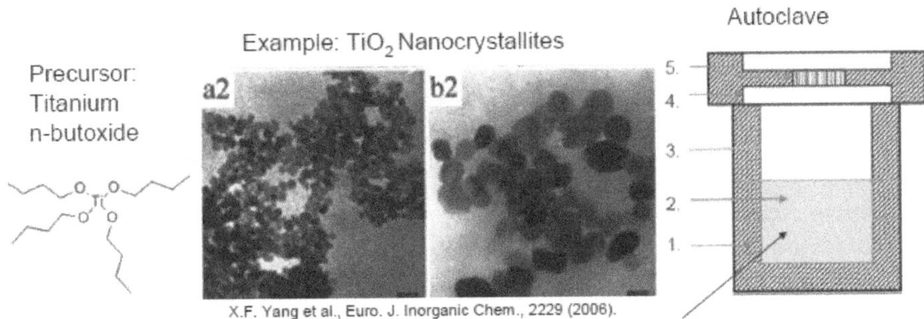

Figure 7.7 TiO_2 nano crystal synthesis by solvo thermal process

7.2.1.2 Sol–Gel Method

Colloids facilitate the formation of nano structured materials and in this process the intermolecular and colloidal forces discussed earlier play important role. Manipulating these forces through varying critical experimental conditions adopting Sol-Gel chemistry concepts greatly help in controlling the morphology of the nano structures.

Sol: Gel process is a two step process involving hydrolysis and condensation. (Figure 7.8).

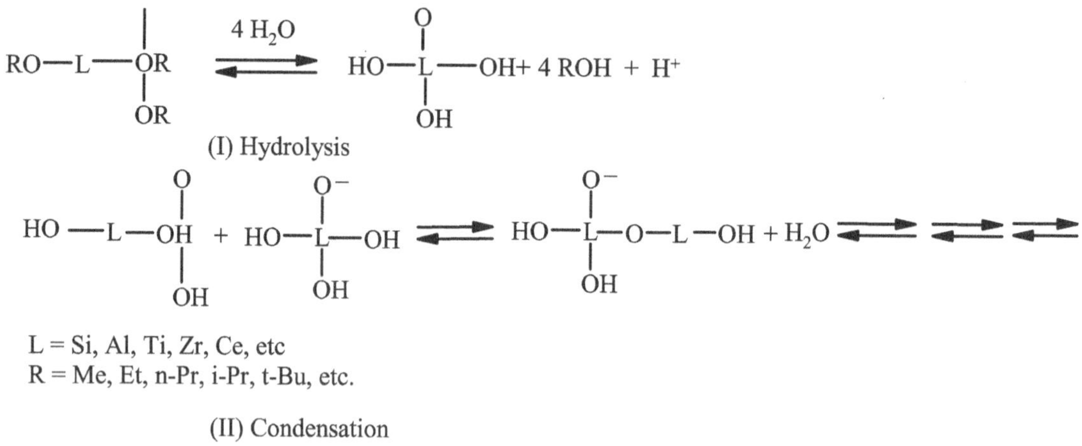

$$RO-\underset{OR}{\overset{|}{L}}-OR \;\xrightleftharpoons{4\,H_2O}\; HO-\underset{OH}{\overset{O}{L}}-OH + 4\,ROH + H^+$$

(I) Hydrolysis

$$HO-\underset{OH}{\overset{O}{L}}-OH + HO-\underset{OH}{\overset{O^-}{L}}-OH \rightleftharpoons HO-\underset{OH}{\overset{O^-}{L}}-O-L-OH + H_2O \rightleftharpoons \rightleftharpoons \rightleftharpoons$$

L = Si, Al, Ti, Zr, Ce, etc
R = Me, Et, n-Pr, i-Pr, t-Bu, etc.

(II) Condensation

Figure 7.8 Sol-gel process for the formation of nanostructured materials

Precursors of metal alkoxides will be first hydrolyzed in the aqueous solution which will be main reactive inorganic species. These will be mainly anioinic unless in the case of extremely low pH. Based on pH of the medium the ratio of anionic group to hydroxide group i.e., the degree of protonation will vary. This will be the monomeric form with possible higher order species in suspension referred as Sol.

The second step involves the condensation between these hydrolyzed forms and other forms also with high degree of cross linking with hydrogen bonded water inside. This results in the formation of bulk inorganic precipitate termed as Gel. Water and solvents will be removed from inside while frame work retains its morphology which facilitates formation of highly porous nano materials with low density.

Critical kinetic control is important in this Sol-Gel process as there may be possibility of considerable of overlap of first and second steps.

Figure 7.9 depicts the critical sequential steps in Sol –Gel method for production of thin films, nano powders and dense ceramics

As can be seen in Figure 7.9, the first step involves formation of stable Sol solution followed by gelation via poly condensation or poly esterification reaction. Then next step is aging the gel into a solid mass. This causes contraction of gel net work, phase transformations and Ostwald ripening.

Drying the gel at 8000 °C will remove M-OH groups for stabilizing the gel, which will help to protect the gel from rehydration.

To collapse the pores in the gel network and to remove remaining organic contaminants heating >8000 °C is recommended to produce dense ceramics.

The typical process for production of nano silioxane network is shown Figure 7.10.

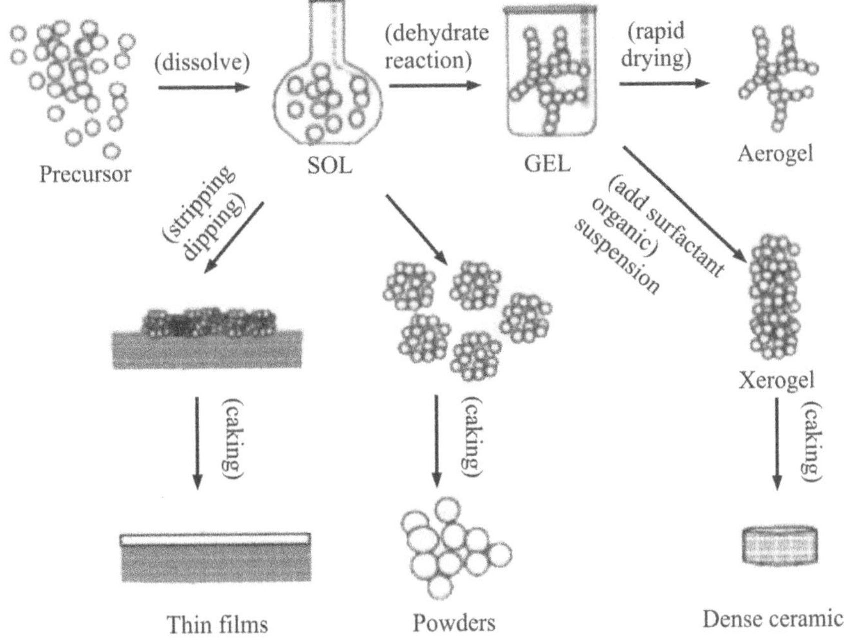

Figure 7.9 Sequential steps in Sol Gel method for production of thin films, nano powders and dense ceramics

Condensation towards network of siloxanes

Figure 7.10 Sol-gel method for production of nano siloxane networks

Hydrolysis: –MOR + H_2O → –MOH + ROH

Ex.: $Si(OR)_4 + 4\ H_2O \rightarrow Si(OH)_4 + 4\ R\text{-}OH$

Condensation: –MOH + ROM- → -MOM- + ROH

$(OR)_3\text{-}Si\text{-}OH + HO\text{-}Si\text{-}(OR)_3 \rightarrow [(OR)_3\ Si\text{-}O\text{-}Si(OR)_3] + R\text{-}OH$

7.2.1.3 Synthesis of Micro Porous Nanomaterials in Structured Media (micro emulsions)

Figure 7.11 shows the mechanism of the preparation of microporous materials using oil-in-water emulsion as structured media (23).

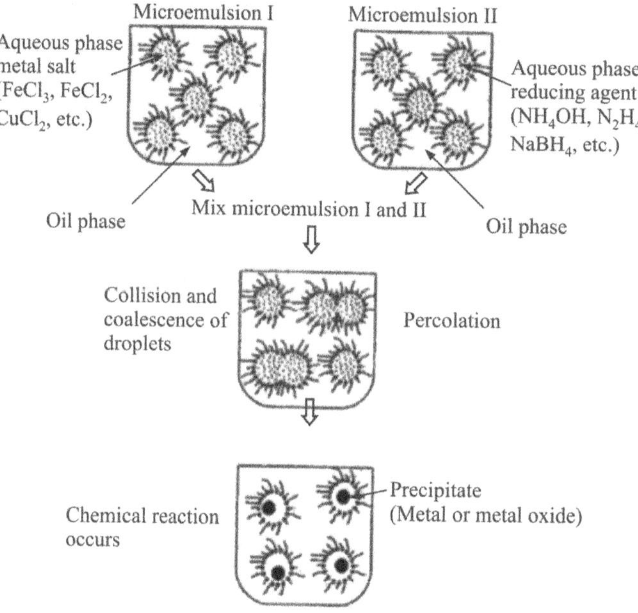

Figure 7.11 Mechanism of nano porous structures formation in structured media (micro emulsions)

This type of emulsion has submicron -size oil droplets dispersed in aqueous solution. The self-assembled surfactant (or polymer) layers at the oil–water interface stabilize this structure.

The reactive inorganic precursors can be dissolved at the aqueous region, and then are attracted onto the surface of oil droplets by electrostatic (for ionic surfactant) or by hydration (for nonionic surfactant) interaction.

Condensation between the inorganics (silanol condensation for silicates) forms the nano structured composites, and the subsequent removal of the organic components provides the final microporous materials.

The growth kinetics can be controlled by regulating inter micellar interchange process via coalescence rate limiting step by imposing constraints from matrices.

The inorganic precursors are hydrolyzed first, and then these hydrolyzed forms co-self-assemble with organic cations to form microstructured composites.

Alkoxides of silicon, aluminum, and titanium are among common inorganic precursors and tetra-alkyl quaternary ammonium cations are typical organic components.

Final zeolites can be obtained as single components such as pure siliceous or multicomponents such as titanosilica, aluminosilica, and so on.

Removal of the organic cations produces the microporous zeolites. Tetra-alkyl quaternary ammoniumcations are the known single molecular structure -directing agents that work for the formation of zeolites.

The pore size, geometry, and structure of mesoporous (10-20 nm) silicas are determined by the size, geometry, and structure of the micelles that are removed. Though there is often some degree of shrinkage during the surfactant removal steps, the diameters of the micelles solely determine the size of pores on a nanometer scale.

These striking structural features make these mesoporous silicas unique in the field of nanotechnology. Not only are the sizes of the pores much larger than those of conventional zeolites (Which limits their applications to small single molecules), but they also can be finely tuned on the nanometer scale of 1.5–15 nm by simply choosing appropriate micelles. Polydispersity of pore size is very narrow.

Wide ranges of mesoporous silicas show 300–1,200 m^2/g of extremely high surface area and 0.3–2.5 cm^3/g of extremely large pore volume.

7.2.1.4 Synthesis of Nano Particles by Kinetic Confinement-by-Adsorption

This is the process wherein the synthesis of nanoparticles is achieved by the direct reaction of the precursors and the subsequent kinetic confinement by the adsorption of organic layers on the surface of growing nanoparticles. This process has 4 critical steps

(i) decomposition,

(ii) nucleation,

(iii) growth and

(iv) adsorption.

Small seed particles that are initially formed by the decomposition of precursors have higher reactivity, and so grow faster. As the reaction goes further into nucleation and growth stages, the size of particles becomes bigger. The organic molecules that preexist with the precursors now can be adsorbed and become saturated on the surface of the particles, which quenches (stops or arrests) particle growth. In most cases, this final step can be controlled on the nanoscale level of the organic-particle complex, and the particle that is obtained also has a nanometer size, which means that a nanoparticle is synthesized. Decomposition of the precursor can be achieved by a variety of methods. This includes the sol– gel process, hydrothermal reaction (thermolysis), electrochemical process, sonication, irradiation by UV, IR, or laser, and a sudden jump in reaction condition such as pH, pressure, or temperature. Depending on the precursor, nucleation can be a homogeneous or heterogeneous type of process.

Control of the size and shape of the nanoparticles can be achieved by the control of various synthetic parameters. For example, the interplay between thermodynamics and kinetics during the nucleation, growth, and quenching steps can provide some degree of control in Ostwald ripening. The non adsorbed species that can change the colloidal force, such as a double- layer or depletion force, will affect the force balance between the nanoparticles. This in turn will affect the coalescence between them, which can have a great effect on their sizes and size distribution. Also, the affinity of amphiphile molecules to any specific facet or specific composition of the nanoparticles can significantly affect their morphology. Instant inhomogeneity on the surface of the nanoparticles can create facets with different surface energy. The amphiphiles or capping agents can be preferentially adsorbed onto this facet, which can cause uneven growth of nanoparticles. This process produces nanoparticles with nonspherical structures.

References

1. Somorjai GA and Li Y (2010). Introduction to Surface Chemistry and Catalysis, Wiley, Hoboken, NJ), 2nd Ed.
2. DH. Everett, Basic Principles of Colloid Science, RSC Publications, London (1988).
3. P. Painter and MM. Coleman, (1994). Fundamentals of Polymer Science, Technomic Publishing Co., Lancaster, PA.
4. G.J. Fleer, MA. Cohen Stuart et al (1993). Polymers at Interfaces, Chapman & Hall, London.
5. J. Lyklema. (1991). Fundamentals of Interface and Colloid Science, Elsevier, Amsterdam.
6. C. Nutzenadel, A. Zuttel, D. Chartouni, G. (2000). Schmidand Schlapbach, Eur. Phys. J D8 245.
7. D. T. Livey and P. Murray. Surface energies of solid oxides and carbides. J. American Chemical Society, Vol 39 N0 9 (156).
8. Parasegian, V.A and B. W. Ninham. (1969). Van der Walls Forces, Nature, 224, 1197.
9. London F. (1937). Transactions of Faraday Society, 33, 19.
10. Hamaker. H. C. (1937). The London-van der Waals Attraction between spherical particles, Physica, IV. No 10, 1058.
11. V. Arunachalam, R. R. Luchese and W.H. Marlow (1998). Development of picture of the van der Waals interaction energy between clusters of nanometer-range particles, Phys. Rev E 58, 3451-3457.
12. Kelvin Cahill and VAP. (2004). Rydberg-London Potential for diatomic molecules and abandonrd atom pairs, J. Chem. Phys. 121: 10839-42.
13. P.J. Flory (1969). Statistical Mechanics of Chain Molecules, Interscience, New York.
14. R.H. Yoon and S.A. Ravishankar, J. (1994). Colloid Interface Sci., 166, 215.
15. R.W. Slater and J.A. Kitchener. (1966). Disc. Faraday Soc., 42, 267.
16. R.M. Pashley et al. (1985). Science, 229, 1088.
17. J.N. Israelachvili and R.M. Pashley, J. (1984). Colloid Interface Sci., 98, 500.
18. C.J. van Oss (1994). Interaction Forces in Aqueous Media, Marcel Dekker, New York.
19. P. Somasundaran, T.W. Healey, and D.W. Fuerstenau, J. (1966). Colloid Interface Sci., 22, 599.
20. H.K. Christenson J. (1985). Colloid Interface Sci., 104, 234.
21. Pollack. G.H. Cells (2001). Gells and the Engines of Life, Ebner & Sons.
22. X. F. Yang et al. (2006). Euro. J. Inorgnic. Chem, 22-29.
23. I. Capek. (2004). Adv. Coll. Interf. Sci.110, 49.

8

Molecular Manufacturing, Future Nano Technologies and Dexler-Smalley Debate

8.1 Molecular Manufacturing

8.1.1 What is Molecular Manufacturing?

Manufacturing today works by cutting or deforming large chunks of matter, then fastening together the remaining pieces into products. Molecular manufacturing plans to be more efficient and make better products by assembling products directly from the smallest pieces: atoms and molecules. The basic idea is to develop a small set of chemical reactions that can be applied repeatedly to build large molecules, then control the sequence and/or position of the reactions by computer in order to build engineered molecular systems.

Though molecular manufacturing not yet realized once scientists get ability to observe and manipulate atoms directly we can expect to have a thorough and inexpensive system for controlling of the structure of matter.

The possibility of building things with atomic precision was first introduced by Richard Feynman (1) in a famous after-dinner talk in 1959 when he said: "The principles of physics, as far as I can see, do not speak against the possibility of maneuvering things atom by atom."

Molecular manufacturing is the term given to "bottom-up" nano technology at molecular or atomic level (2). Molecular manufacturing is the formation of precise molecular structures under direct mechanical guidance. The mechanical component of molecular manufacturing does not require a classical mechanical engineering approach, but such an approach seems to be a good fit for the method especially since the creation of designed molecules will make it relatively easy to specify their shape. Nanoscale structures are often assumed to be inherently floppy and unstable, but this will be less true for strongly crosslinked polymers or covalent solids than for weakly crosslinked biological molecules. It is worth investigating how small a mechanical structure implemented in various materials can be made before thermal, noise or quantum effects spoil its applicability as a mechanical design element.

However the critical question is how to design nanoscale mechanical elements to be reliable machines in the face of thermal noise and surface forces, and how small such designs can be made. The individual control of molecular composition and placement is often cited as a goal of nanotechnology.

Molecular manufacturing is different from biology in that biological systems are not engineered. The functional properties of a cell, or even a protein, are complex and hard to predict. However, the process of building protein molecules from small molecular fragments is quite programmable, and scientists are developing the ability to design and synthesize proteins with desired properties. This will allow protein chemistry to be used in an engineering, rather than a biological, context. This would be one approach to molecular manufacturing. Other approaches using different kinds of chemistry may also work, producing better materials.(3)

Scientists have recently gained the ability to observe and manipulate atoms directly, but this is only one small aspect of a growing array of techniques in nanoscale science and technology. The ability to make commercial products may yet be a few decades away. But theoretical and computational models indicate that molecular manufacturing systems are possible - that they do not violate existing physical law. These models also give us a feel for what a molecular manufacturing system might look like. Today, scientists are devising numerous tools and techniques that will be needed to transform nanotechnology from computer models into reality. While most remain in the realm of theory, there appears to be no fundamental barrier to their development.

Although there are some practical questions remaining to be answered, there is no scientific study demonstrating a limitation or problem with molecular manufacturing theory. Indeed, the chemical mechanisms of life demonstrate that machines of a sort can be constructed out of molecules and can build more machines.

However some scientists dismiss the entire field out of hand have not bothered to study it in any depth. An intermediate position is that some form of molecular manufacturing will be possible, but will be limited. This is usually based on the assumption that engineering cannot improve on biology. This may be true for complex problems, but not for the relatively simple tasks that are tackled by engineering.

Perhaps because of the opposition of these scientists, not many people are working on molecular manufacturing in the lab; most of the work has been theoretical study, which is easier to do without funding. For now, it remains a controversial theory. But after more than a decade, no one has managed to find a real problem with the theory, and evidence is accumulating in favor of it. Some predicted phenomena, such as exceptionally low friction in certain cases, have now been observed. A single atom has been mechanically removed from a crystal and put back in the same place. Although these were not intended as demonstrations of molecular manufacturing theory, they indicate that at least some of the predictions are valid.

Though to date, many nanotechnology efforts have been content to achieve nanoscale—but not atomic—precision, or to build large quantities of small identical molecules, however there are some technologies that are on the verge of achieving the goal.

8.1.2 How the Precise Control the Sequence and/or Position of the Reactions at Atomic or Molecular Lever can be Achieved?

Molecular manufacturing is parallel to conventional automated manufacturing. Both require systems for storing and moving parts, for putting them together, and for transporting finished products. Both require sensors and computers to orchestrate the process. The difference is that the parts in molecular manufacturing will be reactive molecules, while the required conveyor belts, robotic arms, and so forth will be measured in nanometers.

8.1.2.1 Relative Productivity

The productivity of individual control of molecules is limited by the small size of the molecules. A typical scanning probe microscope might weigh two kilograms, have a size of about 10 cm, and carry out ten automated operations per second. If each operation deposits one carbon atom, which masses about 2×10^{-26} kg, then it would take 10^{26} seconds or six billion years for that scanning probe microscope to fabricate its own mass.

The speed with which a molecular manufacturing tool can create its own mass of product may be called "relative productivity." Which depends on several factors like

(a) the mass of the tool,

(b) its frequency of operation, and

(c) the mass deposited per operation.

Roughly speaking, an object's mass will be about the cube of its size; hence, for each factor of ten shrinkage, the mass of the tool will decrease by 1,000. In addition, small things move faster than large things, and the relationship is roughly linear. Taken together, each factor of ten shrinkage of the tool will increase its relative productivity by about 10,000 times; relative productivity increases as the inverse fourth power of the size.

If a tool could be shrunk by a factor of a million, from 10 cm to 100 nm, then its relative throughput could increase by 10^{24}, in which case it would take only 100 seconds to fabricate its own mass. This assumes an operation speed of 10 million per second (which requires a linear speed of only 2 m/s over a 100 nm range). This is about ten times faster than the fastest known enzymes (carbonic anhydrase and superoxide dismutase). But a relative productivity of 1,000 or even 10,000 seconds would be sufficient for a very worthwhile manufacturing technology. (An inkjet printer takes about 10,000 seconds to print its weight in ink.)

8.1.2.2 Possible Physical Techniques for Controlling Factors for Engineered Nano Systems

For controlling molecular joining and/or placement in order to build intricate, engineering functional, nanoscale, kinematic systems that are capable of building more of the same the following techniques may be useful

(i) In general, it is not required that every covalent bond in the final product be formed under direct mechanical control, or even at the time of the deposition process. For example, useful molecular building blocks can be pre-built by bulk chemistry.

(ii) Crosslinking can be triggered optically or chemically after the manufacturing is completed to increase strength and stiffness.

(iii) Many types of control can be broadcast, including chemicals, photons, pressure, and electric or magnetic fields

8.1.3 Molecular Manufacturing Importance in Future Nanotechnologies

Precision at atomic level is the key for any molecular machines and is one of the important factor in understanding why it is necessary to develop molecular manufacturing. In this technology precision means that there is a place for every atom and every atom is in its place and there will be no unnecessary parts anywhere in the design. Machines of precision will be used to create products of equal precision. With this precision, we should be able to recycle all of the waste products produced by the manufacturing processes and put them to good use elsewhere. Manufacturing will also become less expensive as a result.

Technology has never had this kind of precise control; all of our technologies today are bulk technologies. We take a lump of something and add or remove pieces until we're left with whatever object we were trying to create. We assemble our objects from parts, without regard to structure at the molecular level. Precise atomic-level fabrication has previously only been seen in the growth of crystals or in biological molecular machinery, like the ribosome, which assembles all the proteins in living creatures, or DNA, which carries the instructions for creating a living being. If we incorporate similar processes during our development of nanotechnology, we will begin to gain a degree of complexity and control over systems that previously only evolution and nature have had.

Additional benefits arise when we consider the size of devices that we will be able to create. Once we are working on the atomic scale, we can create machines that will go places about which we could once only dream. More information will be packed into smaller and smaller spaces, and we will be able to do much more with much less. Nanotechnology promises unprecedented and efficient control over our environment, but taking advantage of anticipated developments requires forethought and planning.

Exponential molecular manufacturing may significantly impact the semiconductor and pharmaceutical industries, among others; and integrated molecular manufacturing may even be competitive as a general-purpose manufacturing technology. This implies that development of molecular manufacturing may have a very high payoff.

8.1.4 Benefits of Molecular Manufacturing

One of the first obvious benefits is the improvement in manufacturing techniques. By taking familiar manufacturing systems and expanding them to develop precision on the atomic scale will give us greater understanding of the building of things, and greater flexibility in the types and quantity of things we may build. We will be able to expand our control of systems from the macro to the micro and beyond, while simultaneously reducing the cost associated with manufacturing products. Thus gaining molecular-level control over the structure of matter will bring a wide variety of positive applications.

8.1.4.1 Medical Uses

Some of the most dramatic changes are expected in the realm of medicine. Molecular machine systems will be able to sense and rearrange patterns of molecules in the human body, providing the tools needed to bring about a state of health, regardless of a disease's cause. Theorists envision creating machines that will be able to travel through the circulatory system, cleaning the arteries as they go; sending out troops to track down and destroy cancer cells and tumors; or repairing injured tissue at the site of the wound, even to the point of replacing missing limbs or damaged organs. The extent of medical repair systems is expected to be quite broad, with the cumulative impact being equally large.

8.1.4.2 Environmental Applications

Using molecular manufacturing techniques, it will be possible to construct our products with zero chemical pollution, recycling leftover molecules. Environmental restoration could be carried out at the molecular level, detecting and inactivating unwanted chemicals.

8.1.4.3 Water Purification

Once we have the ability to capture, position, and change the configuration of a molecule, we should be able to create filtration systems that will scrub the toxins from the air or remove hazardous organisms from the water we drink. We should be able to begin the long process of cleaning up our environment.

8.1.4.4 Raising Sustainable Living Standards

Molecular manufacturing will be able to cleanly and inexpensively produce high-quality products using common materials (especially carbon, which is in excess in the atmosphere in the form of carbon dioxide) and solar energy.

8.1.4.5 Low Cost to Access to Space

The strong, lightweight materials enabled by molecular manufacturing will greatly lower the cost of access to space and space resources, making their active use affordable for the first time:

Taking all of this into account, it is clear that molecular manufacturing should improve our lives in any area that would benefit from the development of better, faster, stronger, smaller, and cheaper systems.

However these benefits should be attainable though the combined results of

1. a well-funded R&D program,
2. private sector efforts to bring down costs, and
3. public acceptance

8.1.5 Potential Negative Effects of Molecular Manufacturing

Powerful technologies bring problems as well as benefits, and advanced nanotechnologies are expected to bring problems of several sorts:

8.1.5.1 Accidents

Any powerful technology— from fire to biotech— must be controlled to avoid accidents. In the case of molecular manufacturing, rearranging matter at the molecular level can either improve or destroy a system. Molecular machine systems able to build complex objects could build copies of themselves, possibly overdoing this activity from a human point of view, as bacteria do.

An approach to the problem: This issue has to be examined and a set of safety rules need to be framed.

8.1.5.2 Economic Disruption

Technological change continually disrupts employment patterns, but molecular manufacturing is expected to accelerate this significantly: once certain specific points of development in this technology are reached, very rapid change can take place. However this needs increase workforce flexibility through education and training.

8.1.5.3 Deliberate Abuse/Terrorism

Of the potential problems molecular manufacturing may bring, this is regarded as the most serious and most challenging to address. Three main areas of concern have been identified:

1. very rapid construction of conventional weapons, making traditional arms control more difficult,
2. totalitarian control of civilian populations by surveillance using nanoscale sensors, and
3. new weapons made possible by the technology, which can be thought of as "smart" chemical weapons.

8.1.5.4 Reducing Risks from Molecular Manufacturing

Individuals and organizations with legitimate concerns regarding advanced nanotechnology have suggested delays in development, even moratoria or bans. While these reactions are understandable, this approach was examined over a decade ago and rejected as infeasible. Today, both public and private spending on nanotechnology is broadly international. Expected economic and military advantages are driving a technology race already underway. If law-abiding nations choose to delay nanotechnology development, they will relinquish the lead to others.

8.1.6 Different Possible Approaches for Realizing Molecular Manufacturing

8.1.6.1 Polymer Techniques

Ribosomes, which are made of RNA and protein, show that polymers can fold into functional shapes that carry out manufacturing operations. Polymer sequences can be controlled digitally by selecting the building blocks in the right sequence, as ribosomes do, or by flushing different building blocks past while controlling whether or not they stick, as DNA synthesizers do. Some polymers including DNA and RNA fold predictably, so their shapes can be engineered. Seeman's machine selects short sequences to join rather than individual monomers, but a variant that could select monomers or could be constructed out of a few short repeated sequences would be able to build a copy of itself.

Polymer chemistry is known to be quite versatile, and it should be possible to incorporate molecular actuators to select the polymer sequence; this would be faster and probably more reliable than using DNA strands to program the device. Molecular actuators can be controlled and powered by light, electricity, or changes in the composition of the solution (4).

Bulk controlled polymerization techniques, such as DNA synthesis, often use two repeated steps: first they make the end of the polymer reactive by "deprotecting" it, then add a monomer that is protected from further deposition. Nanoscale controlled polymerization could control either the timing of the deprotection step or the monomer selection for the polymerization step. Or the system could protect the addition site by steric hindrance. Alternatively, it could use a polymerization reaction that is exothermic but has a high barrier, and accelerate the desired reaction possibly by many orders of magnitude (5) by holding the monomer in place. The ratio of reaction rates of confined and unconfined monomers will approximate the error rate.

Because there is only one reaction site and one choice to add a monomer, extreme stiffness should not be required in polymer-based molecular manufacturing systems. Even minimally crosslinked polymers (as shown by the ribosome) can implement a controllable, low-error system.

Implementing polymer exponential molecular manufacturing systems appears to be a fairly small step from existing and currently feasible non-exponential polymer-based polymer-building systems.

8.1.6.2 Solids Built in Solution

Instead of building a linear chain that then folds into a shape, it may be possible to build the desired shape directly by depositing small molecular building blocks. These building blocks could either form covalent bonds during deposition, or be held by hydrogen bonds (similar to self-assembly) and perhaps joined later by crosslinking.

Making an atomically precise product does not require atomic precision in block positioning—only enough precision to select between adjacent block deposition sites. Branched molecules, possibly including dendrimers, may be of interest as building blocks.

8.1.6.3 Solids Built in Machine-phase

"Machine-phase" means that all reactive molecules are controlled and moved mechanically rather than diffusing randomly, and reactions take place under mechanical control. Deposition of atoms and inducing of reactions in vacuum has already been accomplished. This approach to building nanoscale tools may be unfamiliar to chemists and molecular biologists, but may be a better fit than most nanoscale techniques for today's mechanical engineering disciplines (6,7). If cleanliness can be maintained, it may be possible to take advantage of superlubricity between stiff surfaces. Investigation to date on machine-phase synthesis of covalent solids has focused on reactions that deposit just one or a few atoms per operation. This would require many operations to build a complete nanoscale manufacturing tool, and this could be difficult to achieve with macro-scale scanning probe microscopes especially given the ultra-clean conditions needed to prevent some of the proposed highly reactive "tool tips" from being spoiled by contact with random molecules. MEMS scanning probes, or even specially built NEMS systems (perhaps made with ion etching), could reduce this problem. An intermediate tool also might be built using one of the solution-phase methods. The smaller the tool, the faster it could work, and the less volume would have to be kept clean. Also, if actuators can be made on the same size scale as the tool tip, then more actuators may be included to make a more flexible and functional tool.

8.1.7 Integrated Molecular Manufacturing

A far-term goal of molecular manufacturing research is to build not just a large mass of products, but large integrated products using chemical and biological techniques (Figure 8.1). Sufficiently large products would create the possibility of manufacture for direct human use, without further expensive manufacturing steps, and the integration of familiar user interfaces. Several problems would have to be solved in order to achieve this.(8)

To produce a heterogeneous product, the manufacturing systems must be individually controlled. The instructions may be delivered through a simple hierarchical network, but probably must be processed locally. This would require nanoscale computers, which would be responsible for significant portions of the mass and power budget. (Commonly used molecular pieces could be built by special-purpose high-speed systems mills substantially reducing the required control inputs.)

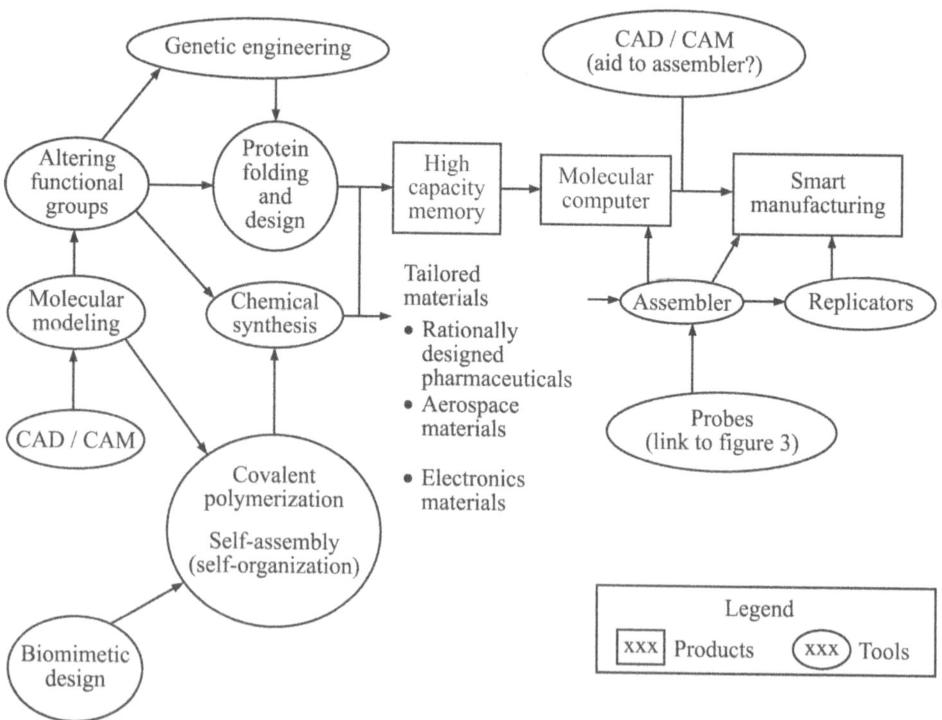

Figure 8.1 Schematic of molecular manufacturing by integrated chemical and biological techniques

8.1.8 Molecular Manufacture - Enabling Tools/Technologies

Several existing and projected technologies discussed below may be useful for development of the various approaches and stages of molecular manufacturing.

8.1.8.1 Computers

Computers have become quite powerful and will continue to improve. Computational chemistry is becoming increasingly trusted. Estimates can be determined of physical properties of molecular structures. Eventually, multilevel integrated simulations might be useful in planning and evaluating larger systems. Computers can also be useful for some design processes (Figure 8.2).

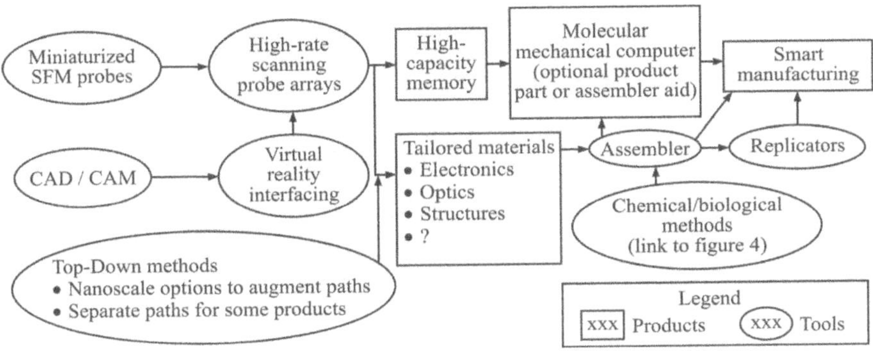

Figure 8.2 Schematic of molecular manufacturing with computer aided design and scanning probes

For example, recent work on protein folding to a desired shape used a repeated simulate-and-modify process to arrive at appropriate sequences. Appropriate DNA sequences for optimum folding to a desired shape also have been generated by computer. As more extensive mechano synthetic capabilities are developed, the ability to automate them will become crucial to carrying out related operations without excess labor.

8.1.8.2 Scanning Probe Microscope

Scanning probe microscopes (SPMs) can image areas with atomic resolution, and also manipulate molecules and do chemistry. Most SPMs have only a single probe with three degrees of freedom, but Xidex Corp. is developing two-probe, six-degree-of-freedom systems that will be able to touch the tips together (9). This may be useful for research into mechano synthetic reactions.

A tip with an active molecule could be brought together with another tip acting as a receptive surface, at various repeatable angles and offsets, to test the effects of displacement and angle on the reaction. SPMs can work in air, water, or vacuum, at room temperature or cryogenic temperature. A deposition system that can create 15 nm lines of molecules has been developed, called dip-pen nanolithography (DPN).

8.1.8.3 Electron Microscopes

Electron microscopes can image with near-atomic resolution. They can be used to cut carbon nanotubes, even to trim outer tubes from multiwalled tubes (10). They can also deposit a variety of materials from gas feedstock (electron beam induced deposition, EBID). These deposits have a feature size as small as 10 nm and can form three-dimensional structures (11).

Sub-wavelength Imaging: FRET (fluorescence resonance energy transfer, which is very sensitive to nanoscale distance) can be used to determine relative positions (12). AngstroVision has claimed to be developing 3D nm-scale imaging using visible light (13). A paper at NASA Tech claims that imaging below the diffraction limit should be possible with incoherent light.

Ion Etching: Ion etching systems can achieve single-atom accuracy and can use tiltable work pieces. This may enable production of freestanding (undercut) kinematic structures from high-performance materials that might be useful for research into nanoscale machinery or even as nanoscale molecular manufacturing systems (19).

8.1.8.4 Engineered Molecules and Molecular Composites

DNA structures can be used to position carbon nanotubes to make a working transistor. Protein folding prediction is still problematic, but the inverse problem—design of sequences that will fold into a desired shape is becoming reliable (15).

There are many possible approaches to molecular manufacturing: many materials in at least four basic processes (polymer, covalent in solvent, noncovalent in solvent, covalent in machine-phase). Most of the possibilities have never been studied. A study of the merits and potentials of a range of approaches and materials would be very useful. Once the potential capabilities and products were better understood, a more detailed roadmap to their development could be produced. Figure 8.3 presents the earlier developments in the Scanning microscopic tools for molecular nano technology process development (16).

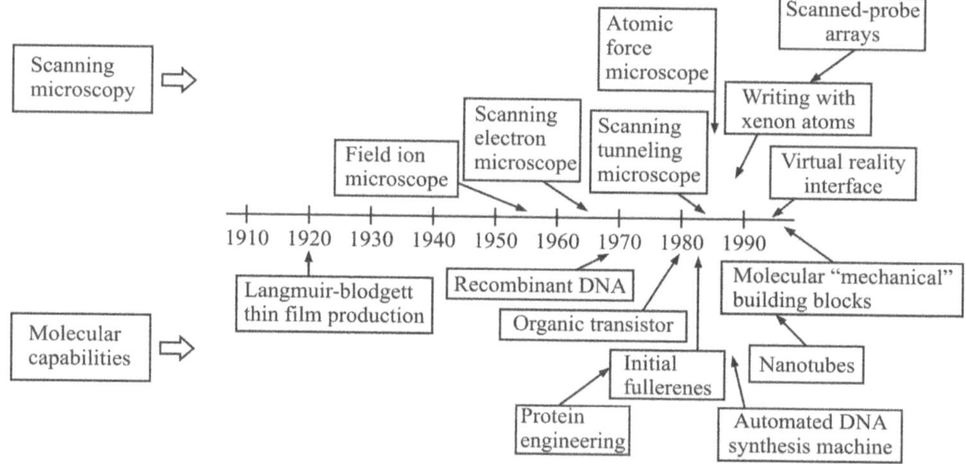

Figure 8.3 History of developments in the scanning microscopic tools for molecular nano technology process development

The development of molecular manufacturing can be an incremental process from today's capabilities. Although the most exciting results rely on the most advanced and integrated capabilities, even the earliest products of basic molecular manufacturing could be useful for basic research and for components such as sensors.

According to its proponents, molecular manufacturing could be revolutionary. Because chemistry is very precise and repeatable, the manufacturing operations should be reliable enough to allow complete automation. Because the chemical steps are simple, a single fabricator could build a wide variety of products by changing the control program. Furthermore, machines built this way could have very precise features. This is a good thing, because building with molecules would be quite slow, and small machines can be built more quickly. In fact, if the calculations are right, a complete nanoscale manufacturing system could build a complete copy of itself in a few hours-and this remains true if many small manufacturing systems are combined into one large one, so that even human-scale factories and products can be built quite rapidly. Finally, it appears that the materials produced by the manufacturing system may be extremely high-performance, allowing products to be far more compact and powerful.

8.2 Self Assembly in Production of Molecular/Nano Structures-Bottom-Up Methods

Bottom-up manufacturing involves the building of structures, atom-by-atom or molecule-by-molecule. The wide variety of approaches towards achieving this goal can be split into three categories:

(i) Chemical synthesis,

(ii) Self-assembly,

(iii) Positional assembly.

8.2.1 Chemical Synthesis

Chemical synthesis is a method of producing raw materials, such as molecules or particles, which can then be used either directly in products in their bulk disordered form, or as the building blocks of more advanced ordered materials. A generic process by which nanoparticles may be produced by chemical synthesis is shown in Figure 8.4.

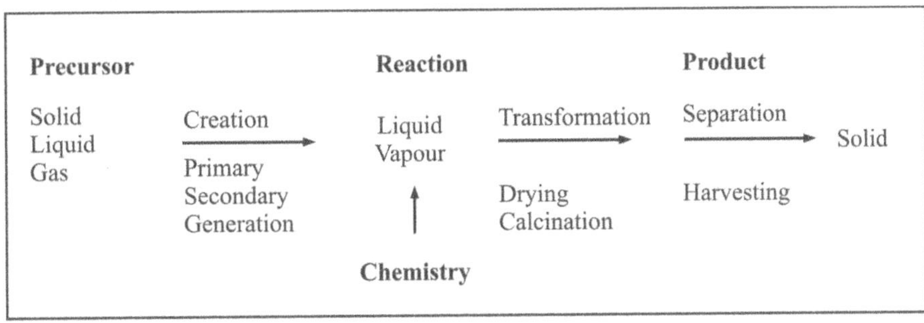

Figure 8.4 Generic process for production of nano structures by chemical methods

The precursor phase is the starting point, and the material can be in any physical state (or multiphase) or spatial arrangement to other components. The first step is the creation of a new phase or state where the nanoparticles either form or can be formed by a chemical step. In other words, the phase change itself could bring about nanoparticle formation (rare but possible) although generally the circumstances are created whereby nanoparticles can be made, for example vaporisation of a precursor mixture. Once in a state where nanoparticles can be made, usually a chemical reaction of some description is performed to generate the desired material. A further phase transformation or even solid-state reaction may be necessary to produce the final product.

8.2.2 Self Assembly

Top-down and bottom-up are the two main approaches (Fig. 8.5) in nano structure systems as discussed earlier. The top-down approach adapts the concept that the construction of nanostructured systems that eventually consist of the length scale of atoms or molecules should begin with bulk materials. By removing the excess portion of bulk materials by physical, chemical, and mechanical means, the desired nanostructured systems can be constructed. Hard lithography using optical laser-light, electron -beam, or x-ray are typical examples (Pl see Chapter 3). The bottom-up approach, on the other hand, starts with the individual building units that are the objects of the nanometer scale, such as atoms, molecules, polymers, and colloids. By assembling these building units with the needed controllability, the desired nanostructured systems can be obtained.

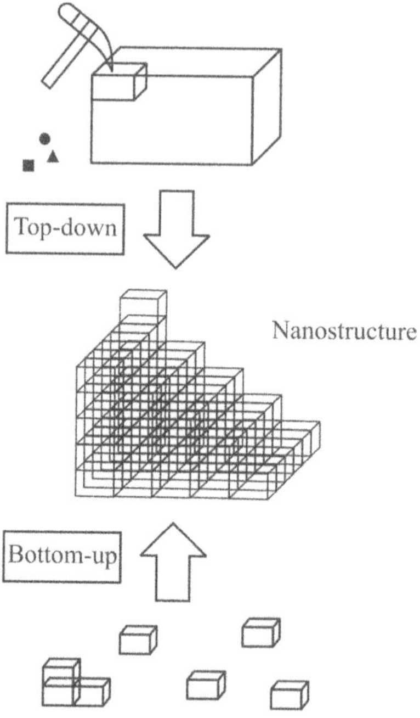

Figure 8.5 Bottom up vs top down approach for the preparation of nano structures

Self assembly is a bottom-up production technique in which atoms or molecules arrange themselves into ordered nanoscale structures by physical or chemical interactions between the units. The formation of salt crystals and snowflakes, with their intricate structure, are examples of self-assembly processes. Although self assembly has occurred in nature for thousands of years, the use of self assembly in industry is relatively new. There is an economic and environmental interest in processes through which materials or product components essentially form themselves, creating less waste and using less energy. However, current understanding extends only to the creation of fairly rudimentary systems. Improved understanding of thermodynamic and kinetic processes at the nanoscale, enabled through advances in the characterization techniques, and improved computer modelling, are expected to aid the development of more complex systems. One potential processing technique involves the use of an external force or field (for example, electric or magnetic) to accelerate the often slow self-assembly process, which is attractive in an industrial context. This is known as directed self assembly.

Depending on the specific system to be constructed, hybrid approaches that try to integrate these two methods are also being sought. Figure 8.6 shows the schematic representation of these hybrid approaches. In the "top- down- after-bottom-up" approach, the rough quasi-nanostructure is assembled first, and the detailed desired structure is obtained by carving out by the top-down method of construction. The "bottom-up-after-top-down" approach is the exact opposite of this process.

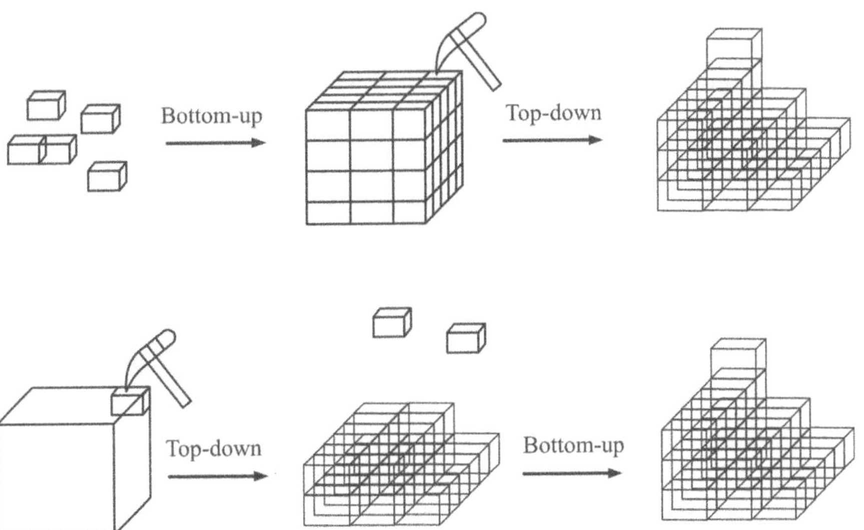

Figure 8.6 Hybrid approaches for nano structure production (a) top down after bottom up (b) bottom up after top down

Nanostructured system is "an integrated system assembled/manipulated from nanometer-scale building units, that the primary building units for the construction of a wide range of nanostructured systems are the same building units as are found in self-assembly systems Table 8.1. They also share the same range of length scale and origins.

Table 8.1 (a) Examples of building units of self assembly (b) Examples of typical nano structured systems and their building blocks

(a)	(b)	
Building Units for Self-Assembly	**Building Units for Nanostructure**	**Nanostructured Systems**
Semiconductor atom Metal atom	Semiconductor atom Metal atom	Quantum dot Nanocrystal
Surfactant molecule	Surfactant molecule	Nanocomposite
Amphiphile molecule	Amphiphile molecule	Film/multilayer nanodevice
Amphiphilic polymer	Amphiphilic polymer	Polymeric nanocomposite
Colloidal object	Colloidal sphere, rod, etc.	Photonic device
Lipid, DNA, RNA	Lipid, DNA, RNA	Model membrane

Molecular self-assembly is a strategy for nanofabrication that involves designing molecules and supramolecular entities so that shape-complementarity causes them to aggregate into desired structures.

Self-assembly has a number of advantages as a strategy: First, it carries out many of the most difficult steps in nanofabrication–those involving atomic-level modification of structure–using the very highly developed techniques of synthetic chemistry.

Second, it draws from the enormous wealth of examples in biology for inspiration: self-assembly is one of the most important strategies used in biology for the development of complex, functional structures.

Third, it can incorporate biological structures directly as components in the final systems.

Fourth, because it requires that the target structures be the thermodynamically most stable ones open to the system, it tends to produce structures that are relatively defect-free and self-healing.

Self-assembly has a special place in nanoscience. Top-down methods of fabrication provide the ability to build patterns, but are capital intensive, two dimensional, and limited in their ability to provide materials in quantity.

Bottom-up methods can make large quantities of nanostructures (including nanostructures too small to be made by any top-down method: e.g., molecules), but without pattern or regularity to their arrangement.

Self-assembly bridges the two: it provides a strategy that makes possible the patterning (in a broad sense) of nanostructures made by bottom-up synthesis; it can also use patterns generated by top-down fabrication to guide the ordering of nanostructures made by bottom-up methods.

The most promising avenues for self-assembly are presently those based on organic compounds, and organic compounds, as a group (although with exceptions), are electrical insulators; thus, many ideas for information processing and electrical/mechanical transduction will require either fundamental redesign in going from the macroscopic systems presently used to self-assembled systems, or the development of new types of organic molecules that show appropriate properties.

There are now a range of different molecular systems that self-assemble, that is, form ordered, monomolecular structures, by the coordination of molecules to surfaces. These systems self-assembled monolayers (SAMs) are reasonably well understood, and increasingly useful technologically. The crucial dimension in SAMs is the thickness perpendicular to the plane of the monolayer: this dimension, and the composition along this axis, can be controlled very simply at the scale of 0.1 nm by controlling the structures of the molecules making up the monolayer. SAMs also provide tailorable functions: for example, by changing the structures of the organic molecules in straightforward ways, interfacial free energies can be controlled. Complementary techniques such as micro contact printing (μCP) allow the in-plane dimensions of structures in SAMs to be controlled easily at the scale of 500 nm, and with difficulty at smaller dimensions.

8.2.3 Positional Assembly

The final bottom-up technique is positional assembly, whereby atoms, molecules or clusters are deliberately manipulated and positioned one-by-one. Techniques such as SPM for work on surfaces, or optical tweezers in free space, are used for this.

Positional assembly is extremely laborious and is currently not suitable as an atomic-scale industrial process. The utility and strength of SPM in industry lie in their ability to characterize and measure surfaces with atomic-level precision. The fact that (albeit very rudimentary) structures can be fabricated atom-by-atom has lead to speculation that tiny nanoscale machines could be made which could be used in parallel to manufacture materials atom-by-atom.

The idea is to fabricate one or a few machines (or assemblers) that would first make copies of themselves, and then go on to make materials in parallel, in principle solving the problem of slow production speed. This speculation has led some individuals to voice fears of uncontrollable self-replication, known as 'grey goo'. Such concerns currently belong in the realm of science fiction without any evidence so far.

Intermolecular and colloidal forces play a critical role throughout the entire process for the formation of nanostructured materials. All possible intermolecular forces have the ability to affect this process at some degree, and their interplay.

Figure 8.7 represents the schematic of possible intermolecular forces and their interplay at the interface between the reactive inorganic species and the self assembly units.

The reactive inorganic species are the precursors to the inorganic frame-work of nanostructured materials. They are mostly anionic, except in rare occasions such as extremely low pH. The self -assembly units are the organic (or polymeric) components.

There is no clear macroscale interface formed between these two components like the oil–water interface. Both components are in solution and the interface is formed at the contact region between the inorganic species and the self-assembly unit (or self-assembled aggregates).

Whether the self-assembly unit as shown in Figure 8.7 is a single molecule (a) or a self-assembled aggregate such as a micelle (b) or an emulsion (or microemulsion) (c) or a colloidal particle (d) there will be always an attractive force between the two components (i).

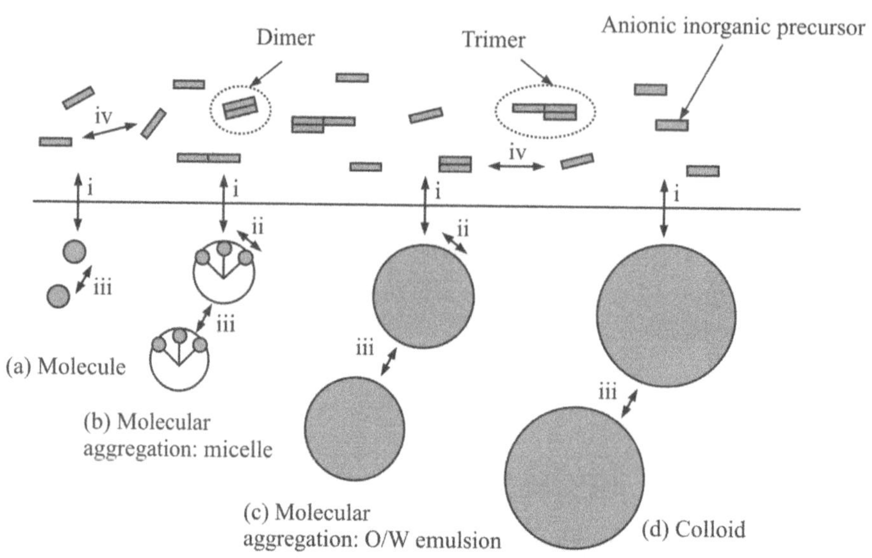

Figure 8.7 Interplay between intermolecular/colloidal forces at the interface between reactive inorganic species and self-assembly units or self-assembled aggregates

This attractive force is the major driving force that brings two components close together; thus, it is responsible for the formation of nanostructured materials.

When the self-assembly unit is cationic, force (i) is the direct electrostatic attraction. When it is anionic, this force is the mediated-electrostatic attraction mediated through the cationic species (or layer) between the two components.

Hydration force is the major player when the self -assembly unit is nonionic.

Force (ii) is the force that is induced as a result of force (i). As the two components come close together, the force balance within the self-assembled aggregates such as micelles or emulsions is forced to change. A shift toward the new balance is inevitable. And this will change the morphology of the self-assembled aggregates and also critical reaction conditions such as charge density, which will eventually affect the nanostructure formation. For the micelle, this occurs on the level of the entire micelle; for the emulsion, it mainly does so on the level of the self-assembled layer that stabilizes the emulsion droplets.

Force (iii) is also induced as a result of force (i) Assume that the self-assembly units are at stable condition before the interaction (either thermodynamically or kinetically). After the interaction between the two components, the reactive inorganic species that are confined on the surface of organic components can induce this attractive interaction. This is usually condensation.

When the inorganic components are nonreactive, force (iii) will be electrostatic repulsion. This will stabilize the entire system, which will result in no formation of nanostructures.

Force (iv) is the force between the inorganic species. For many cases of inorganic precursors, a variety of structural geometries exist in solution. These include ring structure, cage like structure, dimer, trimer, and so on. Understanding the effect of these different structural factors of inorganic species on the self-assembly unit is important in deriving a better explanation of the physical and chemical properties of nanostructured materials formed. Force (iv) is responsible for this to some degree. It can be a rather strong force such as condensation, or can be a weak force such as hydrophobic or van der Waals as well.

8.2.4 Challenges in Self Assembly

The fundamental thermodynamics, kinetics, and mechanisms of self-assembly are surprisingly poorly understood. The basic thermodynamic principles derived for molecules may be significantly different for those that apply (or do not apply) to nanostructures: the numbers of particles involved may be small; the relative influence of thermal motion, gravity, and capillary interactions may be different; the time required to, reach equilibrium may be sufficiently long that equilibrium is not easily achieved (or never reached); the processes that determine the rates of processes influencing many nano systems are not defined. Although the combination of top-down patterning and self-assembly promises to be a powerful one, the development of general strategies, based on a sound understanding of the relationship between the template and the self-assembled nanoscale components, remains to be worked out

8.3 Current Practice and Applications of Nano Bio Systems/Bio Nano Technology

8.3.1 Approaches for Constructing Bio Nano Systems

Bio nanotechnology refers to the harnessing of the vast diversity of self-assembling building blocks and processes for the assembly of nano-scaled structures for the manufacture of highly functional nanomaterials.

Semiconductor and metal nanoparticles in the 2-10 nm size range have aroused great interest due to their unique size-dependent properties, which happen to also have similar dimensions to biological macromolecules (e.g., proteins and nucleic acids) (16). This will have added advantages like the similarities allow integration of nanotechnology and biology, thus promoting great advances in cell biology, molecular biology, targeted therapeutics and medical diagnostics. However, size is just one of many characteristics of nanoparticles that itself is rarely sufficient if one is to use nanoparticles as biological tags.

In order to interact with biological target, a biological or molecular coating or layer acting as a bioinorganic interface should be attached to the nanoparticle. Examples of biological coatings may include antibodies, biopolymers like proteins (17) or monolayers of small molecules that make the nanoparticles biocompatible (18).

In addition, as optical detection techniques are wide spread in biological research, nanoparticles should either fluoresce or change their optical properties. The approaches used in constructing nano-biomaterials are schematically presented in (Figure 8.8).

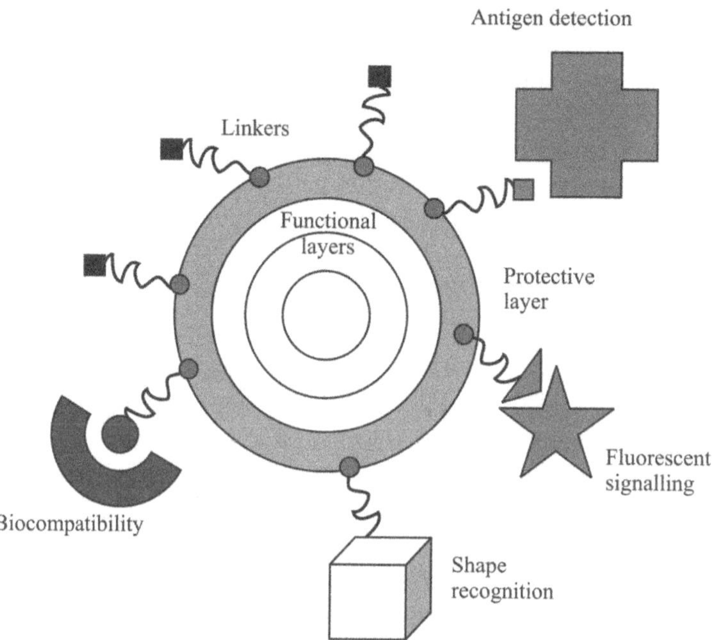

Figure 8.8 Schematic presentation of approaches for constructing bio nano systems

Nano-particle usually forms the core of nano-biomaterial. It can be used as a convenient surface for molecular assembly, and may be composed of inorganic or polymeric materials. It can also be in the form of nano-vesicle surrounded by a membrane or a layer. The shape is more often spherical but cylindrical, plate-like and other shapes are possible. The size and size distribution might be important in some cases, for example if penetration through a pore structure of a cellular membrane is required. The size and size distribution are becoming extremely critical when quantum-sized effects are used to control material properties. A tight control of the average particle size and a narrow distribution of sizes allow creating very efficient fluorescent probes that emit narrow light in a very wide range of wavelengths. This helps with creating biomarkers with many and well distinguished colours. The core itself might have several layers and be multifunctional. For example, combining magnetic and luminescent layers one can both detect and manipulate the particles.

The core particle is often protected by several monolayers of inert material, for example silica. Organic molecules that are adsorbed or chemisorbed on the surface of the particle are also used for this purpose. The same layer might act as a biocompatible material. However, more often an additional layer of linker molecules is required to proceed with further functionalisation. This linear linker molecule has reactive groups at both ends. One group is aimed at attaching the linker to the nanoparticle surface and the other is used to bind various moieties like bio-compatibles (dextran), antibodies, fluorophores etc., depending on the function required by the application (19).

8.3.2 Recent Developments

8.3.2.1 Tissue Engineering

Natural bone surface is quite often contains features that are about 100 nm across. If the surface of an artificial bone implant were left smooth, the body would try to reject it. Because of that smooth surface is likely to cause production of a fibrous tissue covering the surface of the implant. This layer reduces the bone-implant contact, which may result in loosening of the implant and further inflammation. It was demonstrated that by creating nano-sized features on the surface of the hip or knee prosthesis one could reduce the chances of rejection as well as to stimulate the production of osteoblasts. The osteoblasts are the cells responsible for the growth of the bone matrix and are found on the advancing surface of the developing bone.

Titanium is a well-known bone repairing material widely used in orthopedics and dentistry. It has a high fracture resistance, ductility and weight to strength ratio. Unfortunately, it suffers from the lack of bioactivity, as it does not support cell adhesion and growth well. Apatite coatings are known to be bioactive and to bond to the bone. Hence, several techniques were used in the past to produce an apatite coating on titanium. Those coatings suffer from thickness nonuniformity, poor adhesion and low mechanical strength. In addition, a stable porous structure is required to support the nutrients transport through the cell growth. It was shown that using a biomimetic approach a slow growth of nanostructured apatite film from the simulated body fluid resulted in the formation of a strongly adherent, uniform nano porous layer (20). The layer was found to be built of 60 nm crystallites, and possess a stable nano porous structure and bioactivity. A real bone is a nano composite material, composed of hydroxyapatite crystallites in the organic matrix, which is mainly composed of collagen.

An artificial hybrid material was prepared from 15–18 nm ceramic nanoparticles and poly (methyl methacrylate) copolymer (21). Using tribology approach, a visco-elastic behaviour (healing) of the human teeth was demonstrated. An investigated hybrid material, deposited as a coating on the tooth surface, improved scratch resistance as well as possessed a healing behaviour similar to that of the tooth.

8.3.2.2 Cancer Therapy

Photodynamic cancer therapy is based on the destruction of the cancer cells by laser generated atomic oxygen, which is cytotoxic. A greater quantity of a special dye that is used to generate the atomic oxygen is taken in by the cancer cells when compared with a healthy tissue. Hence, only the cancer cells are destroyed then exposed to a laser radiation. Unfortunately, the remaining dye molecules migrate to the skin and the eyes and make the patient very sensitive to the daylight exposure. This effect can last for up to six weeks. To avoid this side effect, the hydrophobic version of the dye molecule was enclosed inside a porous nanoparticle (22). The dye stayed trapped inside the Ormosil nanoparticle and did not spread to the other parts of the body. At the same time, its oxygen generating ability has not been affected and the pore size of about 1 nm freely allowed for the oxygen to diffuse out.

8.3.2.3 Multicolour Optical Coding for Biological Assays (23-25)

The ever increasing research in proteomics and genomic generates escalating number of sequence data and requires development of high throughput screening technologies. Realistically, various array technologies that are currently used in parallel analysis are likely to reach saturation when a number of array elements exceed several millions. A three-dimensional approach, based on optical "bar coding" of polymer particles in solution, is limited only by the number of unique tags one can reliably produce and detect.

Single quantum dots of compound semiconductors were successfully used as a replacement of organic dyes in various bio-tagging applications (25). This idea has been taken one step further by combining differently sized and hence having different fluorescent colours quantum dots, and combining them in polymeric microbeads. A precise control of quantum dot ratios has been achieved. The selection of nanoparticles used in those experiments had 6 different colours as well as 10 intensities. It is enough to encode over 1 million combinations. The uniformity and reproducibility of beads was high letting for the bead identification accuracies of 99.99%.

One of the biggest challenges to the biological application of quantum dots (QDs) is their biocompatibility due to their hydrophobic surface chemistry, which makes them insoluble in aqueous biological media. In order to render QDs biocompatible, they have to be soluble in biological buffer, and over the years many scientists have discovered a wide array of surface coatings for solubilisation of QDs. The techniques used to achieve solubilisation include ligand exchange, surface silanization and phase transfer methods. The "ligand exchange" method is based on the exchange of the hydrophobic surfactant molecules with bi-functional molecules, which are hydrophyllic on one side and hydrophobic on the other, to bind to the ZnS shell on the QD (26) (Alivisatos et al., 2005). Most often thiols (-SH) as functional groups are used to bind to the ZnS and carboxyl (-COOH) groups are used as hydrophyllic ends. The resulting QDs are soluble in both aqueous and polar solvents. This is by far the simplest method to achieve solubilisation. "Surface silanization" involves the growth of a silica shell around the nanocrystal. More details on QDs biological and medical applications are presented in Chapter 9.

8.3.2.4 Manipulation of Cells and Biomolecules (26)

Functionalised magnetic nanoparticles have found many applications including cell separation and probing. Most of the magnetic particles studied so far are spherical, which somewhat limits the possibilities to make these nanoparticles multifunctional. Alternative cylindrically shaped nanoparticles can be created by employing metal electrodeposition into nanoporous alumina template. Depending on the properties of the template, nano-cylinder radius can be selected in the range of 5 to 500 nm while their length can be as big as 60 μm. By sequentially depositing various thicknesses of different metals, the structure and the magnetic properties of individual cylinders can be tuned widely.

As surface chemistry for functionalisation of metal surfaces is well developed, different ligands can be selectively attached to different segments. For example, porphyrins with thiol or carboxyl linkers were simultaneously attached to the gold or nickel segments respectively. Thus, it is possible to produce magnetic nanowires with spatially segregated fluorescent parts. In addition, because of the large aspect ratios, the residual magnetisation of these nanowires can be high. Hence, weaker magnetic field can be used to drive them. It has been shown that a self-assem-bly of magnetic nanowires in suspension can be controlled by weak external magnetic fields. This would potentially allow controlling cell assembly in different shapes and forms. Moreover, an external magnetic field can be combined with a lithographically defined magnetic pattern ("magnetic trapping").

8.3.2.5 Protein Detection (27)

Proteins are the important part of the cell's language, machinery and structure, and understanding their functionalities is extremely important for further progress in human well being. Gold nanoparticles are widely used in immune histochemistry to identify protein-protein interaction. However, the multiple simultaneous detection capabilities of this technique are fairly limited. Surface-enhanced Raman scattering spectroscopy is a well-established technique for detection and identification of single dye molecules. By combining both methods in a single nanoparticle probe one can drastically improve the multiplexing capabilities of protein probes.

A number of nano materials being commercialized by various companies for bio and medical applications are presented in Table 8.2.

Table 8.2 List of some companies commercializing nano materials for bio and medical applications

Company	Major area of activity	Technology
Adventys Life Sciences Inc.	Drug delivery	Polymeric nonoparticles engineered to carry at tumour drug across the blood-brain barrier
Alnis Biosciences, Inc.	Bio-pharmaceutical	Biodegradable polymeric nonoparticles polymers nonoparticules for drug delivery
Argonide	Membrane filtration	Nonoporous ceramic materials for endotoxin filtration, orthopaedic and dental implants, DNA protein separation
BASF	Toothpaste	Hydroxyaoatite nano particles seems to improve dental surface
Biophan Technologies,Inc,	MRI shielding	Nanomagnetic/carbon composite materials to medical devices from RF-fields
Capsulution NanoScience AG Dynal Biotech	Pharmaceutical coatings to improve solubility of drugs	Layer-by-layer poly-electro coatings, 8-50 Magnetic beads
Eiffel Technology	Drug delivery	Reducing size of the drug particles to 50-100
Environments. Inc	Surface desinfectsant	Nanoemulsions
Evident Technologies	Luminescent biomarkers	Semiconductor quantum dots with amine or Ca groups on the surface, emission from 350 to 250

Table 8.2 ***Contd...***

Company	Major area of activity	Technology
Immunicon	Tarcking and separation of different cell types	Magnetic core surrounded by a polymeric layer coated with antibodies for capturing cells
KES Science and Technology, Inc	AiroCide filters	Nano-TiO_2 to destroy airborne pathogens
NanoBio Cortporation	Pharmaceutical	Antimicrobal non-emulsions
NanoCarrier Co., Ltd	Drug delivery	Micellar nanoparticles for encapsulation of drug proteins, DNA
NanoPharm AG	Drug delivery	Polybutilcynoacrylate nonoparticles are coate drugs and then with surfactant, can go across I blood-brain barrier
Nonoplex Technologies, Inc Nanoprobes, Inc.	Nanobarcodes for bioanalysis Gold nanoparticles for biological markers	Gold nanoparticles bio-conjugates for TEM and fluorescent microscopy
Nanoshpere, Inc.	Glod biomarkers	DNA barcode attached to each nanoprobe for identification purposes, PCR is used to amplify signal; also catalytic silver deposition to amplify signal using surface plasmon resonance
NanoMed Pharmaceutical, Inc.	Drug delivery	Nanoparticles for drug delivery
Oxonica Ltd	Sunscreens	Doped transparent nanoparticles to effectively absorb harmful UV and convert it into heat
PSiVida Ltd	Tissue engineering, implants, drugs and gene delivery bio-filtration	Exploiting material properties of nanostructure porous silicone
Smith & Nephew	Acticoat bandages	Nanocrystal silver is highly toxic to pathogene
QuantumDot Corporatin	Luminescent biomarkers	Bioconjugated semiconductor quantum dots

8.4 Drexler-Smalley Debate - Realistic Projections on Molecular Nano Technology

8.4.1 Essence of Drexler – Smalley Debate on Molecular Manufacturing

The debate between K. Eric Drexler and Richard Smalley on the feasibility of molecular assemblers began in 2001 and concluded in a cover story in Chemical & Engineering News in 2003 (28, 29).

The Drexler–Smalley debate on molecular nanotechnology was a public dispute between K. Eric Drexler, the originator of the conceptual basis of molecular nanotechnology, and Richard Smalley, a recipient of the 1996 Nobel prize in Chemistry for the discovery of the nanomaterial buck minster fullerene. The dispute was about the feasibility of constructing molecular assemblers, which are molecular machines which could robotically assemble molecular materials and devices by manipulating individual atoms or molecules.

The concept of molecular assemblers was central to Drexler's conception of molecular nanotechnology, but Smalley argued that fundamental physical principles would prevent them from ever being possible. The two also traded accusations that the other's conception of nanotechnology was harmful to public perception of the field and threatened continued public support for nanotechnology research.

The debate was carried out from 2001 to 2003 through a series of published articles and open letters. It began with a 2001 article by Smalley in Scientific American, which was followed by a rebuttal published by Drexler and coworkers later that year, and two open letters by Drexler in early 2003. The debate was concluded in late 2003 in a "Point–Counterpoint" feature in Chemical & Engineering News in which both parties participated.

The debate has been often cited in the history of nanotechnology due to the fame of its participants and its commentary on both the technical and social aspects of nanotechnology. It has also been widely criticized for its adversarial tone, with Drexler accusing Smalley of publicly misrepresenting his work, and Smalley accusing Drexler of failing to understand basic science, causing commentators to go so far as to characterize the tone of the debate as similar to "a pissing match" and "reminiscent of Saturday Night Live sketch".

Drexler's vision of nanotechnology, now called molecular nanotechnology, is based on the concept of the molecular assembler, a molecular machine which would manufacture molecules and molecular devices atom-by-atom. Drexler drew a distinction between wet nanotechnology based on biological systems, and "second-generation" dry nanotechnology which would be based on mechanosynthesis, positional control of molecules through principles more related to mechanical engineering. Drexler and his followers have focused almost exclusively on the latter form of molecular nanotechnology, but Drexler has stated that both are valid pathways to creating molecular machine systems. They made following projections.

8.4.2 K. Eric. Drexler's Future Projections on Molecular Manufacturing

Drexler in his article on Engines of Creation- The coming Era of Nano Technology (30) projected that by integrating advances in nano technologies and molecular engineering concepts the following new developments which will have great impact on the living nature of mankind can happen in future.

1. Engineered proteins can bind at precise locations on million atom frame works build through DNA engineering with strong stiff, electrically or chemically active force which can serve as self constructing machines. This can lead to step by step advanced molecular manufacturing.
2. Nano factories ranging from desktop scale and larger will be able to convert simple chemical feed stock into large atomically precise products cleanly, inexpensively and with moderate energy consumption.
3. By Molecular manufacturing A 10 kg factory will be able to produce 10 kg products in hours or less, a stack of billion processor laptops, a package containing a trillion cell sized medical devices or a roll containing hundreds of square meters of tough, flexible stuff that

converts sunlight to electric power. In these process the cost of raw materials will be only the cost of production.

At a $ per kg of industrial feed stock the soloar electric material will cost only one cent per square meter and the computers would cost only a dime with shipping and transport costs will be extra.

4. Just as present day computers are able to replace a type writer, a TV, a drawing board, a dark room, a record player, a pin ball machine, a library which people would not have belied to be possible, the effects of molecular manufacturing will be far away any one can imagine. Just as computers can do all this with their tiny, fast cycling parts can be directed to form complex patterns of elementary building blocks of information bits, likewise molecular manufacturing system will be able to use tiny fast cycling parts of elementary building blocks of matter.
5. Molecular engines may be created which can extract excess carbon dioxide from air and keep it in liquid form which can address the present day global warming problem.
6. Dexler predicts due to molecular manufacturing what we can do depends on what we can build. For example we can build assembly machines much smaller than a living cell and make materials stronger and lighter than available today which helps mankind to build better space crafts and tiny nano devices which can travel along capillaries to enter and repair living cells to heal diseases and reverse the ravages of age or make our bodies are speedier and stronger than before.
7. Molecular Manufacturing can facilitate to produce machines of the size of viruses and machines that would work at speeds that none of us can appreciate. Once we learn how to produce these tiny molecular machines we can have the option of assembling them into large frame work of intelligent machines based on using trillions of nano scopic parallel processing devices which make descriptions compare them to recorded patterns, and exploit their memories of previous experiments. These new molecular technologies not only change the matrerials we use in the physical environment but also the activities we will be then able to pursue inside which ever kind of world we make.
8. Dexler projects that the Engines of creation generated from molecular engineering concepts could be directed to address the major concern of mankind like wealth or poverty, health or sickness, peace or war. Engines of creation are best attempt so far to prepare us to think of what we might become, should we persist in making new technologies.

8.4.3 Exchange of Letters between Smalley and Drexler in Chemical and Engineering News

Smalley questioned whether devices like these could be constructed using an enzyme-like mechanical process, either in an aqueous solution or with some other chemistry, and whether such devices would be operational at all.

The debate was concluded in a "Point–Counterpoint" feature that was the 1 December 2003 cover story of Chemical & Engineering News, the news magazine of the American Chemical Society. The feature first reproduced Drexler's April 2003 open letter to Smalley. Smalley's response began by apologizing for any offense his September 2001 article had caused, and stating that Drexler's book Engines of Creation had triggered Smalley's own interest in nanotechnology. He agreed that "Smalley fingers" could not work, and then asserted that the same reasons that would preclude atomic control of reactions would also preclude the manipulation of larger building blocks, since each molecule would have multiple atoms which would need to be controlled.

He then agreed that something like an enzyme or ribosome would be capable of precise chemistry, but asked how the nanorobot would be able to obtain, control, and repair such an enzyme, and noted the incompatibility of many reactions with water-based biological systems, stating that "biology is wonderous in the vast diversity of what it can build, but it can't make a crystal of silicon, or steel, or copper, or aluminum, or titanium, or virtually any of the key materials on which modern technology is built." Smalley asked what kind of "nonaqueous enzyme like chemistry" Drexler would envision for his molecular assemblers to operate upon, calling this "a vast area of chemistry that has eluded us for centuries."

Drexler's counter response began by returning to Feynman's 1959 talk, stating that "although inspired by biology... Feynman's vision of nanotechnology is fundamentally mechanical, not biological." He characterized the challenges as being that of systems engineering rather than solely chemistry, and referred Smalley to Nano systems, with its vision of mechanical control of chemical reactions with no enzymes and no reliance on solvents or thermal motion. He stated:

Positional control naturally avoids most side reactions by preventing unwanted encounters between potential reactants. Transition-state theory indicates that, for suitably chosen reactants, positional control will enable synthetic steps at mega hertz frequencies with the reliability of digital switching operations in a computer. The supporting analysis for this conclusion appears in "Nanosystems" and has withstood a decade of scientific scrutiny.

Drexler reiterated that these molecular assemblers would require no impossible fingers, and would augment solution-phase chemistry to produce macroscopic products with precise arrangements of chemical building blocks, using solution-phase molecular assemblers to boot strap the construction of more sophisticated assemblers. He concluded by writing:

U.S. progress in molecular manufacturing has been impeded by the dangerous illusion that it is infeasible. I hope you will agree that the actual physical principles of molecular manufacturing are sound and quite unlike the various notions, many widespread in the press, that you have correctly rejected. I invite you to join me and others in the call to augment today's nanoscale research with a systems engineering effort aimed at achieving the grand vision articulated by Richard Feynman.

Smalley began his concluding letter: I see you have now walked out of the room where I had led you to talk about real chemistry, and you are now back in your mechanical world. I am sorry we have ended up like this. For a moment I thought we were making progress. You still do not appear to understand the impact of my short piece in Scientific American. Much like you can't make a boy and a girl fall in love with each other simply by pushing them together, you cannot make precise chemistry occur as desired between two molecular objects with simple mechanical motion along a few degrees of freedom in the assembler-fixed frame of reference. Chemistry, like love, is more subtle than that.

Smalley stated his belief that most reactions using mechano synthesis would simply give the wrong product, and that very few reactions and target molecules would likely be compatible with such an approach. He asserted that any robotic assembler arm would need an enzyme-like tool at its end such would require a liquid medium, and as all known enzymes use water as that medium, the range of products must be limited to the "meat and bone of biology." He accused Drexler of creating "a pretend world where atoms go where you want because your computer program directs them to go there."

Lastly, Smalley recounted his recent experience reading essays written by middle and high school students after an outreach visit, saying that nearly half of them thought that self-replicating nanorobots were possible and that most were worried about the results of them spreading across the world.

Smalley called this a deeply troubling bedside story that he did his best to allay. Smalley concluded his letter:

You and people around you have scared our children. I don't expect you to stop, but I hope others in the chemical community will join with me in turning on the light, and showing our children that, while our future in the real world will be challenging and there are real risks, there will be no such monster as the self-replicating mechanical nanorobot of your dreams.

The debate has received technical criticism as well. Steven A. Edwards in The Nanotech Pioneers noted that the ambiguity of the specifications and even definition of a molecular assembler makes an evaluation of the argument difficult and minimizes its scientific implications. He remarked that "nowhere in it does Nanosystems contain a blueprint for a molecular assembler.... We are told, for instance, that a manipulator arm would involve 4,000,000 atoms, but we are not told which atoms, or how they would be put together." He concludes that "the debate over mechano synthesis so far is huge to the participants, but mainly an entertaining academic diversion to most nanotechnologists."

On the other hand, futurist Ray Kurzweil in his book The Singularity is Near declared Drexler as the winner of the debate, reiterating the view that Smalley distorted Drexler's ideas and calling Smalley's responses "short on specific citations and current research and long on imprecise metaphors" and asserting that "Smalley is ignoring the past decade of research on alternative means of positioning molecular fragments using precisely guided molecular reactions... [which have] been extensively studied." He quoted experimental results on enzyme function in nonaqueous solutions, and pointed out that modern non-biological technology such as airplanes and computers have exceeded the capabilities of natural biological systems. He also noted that "earlier critics also expressed skepticism that either worldwide communication networks or software viruses that would spread across them were feasible.... [but today] we are obtaining far more gain than harm from this latest example of intertwined promise and peril."

8.4.4 Commentary on Public Perception of Nanotechnology

The debate's focus on the public perception of nanotechnology has also received commentary. Political blogger Glenn Reynolds stated that "the business community is afraid that advanced nanotechnology just seems too, well, spooky—and worse, that discussions of potentially spooky implications will lead to public fears that might get into the way of bringing products to market." Lawrence Lessig criticized The emerging science of bionanotechnology refers to the harnessing of the vast diversity of self-assembling building blocks and processes for the assembly of nano-scaled structures for the manufacture of highly functional nanomaterials.

8.4.5 Molecular Manufacturing Present: Overview

Molecular manufacturing (MM) can solve many of the world's current problems. For example, water shortage is a serious and growing problem. Most water is used for industry and agriculture; both of these requirements would be greatly reduced by products made by molecular manufacturing. Infectious disease is a continuing scourge in many parts of the world. Simple products like pipes, filters, and mosquito nets can greatly reduce this problem. Information and communication are valuable, but lacking in many places. Computers and display devices would become stunningly cheap. Electrical power is still not available in many areas. The efficient, cheap building of light, strong structures, electrical equipment, and power storage devices would allow the use of solar thermal power as a primary and abundant energy source. Environmental degradation is a serious problem worldwide. High-tech products can allow people to live with much less environmental impact. Many areas of the world cannot rapidly bootstrap a 20th century manufacturing infrastructure. Molecular manufacturing technology can be self-contained and clean; a single packing crate or suitcase could contain all equipment required for a village-scale industrial revolution. Finally, MM will provide cheap and advanced equipment for medical research and health care, making improved medicine widely available. Much social unrest can be traced directly to material poverty, ill health, and ignorance. MM can contribute to great reductions in all of these problems, and in the associated human suffering.

The following are some future areas of research and development for realization of molecular manufacturing in large scale

- Molecular design and folding
- Innovative self assembly methods
- New catalysts (Inorganic,enzyme etc)
- Dendimers, fullerenes and other novel chemical structures
- Bioenergitic, nano batteries and ultrasound driven chemistry
- Semiconductor-organic/biological interfaces
- Minaturization and massive parallelism of SFM
- Molecular modelling tools

References

1. Feynman R (1991). There's plenty of room at the bottom. Science.
2. Philip Ball (1994). Designing the Molecular world: Chemistry at the Frontier. Princeton University Press.
3. Taton TA (2002). Nanostructures as tailored biological Probes. Trends Biotechnol, 20, 277-279.

4. C. Mavroidis, A. Dubey, and M.L. Yarmush. (2004). Annu. Rev. Biomed. Eng.. 6, 10.1–10.33. 20-BE06-10_001-033.
5. Creighton, T. E. (1984). Proteins. New York: W. H. Freeman and Company.
6. http://focus.aps.org/story/v13/st14 links to papers. http://www.berkeley.edu/news/media/releases/2000/07/27_nano.html for mention of near-frictionless sliding nanotubes.
7. Drexler (1992). Design of a Primitive Nanofactory, Nano Systems, Wiley.
8. Merkle, R. (1999). Casing an assembler. Nanotechnology 10, 315-322.
9. http://www.xidex.com/Xidex%20Caliper%20CD%20AFM%20SPIE%202002.pdf
10. http://www.berkeley.edu/news/media/releases/2000/07/27_nano.html
11. http://www.nims.go.jp/hvems/nano_char/facilities/facility6/NanofabSEM_e.html
12. http://www.iscid.org/encyclopedia/Fluorescense_Resonance_Energy_Transfer
13. http://www.nanosight.ukideas.com/
14. http://www.parc.com/cms/get_article.php?id=223
15. George M. Whitesides, John P. Mathias, and Christopher, T. Seto. (1991). Molecular Self Assembly and Nano Chemistry, A Chemical Strategy for the synthesis of nano structures: Science, Vol 254, 1312-1319.
16. Sinani VA, Koktysh DS, Yun BG, Matts RL, Pappas TC, Motamedi M, Thomas SN and Kotov. Collagen coating promotes biocompatibility of semiconductor nanoparticles in stratified LBL films. Nano Letter 3(9)-1177-1182.
17. Zhang Y, Kohler N and Zhang M (2002). Surface modification of superpar- amagnetic magnetite nanoparticles and their intracellular uptake. Biomaterials, 23, 1553-1561.
18. Gutwein LG and Webster TJ (2003). Affects of alumina and titania nano-particulates on bone cell function. American Ceramic Society 26th Annual Meeting Conference Proceedings.
19. De Mel, A., Oh, J. T., Ramesh, B. and Seifalian, A. M. (2012). Bio functionalized quantum dots for live monitoring of stem cells: applications in regenerative medicine. Regen Med, 7, 335-47, 254: 1300-1301.
20. Ma J, Wong H, Kong LB and Peng KW (2003). Biomimetic processing of nanocrystallite bioactive apatite coating on titanium. Nanotechnology, 14, 619-623.
21. de la Isla A, Brostow W, Bujard B, Estevez M, Rodriguez JR, Vargas S and Castano VM (2003). Nanohybrid scratch resistant coating for teeth and bone visco elasticity manifested in tribology. Mat Resr Innovat, 7, 110-114.
22. Roy I, Ohulchanskyy TY, Pudavar HE, Bergey EJ, Oseroff AR, Morgan J, Dougherty TJ and Prasad PN. (2003). Ceramic-based nanoparticles entrapping water-insoluble photosensitizing anticancer drugs: a novel drug-carrier system for photodynamic therapy. J Am Chem Soc, 125, 7860-7865.
23. Han M, Gao X, Su JZ and Nie S (2001). Quantum-dot-tagged microbeads for multiplexed optical coding of biomolecules. Nature Biotechnology, 19, 631-635.
24. Parak WJ, Gerion D, Pellegrino T, Zanchet D, Micheel C, Williams CS, Boudreau R, Le Gros MA, Larabell CA and Alivisatos AP. (2003). Biological applications of colloidal nanocrystals. Nanotechnology, 14, R15-R27.
25. Alivisatos, A. P., Gu, W. and Larabell, C. (2005). Quantum dots as cellular probes. Annu Rev Biomed Eng, 7, 55-76.

26. Reich DH, Tanase M, Hultgren A, Bauer LA, Chen CS and Meyer GJ (2003) Bio-logical applications of multifunctional magnetic nanowires. J Appl Phys, 93, 7275-7280.

27. Cao YC, Jin R, Nam JM, Thaxton CS and Mirkin CA (2003). Raman dye- labeled nanoparticle probes for proteins. JACS, 125, 14676-14677.

28. K.Eric. Drexler (1986). Engines of Creation: The coming Era of Nano Technology, Doubleday, Newyork.

29. K. Grexler and C. Peterson (1991). with Pergamit, Unbounding the Future: The Nanotechnology Revolution, William Morrow and Company. Inc.

9

Semiconductors, Nano Size Effects and Moore's Law

9.1 Semiconductor Materials and their Properties

Materials can be categorized into *conductors*, *semiconductors* or *insulators* by their ability to conduct electricity.

Conductors

Generally speaking, most metals are good conductors of electricity, as they have very small resistance values, usually in the region of micro-ohms per meter. Metals conduct electricity easily because the energy levels between the conduction and valence band are closely spaced, or there are more energy levels available than there are electrons to fill them so very little energy is required to find new energies for electrons to occupy. The resistivity of a material is a measure of how difficult it is for a current to flow.

While metals such as copper and aluminum are very good conductors of electricity, they still have some resistance to the flow of electrons and consequently do not conduct perfectly. The energy which is lost in the process of passing an electrical current, appears in the form of heat which is why conductors and especially resistors become hot. Also the resistivity of conductors increases with ambient temperature because metals are also generally good conductors of heat.

Insulators

Insulators on the other hand are the exact opposite of conductors. They are made of materials, generally non-metals, that have very few or no "free electrons" floating about within their basic atom structure because the electrons in the outer valence shell (Figure 9.1) are strongly attracted by the positively charged inner nucleus. So if a potential voltage is applied to the material no current will flow as there are no electrons to move and which gives these materials their insulating properties.

Insulators also have very high resistances, millions of ohms per meter, and are generally not affected by normal temperature changes (although at very high temperatures wood becomes charcoal and changes from an insulator to a conductor). Examples of good insulators are marble, fused quartz, p.v.c. plastics, rubber etc.

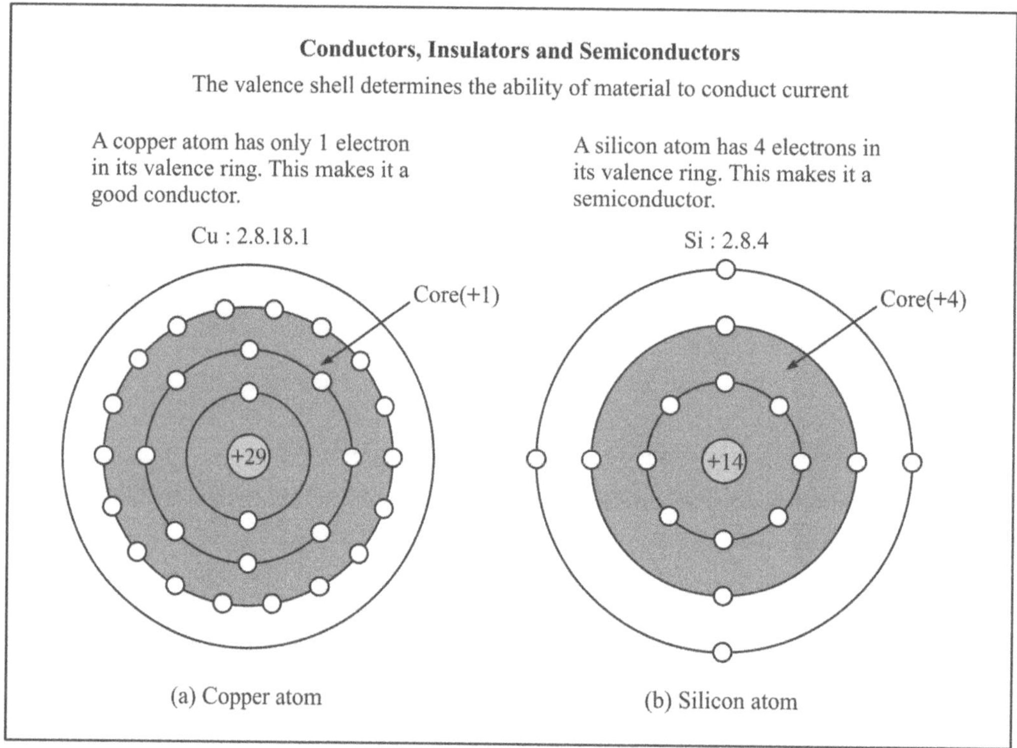

Figure 9.1 Valence schell of Cu and Si

Insulators play a very important role within electrical and electronic circuits, because without them electrical circuits would short together and not work. For example, insulators made of glass or porcelain are used for insulating and supporting overhead transmission cables while epoxy-glass resin materials are used to make printed circuit boards, PCB's etc., while PVC is used to insulate electrical cables.

9.1.1 Semiconductors

Semiconductors materials such as silicon (Si), germanium (Ge) and gallium arsenide (GaAs), have electrical properties somewhere in the middle, between those of a "conductor" and an "insulator". They are not good conductors nor good insulators (hence their name "semi"-conductors). They have very few "fee electrons" because their atoms are closely grouped together in a crystalline pattern called a "crystal lattice".

Semiconductors can be grouped into two categories

(i) Elemental semiconductors: Silicon (Si) most common semiconductor used today Germanium (Ge): First semiconductor used in p-n diodes

(ii) Compound Semiconductors: Gallium Arsenide (GaAs), Gallium Nitride (GaN), Silicon carbide (SiC)

By far, silicon (Si) is the most widely used material in semiconductor devices. Its combination of low raw material cost, relatively simple processing, and a useful temperature range make it currently the best compromise among the various competing materials. Silicon used in semiconductor device manufacturing is currently fabricated into boules that are large enough in diameter to allow the production of 300 mm (12 in.) wafers.

Germanium (Ge) was a widely used early semiconductor material but its thermal sensitivity makes it less useful than silicon. Today, germanium is often alloyed with silicon for use in very-high-speed SiGe devices; IBM is a major producer of such devices. Gallium arsenide (GaAs) is also widely used in high-speed devices but so far, it has been difficult to form large-diameter boules of this material, limiting the wafer diameter to sizes significantly smaller than silicon wafers thus making mass production of GaAs devices significantly more expensive than silicon. Silicon carbide (SiC) has found some application as the raw material for blue light-emitting diodes (LEDs) and is being investigated for use in semiconductor devices that could withstand very high operating temperatures and environments with the presence of significant levels of ionizing radiation. IMPATT diodes have also been fabricated from SiC.

Various indium compounds (indium arsenide, indium antimonide, and indium phosphide) are also being used in LEDs and solidstate laser diodes. Selenium sulfide is being studied in the manufacture of photovoltaic solar cells.

9.1.1.1 Elemental Semiconductors

These are semiconductors where each atom is of the same type such as Ge, Si. are bound together by covalent bonds, so that each atom shares an electron with its nearest neighbor, forming strong bonds.

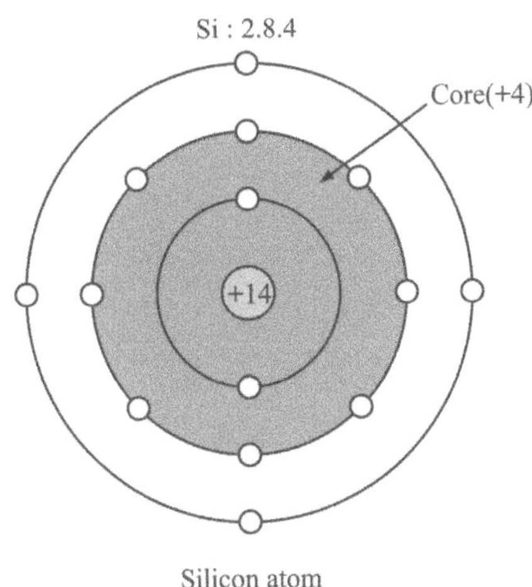

Figure 9.2 Valence electrons of Si

In Silicon (Si) there are 4 electrons in the valence shell (Figure 9.2). The potential required to remove any one of these 4 valence is lower than that required for any other electron in the structure In silicon crystal (Figure 9.3) these 4 valence electrons are bonded to 4 adjoining atoms. Materials in group IV in Periodic Table are referred as tetravalent atoms because they each have 4 valence electrons. A bonding of atoms, strengthened by the sharing of electrons is called covalent bonding. Although the covalent bond will result in a stronger bond between the valence electrons and their parent atom, it is still possible for the valence electrons to absorb sufficient kinetic energy from natural causes to break the covalent bond and assume the free state.

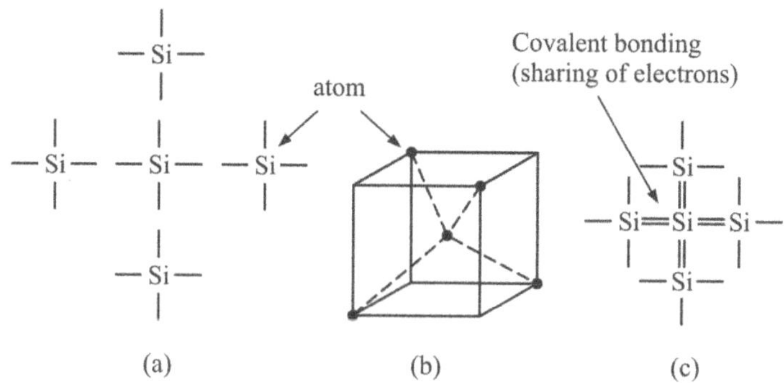

Figure 9.3 Silicon crystal structure

9.1.1.2 Compound Semiconductors

These are made of two or more elements. Common examples are GaAs or InP. These compound semiconductors belong to the III-V semiconductors so called because first and second elements can be found in group III and group V of the periodic table respectively. In compound semiconductors, the difference in electro-negativity leads to a combination of covalent and ionic bonding. Ternary semiconductors are formed by the addition of a small quantity of a third element to the mixture, for example Al x Ga In-x As. The subscript x refers to the alloy content of the material, what proportion of the material is added and what proportion is replaced by the alloy material. The addition of alloys to semiconductors can be extended to include quaternary materials such as Ga x In (1-x) As y P (1-y) or GaInNAs and even quaternary materials such as GaInNAsSb. The subscripts denote the proportion elements that constitute the mixture of elements. Alloying semiconductors in this way allows the energy gap and lattice spacing of the crystal to be chosen to suit the application

9.1.1.3 Energy Band Theory (1-9)

Semiconductors have a resistivity between $10^{-4} < \rho < 10^{8}\ \Omega\text{m}$ although these are rough limits. The band theory of materials explains qualitatively the difference between these types of materials. Electrons occupy energy levels from the lowest energies upwards (Figure 9.4). However, some energy levels are forbidden because of the wave like properties of atoms in the material. The allowed energy levels tend to form bands. The highest filled level at T=0 K is known as the *valence band* . Electrons in the valence band do not participate in the conduction process. The first unfilled level above the valence band is known as the *conduction band.* In metals, there is no forbidden gap; the conduction band and the valence band overlap, allowing free electrons to participate in the conduction process. Insulators have an energy gap that is far greater than the thermal energy of the electron, while semiconductor materials the energy gap is typically around 1 eV. The (Figure 9.4) shows the differences in metals, semiconductors and insulators in terms of the how the energy bands are separated.

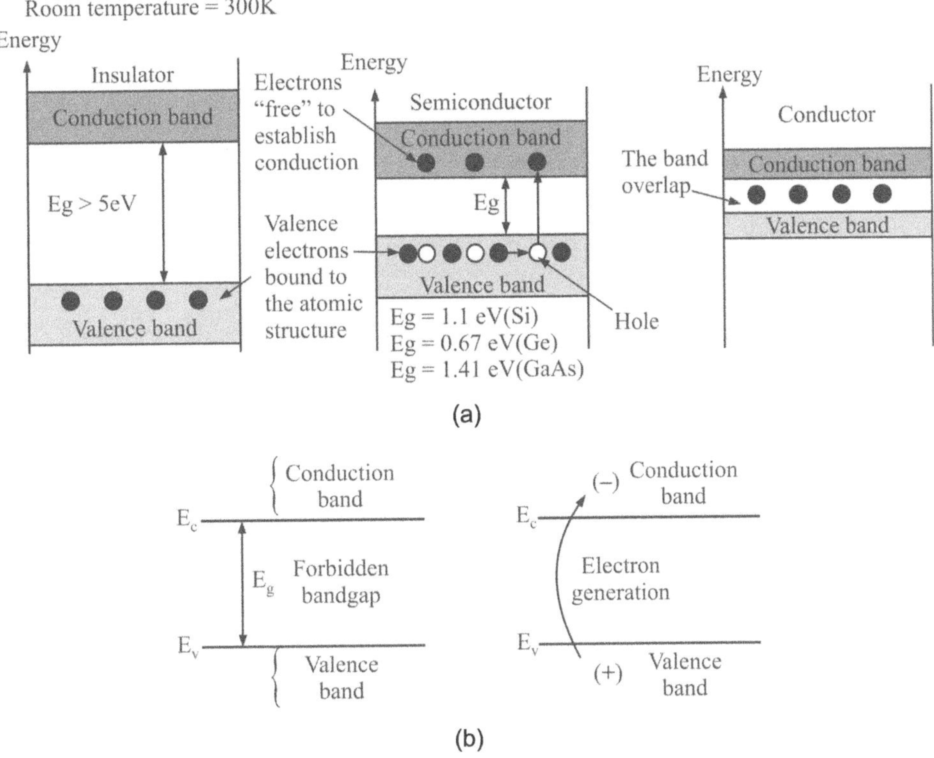

Figure 9.4 Energy band diagram (a) Energy bands in conductors, semiconductors and insulators; energy band diagram (b) EV = Maximum Energy of a valency electron or hole; EC = Minimum energy free electron, Eg = Energy required to break the covalent bond

9.1.1.4 Intrinsic Semiconductors

Intrinsic materials are those semiconductors that have carefully refined to reduce the impurities to very low level i.e., 100% pure material. The semiconductor material structure should contain no impurity atoms. Elemental and compound semiconductors can be intrinsic semiconductors. At room temperature, the thermal energy of the atoms may allow a small number of the electrons to participate in the conduction process. Unlike metals, where the resistance of the material decreases with temperature (Figure 9.5). For semiconductors, as the temperature increases, the thermal energy of the valence electrons increases, allowing more of them to breach the energy gap into the conduction band. When an electron gains enough energy to escape the electrostatic attraction of its parent atom, it leaves behind a vacancy which may be filled be another electron. The vacancy produced can be thought of as a second carrier of positive charge. It is known as a hole (Figure 9.6). As electrons flow through the semiconductor, holes flow in the opposite direction (Figure 9.7). If there are n free electrons in an intrinsic semiconductor, then there must also be n holes. Holes and electrons created in this way are known as intrinsic charge carriers. The carrier concentration, or charge density, defines the number of charge carriers per unit volume. This relationship can be expressed as n=p where n is the number of electrons and p the number of holes per unit volume. The variation in the energy gap between different semiconductor materials means that the intrinsic carrier concentration at a given temperature also varies.

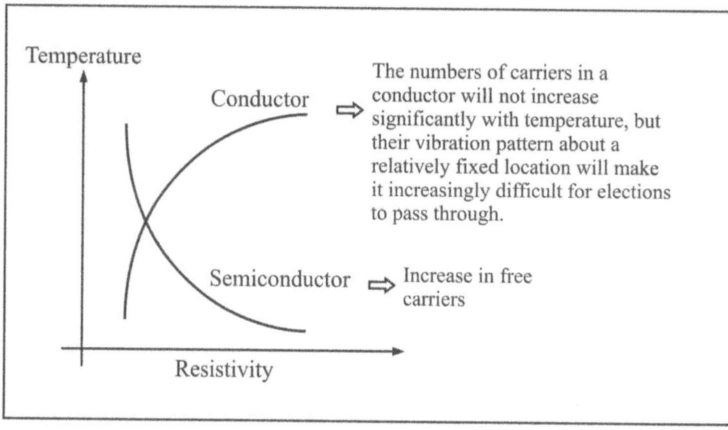

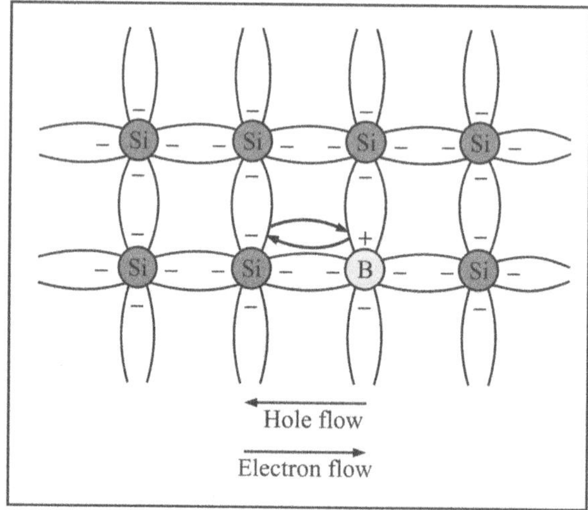

Figure 9.5 Effect of temperature on the resistivity

Effect of the hole on conduction: If a valence electron acquires sufficient kinetic energy to break its covalent bond and fills the void created by a hole, then a vacancy or hole will be created in the covalent bond that released the electron. There is therefore a transfer of holes to the left and electrons to the right.

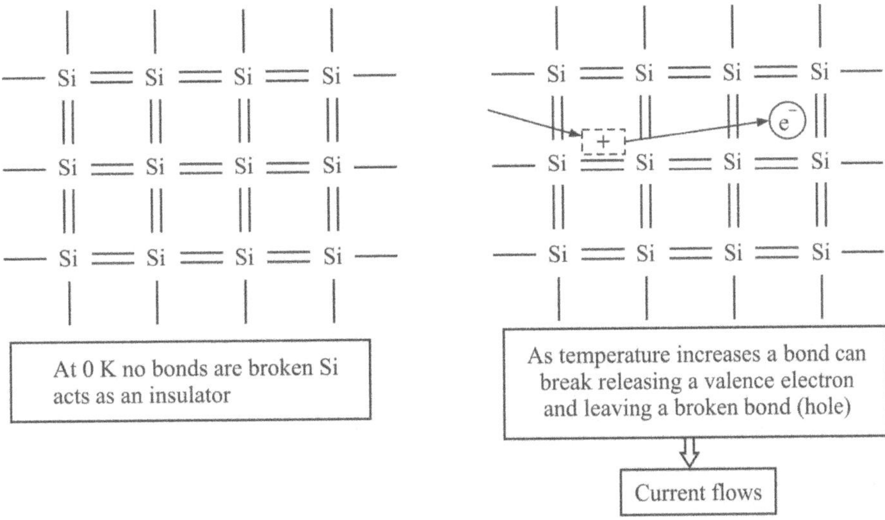

Figure 9.6 Effect of temperature on hole formation in semiconductors

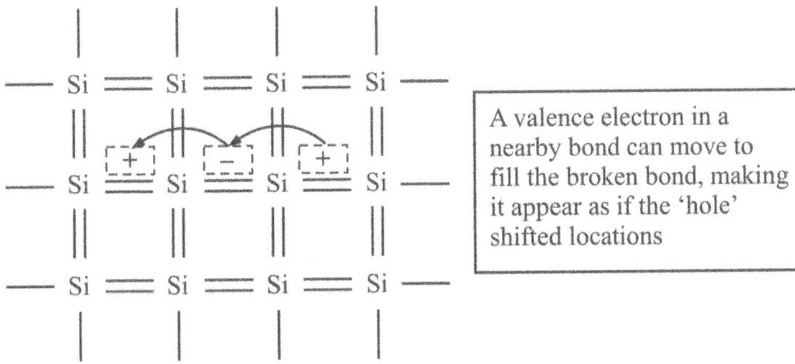

Figure 9.7 Movement of holes and electron vs hole flow

9.1.1.5 Extrinsic Semiconductors

The characteristic of semiconductor materials can be altered significantly by the addition of certain impurity atoms into the relatively pure semiconductor materials. These impurities although only added at 1 part in 10 million ,can alter the band structure sufficiently to totally change the electrical properties of the material. A semiconductor material subjected to doping process is called extrinsic semiconductor

If certain impurities are added to the intrinsic semiconductor materials energy states in the forbidden bands will occur which will cause net reduction in E_g for both semiconductor materials, consequently increased carrier density in the conduction band at room temperature. There are two types of extrinsic semiconductors: **n-type** (Figure 9.8a and **p-type** (Figure 9.8b).

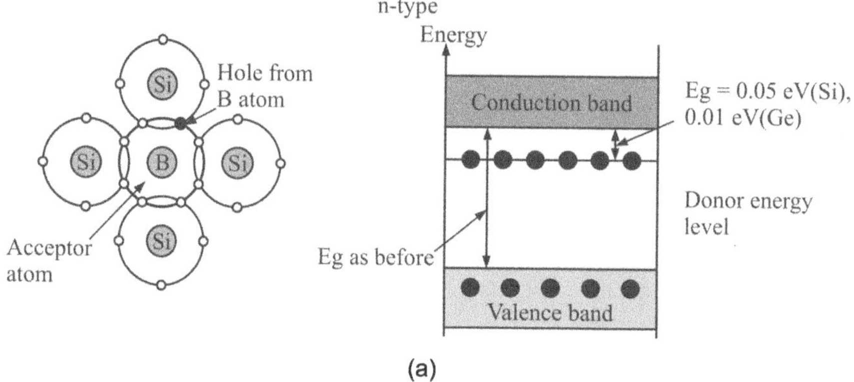

(a)

Figure 9.8 (a) To create n-Type semiconductors-dope other atoms with 5 valence electrons such as Sb, As and P to silicon to increase the free electrons

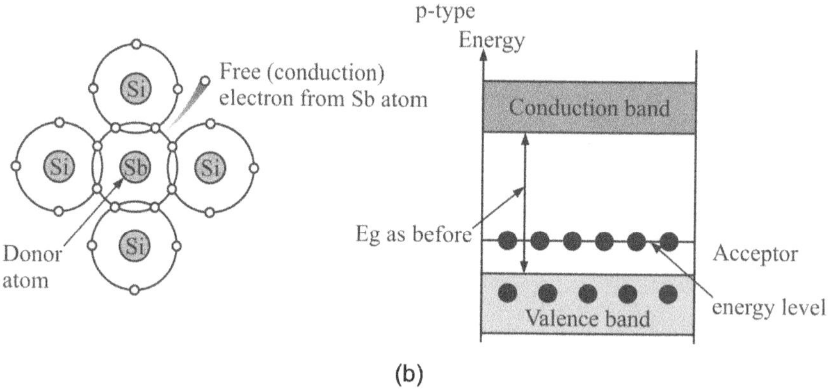

(b)

Figure 9.8 (b) To create p-Dope other atoms with 3 valence electrons such as Boron(B), Gallium (Ga) and Indium (In) to Silicon to create deficiency of electrons or hole charges

9.1.1.6 Free Carriers

The discrete energy level (energy level for donor/acceptor) appears in the forbidden band with an E_g significantly lessor more than that of the intrinsic material. Those free electrons due to the added impurity sit at this energy level and will have less difficulty in absorbing sufficient thermal energy to move into the conduction band at room temperature due to which there will be large number of carriers (electrons) in the conduction level and the conductivity of the material increases significantly. The natural causes for generation of free carriers are

(i) effects as light energy in the form of photons,

(ii) thermal energy from the surrounding medium.

The free state refers to existence of free carriers which refers to electrons and holes. At room temperature there are approximately 1.5×10^{10} free carriers in a cubic centimeter of intrinsic silicon material. The free carriers are sensitive to applied electric fields such as established by voltage sources or any difference in potential leading to current flow.

Majority and Minority carriers In the intrinsic state: The number of free electrons in Ge is only to those few electrons in the valence band that have acquired sufficient energy from thermal or light sources to break the covalent bond or to the few impurities that could not be removed. The vacancies left behind in the covalent bonding structure give limited supply of holes. In an n-type material the number of holes has not changed significantly from this intrinsic level. The net result is therefore that the number of electrons far outweighs the number of holes.

In an n-type of material the electron is called the majority carrier and the hole is called minority carrier. For the p –type material the number of holes far outweighs the number of electrons. In a p –type the hole is the majority carrier.

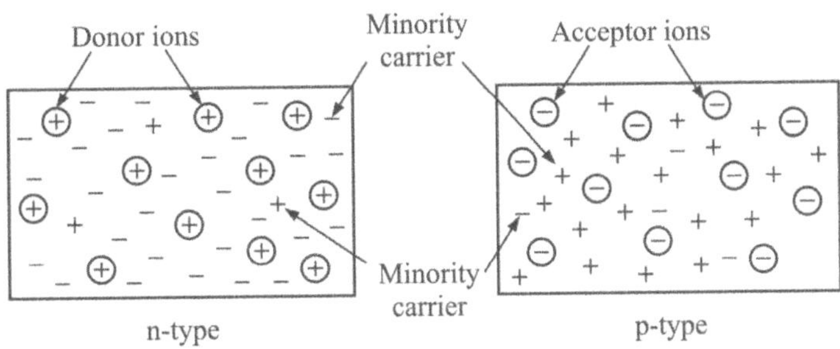

Figure 9.9 Majority and minority carrier

When the fifth electron of a donor atom leaves the parent atom, the atom remaining acquires a net positive charge: hence the positive sign in the donor-ion representation. For Similar reasons the negative sign appears in the acceptor ion (Figure 9.9).

9.1.1.7 Calculation of Carrier Concentrations

For the calculation of semiconductor electrical properties and analyzing device behavior, it is necessary to know the number of charge carriers/cm^3 in the material. The majority carrier concentration in a heavily doped material is obvious, since for each impurity atom, one majority carrier is obtained. However, the minority carrier concentration and the dependence of carrier concentrations on temperature are not obvious. To obtain the carrier concentrations, their distribution over the available energy states is required. These distributions are calculated using statistical methods. The following equations provides a mathematical expression for calculating electron and hole concentrations, intrinsic carrier concentration.

Electron and Hole Concentrations

n = electron concentration

p = hole concentration

$$n_i^2 = n.p \qquad .....(9.1)$$

n-type:

$n = N_D$, the donor concentration $p = p = n_i^2 / N_D$

p-type:

$p = N_A$, the acceptor concentration $n = n_i^2 / N_A$

Intrinsic Carrier Concentration

$$n_i = BT^{3/2} e^{\frac{-E_g}{2kT}} \quad .....(9.2)$$

B = Coefficient related to specific semiconductor

T = Temperature in Kelvin

E_g = Semiconductor bandgap energy

K = Bolzmann's constant

$$n_i(\text{Si}, 300\,\text{K}) = 1.5 \times 10^{10}\,\text{cm}^{-3} \quad .....(9.3)$$

9.2 Semiconductor Devices

These are electronic components that exploit the electronic properties of semiconductor materials, principally silicon, germanium, and gallium arsenide, as well as organic semiconductors. Semiconductor devices have replaced thermionic devices (vacuum tubes) in most applications. They use electronic conduction in the solid state as opposed to the gaseous state or thermionic emission in a high vacuum. Semiconductor devices are manufactured both as single discrete devices and as integrated circuits (ICs), which consist of a number from a few (as low as two) to billions of devices manufactured and interconnected on a single semiconductor substrate, or wafer.

Semiconductor materials are so useful because their behavior can be easily manipulated by the addition of impurities, known as doping.

Semiconductor conductivity can be controlled by introduction of an electric or magnetic field, by exposure to light or heat, or by mechanical deformation of a doped monocrystalline grid; thus, semiconductors can make excellent sensors. The semiconductor material used in devices is doped under highly controlled conditions in a fabrication facility, or fab, to precisely control the location and concentration of p- and n-type dopants. The junctions which form where n-type and p-type semiconductors join together are called p–n junctions. Exposing a semiconductor to light can generate electron–hole pairs, which increases the number of free carriers and its conductivity. Diodes optimized to take advantage of this phenomenon are known as photodiodes.

The following are some semiconductor devices which are being used presently for various applications:

Two-terminal devices: DIAC, Diode (rectifier diode), Gunn diode, IMPATT diode, Laser diode, Light-emitting diode (LED), Photocell, PIN diode, Schottky diode, Solar cell, Tunnel diode, VCSEL, VECSEL, Zener diode.

Three-terminal devices: Bipolar transistor Darlington transistor, Field-effect transistor, Insulated-gate bipolar transistor (IGBT), Silicon-controlled rectifier, Thyristor, TRIAC, Unijunction transistor.

Four-terminal devices: Hall effect sensor (magnetic field sensor).

Multi-terminal devices: Integrated circuit (ICs), Charge-coupled device (CCD), Microprocessor Random-access memory (RAM), Read-only memory (ROM).

9.3 Growing Silicon Demand for Semiconductors

Polycrystalline silicon (Si), or poly-silicon mainly contribute to the raw material use in the production of single crystal wafers, the substrate for silicon-based solar cells and semiconductors. The raw material has a dark gray color, with a bronze to bluish metallic sheen. Meeting the demands of end-users for solar power systems begin with the manufacturing of high quality polycrystalline silicon, in short poly-silicon.

Silicon-based solar panels make up 90 percent of the photovoltaic market. Thin-film technologies, which consist mainly of cadmium tellurium (CdTe) and copper indium gallium selenium (CIS/CIGS), cover the other ten percent. The thin film panel manufacturing process requires about 1/100th of the charging material needed for silicon-based modules. However, these panels have a lower efficient than standard solar modules.

9.3.1 Poly-silicon Manufacturers

In 2000, the photovoltaic (PV) industry accounted for ten percent of the market for polysilicon. Semiconductor companies accounted for the other 90 percent of the market. These chip companies produced the semiconductor chips required to power calculators, cellular phones, wristwatches, stereos, and other electronic gadgets.

Before the solar industry set off on its historic path of growth over the last decade, silicon manufacturers were slow to ramp up production capacity. About the same time, Germany's feed-in-tariff policy (FIT), Erneuerbare-Energien-Gesetz (EEG)--German for the "Renewable Energy Sources Act," started to gather momentum.

By late 2004, as the number of solar module manufacturers increased, solar cell manufacturers began competing with the semiconductor industry for the limited supply of poly-silicon feedstock more aggressively.

National Renewable Energy Laboratory (NREL) statistics show the PV industry surpassing the semiconductor sector as the largest consumer of polysilicon feedstock, in 2008. From 2007 to 2008, solar cell manufacturers' purchase of polysilicon increased from 17,000, to 44,000 metric tons, which resulted in significant supply shortages. In 2003, silicon feedstock had a spot price of $24 per kilogram (kg). Explosive demand pushed spot prices to $200 kg by 2007. At its peak in 2008, the spot price for polysilicon skyrocketed to $500 per kg.

Manufacturers scrambled to secure product—most market participants entered long-term supply agreements with durations of 6 to 10 years. Polysilicon manufacturers required buyers to pay in advance. Firms that lack sufficient capital to secure a stable supply of polysilicon pre-paid the contract by selling equity in their companies.

Historic high prices motivated many new companies to enter the polysilicon manufacturing segment, especially in the United States - where state and federal subsidies for the solar industry began to take shape. Companies in Europe, China, Russia, and Qatar have also entered the industry.

Some companies, already involved in the solar industry, simply vertically integrated their operations into the polysilicon manufacturing segment.

LDK constructed a production plant near its solar wafer manufacturing facilities in Xinyu Hi-Tech Industrial Park. The two factories - 1000 and 15,000 metric ton capacity came online in 2009.

Polysilicon manufacturing plants entail a time-consuming construction process, which may cost $1-1.5 billion to complete.

However, the financial collapse of 2008 led to a crash in polysilicon prices, led to most of these enterprises going under, in 2009, as the drop in demand sent polysilicon prices plummeting below their marginal costs.

9.3.2 Global Demand for Silicon

Global demand for silicon grew at an average rate of more than 8% between 2002 and 2008, but fell by almost 20% in 2009, and recovered by the same amount in 2010. Ten countries accounted for about 80% of the world market in 2010, with the USA, China, Germany and Japan as the major markets (Figure 9.10).

Apparent consumption in China does not indicate the rapid market growth for silicon metal in the mid-2000s that is seen with most other markets. One explanation is that production is underestimated between 2004 and 2008. In 2009, however, a sharp decline in exports resulted in apparent consumption more than doubling, assuming domestic production of about 780 kt in 2009. Considerable volumes also probably went into stocks.

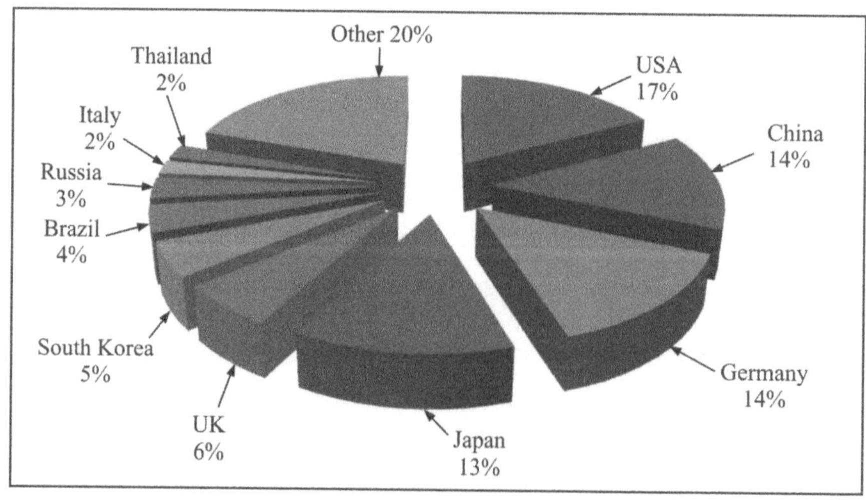

Figure 9.10 World division of apparent consumption of silicon by various countries (2010)

9.3.3 Poly Silicon Production

In 2009, China increased its polysilicon production output threefold, over 2008 levels, to 18,000 tons. Overall, the strong growth in polysilicon production and shows no signs of subsiding. Production reached 209,000 tons in 2010 and may approach 350,000 tons by 2012 (Figure 9.11).

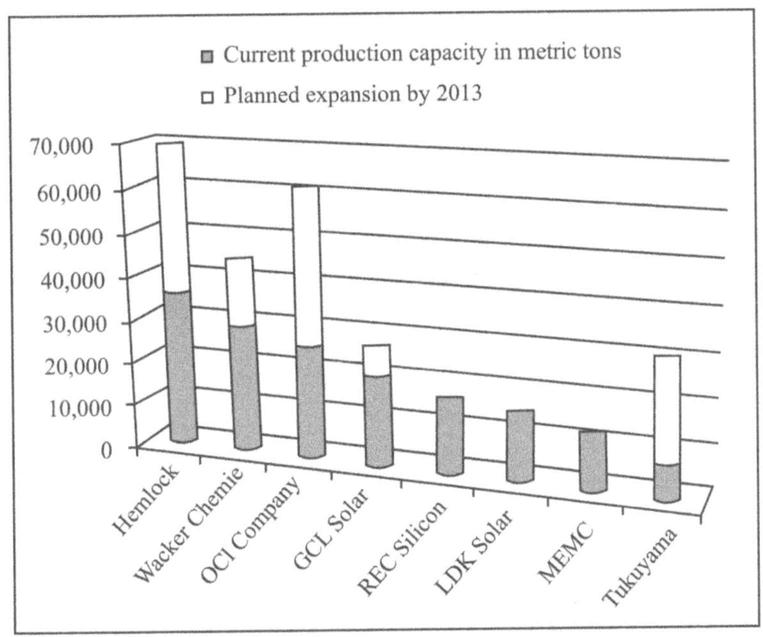

Figure 9.11 Poly silicon production

9.4 Nano Size (Quantum) Effects on Semiconductors

9.4.1 Difference in Energy Band Gap in Bulk Semiconductors and Nano Sized Semiconductors

According to the laws of **quantum mechanics**, free carriers in a **metal** or semiconductor can only take on specific values of energy, as defined by the crystal structure; that is, the energy is quantized. For most practical purposes, there are so many closely spaced energy levels (Figure 9.12) in bulk semiconductor materials and it appears that the carriers have a continuum of possible energies, except for the well-defined gaps characteristic of semiconductors. When the carrier is confined to a region where one or more of the dimensions reach the range of less than 100 nm, the quantum energy levels begin to spread out and the quantum nature becomes detectable. This reduction in size can take place in one, two, or three dimensions, using the fabrication techniques discussed earlier, yielding structures known respectively as super lattices, quantum wires, and quantum dots (QDs).

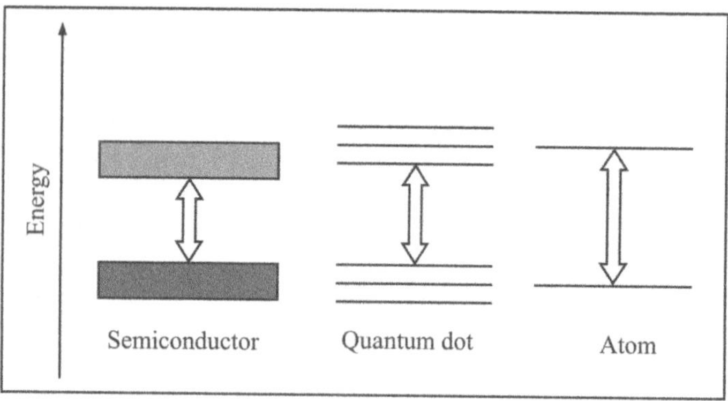

Figure 9.12 Energy band gap between bulk semiconductors, quantum dots and atoms

Quantisation arises because the size of a nanoparticle is comparable to the de Broglie wavelength of its charge carriers (i.e., electrons and holes) (discussed in detail in Chapter 1). Considering the thermal motion of a particle of mass m along a single direction in a crystal with lattice temperature T, then the de Broglie wavelength is given by (10).

$$\lambda_{deB} = \frac{h}{\sqrt{mk_B T}} \qquad .....(9.4)$$

where k_B is the Boltzmann constant. For an electron in GaAs at room temperature (T = 300 K), $\lambda_{deB} = 42$ nm

Due to the spatial confinement of the charge carriers, the edge of the valance and conduction bands split into discrete, quantized, electronic levels. These electronic levels are similar to those in atoms and molecules.

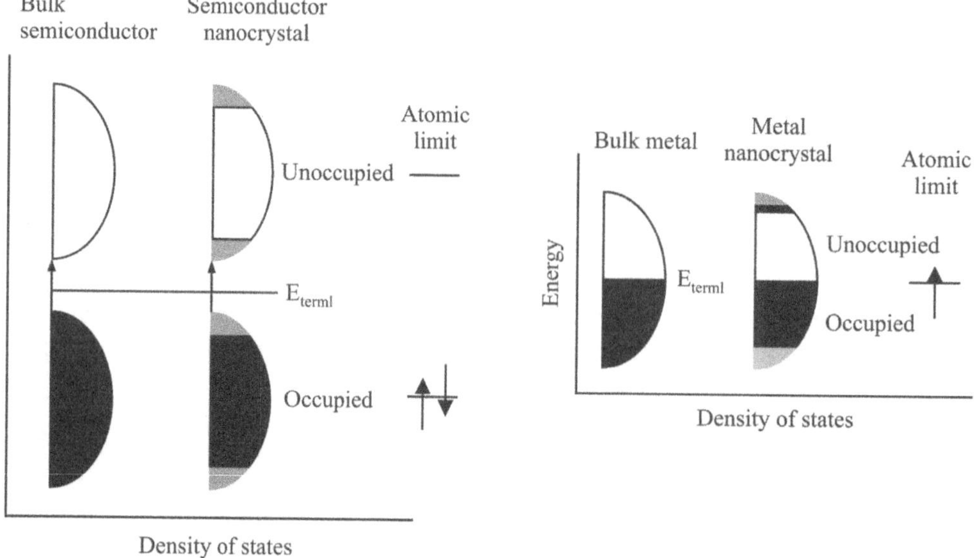

Figure 9.13 Conduction and valence bands in bulk semiconductor and nano particles

Figure 9.13 shows the conduction and valence bands in the solid, nanoparticle and atom state of a semiconductor and a metal are compared. It can be seen here again that nanoparticles represent a state of matter in the transition state between bulk solids and individual atoms.

The spacing of the electronic levels and the bandgap increases with decreasing particle size. This is because the electron hole pairs are now much closer together and the Coulombic interaction between them can no longer be neglected giving an overall higher kinetic energy.

This means that for nanostructures with spatial dimensions on the order of tens of nm, which are routinely grown using epitaxial techniques, quantum size effects will be important from room temperature down to liquid helium temperatures. In semiconductors, quantum confinement is provided by the discontinuity of the bandgap at interfaces. For structures that satisfy the confinement conditions, the motion of electrons and holes is restricted and quantized along the confinement direction, giving rise to a series of discrete energy levels. This behavior is depicted in Figure 9.14., which compares the density of states between the bulk, QWs, quantum wires, and QDs.

In general Quantization of the electron and hole spatial motion has three characteristics effects:

1. The effective band edge is blue shifted,
2. Confinement forces the electron and hole to overlap
3. The density of state becomes discrete

Confinement of electron-hole pairs (excitons) in semiconductor nano crystals (quantum dots (QDs) leads to novel quantum phenomena, tunable optical properties and enhanced Coulomb interactions, all of which are sensitive to the size, shape and material composition of the QDs. The Quantum confinement effects on the electronic and optical properties become important when the physical size of the confining potential becomes comparable to the de Broglie wavelength of the electrons or holes.

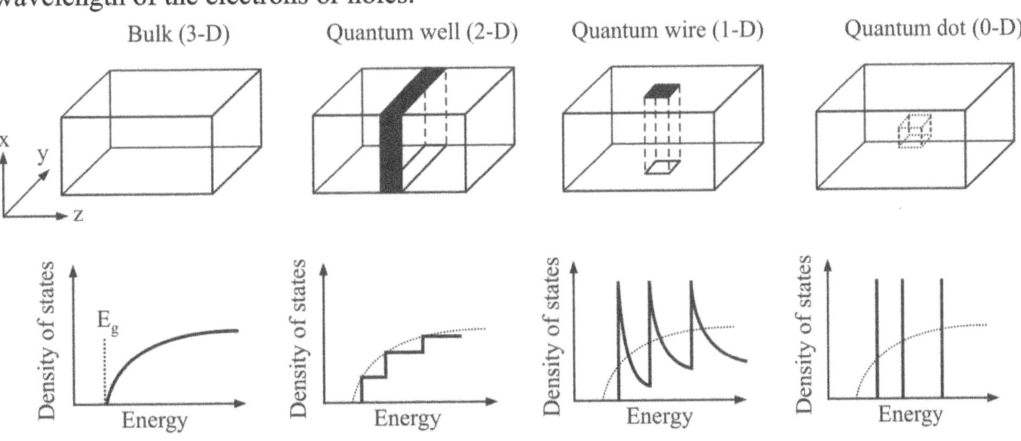

Figure 9.14 Density of states for particles with three (bulk), two (quantum well), one (quantum wire) and zero (quantum Dots) spatial degree of freedom

This increase in bandgap can be observed experimentally by the blue-shift in the absorption spectrum or sometimes even visually by the colour of the samples. A larger bandgap means that more energy is required to excite an electron from the valance band to the conduction band and hence light of a higher frequency and lower wavelength would be absorbed.

The increase in molar absorption coefficient shows that the oscillator strength also increases as particle size decreases. This is due to the strong overlapping of the wave functions of confined charge carriers. Experiments carried out in the past have found the absorption spectrum to be potential dependent, with decreasing wavelengths as applied potential decreases. This change is fully reversible before the threshold potential. More details on the quantum effects of decrease in their size are presented in Chapter 11.

By changing the particle size of the same material, and hence the energy bandgap, different coloured samples can be obtained.

Control over dimensions as well as composition of structures thus makes it possible to tailor material properties to specific applications. Both semiconductor and metal nanostructures have been investigated over the years. The ability to engineer artificial atoms and molecules in semiconductors using nanofabrication techniques has resulted in a powerful new tool in creating novel semiconductor devices, such as quantum dots where the number of carriers trapped by the dot can be controlled by an external voltage. While the applications of metal nanostructures are still very limited semiconductor electronic and optoelectronic components based on structures with quantum size effects have been on the market for several years.

It appears possible that nanoscale quantum-effect devices may become widely used in complex electronic systems, such as a neural array of quantum dots spaced only a few 100 nm apart, but this will only take place after significant progress has been made in fabrication and tolerance

9.5 Moore's Law

Moore's Law is the observation that, over the history of computing hardware, the number of transistors in a dense integrated circuit doubles approximately every two years. The law is named after Gordon E. Moore, co-founder of the Intel Corporation, who described the trend in his 1965 paper (11-14). His prediction has proven to be accurate, in part because the law is now used in the semiconductor industry to guide long-term planning and to set targets for research and development (15). The capabilities of many digital electronic devices are strongly linked to Moore's law: quality-adjusted microprocessor prices, (16) memory capacity, sensors and even the number and size of pixels in digital cameras (17). All of these are improving at roughly exponential rates as well. This exponential improvement has dramatically enhanced the impact of digital electronics in nearly every segment of the world economy (18).

To break down the law even further, it specifically stated that the number of transistors on an affordable CPU would double every two years (which is essentially the same thing that was stated before) but 'more transistors' is more accurate. If you were to look at processor speeds from the 1970's to 2009 and then again in 2010, one may think that the law has reached its limit or is nearing the limit. In the 1970's processor speeds ranged from 740 KHz to 8 MHz; notice that the 740 is KHz, which is Kilo Hertz – while the 8 is MHz, which is Mega Hertz. From 2000 – 2009 there has not really been much of a speed difference as the speeds range from 1.3 GHz to 2.8 GHz, which suggests that the speeds have barely doubled within a 10 year span. This is because we are looking at the speeds and not the number of transistors; in 2000 the number of transistors in the CPU numbered 37.5 million, while in 2009 the number went up to an outstanding 904 million; (Fig 9.15) this is why it is more accurate to apply the law to transistors than to speed. Moore's law describes a driving force of technological and social change, productivity and economic growth in the late 20th and early 21st centuries.

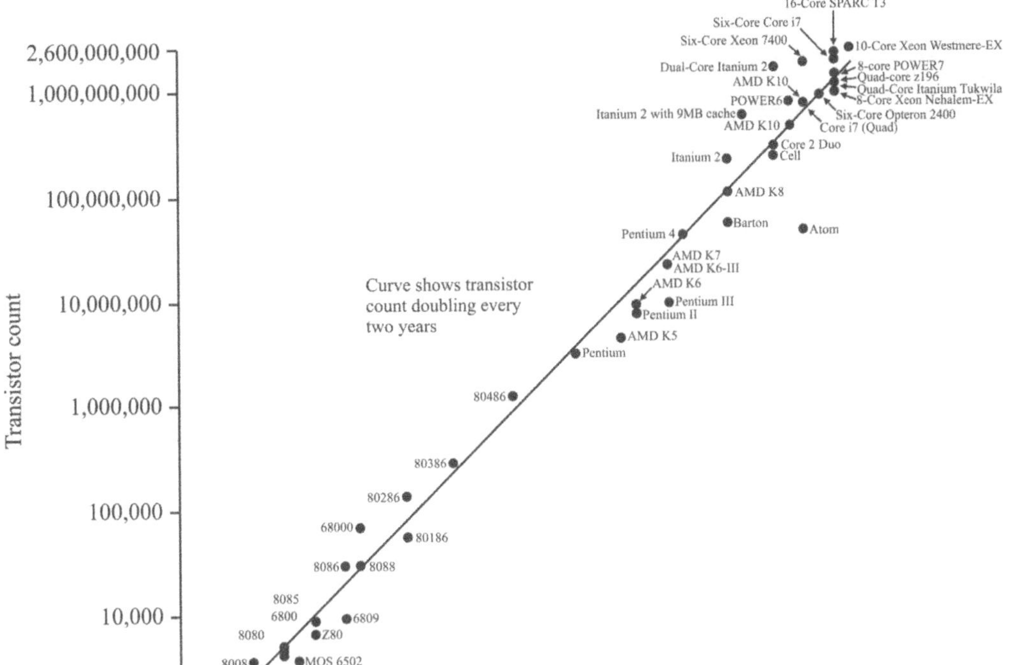

Figure 9.15 Microprocessor count from 1971-2011

In 1975 Moore reduced his forecast regarding the rate of density-doubling, stating circuit density-doubling would occur every 24 months. During the 1975 IEEE International Electron Devices Meeting he outlined his analysis of the contributing factors to this exponential behavior as:

- Die sizes were increasing at an exponential rate and as defective densities decreased, chip manufacturers could work with larger areas without losing reduction yields.

- Simultaneous evolution to finer minimum dimensions.
- and what Moore called "circuit and device cleverness".

Shortly after the 1975 IEEE Meeting, Caltech professor Carver Mead popularized the term "Moore's law".

The period is often quoted as 18 months because of Intel executive David House, who predicted that chip performance would double every 18 months (being a combination of the effect of more transistors and their being faster) (19-22).

Although this trend has continued for more than half a century, Moore's law should be considered an observation or conjecture and not a physical or natural law. Sources in 2005 expected it to continue until at least 2015 or 2020(22). However, the 2010 update to the International Technology Roadmap for Semiconductors predicted that growth will slow at the end of 2013, when transistor counts and densities are to double only every three years (23, 24).

9.5.1 Nano Fabrication Techniques and Moore's Law

In a competitive market scaling solid-state devices has the peculiar property of improving cost, performance, and power, for any electronic Industry. As a result, the microelectronics industry has driven transistor feature size scaling from 10 μm to ~30 nm during the past 40 years. During most of this time, scaling simply consisted of reducing the feature size. However, during certain periods, there were major changes as with the industry move from Si bipolar to p-channel metal-oxide-semiconductor (MOS), then to n-channel MOS, and finally to complementary MOS (CMOS) planar transistors in the 1980s, which has remained the dominate technology for the past two decades. The big challenge going forward is that the end of planar CMOS transistor scaling is near as the transistor size approaches tens of nanometers. How the industry evolves after this limit is reached is unclear.

To address these challenges, present day research is focused on identifying new materials and devices that can augment and/or potentially replace the aging ~50-year-old Si transistor. Two approaches under investigation are:

1. nonclassical CMOS, which consists of new channel materials and/or multigate fully depleted device structures; and
2. alternatives to CMOS, such as spintronics, single electron devices, and molecular computing.

While some of these non-Si research areas are important and will be successful in new applications and markets, it seems unlikely any of the non-Si options can replace the Si transistor for the $300 billion microelectronics industry in the foreseeable future (perhaps for coming as 30 years).

9.5.2 Status of Moore's Law

Moore's law has continued unabated for 40 years. Driven by tremendous advances in lithography, the 65 nm logic technology node featuring ~30 nm transistors is currently in high volume production and furthermore, 45 nm and 32 nm technologies with process targets defined to maintain Moore's law are currently under development at several companies. With such small feature sizes in high volume production and under development, Si CMOS technologies are now leading the field of nanotechnology and will continue to do so (25-27).

Si MOSFETs entered the nanometer era around 2000, as seen in Fig. 9.16 that shows technology node and transistor feature size versus year for the semiconductor industry. For the 0.13 μm technology node, the industry incorporated ~70 nm gate length transistors on average. What is interesting to note is that traditional top-down microelectronics have not only become nanoelectronics but the device dimensions are now comparable to those being explored in the new field of bottom-up nanotechnology and molecular electronics.

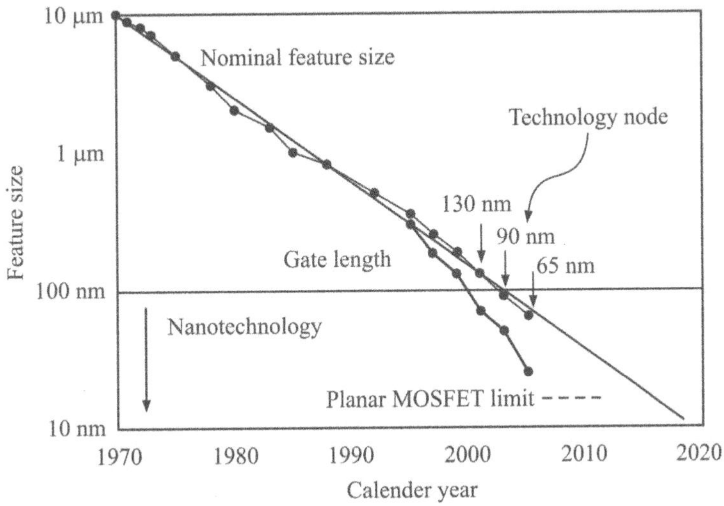

Figure 9.16 Logic technology node and transistor gate length versus calendar year. Note mainstream Si technology is nanotechnology (Published at the International Electron Devices Meeting)

The key driver behind these trends is economics, as pointed out by Moore in 1965. According to Moore, integrated circuits and scaling are "the cheap way to do electronics". Even with large increases in lithography tool cost to fabricate nanoscale CMOS transistors the cost per transistor has decreased by seven orders of magnitude during the last 40 years and is likely to continue to decrease for another decade. However, CMOS transistor scaling must inevitably slow down and finally halt, at least in the traditional sense, as the lithography scale approaches atomic dimensions.

9.5.3 Prospects to Replace Si Electronics

With Si CMOS scaling limits in sight, the obvious questions are How nanotechnology is on the horizon to replace planar Si CMOS transistors and in what time frame could this happen?"

Though these are difficult questions to try to answer, it is important to attempt to do so since this affects a $300 billion worldwide industry and the careers of most engineers.

Addressing the time frame question first, how quickly the industry can adopt a radically different device type depends on how many research and development levels it impacts.

Furthermore, radically new device types will require changes along many research and development levels from materials to software. The time frame to implement a radically new device is estimated to be ~30 years. Thus Si CMOS will be the dominant form of nanotechnology for the foreseeable future.

Finally, we must remember Moore's law is not a physical law but a law about economics. Consumer products and emerging markets have become the dominate end markets for semiconductors and will continue to be so for the next decade (versus the military in the 1960s-70s). A key attribute in these markets is price. Si technology is on course in the next decade to offer a billion transistor chips for ~$1, which will be a very difficult technology to displace.

References

1. Ball and Philip (1997). Made to Measure: new materials for the 21st century, Princeton: Princeton University Press.
2. Braun, Ernest and Macdonald (1982). Stuart, Revolution in Miniature, 2nd ed. Cambridge: Cambridge University Press.
3. Cahn, R.W. (2001). The Coming of Materials Science, Amsterdam: Pergamon.
4. Dummer, G.W.A. (1983). Electronic Inventions and Discoveries: Electronics from its earliest beginnings to the present day, 3rd ed. Oxford: Pergamon.
5. Seitz, Frederick and Enspruch, Norman G. (1998). Electronic Genie: The Tangled History of Silicon. Urbana: University of Illinois Press.
6. Gualtieri J.G., Kosinski J.A. and Ballato A. (1994). Piezoelectric materials for acoustic wave applications. IEEE Transactions on Ultrasonics Ferroelectrics & Frequency Control, 41, 53-9.
7. Eber, K., Schmidt O.G. and Duschl R. (2000). Structural Properties of SiC and SiGeC Alloy Klara Lyutovich, eds., London, INSPEC.
8. Neuberger, M. (1971). Handbook of Electronic Materials, III-V Semiconducting Compounds New York, Plenum.
9. Advanced Semiconductor Materials. New York, John Wiley & Sons, 2001.
10. Bube and Richard H (1992). Electrons in Solids: an Introductory Survey. Boston: Academic Press.
11. Moore and Gordon E. (1965). Cramming more components onto integrated circuits (PDF). Electronics Magazine. 4.
12. Excerpts from A Conversation with Gordon Moore: Moore's Law (PDF). Intel Corporation. 2005. 1.
13. 1965 – Moore's Law Predicts the Future of Integrated Circuits. Computer History Museum. 2007.
14. Disco, Cornelius, van der Meulen and Barend (1998). Getting new technologies together. New York: Walter de Gruyter. 206–207. ISBN 3-11-015630-X. OCLC 39391108.
15. Byrne David M., Oliner Stephen D., Sichel Daniel E. (2013-03). Is the Information Technology Revolution Over? (PDF). Finance and Economics Discussion Series Divisions of Research & Statistics and Monetary Affairs Federal Reserve Board. Washington, D.C.
16. Nathan Myhrvold (2006). Moore's Law Corollary: Pixel Power. New York Times.

17. Rauch and Jonathan (2001). The New Old Economy: Oil, Computers, and the Reinvention of the Earth. The Atlantic Monthly.
18. Keyes and Robert W. (2006). The Impact of Moore's Law. Solid State Circuits Newsletter.
19. Liddle and David E. (2006). The Wider Impact of Moore's Law. Solid State Circuits Newsletter.
20. Kendrick and John W. (1961). Productivity Trends in the United States. Princeton University Press for NBER. 3.
21. Dale W. Jorgenson, Mun S. Ho and Jon D. Samuels (2014). Long-term Estimates of U.S. Productivity and Growth. World KLEMS Conference.
22. Moore's Law to roll on for another decade. Retrieved 2011, 11, 27.
23. Kanellos and Michael (2005). New Life for Moore's Law. cnet.
24. Overall Technology Roadmap Characteristics. International Technology Roadmap for Semiconductors. 2010.
25. Evans, Dean. Moore's Law: how long will it last?. http://www.techradar.com
26. Schaller and Bob (1996). The Origin, Nature, and Implications of Moore's Law. Microsoft. Retrieved 10 September 2014.
27. Tuomi, I. (2002). The Lives and Death of Moore's Law. First Monday 7 (11). doi: 10. 5210/fm. v7i11. 1000.

10

Quantum Computing

10.1 Quantum Theory

Modern understanding of quantum theory is based on Max Plancks theory of energy existence in individual quantas similar to matter which he presented to German Physical Society in 1900.

10.1.1 The Essential Elements of Quantum Theory are

- Energy, like matter, consists of discrete units, rather than solely as a continuous wave.
- Elementary particles of both energy and matter, depending on the conditions, may behave like either particles or waves.
- The movement of elementary particles is inherently random, and thus, unpredictable.
- The simultaneous measurement of two complementary values, such as the position and momentum of an elementary particle, is inescapably flawed; the more precisely one value is measured, the more flawed will be the measurement of the other value.

10.2 Quantum Devices

These function to create, control, measure, and protect quantum coherence and entanglement in solid-state electronic device. For example Electrical circuits can be made to behave quantum mechanically, when fabricated from superconductors and operated at microwave frequencies and very low temperatures. This allows them to be used to both study quantum mechanical effects in a uniquely controllable manner, and also to be explored for use in quantum technology.

10.3 Quantum Computing

Quantum computing brings together ideas from classical information theory, computer science, and quantum physics. First proposed in the 1970s, quantum computing relies on quantum physics by taking advantage of certain quantum physics properties of atoms or nuclei that allow them to work together as quantum bits, or qubits, to be the computer's processor and memory. By interacting with each other while being isolated from the external environment, qubits can perform certain calculations exponentially faster than conventional computers (1-9).

Thus Quantum computing is essentially harnessing and exploiting the amazing laws of quantum mechanics to process information. A traditional computer uses long strings of "bits," which encode either a zero or a one. A quantum computer, on the other hand, uses quantum bits, or qubits. A qubit is a quantum system that encodes the zero and the one into two distinguishable quantum states. But, because qubits behave quantumly, we can exploit the phenomena of "superposition" and "entanglement

Though Paul Benioff, working at Argonne National Labs, in 1981 has originated the essential elements of quantum computing, it is generally accepted that David Deutsch of Oxford University is the man who provided the critical impetus for quantum computing research. David Deutsch in 1984 at a computation theory conference introduced about the possibility of designing a computer that was based exclusively on quantum rules, then published his breakthrough paper a few months later. From then a number of scientists began to exploit his ideas.

10.3.1 Quantum Computer

Quantum computer is any device essentially harnessing and exploiting the amazing laws of quantum mechanics to process information. A traditional computer uses long strings of "bits," which encode either a zero or a one. A quantum computer, on the other hand, uses quantum bits, or qubits. A qubit is a quantum system that encodes the zero and the one into two distinguishable quantum states. But, because qubits behave quantumly, we can capitalize on the phenomena of "superposition" and "entanglement."

In Digital computers Data will be encoded into binary digits (bits), each of which is always in one of two definite states (0 or 1), while in quantum computation uses qubits (quantum bits), which can be in superpositions of states.

Superposition is essentially the ability of a quantum system to be in multiple states at the same time that is, something can be "here" and "there," or "up" and "down" at the same time.

Entanglement is an extremely strong correlation that exists between quantum particles — so strong, in fact, that two or more quantum particles can be inextricably linked in perfect unison, even if separated by great distances. The particles remain perfectly correlated even if separated by great distances. The particles are so intrinsically connected, they can be said to "dance" in instantaneous, perfect unison, even when placed at opposite ends of the universe. This seemingly impossible connection inspired Einstein to describe entanglement as "spooky action at a distance."

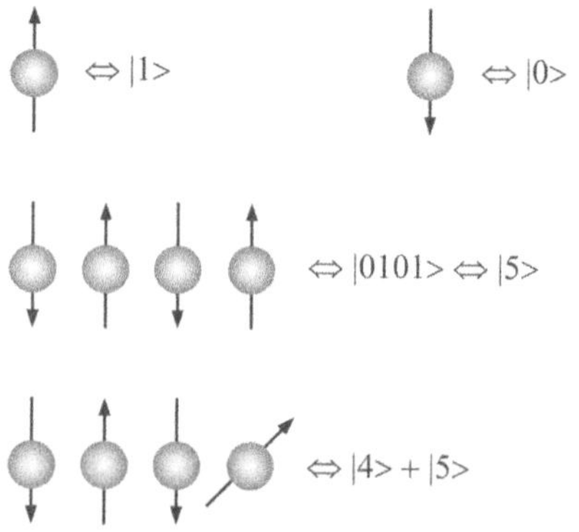

Figure 10.1 Qubits-representation of super position of all the clasically allowed states

Digital computers based on transistors can hold not only the state '0' and '1' but a linear super position of both states $\alpha|0> + \beta|1>$. In physics terms we denote the states |0> and |1> respectively.

A quantum computer maintains a sequence of qubits. A single qubit can represent a one, a zero, or any quantum superposition of these two qubit states; moreover, a pair of qubits can be in any quantum superposition of 4 states, and three qubits in any superposition of 8 (Figure 10.1). Quantum computers share theoretical similarities with non deterministic and probabilistic computers; one example is the ability to be in more than one state simultaneously.

In general, a quantum computer with n qubits can be in an arbitrary superposition of up to 2^n different states simultaneously (this compares to a normal computer that can only be in one of these 2^n states at any one time). A quantum computer operates by setting the qubits in a controlled initial state that represents the problem at hand and by manipulating those qubits with a fixed sequence of quantum logic gates. The sequence of gates to be applied is called a quantum algorithm. The calculation ends with a measurement, collapsing the system of qubits into one of the 2^n pure states, where each qubit is purely zero or one. The outcome can therefore be at most n classical bits of information. Quantum algorithms are often non-deterministic, in that they provide the correct solution only with a certain known probability.

Implementation of qubits for a quantum computer could start with the use of particles with two spin states: "down" and "up" (typically written |↓> and |↑>, or |0> and |0>). But in fact any system possessing an observable quantity *A*, which is *conserved* under time evolution such that *A* has at least two discrete and sufficiently spaced consecutive eigen values, is a suitable candidate for implementing a qubit. This is true because any such system can be mapped onto an effective spin-1/2 system.

A quantum computer with a given number of qubits is fundamentally different from a classical computer composed of the same number of classical bits. For example, to represent the state of an n-qubit system on a classical computer would require the storage of 2^n complex coefficients. Although this fact may seem to indicate that qubits can hold exponentially more information than their classical counter parts, care must be taken not to overlook the fact that the qubits are only in a probabilistic superposition of all of their states. This means that when the final state of the qubits is measured, they will be found in one of the possible configurations they were in before measurement. Moreover, it is incorrect to think of the qubits as only being in one particular state before measurement since the fact that they were in a superposition of states before the measurement was made directly affects the possible outcomes of the computation.

Qubits are made up of controlled particles and the means of control (e.g., devices that trap particles and switch them from one state to another). For example (10), consider first a classical computer that operates on a three-bit register. The state of the computer at any time is a probability distribution over the $2^3 = 8$ different three-bit strings 000, 001, 010, 011, 100, 101, 110, 111. If it is a deterministic computer, then it is in exactly one of these states with probability 1. However, if it is a probabilistic computer, then there is a possibility of it being in any one of a number of different states. We can describe this probabilistic state by eight nonnegative numbers A, B, C, D, E, F, G, H (where A = probability computer is in state 000, B = probability computer is in state 001, etc.). There is a restriction that these probabilities sum to 1.

The state of a three-qubit quantum computer is similarly described by an eight-dimensional vector (*a*, *b*, *c*, *d*, *e*, *f*, *g*, *h*), called a ket. Here, however, the coefficients can have complex values, and it is the sum of the *squares* of the coefficients' magnitudes, $|a|^2+|b|^2+...+|h|^2$ that must equal 1. These square magnitudes represent the probability amplitudes of given states. However, because a complex number encodes not just a magnitude but also a direction in the complex plane, the phase difference between any two coefficients (states) represents a meaningful parameter. This is a fundamental difference between quantum computing and probabilistic classical computing. (11) (Figure 10.2).

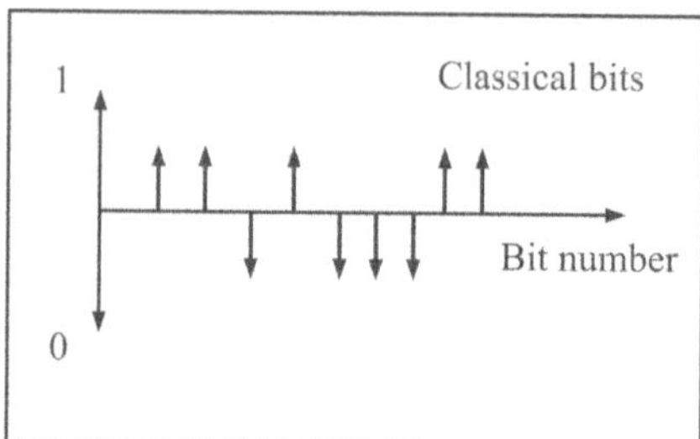

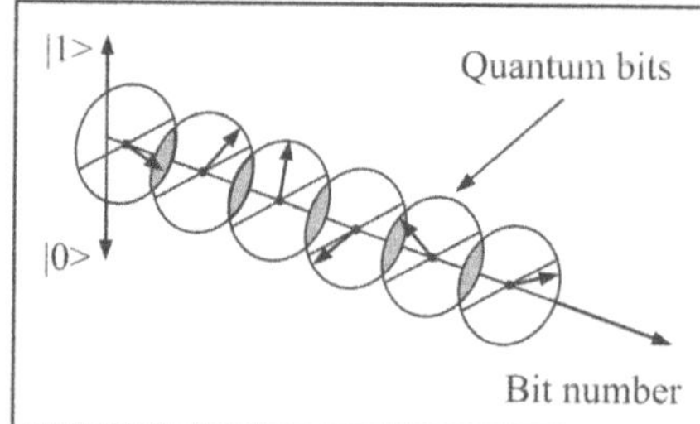

Figure 10.2 Qubit vs classical bit

If you measure the three qubits, you will observe a three-bit string. The probability of measuring a given string is the squared magnitude of that string's coefficient (i.e., the probability of measuring 000 = $|a|^2$, the probability of measuring 001 = $|b|^2$ etc). Thus, measuring a quantum state described by complex coefficients (*a*, *b*, ..., *h*) gives the classical probability distribution $\left(|a|^2,|b|^2,...,|h|^2\right)$ and we say that the quantum state "collapses" to a classical state as a result of making the measurement.

10.3.2 Quantum Decoherence

One of the greatest challenges is controlling or removing quantum decoherence. This usually means isolating the system from its environment as interactions with the external world cause the system to decohere. However, other sources of decoherence also exist. Examples include the quantum gates, and the lattice vibrations and background nuclear spin of the physical system used to implement the qubits. Decoherence is irreversible, as it is non-unitary, and is usually something that should be highly controlled, if not avoided. Decoherence times for candidate systems, in particular the transverse relaxation time T_2 (for NMR and MRI technology, also called the *dephasing time*), typically range between nanoseconds and seconds at low temperature.[11]

These issues are more difficult for optical approaches as the time scales are orders of magnitude shorter and an often-cited approach to overcoming them is optical pulse shaping. Error rates are typically proportional to the ratio of operating time to decoherence time, hence any operation must be completed much more quickly than the decoherence time.

If the error rate is small enough, it is thought to be possible to use quantum error correction, which corrects errors due to decoherence, thereby allowing the total calculation time to be longer than the decoherence time. An often cited figure for required error rate in each gate is 10^{-4}. This implies that each gate must be able to perform its task in one 10,000th of the decoherence time of the system.

Meeting this scalability condition is possible for a wide range of systems. However, the use of error correction brings with it the cost of a greatly increased number of required qubits. The number required to factor integers using Shor's algorithm is still polynomial, and thought to be between L and L^2, where L is the number of bits in the number to be factored; error correction algorithms would inflate this figure by an additional factor of L. For a 1000-bit number, this implies a need for about 10^4 qubits without error correction (12). With error correction, the figure would rise to about 10^7 qubits. Note that computation time is about L^2 or about 10^7 steps and on 1 MHz, about 10 seconds.

A very different approach to the stability-decoherence problem is to create a topological quantum computer with anyone, quasi-particles used as threads and relying on braid theory to form stable logic gates (13, 14).

10.3.3 A Comparison of Classical and Quantum Computing

Classical computing relies, at its ultimate level, on principles expressed by Boolean algebra, operating with a (usually) 7-mode logic gate principle, though it is possible to exist with only three modes (which are AND, NOT, and COPY). Data must be processed in an exclusive binary state at any point in time - that is, either 0 (off/false) or 1 (on/true). These values are binary digits, or bits. The millions of transistors and capacitors at the heart of computers can only be in one state at any point. While the time that the each transistor or capacitor need be either in 0 or 1 before switching states is now measurable in billionths of a second, there is still a limit as to how quickly these devices can be made to switch state. As we progress to smaller and faster circuits, we begin to reach the physical limits of materials and the threshold for classical laws of physics to apply. Beyond this, the quantum world takes over, which opens a potential as great as the challenges that are presented?

The Quantum computer, by contrast, can work with a two-mode logic gate: XOR and a mode we'll call QO1 (the ability to change 0 into a superposition of 0 and 1, a logic gate which cannot exist in classical computing). In a quantum computer, a number of elemental particles such as electrons or photons can be used (in practice, success has also been achieved with ions), with either their charge or polarization acting as a representation of 0 and/or 1. Each of these particles is known as a quantum bit, or qubit, the nature and behavior of these particles form the basis of quantum computing. The two most relevant aspects of quantum physics are the principles of superposition and *entanglement*.

10.3.4 Superposition

Think of a qubit as an electron in a magnetic field. The electron's spin may be either in alignment with the field, which is known as a *spin-up* state, or opposite to the field, which is known as a *spin-down* state. Changing the electron's spin from one state to another is achieved by using a pulse of energy, such as from a laser - let's say that we use 1 unit of laser energy. But what if we only use half a unit of laser energy and completely isolate the particle from all external influences? According to quantum law, the particle then enters a superposition of states, in which it behaves as if it were in both states simultaneously. Each qubit utilized could take a superposition of both 0 and 1. Thus, the number of computations that a quantum computer could undertake is 2^n, where n is the number of qubits used. Thus Qubits are useful for quantum computation. Useful when lots of permutations need to be tried. Suppose we want to calculate and store all possible outcomes of particular calculation we can achieve in single computation using quantum computation while we need 4 computation steps in classical computation (Figure10.3).

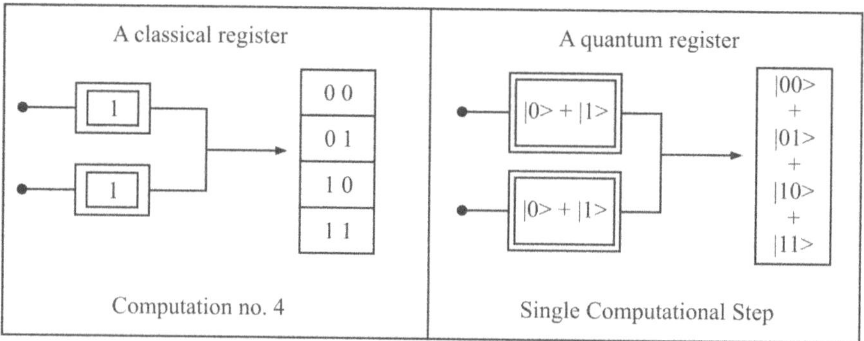

Figure 10.3 Computation steps classical vs quantum computing

A quantum computer comprised of 500 qubits would have a potential to do 2^{500} calculations in a single step. This is an awesome number - 2^{500} is infinitely more atoms than there are in the known universe (this is true parallel processing - classical computers today, even so called parallel processors, still only truly do one thing at a time: there are just two or more of them doing it). But how will these particles interact with each other? They would do so via quantum entanglement.

10.3.5 Entanglement

Particles (such as photons, electrons, or qubits) that have interacted at some point retain a type of connection and can be entangled with each other in pairs, in a process known as *correlation*. Knowing the spin state of one entangled particle - up or down - allows one to know that the spin of its mate is in the opposite direction. Even more amazing is the knowledge that, due to the phenomenon of superposition, the measured particle has no single spin direction before being measured, but is simultaneously in both a spin-up and spin-down state. The spin state of the particle being measured is decided at the time of measurement and communicated to the correlated particle, which simultaneously assumes the opposite spin direction to that of the measured particle. This is a real phenomenon (Einstein called it "spooky action at a distance"), the mechanism of which cannot, as yet, be explained by any theory - it simply must be taken as given. Quantum entanglement allows qubits that are separated by incredible distances to interact with each other instantaneously (not limited to the speed of light). No matter how great the distance between the correlated particles, they will remain entangled as long as they are isolated.

Taken together, quantum superposition and entanglement create an enormously enhanced computing power. Where a 2-bit register in an ordinary computer can store only one of four binary configurations (00, 01, 10, or 11) at any given time, a 2-qubit register in a quantum computer can store all four numbers simultaneously, because each qubit represents two values. If more qubits are added, the increased capacity is expanded exponentially.

10.3.6 Choosing your Qubit

Qubits can be realized using many different methods.

Photonic Systems / Laser Ion Traps / NMR systems, Quantum Dots.

At present researchers are working towards solid state Qubits using super coductor technology (Figure 10.4).

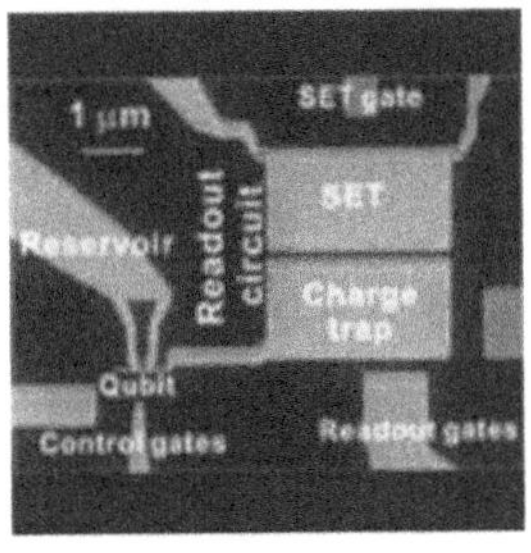

Charge qubit

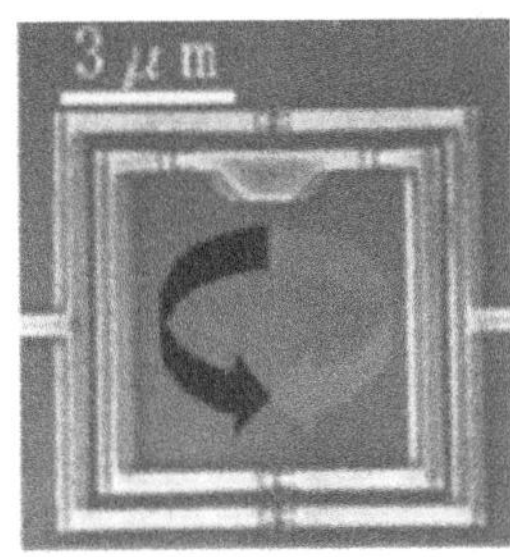

Flux qubit

Figure 10.4 Super conducting qubits

The basic Principles of super conducting Qubits

The Josephson Junction is the basic building block of a superconducting Qubit and thus quantum computer. This Junction consists of 2 super conducting regions separated by a weak link (usually an Insulator) Figure 10.5.

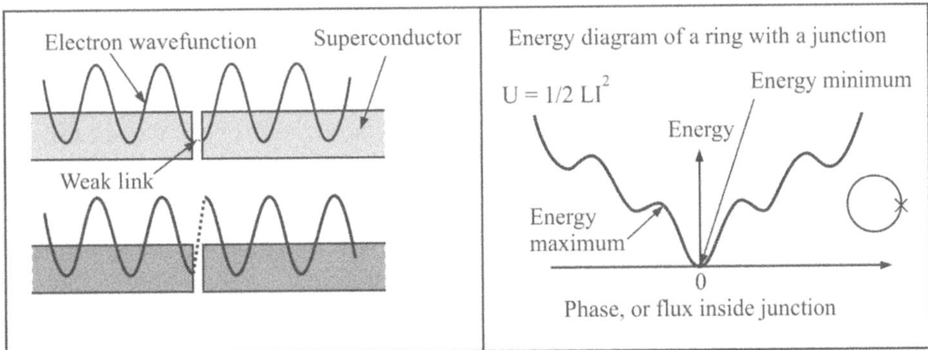

Figure 10.5 Mechanism of functioning of super conducting qubits

There are a number of technical challenges in building a large-scale quantum computer, and thus far quantum computers have yet to solve a problem faster than a classical computer. David DiVincenzo, of IBM (11), listed the following requirements for a practical quantum computer.

- Scalable physically to increase the number of qubits
- Qubits can be initialized to arbitrary values
- Quantum gates faster than decoherence time
- Universal gate set
- Qubits can be read easily

10.3.7 Quantum Programming

Perhaps even more intriguing than the sheer power of quantum computing is the ability that it offers to write programs in a completely new way. For example, a quantum computer could incorporate a programming sequence that would be along the lines of "take all the superpositions of all the prior computations" - something which is meaningless with a classical computer - which would permit extremely fast ways of solving certain mathematical problems, such as factorization of large numbers, one example of which we discuss below.

There have been two notable successes thus far with quantum programming.

Factoring large numbers: Multiplying two large numbers is easy for any computer. But calculating the factors of a very large (say, 500-digit) number, on the other hand, is considered impossible for any classical computer. In 1994, a mathematician from the Massachusetts Institute of Technology (MIT) Peter Shor, who was working at AT&T at the time, unveiled that if a fully working quantum computer was available, it could factor large numbers easily. Nobody wants to factor very large numbers! That's because it's so difficult – even for the best computers in the world today.

In fact, the difficulty of factoring big numbers is the basis for much of our present day cryptography. It's based on math problems that are too tough to solve. RSA encryption, the method used to encrypt your credit card number when you're shopping online, relies completely on the factoring problem. The website you want to purchase from gives you a large "public" key (which anyone can access) to encode your credit card information.

This key actually is the product of two very large prime numbers, known only to the seller. The only way anyone could intercept your information is to know those two prime numbers that multiply to create the key. Since factoring is very hard, no eavesdropper will be able to access your credit card number and your bank account is safe. Unless, that is, somebody has built a quantum computer and is running Peter Shor's algorithm!

Quantum key distribution relies on another interesting property of quantum mechanics: any attempt to observe or measure a quantum system will disturb it.

The Institute for Quantum Computing (IQC) is home of one of the few QKD prototypes in the world. "Alice," a device located at IQC headquarters, receives half of the entangled (highly correlated) one of the photons generated by a laser on the roof of a building at the University of Waterloo. "Bob" is housed at the nearby Perimeter Institute, and receives the other half of the entangled photons.

Photons have a unique measurable property called polarization (which should sound familiar to any connoisseur of sunglasses).

Since the polarization of each individual photon is random, there's no way of knowing the unique properties of each photon in advance. But here is where entanglement becomes interesting: if Alice and Bob measure the polarization of the entangled photons they receive, their results will be the same (remember, "entangled" means the particles are highly correlated with each other, even at great distances). Depending on the polarization of each photon, Alice and Bob ascribe either a "one" or a "zero" to each photon they receive. Therefore, if Alice gets a string like 010110, Bob also gets a 010110. Unless, that is, an eavesdropper has been attempting to spy on the signal. This will disturb the system, and Alice and Bob will instantly notice that their keys don't match.

Alice and Bob keep receiving photons until their identical keys are long and identical enough and, presto, they've got ultra-secure keys for encrypting communications

The other major breakthrough happened with Lov Grover of Bell Labs in 1996, with a very fast algorithm that is proven to be the fastest possible for searching through unstructured databases. The algorithm is so efficient that it requires only, on average, roughly N square root (where N is the total number of elements) searches to find the desired result, as opposed to a search in classical computing, which on average needs N/2 searches.

10.3.8 The Problems - and Some Solutions

The above sounds promising, but there are tremendous obstacles still to be overcome. Some of the problems with quantum computing are as follows:

- **Interference:** During the computation phase of a quantum calculation, the slightest disturbance in a quantum system (say a stray photon or wave of EM radiation) causes the quantum computation to collapse, a process known as de-coherence. A quantum computer must be totally isolated from all external interference during the computation phase. Some success has been achieved with the use of qubits in intense magnetic fields, with the use of ions.
- **Error correction:** Because truly isolating a quantum system has proven so difficult, error correction systems for quantum computations have been developed. Qubits are not digital bits of data, thus they cannot use conventional (and very effective) error correction, such as the triple redundant method. Given the nature of quantum computing, error correction is ultra critical - even a single error in a calculation can cause the validity of the entire computation to collapse. There has been considerable progress in this area, with an error correction algorithm developed that utilizes 9 qubits (1 computational and 8 correctional). More recently, there was a breakthrough by IBM that makes do with a total of 5 qubits (1 computational and 4 correctional).
- **Output observance:** Closely related to the above two, retrieving output data after a quantum calculation is complete risks corrupting the data. In an example of a quantum computer with 500 qubits, we have a 1 in 2^{500} chance of observing the right output if we quantify the output. Thus, what is needed is a method to ensure that, as soon as all calculations are made and the act of observation takes place, the observed value will correspond to the correct answer. How can this be done? It has been achieved by Grover with his database search algorithm, that relies on the special "wave" shape of the

probability curve inherent in quantum computers, that ensures, once all calculations are done, the act of measurement will see the quantum state decohere into the correct answer.

Even though there are many problems to overcome, the breakthroughs in the last 15 years, and especially in the last 3, have made some form of practical quantum computing not unfeasible, but there is much debate as to whether this is less than a decade away or a hundred years into the future. However, the potential that this technology offers is attracting tremendous interest from both the government and the private sector. Military applications include the ability to break encryptions keys via brute force searches, while civilian applications range from DNA modeling to complex material science analysis. It is this potential that is rapidly breaking down the barriers to this technology, but whether all barriers can be broken, and when, is very much an open question.

10.3.9 Quantum Dots as Qubits in Quantum Computing

Coupled Quantum Dots can be used as two-Qubit gates in Quantum computing. In this Qubit system is realized as a spin of excess electron on a single electron quantum dot. The qubit is realized as the spin of the excess electron on a single-electron quantum dot; (Fig. 10.6).

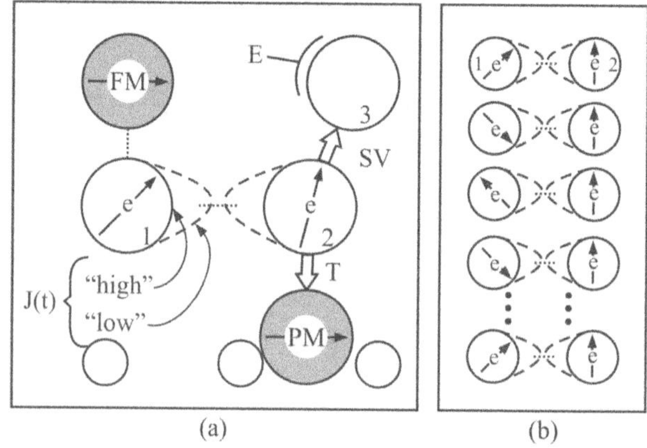

Figure 10.6 Schematic top view of two coupled quantum dots labeled 1 and 2, each containing one excess electron (e) with spin 1/2. The tunnel barrier between the dots can be raised or lowered by setting a gate voltage "high" (solid equipotential contour) or "low" (dashed equipotential) contour

The mechanism for two-qubit quantum-gate operation is that it operates by a purely electrical gating of the tunneling barrier between neighboring quantum dots rather than by spectroscopic manipulation as in other model. It is expected that quantum dots as qubits the spin-1/2 degrees of freedom will have longer decoherence times than charge degrees of freedom since they are insensitive to any environmental fluctuations of the electric potential which facilitates scalability.

Quantum computing is still in its infancy but experiments have been carried out in which quantum computational operations were executed on a very small number of qubits. Both practical and theoretical research continues, and many national governments and military funding agencies support quantum computing research to develop quantum computers for both civilian and national security purposes, such as crypt analysis (7).

Large-scale quantum computers will be able to solve certain problems much quicker than any classical computer using the best currently known algorithms, like integer factorization using Shor's algorithm or the simulation of quantum many-body systems. There exist quantum algorithms, such as Simon's algorithm, which run faster than any possible probabilistic classical algorithm (8). Given sufficient computational resources, however, a classical computer could be made to simulate any quantum algorithm.

Quantum computation helps greatly in calculating the positions of individual atoms in very large molecules like polymers and in viruses. The way that the particles interact with each other – there's so many different possibilities that normally they say that you can't calculate anything properly with more than about 10 atoms inside the molecule using classical computation So if you have a quantum computer you could use it to develop drugs and understand how molecules work a bit better.

References

1. Gershenfeld, N. and Chuang, I. L. (1998). Quantum Computing with Molecules. Scientific American.
2. Manin, Yu. I. (1980). Vychislimoei, Nevychislimoe [Computable and Noncomputable] (in Russian). Sov. Radio. 13–15. Retrieved 2013-03-04.
3. Feynman, R.P. (1982). Simulating physics with computers. International Journal of Theoretical Physics, 21 (6), 467–488. doi:10.1007/BF02650179.
4. Deutsch, D. (1992). Quantum computation. Physics World.
5. Finkelstein, D. (1968). Space-Time Structure in High Energy Interactions. In Gudehus, T., Kaiser, G. Fundamental Interactions at High Energy. New York: Gordon & Breach.
6. Gershon, E. (2013). New qubit control bodes well for future of quantum computing. Phys.org. Retrieved 2014-10-26.
7. Quantum Information Science and Technology Roadmap for a sense of where the research is heading. Report of the Quantum Information Science and Technology, Experts Panel, April, 2, 2004.
8. Simon, D.R. (1994). On the power of quantum computation. Foundations of Computer Science, 1994 Proceedings, 35th Annual Symposium on: 116–123. doi:10.1109/SFCS.1994.365701. ISBN 0-8186-6580-7.
9. Nielsen, M.A. and Chuang, I.L. (2000), Quantum Computation and Quantum Information. Cambridge University Press, UK, 202.
10. Waldner J. (2007), Nanocomputers and Swarm Intelligence. London: ISTE. 157. ISBN 2-7462-1516-0.
11. DiVincenzo D.P. (1995). Quantum Computation. Science, 270(5234), 255–261.
12. DiVincenzo D.P., (2000). The Physical Implementation of Quantum Computation. Ar Xiv: quant-ph/0002077 [quant-ph].

13. Dyakonov M. I. (2006). Is Fault-Tolerant Quantum Computation Really Possible?. In: Future Trends in Microelectronics. Up the Nano Creek, S. Luryi, J. Xu, and A. Zaslavsky (eds), Wiley, 4–18. arXiv:quant-ph/0610117.

14. Freedman, M.H., Kitaev, A., Larsen, M.J. and Wang, Z. (2003). Topological quantum computation. Bulletin of the American Mathematical Society 40(1): 31–38. arXiv: quant-ph/0101025. doi:10.1090/S0273-0979-02-00964-3. MR 1943131.

11

Quantum Dots, Nano-Photonics and their Applications

11.1 Quantum Dots

11.1.1 Introduction

Semiconductor nanocrystals in the 2-10 nm size range have aroused great interest due to their unique size-dependent optical and electrical properties. These are known as Quantum dots (QDs) and are often assembled from atoms from groups II-VI or III-V elements in the periodic Table (For example cadmium telluride (Cd from group II and Te from group VI) and Indium Phosphamide (In from group III and P from group V). These nano particles when compared to their bulk have smaller exciton Bohr radius (EBR) (1) which imposes quantum confinement (QC) imparting special optical and electronic properties for them which find wide defense applications. Further as they have similar dimensions to biological macromolecules (e.g., proteins and nucleic acids) (2) which allow integration of nanotechnology and biology, thus promoting great advances in cell biology, molecular biology, targeted therapeutics and medical diagnostics. These nano-sized bio-conjugates are used in the construction of hybrid materials for the development of multicolour fluorescent probes and bioassays (3) for ultrasensitive detection and imaging (4).

In nanotechnology, the main precursors used in the assembly of QDs is cadmium and tellurium forming the core which they are in range of sizes from 2-10 nm in diameter containing approximately 200-10000 atoms (5). Due to the effects of quantum confinement QDs posses unique optical properties such brighter, highly photo and chemical stability, with broad absorption, narrow and symmetric emission spectrum. An outstanding feature of QDs is that their emission wavelength can be tuned by adjusting their size and chemical composition.

11.1.2 Structure of Quantum Dots

Structurally, QD possess a metalloid crystalline core, which ultimately depending on its composition and size will confer the type of fluorescence it will emit. A bare nanocrystal core is highly reactive and toxic resulting in a very unstable structure that is prone to photochemical degradation. Also the core crystalline structure has surface irregularities that may lead to emission irregularities like blinking. The core is composed of materials such as cadmium–

selenium (CdSe), cadmium-tellurium (CdTe), indium-phosphate (InP) or indium-arsenate (InAs) (6) (Figure 11.1).

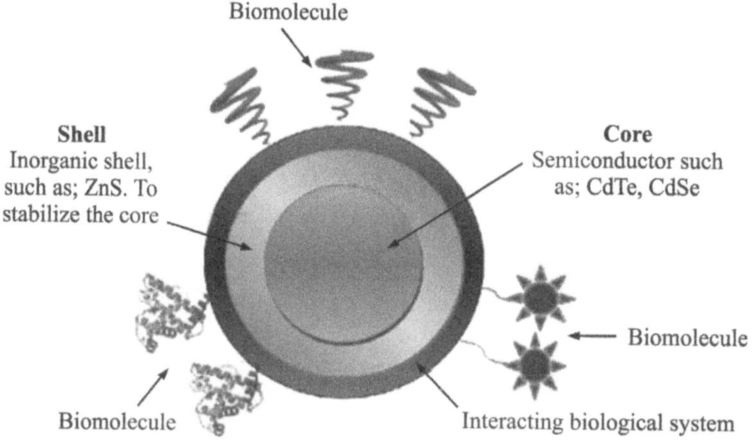

Figure 11.1 Structure of QDs modified with bioactive agents materials such as zinc sulphide (ZnS) used to coat QDs to reduce its toxicity and facilitate covalent conjugation to bioactive molecules such as antibodies and peptides

Semiconductor shells or coatings typically using zinc sulphide (ZnS) or mercapto-succinic acid (MSA) which not only stabilize the core, but further improves the optical, physical properties and bioavailability of the material. They also further reduce the toxicity of the core by shielding reactive Cd^{2+} and Te^{2-} ions from being exposed to photo-oxidative environments *e.g.*, exposure to UV and air (6). Additional capping or coatings with biocompatible material or polymer layers such as functionalized silica, phospholipid micelles, amphiphylic polymers (polyacrylic acid) have been introduced to render them soluble in aqueous media. Further coatings with POSS-PCU (Polyhedral Oligomeric Silsesquioxane-Polycarbonate urethane, (7)) or PEG (Polyethylene Glycol) to the QD core-shell promote water solubility,

11.1.3 Quantum Confinement in QDs

In natural semiconductor bulk material, the electrons (larger than 10 nm) exist in a range of energy levels. The energy levels are so compact that they can be described as continuous, meaning there is negligible energy difference between them. Also there is region where electrons do not reside or go and is known as "band gap". This band gap is different for each type of bulk semiconductor material.

Electrons in energy level below this band gap are known to be in the valence band while those occupying above are known to be in the conduction band. Also, the maximum amount of electrons resides in the valence band whilst a minimum amount of electrons in the conduction band. On acquiring a stimulus such as photon, voltage or heat can push the electrons from the valence band beyond the band gap to conduction band.

By leaving the valence band the space left vacant is known as a "hole" which happens to be positively charged. This elevated electron and the hole as a pair, is known as an "exciton".

Conduction electrons and valence holes have an average physical distance and this is referred to as exciton Bohr radius (EBR). These can vary with different materials. In bulk the distance is much larger than the EBR, which expands beyond its limit. This property also applies to QDs at the nanoscale level when it becomes equal to the EBR or less when the energy level is said to be discrete and not continuous anymore. This imposition of discrete energy is known as **quantum confinement (QC)** and at this stage the semiconductor material loses all resemblance to the bulk and adopts the structure of QD.

This alters the absorptive and emissive properties of the semiconductor material (8). Once excited to the conduction band, the electron falls back down across the band gap towards the valence band, in doing so electromagnetic radiation with a wavelength corresponding to the energy it loses in the transition is emitted (Figure 11. 2.).

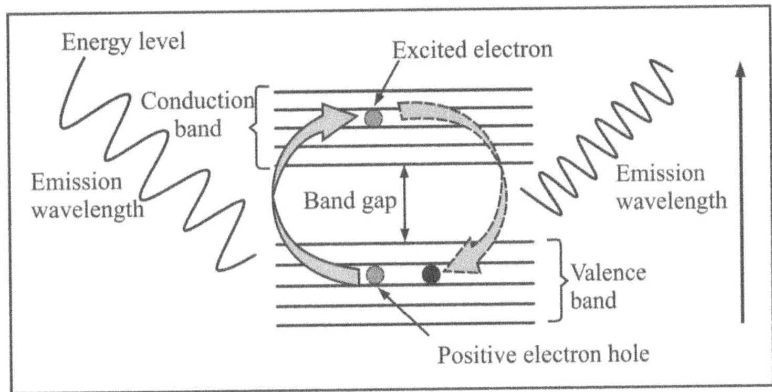

Figure 11.2 Schematic representation of the mechanism of emission by QDs

Excited electron in the valence band is elevated across the band gap onto the conduction band. As the raised electron in the conduction band falls back across the band gap to its valence energy level, a wavelength of light is emitted in this process.

11.1.4 Size dependent Fluorescence in QDs

The emission frequency due to the drop of electron from conduction band to valence band is perceived as fluorescence and depends on the size of the band gap, which can be altered by changing the size of the QD through changing the surface chemistry. It is important to note that the band gap in a QD will always be energetically higher and the radiation emitted is referred to being "blue shifted" thus producing a shorter wavelength. Hence, with QD, the size of the band gap is dependent on the size of the dot and wavelength of the emitted photons. Because the emission frequency of QD depends on the band gap, it is possible to control the output wavelength of the QD with extreme precision (9).

At this scale, quantum dots act similarly to large molecules; adding or subtracting single orbitals can shift the energy levels in the material, changing the bandgap and making their emission tunable. This occurs when all three dimensions of a particle are smaller than the exciton Bohr radius (Figure. 11.3).

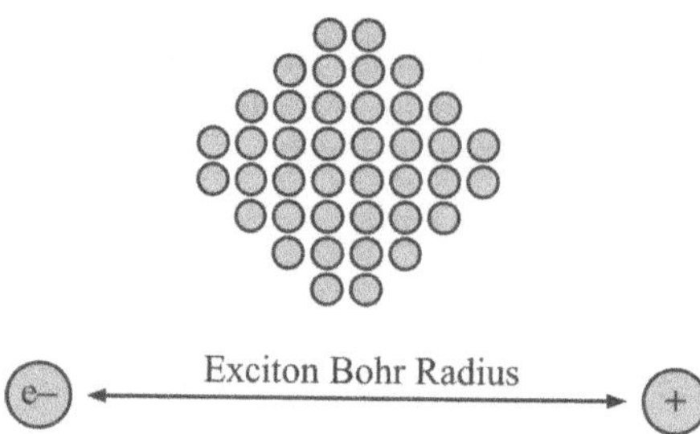

Figure 11.3 A quantum dot exhibits bandgap tunability because it is smaller than the spatial separation between the electron and its hole, known as the exciton Bohr radius

11.1.5 Modeling Quantum Confinement

We can model the confinement of the exciton to the edges of the quantum dot by viewing it as a particle-in-a-box. Brus developed an approximate relationship between the particle size and its resultant bandgap, based on the material being used and its bandgap in the bulk form (*Equation 11.1*) (10). In the equation, E_g^{QD} is the theoretical bandgap of the quantum dot, E_g^{bulk} is the bandgap of the bulk material, *h* is Planck's constant, *r* is the radius of the nanoparticle, m_0 is the electron mass, m_{e*} is the effective mass of the electron for the material, m_{h*} is the effective mass of the hole for the material, *e* is the charge of the electron, ε_0 is the permittivity of free space, and ε is the permittivity of the material.

$$E_g^{QD} = E_g^{bulk} + \frac{h^2}{8m_0r^2}\left(\frac{1}{m_e^*} + \frac{1}{m_h^*}\right) - \frac{1.8e^2}{4\pi\varepsilon_0\varepsilon_r} \quad .....(11.1)$$

The first term is based on the properties of the bulk material, the second term is based on the particle-in-a-box confinement of the exciton, and the third term is based on the Coulombic attraction between the electron and the hole. While it is not a perfect fit to experimental values, what we can see from this equation is that the bandgap, and therefore the wavelength of light emitted, changes significantly with small changes in particle size.

11.1.6 Optical and Electronic Properties of Quantum Dots

This size tunable absorption and emission property of QDs is an extremely valuable property for Electronic, Defense and Biological applications as they can be tuned all the way from the ultraviolet (UV) to the NIR of the spectrum such that smaller dots emit in the blue range and the larger dots in the red and NIR region (**Figure 11.4**). QDs have a broad excitation and narrow, discrete emission spectra. The peak emission wavelength of the QD is slightly longer than the first exciton peak or absorption onset, and this energy separation is known as the "Stoke's shift". Also their peak emission wavelength is independent of the wavelength of excitation light. This means that variable sized QDs can be excited by a single wavelength of light, as long as this wavelength is shorter than the absorption onset. This can be used in multiplexed imaging where a number of different sized QDs with discrete emission peaks and hence different colours can be excited by a single wavelength of light.

With such small sizes, QDs exhibit some unique and fascinating optical properties, such as sharp, symmetrical emission spectra, high quantum yield (QY), chemical and photo stability and size dependent emission wavelength tunability. Their intrinsic brightness is often of many orders that observed for traditional organic fluorophores. Their large surface to mass ratio confers the particles the ability to absorb and carry many compounds. QDs have relatively large (functional) surface, able to bind, adsorb and carry compounds such as drugs, probes and proteins. Based on these desirable properties, QDs have the potential to improve the sensitivity of biological imaging at the cellular level which includes cancer detection, progression and treatment, radio-and chemo- sensitizing agents, electron and x-ray contrast agent and targeted drug delivery. However, caution has to be exercised with these attractive applications as QDs contain substantial quantities of cadmium, which are reactive in an ionic state, and limited studies are only available concerning the health risk of exposure to nanoparticles (11), (12).

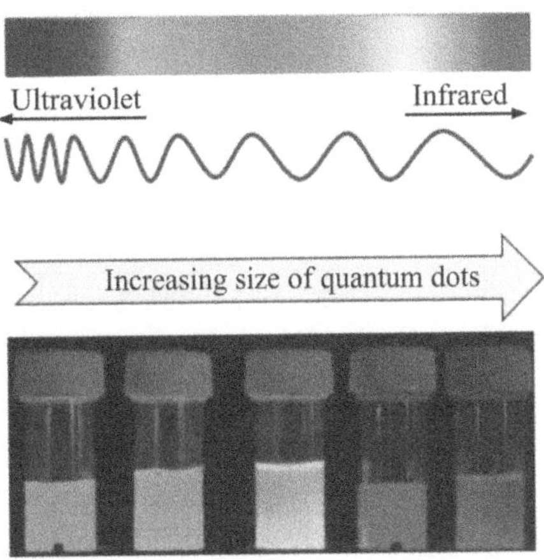

Figure 11.4 Size dependent multi colour fluorescence of QDs in aqueous solution

11.1.7 Quantum Dot Materials

Quantum dots are made from semiconducting materials typically made from III-V and II-VI semiconductors, such as CdSe, CdS, InP, and ZnS (Table 11.1). As the bandgap of the material from which a quantum dot is made is very important to its properties, different materials are used when different properties are needed for an application. The first quantum dots were made primarily from II-VI semiconductors, such as cadmium and zinc chalcogenides.

Table 11.1 Important parameters of bulk semiconductors commonly used for quantum dots. (13)

Material	Structure (300 K)	Type	Egap (ev)	Lattice Parameter [A]	Density (kg m–2)
ZnS	Zinc blende	II-VI	3.61	5.41	4090
ZnSe	Zinc blende	II-VI	2.69	5.668	5266
ZnTe	Zinc blende	II-VI	2.39	6.104	5636
CdS	Wurtzite	II-VI	2.49	4.136/6.714	4820
CdSe	Wurtzite	II-VI	1.74	4.3/7.01	5810
CdTe	Zinc blende	II-VI	1.43	6.482	5870
GaN	Wurtzite	III-V	3.44	3.188/5.185	6095
GaP	Zinc blende	III-V	2.27	5.45	4138
GaAs	Zinc blende	III-V	1.42	5.653	5318
GaSb	Zinc blende	III-V	0.75	6.096	5614
InN	Wurtzite	III-V	0.8	3.545/5.703	6810

Table 11.1 ***Contd...***

Material	Structure (300 K)	Type	Egap (ev)	Lattice Parameter [A]	Density (kg m^{-2})
InP	Zinc blende	III-V	1.35	5.869	4787
InAs	Zinc blende	III-V	0.35	6.058	5667
InSb	Zinc blende	III-V	0.23	6.479	5774
PbS	Rocksalt	IV-VI	0.41	5.936	7597
PbSe	Rocksalt	IV-VI	0.28	6.117	8260
PbTe	Rocksalt	IV-VI	0.31	6.462	8219

Most II-VI and III-V semiconductor materials crystallize in either the hexagonal wurtzite or cubic zincblende form (Figure 11.5). For some materials, such as ZnSe and CdTe, there is very little difference in energy between the zincblende and wurtzite structures, and so they can exhibit wurtzite-zincblende polytypism (14). Depending on the synthesis conditions, these nanocrystals may crystallize in either structure or both may coexist in the same nanoparticle. Lead chalcogenides crystallize in the rocksalt structure, although it has been shown that CdSe quantum dots can also crystallize in this structure if the diameter exceeds 11 nm.

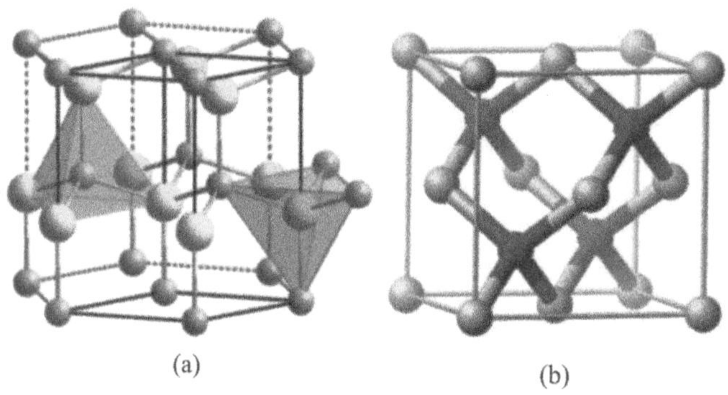

Figure 11.5 **(a)** Wurtzite and **(b)** Zincblende crystal structures

For the most part, the choice of material for quantum dots is primarily focused on the optical properties of the material, but consideration also should be made for the preferred structure for the application, toxicity (such as being free of heavy metals), and ability to coordinate ligands and functional groups to the surface.

11.1.8 Quantum Dot Synthesis Techniques

A lot of synthetic methods have been developed for QD preparation, which can be divided into two classifications: physical approach and chemical approach.

11.1.8.1 Physical Approach

This approach was mainly referred to epitaxial growth and/or Quantum Dots nanoscale patterning, has been widely used to provide QDs predominantly by the combination of high-resolution electron beam lithography and subsequent etching. However, the production of defection formation, size non uniformity, poor interface quality, and even damage to the bulk of the crystal itself became the main disadvantages of this method.

11.1.8.2 Chemical Approach

In contrast, the size and composition of QDs prepared by chemical approach are easily controlled and the average size distribution varies within 5-10%. The basic process of this method relies on the pyrolysis of organometallic and chalcogen precursors, where rapid nucleation followed by slower and steady growth is desired (15).

By depositing semiconductors on substrates with a large degree of lattice mismatch, it was found that the layer would bead up into droplets, forming quantum dots. However, this approach limited size dispersions to greater than 10%.

Another approach for production of the quantum dots is by controlling QDs size in colloids-micelles medium limiting their growth to the size of the micelle. While this method did not require high temperature, organic solvents, or complicated equipment, the size distribution of QDs is found to be poor and the concentration was limited, as well as the quantum dots exhibiting poor crystallinity and a large degree of defects (10).

11.1.8.3 Nucleation and Controlled Growth Techniques

The major breakthrough that made in QDs synthesis easier and more controllable is from nucleation and growth techniques to synthesize them in high temperature organic solvents. In nucleation and growth processes to make quantum dots, ionic sources of the constituent materials such as Cd^{2+} are used. These methods utilize the pyrolysis of organometallic precursors to produce mono disperse (less than 5% size dispersion) quantum dots made of cadmium chalcogenides. In this nucleation and growth process, an excess of organometallic precursors, such as dimethyl cadmium and selenium-trioctylphosphine (SeTOP) will be injected into a hot solution of coordinating solvent, such as a mixture of trioctylphosphine and trioctylphosphine oxide (TOP/TOPO) at over 280 °C, supersaturating the solution. During the first few seconds following the injection, particles nucleate homogeneously depleting the reactants, followed by particle growth, Ostwald ripening, and eventually saturation of the solution (Figure 11. 6).

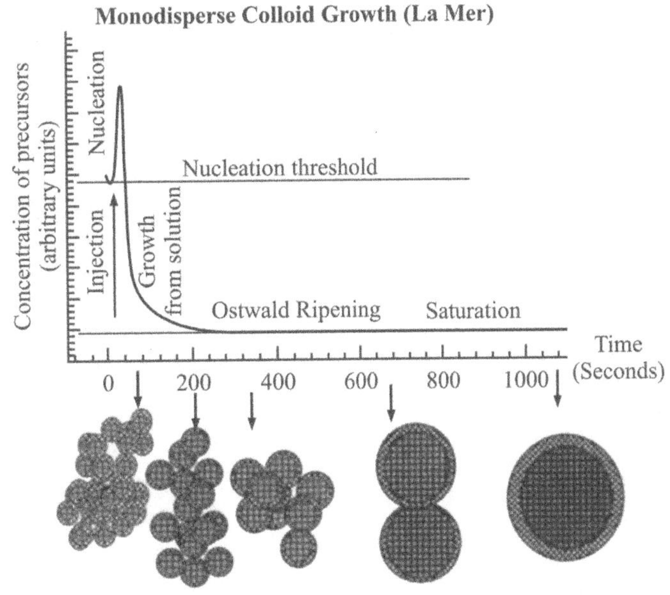

Figure. 11. 6 Nucleation and growth of nanoparticles in a solution of hot organic solvents (16)

This procedure was the first to result in quantum dots with sufficiently high quantum yield, between 10 and 20%, coordinated with organic ligands stabilizing the colloid, as well as producing mono dispersity.

Since the development of a nucleation and growth technique for synthesizing quantum dots, almost all newer techniques have built on it, changing solvents and precursors and working to increase the quantum yield and mono dispersity, as well as introducing greater control in the process.

In 2002, a major development was made towards using "green chemistry" to synthesize quantum dots (17). While the pyrolysis of organometallic precursors produces high quality quantum dots, the precursors are not air-stable, are pyrophoric, and very toxic. In addition, the reaction was not very tunable, so the balance between nucleation and growth could not be controlled well. The new "green" method, developed by the Peng group, used the non-coordinating organic solvent octadecene (ODE) in conjunction with the surfactant oleic acid (OA) and cadmium oxide as a cadmium ion source, and a solution of elemental sulfur and ODE as the sulfur source. Not only were the precursors air-stable and less toxic than organometallic precursors, but the reaction could be tuned by changing the concentration of Oleic acid (OA) (Figure. 11.7).

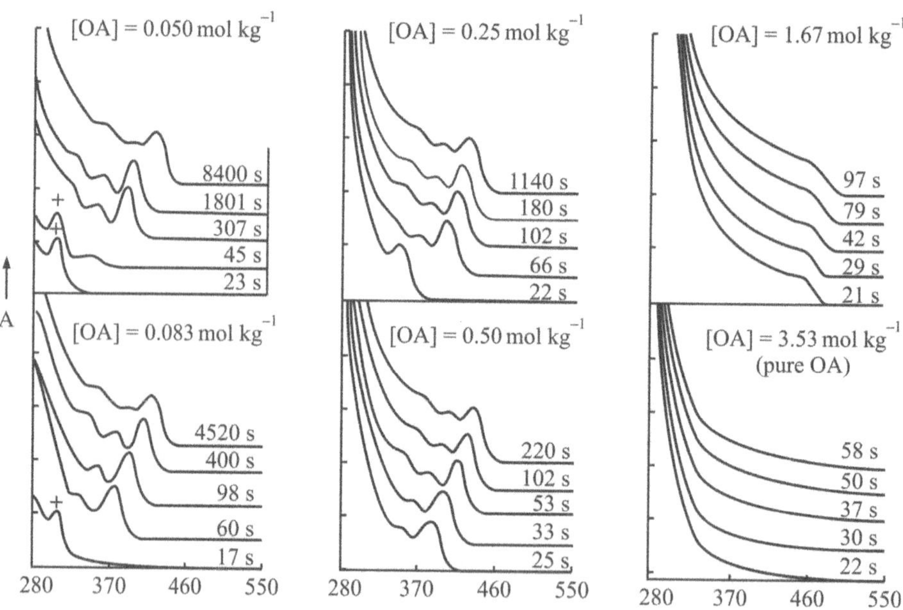

Figure. 11.7 Absorbance of quantum dots produced using the CdO/ODE/OA method, showing tunable reactivity of the precursors through adjustment of the ligand concentration

11.1.8.4 Synthesis of Quantum Dots for Biological Applications

The above organometallic way produces QDs, which are generally capped with hydrophobic ligands (e.g., trioctylphosphine oxide – TOPO or trioctylphosphine - TOP) and hence cannot be directly employed in bioapplications. To be used in biological applications, QDs need to be soluble in aqueous solutions and require surface modifications to achieve biocompatibility and stability [18]. Coordinating solvents stabilize the bulk semiconductors and prevent aggregation as the QDs grow. The semiconductor core material must be protected from degradation and oxidation to optimize QDs performance. Shell growth and surface modification enhance stability and increase photoluminescence of the core. The inorganic core-shell semiconductor nanoparticles are soluble in nonpolar solvents only. However, a further process is needed for the QDs used in biological system. Therefore numerous effective methods have been developed for creating of hydrophilic QDs, which can be divided into two main categories. The first route is commonly designated as "cap exchange". The hydrophobic layer of organic solvent can be replaced with bifunctional molecules containing a soft acidic group (usually a thiol, e.g., sodium thiolycolate) and hydrophilic groups (e.g., carboxylic or amino groups) which point outwards from the QDs surfaces to bulk water molecules [19, 20]. In fact, substitution of monothiols by polythiols or phosphines usually improves stability. The second route is native surface modification, for example adding a silica shell to the nanoparticles using a silica precursor during the polycondensation [21]. Amorphous silica shells can be further functionalized with other molecules or polymers. The method of QDs encapsulation into solid lipid nanoparticles, which are composed of high biocompatible lipids with physical and chemical long-term stability, was also successfully tested. These lipid nanoparticles are more convenient than small molecules (e.g., mercaptopropionic acid) traditionally used for QDs surface modification, which are rather unstable since they can be easily degraded by hydrolysis or oxidation of the capping ligand.

11.1.8.4.1 The Water Phase Method

The second way is the aqueous synthesis route, producing QDs with excellent water solubility, biological compatibility, and stability. One of the most widespread approaches for creating water-soluble QDs is ligand exchange with thioalkyl acids such as mercaptoacetic acid, mercaptopropionic acid, mercaptoundecanoic acid or reduced glutathione (GSH). These QDs have lower quantum yields than the above mentioned ones (up to 10%) without any following treatment. From these ligands, GSH seems to be very perspective molecule, since provides an additional functionality to the QDs due to its key function in detoxification of heavy metals in organisms. The fluorescence is considerably quenched at the presence of heavy metals and therefore GSH-QDs were successfully employed for determination of heavy metals. In addition, GSH-QDs exhibit high sensitivity to H_2O_2 produced from the glucose oxidase catalyzing oxidation of glucose and therefore glucose can be sensitively detected by the quenching of the GSH-QDs florescence [22]. In recent years, more work has been done to develop a large variety of methods for producing colloidal quantum dots in organic solvents, giving researchers a wide variety of chemical systems in which to work depending the on the properties they desire which are summarized in Table 11.2.

Table 11.2 Available synthesis methods for producing II-VI semiconductor quantum dots. (17)

Material	Precursors and stabilizers	Solvent (S)	Method[a]
CdS, CdSe, CdTe	$CdMe_2$ (TOP, $(TMS)_2$ Se or $(TMS)_2$S or $(BDMS)_2$ Te	TOPO	HI
CdSe	$CdMe_2$/TOP, TOP-Se	TOPO, HAD	HI
CdSe, CdTe	CdO, TDPA, TOP-Se or TOP-Te	TOPO	HI
CdSe	CdO or $Cd(ac)_2$ or $CdCO_3$, TOP-Se, TDPA or SA or LA	TOPO	HI
CdS, CdSe	CdO, SAODE or TBP-Se/ODE, OA	ODE	HI
CdSe	$Cd(st)_2$, TOP-Se, HH or BP	HAD, octadecane	HI
CdSe	$Cd(my)_2$, Se, OA/ODE	ODE	HU
CdSe	CdO, Se/ODE, OA	ODE	HI
CdSe	CdO, Se, OA	Olive oil	HI
CdSe	$Cd(st)_2$, TBP-Se, SA, DDA	ODE	HI
CdS	$Cd(ac)_2$, S, MA	ODE	HU
CdS, ZnS	$Cd(hdx)_2$ or $Cd(ex)_2$ or $Cd(dx)_2$ or $Zn(hdx)_2$	HAD	HU
CdS, ZnS	CdCl2/OAm or ZnCl2/OAm/TOPO, S/OAm	OAm	HU
CdTe	$CdMe_2$, TOP-Te	DDA	HI
CdTe	CdO, TBP-Te/ODE or TOP-Te/ODE, OA	ODE	HI
CdTe	CdO, TBP-Te, ODPA	ODE	HU
ZnS	$Zn(st)_2$, SAODE	ODE, Tetracosane	HI
ZnS	$ZnEt_2$, S	HAD/ODE	HU
ZnSe	$ZnEt_2$, TOP-Se	HDA	HI
ZnSe	$Zn(st)_2$, TOP-Se	Ocladecane	HI
ZnTe	Te and ZnCl2, in TOP	ODA, ODE	HI
HgTe	$HgBt_2$, TOP-Te	TOPO	HI
$Cd_{1\_x}Zn_xSe$	$ZnEt_2$/TOP, $CdMe_2$,/TOP	TOPO,HAD	HI
$Cd_{1\_x}Zn_xSe$	$Zn(st)_2$, $Cd(st)_2$, TOP-Se	ODE	HI
$Cd_{1\_x}Zn_xS$	CdO, ZnO, S/ODE, OA	ODE	HI
$CdSe_{1\_x}Te_x$	CdO, TOP-Se, TOP-Te	TOPO, HAD	HI

HI = Hot injection method; HU = Heating up method; $CdMe_2$ = dimethylcadmium; Zn Et_2 = diethylzinc; TMS = Trimethylsilyl; $(BDMs)_2$ Te = Bis (tert-butyldimesilyl) telluride; TDPA= tetradecylphosphonic acid; ODPA = octadecylphosphonic acid; SA = stearic acid; LA = Lauric acid; OA = Oleic acid; MA = Myristic acid; ac = acetate; my = myristate; st = stearate, hdx = hexadecylxanthate; dx = decylxanthate; TOPO = Trioctyl phosphine oxide; HAD = Hexadecylamine; DDA = Dodecylamine; ODA = Octadecylamine; TOP = Trioctyl phosphine, TBA = Tributylphosphine; ODE = 1-octadecene; HH = Hexadecyl Hexadecanoate; BP = benzophenone

11.1.9 Core-Shell Quantum Dots

Since quantum dots are only a few nanometers in diameter, they have a very high surface-to-volume ratio, as much as 80% of the atoms reside on the surface. Having such a high surface-to-volume ratio suggests that the properties of the surface have significant effects on the optical and structural properties of the particles. Surface defects, such as dangling bonds are surface-related trap states that act as non-radiative recombination sites, which degrade the fluorescence quantum yield of quantum dots (23). The organic ligands that surround colloidal quantum dots lend some degree of surface passivation, but do not provide sufficient protection from the surrounding environment or complete passivation of surface defects. To better passivate the surface, a secondary semiconductor can be epitaxially grown surrounding the core particle. After coating the core with such a shell, the quantum yield has been shown to greatly increase up to ten times, as well as displaying increased stability against photo-oxidation and environmental attack (24).

11.1.10 Types of Core-Shell Quantum Dots

Choosing the material for the shell layer depends on the properties that we desire after coating. To understand this a little better, we need to look at the different "types" of core-shell systems. There are three main types, characterized by the alignment of the valence and conduction bands between the core and shell (Figure 11. 8) (23).

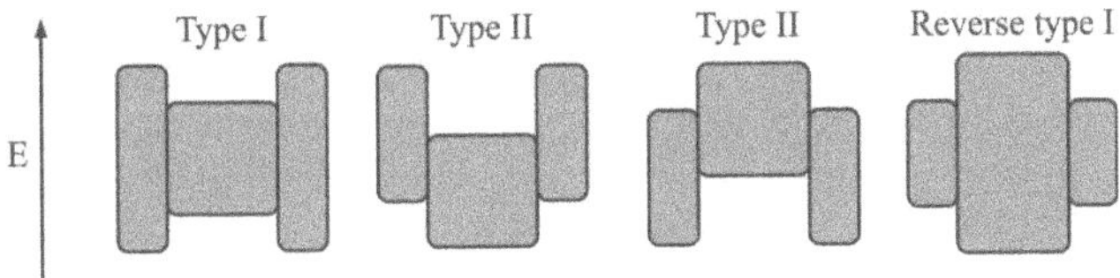

Figure 11.8 Band (valence and conduction bands) alignment of different core-shell systems

The first and most common core-shell system is type-I in which a higher bandgap semiconductor shell is formed on the core, confining the exciton to the core. The primary purpose of the type-I core-shell system is increasing fluorescence quantum yield by passivating the surface of the core, as well as isolating the core from the environment and reducing degradation. One of the first core-shell systems was CdSe-ZnS (25). In type-I systems, there is a characteristic slight red shift, usually around 10 nm, of the fluorescence due to some leakage of the exciton from the core into the shell. In reverse type-I systems, a narrower bandgap semiconductor is grown onto a higher bandgap core, partially delocalizing charge carriers from the core to the shell. Reverse type-I core-shell quantum dots are used when control is wanted over the red shifting of the fluorescence spectrum, as the shift can be controlled by changing the coating thickness. The most common reverse type-I systems are CdS-CdSe and ZnSe-CdSe. Type-II core-shell systems aim to significantly red shift the fluorescence, often into wavelengths that are otherwise unattainable with the same materials. This can be done by coating the core with a shell that has a staggered bandgap from its own, creating a smaller effective bandgap than either the core or shell. These core-shell types can be achieved using a wide variety of combinations of materials depending on the desired band alignment (Figure 11. 9).

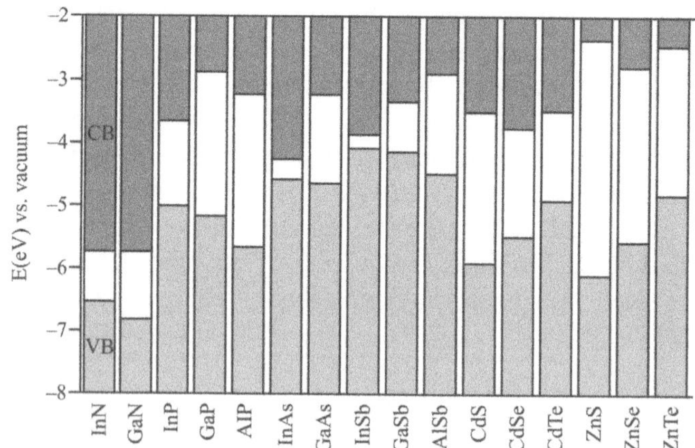

Figure 11.9 Electronic energy levels of selected III-V and II-VI semiconductors based valence band offsets (CB = conduction band, VB = valence band) (23)

11.1.11 Choosing a Shell Material for Type-I Systems

Type-I systems, will be suitable for achieving increase in the fluorescence quantum yield. Choosing a shell material involves both band alignment and crystal structure. Since the shell can be grown epitaxially on the core, a balance need to be made between bandgap alignment and lattice mismatch between the core and the shell. If the lattice mismatch between the core and shell is too great, new defects can form at the interface, effectively reducing the desired increase in fluorescence. When forming a shell on a core particle, the shell will tend to take the crystal structure of the core to minimize lattice mismatch if the shell can form the crystal structure of the core material. The other factor to account for in choosing a shell material is the possibility of alloying between the core and shell. Since we want to fully confine the exciton to the core, there should be a distinct change of electronic properties at the interface, so alloying should not be present.

Considerations also have to be made for the deposition of the shell material onto the core. The shell material should be able to be deposited in a colloidal system, at a lower temperature than was necessary to nucleate the core. Using a lower temperature allows the shell to be formed without growing the core significantly or nucleating separate particles of the shell material.

11.1.12 CdSe-ZnS Core-Shell System

The CdSe-ZnS core-shell system was one of the first type-I systems to be studied, and has been studied the most extensively (Figure 11. 10) (25). Due to the large difference in bandgap between the CdSe core (1.74 eV) and the ZnS shell (3.61 eV), the exciton is well confined to the core. The ZnS shell also passivates surface defects very well, greatly increasing the fluorescence quantum yield.

Zn sulfide can be deposited on CdSe cores from a variety of chemical precursors in a colloidal system, such as pyrolysis of the organometallic precursors diethylzinc (or dimethylzinc) and hexamethydisilathiane (26). These precursors will decompose at a lower temperature than is necessary for CdSe nucleation, as low as 140 °C and as high as 200 °C (27). In addition, ZnS will crystallize in the zinc blende structure on its own, but wurtzite is also thermodynamically stable at room temperature and atmospheric pressure, allowing epitaxial growth of wurtzite ZnS on CdSe cores. There is however ~12% lattice mismatch between the CdSe and ZnS, so coatings thicker than 2 to 3 monolayers tend to have decreased quantum yield due to the formation of new defects at the interface.

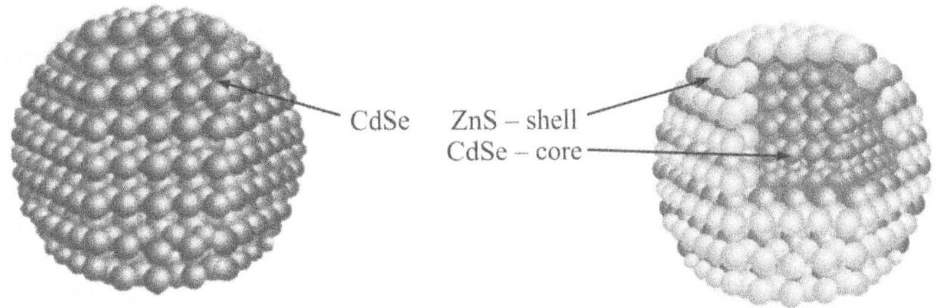

Figure 11.10 Efficiency of light produced by incandescent, compact fluorescent, and LED lamps, expressed in lumens per watt (28)

11.1.13 General Applications of Quantum Dots

Quantum dots find use in many applications that need strong, stable fluorescence with tunable emission. The primary applications of quantum dots are in energy efficient lighting, photovoltaics, and biological imaging. They are also finding a variety of potential applications in photonic devices such as (a) emitters for color displays (b) color modifiers for light emitting diodes (LEDs), (c) optical fiber amplifiers (d) low threshold lasers (e) self-assembled photonic sphere arrays (f) polymer-based photovoltaic cells (g) optical temperature probes (h) chemical sensors (i) high-speed signal-processing filters.

11.1.13.1 Lighting

Lighting accounts for up to 25% of energy usage in any country, so introducing more energy efficient lighting is of key importance. Lighting has progressed from black body radiators, such as incandescent lamps, to fluorescent lamps to more efficient forms of lighting such as light emitting diodes (LEDs) (Figure 11.11). Throughout this transition, though, it has become very important to retain or improve the quality of light produced.

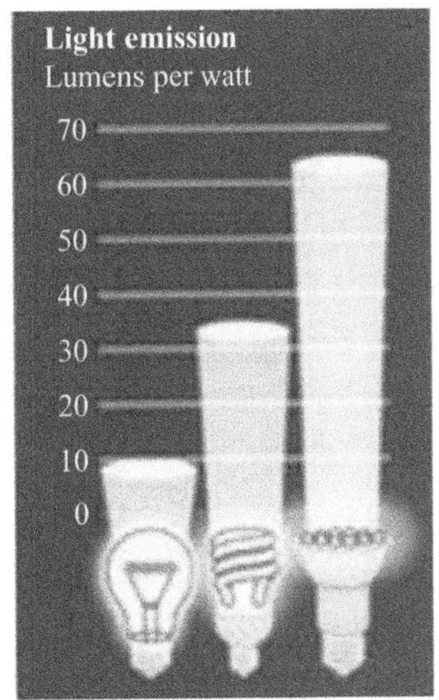

Figure 11.11 Efficiency of light produced by incandescent, compact fluorescent, and LED lamps, expressed in lumens per watt (28)

Quantum dots can be used in lighting either in conjunction with inorganic semiconductor light emitting diodes (LEDs), such as GaAs or InGaN, or as a replacement for, or complement to, conductive polymer junctions in thin film LEDs, such as organic LEDs. (28)

11.1.13.1.2 Light Sources with Spontaneous Emission

Based on narrow emission width possible from quantum dots which can be used for producing new emitters for applications that require carefully controlled color the animated billboards on buildings and in sports arenas are developed. Presently, these large color displays contain millions of light emitting diodes (LEDs); the intensities of red, green, and blue LEDs are adjusted to produce a color image, much like the elements of a color television. The cost of such displays could potentially be reduced if discrete LEDs could be replaced by polymer films containing quantum dots with different emission wavelengths. Such composites have demonstrated photoluminescence quantum yield (PLQY), (defined as the ratio of photons emitted to photons absorbed) as high as 22% for red emission at 426 nm, and 40% for green emission at 514 nm.

The absorbance and photoluminescence spectra of quantum dots also allow them to act as wavelength conversion materials for light emitting diodes. Quantum dot emission is typically narrower than that of inorganic phosphorus currently used as wavelength conversion materials. Quantum dots absorb light from wavelengths shorter than their absorbance edge and then emit light at their luminescence peak wavelength. For example, commercially available gallium nitrite (GaN, λ = 425 nm) blue LEDs provide a convenient excitation source that can be converted to any wavelength in the visible spectrum using appropriately designed CdSe quantum dots.

11.1.13.1.3 Optical Amplifiers

Theoretically, quantum dot lasers offer the potential of better performance than bulk semiconductor lasers or quantum well lasers. This potential arises from tighter confinement of carriers. "In small, strongly confined dots, the separation between energy states is greater than the thermal carrier energies, inhibiting thermal depopulation of the lowest, 'emitting' transition and resulting in a low, temperature-insensitive optical-gain threshold, a narrow emission line, and a high-temperature stability." To realize this dream, quantum dots must have a high PLQY, a narrow size distribution, and a close packing density.

For amplification, the electrons and holes in quantum dots could be excited by pumping at the absorption peak or shorter wavelengths, so that light propagating near the quantum dots' peak emission wavelength would stimulate coherent emission that amplifies the propagating signal. To use quantum dots as the gain media of a laser, a cavity must be formed, which multiplies the stimulated emission within a narrow wavelength band. Some of the techniques for creating laser cavities using optical fiber include feedback rings, etched Bragg gratings, splices, and whispering gallery modes.

11.1.13.2 Solar and Photovoltaics

Since quantum dots absorb all wavelengths higher in energy than their bandgap and convert them to a single color, they can be used to increase the range of wavelengths absorbed by photo voltaics, increasing their efficiency (Figure 11.12). There are a couple of different approaches to use this capability (Figure 11.13) (29,30).

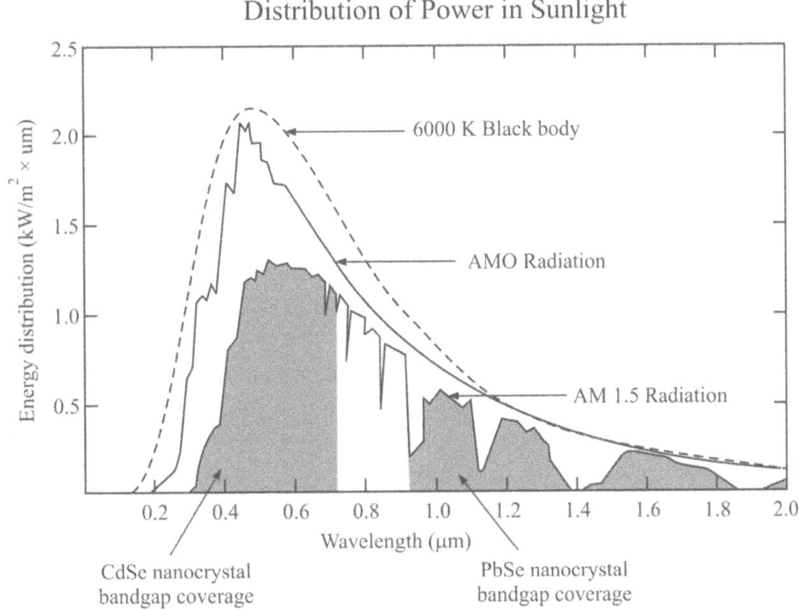

Figure 11.12 Comparison of solar spectrum with wavelengths that nanocrystals can efficiently absorb (31)

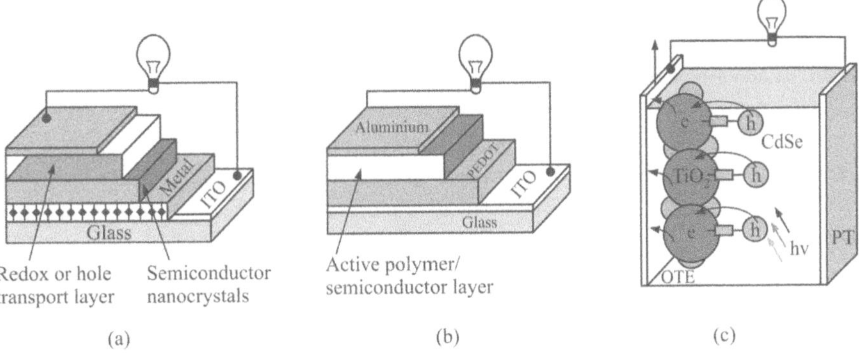

Figure 11.13 Current strategies to create quantum dot based solar cells. **(a)** metal-QD junction, **(b)** polymer-QD junction, **(c)** QD-dye sensitized solar cells

Organic photovoltaics (OPVs) are a growing industry in the same way as OLEDs for many of the same reasons. OPVs function in a very similar fashion as OLEDs. As in QLEDs, quantum dots can be substituted for or used in conjunction with organic molecules in thin film, printable solar cells. Another method for using quantum dots to harvest solar energy uses quantum dots for dye-sensitization with TiO_2 nanoparticles.

11.1.14 Potential Defense Applications of Quantum Dots

(a) **Optical, Thermal and Chemical Sensors:** Optical temperature probes which have a number of defense applications have been made using CdSe quantum dots in a polylaurylmethacrylate matrix. The photoluminescence intensity decreases with increasing temperature from 100 to 300 K. The decrease in PL intensity is linear near room temperature. There is also a redshift in the emission peak with increasing temperature.

(b) **Application for surveillance:** A viscous coating of the QDS that can be sprayed over an area to be monitored. The QDS preferably comprises an infrared fluorescing polymer material in a low-cost carrier such as SolGel. The QDS will have a unique optical phosphor luminescence, With a strong signature that can be observed by the camera or sensor system. The material will absorb optical energy, either solar or from an illuminator device, and re-emit this energy with the specified signature. The signature can be multi-spectral, or hyperspectral if required for anti-spoofing reasons. Since the agent will be deployed over a large area, the emission will provide a method of detecting changes will be relatively uniform throughout the region being monitored.

After Spraying the QDs the changes in area under surveillance can also be monitored for changes by aircraft or satellite by illuminating the scene with laser . An airborne laser illuminator can be employed to ensure that the QDs will have energy to emit, even at night. Any articles introduced into the area, items moved can be observed which will immediately be apparent in the imagery.

(c) **Optical Signature:** The QDs which exhibit unique optical fluorescence/ phosphoresce in infrared region will be mixed with environmentally friendly paint, making it invisible to the human eye. To produce specific fluorescence, mixtures of quantum dots can be used to produce random patterns of spectrally varying fluorescence.These QDs with strong optical signature can be tracked using infrared surveillance camera system that can track any object that has been marked with these compounds. The signature can be multi-spectral, or hyper-spectral if required for anti-spoofing reasons.

(d) **Sensors:** The properties of quantum dots are such that they can be functionalized to emit.

Light on binding target molecule which can be used to develop sensors for direct specific airborne pollutants.

(e) **Coding:** Quantum Dots can be used as fluorescent taggants in security in inks, plastics, papers, explosives or any other item or substance in which it is required to provide a distinct signature or mark. For a variety consumer goods like drugs, software, electronic products etc. QDs provide a method of creating unique, optical barcodes using the precise combinations of different quantum dots.

11.1.15 Quantum Dots for Biological and Medical Applications

Quantum Dots find great potential for biological and medical applications (32,33).A wide array of potential biomedical applications are increasingly motivating research on quantum dots. Many pathogens can be identified by characteristic proteins. Furthermore, antibodies have been developed that bind to many of these target proteins. With recent advances in immune histo chemistry, these antibodies can be joined to quantum dots to form new luminescent tags. As a medical tool for diagnosis, luminescent tags can help detect the presence of selected diseased tissues or illuminate the structure of diseased areas.

Quantum dots can help read DNA sequences. By attaching quantum dots and antibodies to polymer microspheres, the prevalence of targeted DNA sequences can be read by a standard flow cytometer. The surface of each sphere would also be covered with specific antibodies that would bind to specific protein sequences. This method is being compared to immuno-assay fluorescence arrays.

One of the keys to successful cancer treatment is accurate identification. To diagnose a specific type of cancer, slides are traditionally made from tumor tissue samples, and selective chemical stains can highlight one specific feature at a time, such as nuclei or aster cells. Quantum dots can enhance the capability of simultaneously marking numerous features with different colors, thus rapidly providing oncologists with a wealth of accurate and valuable information.

Two main areas of biological applications of QDs are biosensors and labels in biological imaging as shown in Figure 11.14.

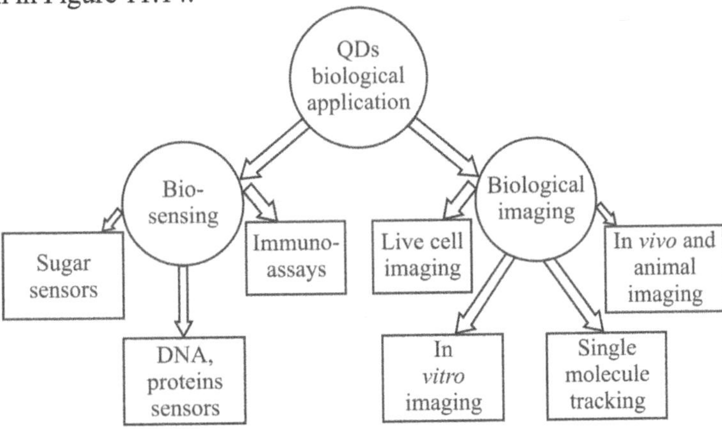

Figure 11.14 Biological applications of quantum dots

For in vivo biological applications of QDs the fluorescent emission wave length ideally should be in a region of the spectrum where blood and tissue absorb minimally but still detectable by the instruments. Thus the QDs should emit at approximately 700-900 nm in the NIR region to minimize the problems of indigenous fluorescence of tissues. Moreover, the spectroscopic properties of NIR QDs would allow imaging deeper penetration than conventional near infrared dyes (34).

Besides the application as simple sensors QDs have much higher impact as unique fluorescent labels. Various specific labeling strategies are known and most of these approaches are based on bio conjugation with other bio molecule exhibiting some specific affinity to the target compound.

Traditional synthesis of QDs in organic solvents such as toluene or chloroform at high temperatures in the presence of surfactants produce hydrophobic nanocrystals due to surface modifications, which are incompatible with aqueous biological environment (35). In biological environment, water is the main solvent and water-soluble nanocrystals will be an attractive proposition. To take full advantage of their optical properties for biosensing and bioimaging applications, they must possess minimum toxicity and monodispersible in aqueous biological environments. Due to the advances in the synthesis of highly luminescent QDs with high yields, favorable surface chemistry for surface modification, adaptation to water solubility, biocompatibility and multi colour emission for multiplex optical detection of biomolecules biological applications of QDs is taking a great momentum.. Recent developments have introduced multifunctional QDs for molecular and cellular imaging, nanoparticle drugs for targeted therapy, disease screening, detection and diagnosis.

Novel strategies have been developed adopting colloidal methods (Figure11.15) to make them water soluble and biocompatible, whereby the surface layer is exchanged or additional layer added with hydrophilic properties. These include the adsorption of small molecules such as mercaptoacetic acid or cysteine to QD surface (36) silica (37) or wrapping within micelles (38), liposomes (39) and amphiphilic polymers. To overcome, the bioincompatibility and instability, development in hybrid coatings have been investigated in the followings studies. The hydrophilic coatings may have reactive functional groups for binding to the core at one end and for conjugation at the other end. These allow conjugation to specific molecules for *in vitro* and *in vivo* targeted delivery and imaging.

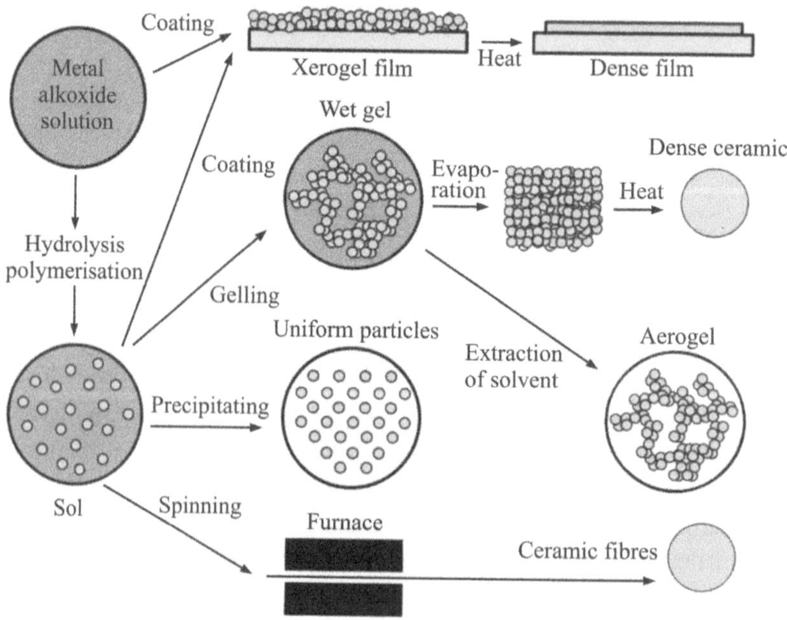

Figure 11.15 Colloidal methods for QD synthesis

11.1.15.1 Surface Coatings and Water-solubility of QDs

One of the biggest challenges to the biological application of QDs is their Bio compatibility. This dilemma is due to their hydrophobic surface chemistry, which makes them insoluble in aqueous biological media. In order to render QDs biocompatible, they have to be soluble in biological buffer, and over the years many scientists have discovered a wide array of surface coatings for solubilisation of QDs (35). The aim is to achieve an ideal surface chemistry that is stable in biological media and not alter the photophysical properties of the QDs, while at the same time retaining a small QD size and providing free reactive surface groups for binding and recognition of biomolecules. The techniques used to achieve solubilisation include ligand exchange, surface silanization and phase transfer methods. The "ligand exchange" method is based on the exchange of the hydrophobic surfactant molecules with bi-functional molecules, which are hydrophyllic on one side and hydrophobic on the other, to bind to the ZnS shell on the QD (40). Most often thiols (-SH) as functional groups are used to bind to the ZnS and carboxyl (-COOH) groups are used as hydrophyllic ends. The resulting QDs are soluble in both aqueous and polar solvents. This is by far the simplest method to achieve solubilisation. "Surface silanization" involves the growth of a silica shell around the nanocrystal. As silica shells are highly cross-linked they are very stable. However, the drawback is that the process is laborious and the shell may be hydrolysed (40)). The "phase transfer" method uses amphiphyllic polymers to coat the QDs surface (41,) The hydrophobic alkyl chains of the polymer interdigitate with the alkyl groups on the QDs surface while the hydrophyllic groups orientate outwards to attain water solubility. However, coating with a polymer may increase the overall diameter of the QDs and this may reduce emission and limits their use biological applications (42). Various techniques of coating using phospholipids micelles, dithioretol, organic dendrons, and oligomeric ligands can be found in numerous publications (43,44).

11.1.15.2 Surface Functionalization

11.1.15.2.1 Functionlazation of QDs by conjugation with biological molecules

Once solubilisation is achieved, QDs can be functionalized by conjugation to a number of biological molecules including avidin, biotin, oligonucleotides, peptides, antibodies, DNA and albumin (45) through surface reactive groups for specific targeted action. Methods of bioconjugation broadly fall into two categories: noncovalent and covalent conjugation (46). Noncovalent interactions include absorption, electrostatic interaction and mercapto-exchange. Biomolecules like oligonucleotide and various serum albumins (47) can be absorbed on the surface of the water-soluble QDs. This process is non-specific and depends on pH, temperature, ionic strength and surface charge of the molecule (48). QDs may be cationic or anionic and may interact with biomolecules through electrostatic interaction. The surface charge plays an important role in the cellular interaction of QDs and is determined by the free surface reactive groups. It has been demonstrated that proteins engineered with positively charged domains, can interact with the negative charges on the QDs surface coated with dihydrolipoic acid (DHLA) through electrostatic interaction (49). These conjugates have greater photoluminescence but are also more stable than unconjugated dots. However, electrostatic interactions are nonspecific and relatively weaker compared to covalent bonding and this may pose a problem in the biological environment. Many biological molecules have a thiol group, which can be tagged on to the surface of a QD by a mercapto-exchange process (49). However, the resulting bond between thiol and Zn is not only weak but also dynamic and this may lead to precipitation of the biomolecules in solution as they easily detach from the QD surface. Covalent linkage is the most stable of all the bioconjugation methods and utilises functional groups on the QDs surface like primary amine, carboxylic acids, and thiols to form a covalent bond with similar groups present on biomolecules or through the use of cross-linker molecules. The commonest and convenient method of bioconjugation is via the avidin-biotin interaction. This is based on the high affinity interaction between avidin and biotin through a generic key-lock mechanism. Avidin is attached to antibodies and biotin can covalently bind to the surface of QDs and vice versa. As most of the commercially available biomolecules are avidin or biotin linked, it makes this process most convenient as well Figure 11.16.

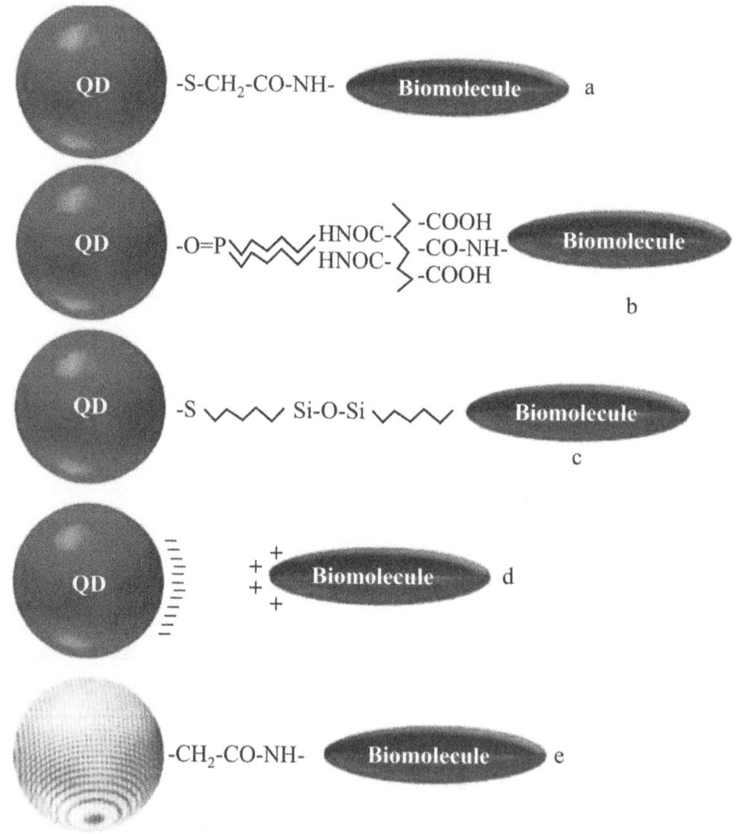

Figure 11.16 Schematic representation of different approaches to conjugation of QDs to biomolecules

(a) Use of a bifunctional ligand such as mercaptoacetic acid for linking QDs to biomolecules. (b) TOPO- capped QDs bound to a modified acrylic acid polymer by hydrophobic forces. (c) QD solubilisation and bioconjugation using a mercaptosilane compound. (d) Positively charged biomolecules linked to negatively charged QDs by electrostatic attraction. (e) Incorporation of QDs into microbeads and nanobeads.

11.1.15.2.2 Functonalzation of QDs with chemical -polymer groups

The functionalization of QDs with polimerizable groups can induce graft copolymerization with vinyl monomers, thus leading to a homogeneous distribution of QDs inside polymer latexes. The introduction of vinyl-functionalized CdSe/ZnS in styrene droplets, by exchange of the native capping by 4-mercaptovinylbenzene, has been investigated in miniemulsion polymerization but still migration of the dots to the surface was observed along with the deterioration of the photoluminescence of the final material. An hybrid method that comprises in a first step the silica encapsulation of the CdSe/ZnS dots, followed by functionalization of the silica at the surface with methacryl oxypropyl trimethoxysilane (MAS) and then graft copolymerization, gave core-shell QDs-polymer nanocomposites, but again there was a decrease on the luminescence of the nanomaterials (50).

Figure 11.17 summarizes some of the innovative chemical strategies that have been used to produce organically capped and/or polymer encapsulated QDs, as discussed above. However, there is still a number of practical issues which need improvement interface of the quantum dot and the polymer matrix, and because the polymer may also exhibit electroluminescence.

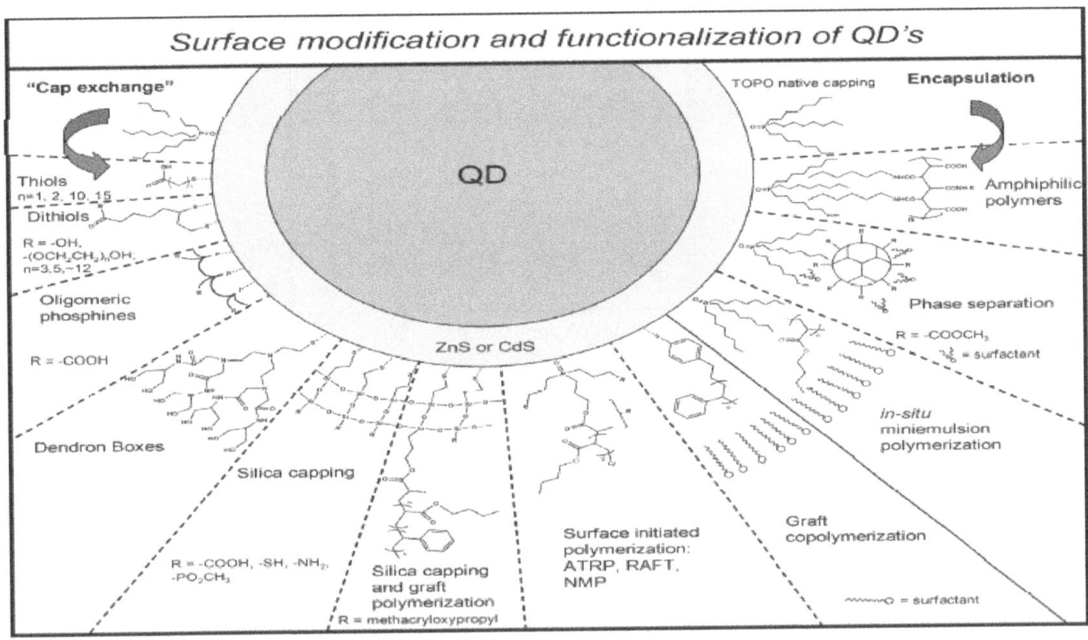

Figure 11.17 Methods for Chemical Surface Modification of QDs

11.1.15.3 Bio Imaging Applications

One of the primary areas of research and commercialization of quantum dots is in biological imaging. Quantum dots are approximately the same size as a protein, thus allowing them to enter cells in a similar manner. Most fluorescent dyes are based on organic molecules, often xanthenes such as rhodamine and fluorescein. There are a couple of key issues with organic dyes that can be remediated with quantum dots. The absorbance and fluorescence of organic dyes are tied to their molecular structure, requiring excitation and detection at specific wavelengths. Unlike organic fluorophores, quantum dots absorb a broad spectrum and emit symmetric, narrow spectra (Figure 11.18).

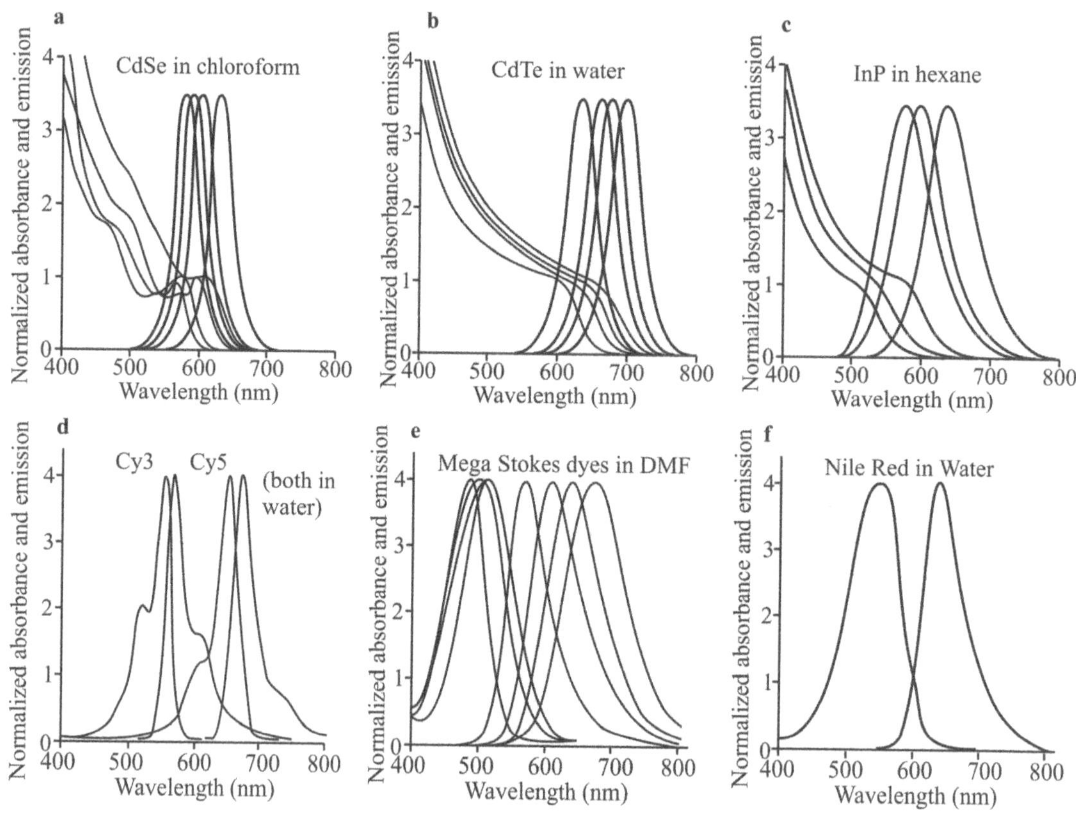

Figure 11.18 Absorbance and fluorescence spectra of quantum dots (a-c) in comparison to organic dyes (d-f). (50)

This feature of quantum dots give them advantages over organic fluorophores because the excitation wavelength can be anywhere within a broad range, making it easier to avoid excitation of background tissues, as well as simple separation of excitation and emission.

11.1.15.3.1 Quantum Dots vs. Organic Dyes in Biomedical Imaging

For visible imaging, quantum dots have several potential advantages over organic dyes. First, traditional organic dyes have narrow absorption peaks which are very close to their photoluminescence peak. Therefore each dye requires its own expensive, carefully-tuned, sharp-cutoff filter to block the excitation background from the imaging camera. In contrast, quantum dots have a broad absorption tail over wavelengths shorter than the peak emission wavelength. So a single common long pass filter and a single excitation source can pump quantum dots that emit in several distinct wavelength ranges. Although the PLQY of some common visible dyes, such as rhodamine 6G and fluorescein, can be higher than 90% in plain dye-solvent solutions, conjugation to antibodies is required to form luminescent protein tags. Proximity to other moleculesoften degrades the PLQY of organic dyes (*e.g.* the PLQY of fluorescein may be cut in half when bound to antibodies). Quenching of this magnitude can be avoided with quantum dots. Additionally the luminescence from organic dyes decreases rapidly with exposure time, which leads to the use of more expensive pulsed excitation sources and detectors. This "bleaching" problem is reduced by using quantum dots.

11.1.15.4 In Vitro and In Vivo Imaging

The above-described properties of QDs are most promising for enhancing the sensitivity and the multiplexing capabilities of molecular histopathology and disease diagnosis. Recent developments in QDs have produced very bright and stable probes that have found applications in profiling genetic and protein biomarkers in intact cells and clinical tissue specimens. Because of inherent potential toxicity QDs have been generally excluded from a *in vivo* imaging applications. However, the use in immunohistological studies has found favour on *in vitro* or *ex vivo* clinical specimens. The utilization of multicolour QD probes in immunohistochemistry (IHC) will be a major contribution in clinical diagnostics in the coming years.

In vivo imaging has traditionally used contrast agents as imaging probes that include radioisotope small molecules in positron emission tomography (PET) and single photoemissions computed tomography (SPECT), gadolinium compounds in magnetic resonance imaging (MRI), and isotope conjugated antibodies. The unique features of bioconjugated targeted QDs however can provide extraordinary capabilities that could enhance the sensitivity and specificity of diseased tissues and improve diagnosis. The inherent, size dependent optical and electronic features of QDs can be manipulated continuously by altering the particle size. This tunability feature facilitates the use of a broadband range of QDs, for simultaneous detection of multiple cancer cell surface biomarkers. Additionally the large surface to volume ratio of QDs allows multifunctional groups suitable for conjugation with multiple diagnostic (magnetic, NIR and radioisotope) and therapeutic (e.g. anticancer) ligands. This allows the development and design of multifunctional "intelligent" nanocrystals for multi-modal imaging combining therapy. There is evidence that nanoparticle in the range of 10-100 nm preferentially concentrate at tumour sites through a phenomena defined as enhanced permeability retention (EPR) (48-49).

11.1.15.5 Nano-drugs

Developing QDs as drugs and additional to drug delivery carriers and targeting probes is challenging and important in the arena of nanotechnology. Introducing nanosize range has improved the efficacy and toxicity profile of chemotherapeutic drugs by virtue of their desirable surface properties. Additionally it facilitates dispersibility of drugs that are otherwise poorly soluble in aqueous medium. Traditional drug formulations suffer from bioavailability with poor release mechanisms.

Various Biological applications of QDs after surface functionalization or conjugation with bio molecules is summarized in Table 11.3.

Table 11.3 Summary of biological applications of quantum dots

Biological Applications of Quantum Dots			
Imaging	*In vitro* imaging	Fixed cell imaging	Immunofluorescent labeling of proteins, biomarkers
		Live cell imaging	Immunohistochemical detection Fluorescence in situ hybridization
	In vivo imaging		Cancer studies and cancer localization
Labeling			As well as labeling quantum dots have potential for detecting
Drug Delivery			Loading biological compounds on QDs such as any types of drugs
Other Applications			Photodynamic Therapy Pathogen toxin detection

11.2 Nano Photonics

11.2.1 Introduction

Photonics is the study of the interaction of light with matter. The field was opened up in 1960s with the invention of the laser. Ten years later, the invention of the optical fibre as a means of transmitting information via light formed the basis for optical communication. The field is now enormous and consists of many sub-disciplines and applications, such as laser technology, biological and chemical sensing, display technology, optical computing, fibre optics, photonic crystals and more. In 1987, Eli Yablonovitch at Bell Communications Research Centre created an array of 1 mm holes in a material with a refractive index of 3.6. It was found that the array prevented microwave radiation from propagating in any direction. This discovery launched the research on nano photonic crystals, but it took more than a decade to fabricate photonic crystals that do the same in the near-IR and visible range. Photonic crystals are now receiving much attention because of their potential in particular in the optical-communication industry. Researchers are considering using light and photonic crystals (in alternative to electrons travelling in wires) for the new generation integrated circuits. Light can travel much faster in a dielectric medium than an electron in a wire, and it can carry a larger amount of information per second. Given the impact that semiconductor materials have had on every sector of society, photonic crystals could play an even greater role in the 21st century.

11.2.2 Photonic Crystals

Photonic crystals are designed to confine, manipulate and control the propagation of photons in three dimensions. The properties of a photonic crystal are determined by the radius of the holes (or other features such as dielectric rods) forming the array, the periodicity of the holes (or rods), the lattice structure, the thickness of the material and the refractive index. In a semiconductor, it is possible to break the perfect periodicity of a silicon-crystal lattice by introducing defects (doping), and to have electrons in the forbidden energy gap region. Similarly, it is possible to create energy levels in the photonic band gap by changing the size of the holes (or rods) forming the photonic crystals. This introduces an 'allowed frequency' (i.e. an allowed wavelength) in the photonic band gap, and the photonic crystal then acts like a wave guide

In reality a photonic crystal consists of a periodic arrangement of alternating layers of two different dielectric materials or metal-dielectric nanostructures that affects the propagation of electromagnetic waves. Essentially, photonic crystals contain regularly repeating internal regions of high and low dielectric constant. Photons (behaving as waves) propagate through this structure - or not - depending on their wavelength. The periodicity of the photonic crystal structure has to be of the same length-scale as half the wavelength of the electromagnetic waves, i.e. ~200 nm (blue) to 350 nm (red) for photonic crystals operating in the visible part of the spectrum. Such crystals have to be artificially fabricated by methods such as electron-beam lithography and X-ray lithography.

A schematic diagram for two dimensional and three dimensional photonic crystals is presented in Figure 11.19 which shows the two dielectric materials are shown as being red and yellow, with indices of refraction of nA and nB. The two materials are shown as having the same thickness in each direction, but these can be different. The two or three lengths involved could also be different, but then the band gaps would be different for each direction you consider. If the reflections interfere constructively, light will be unable to propagate through the crystal, and total reflection will be observed. In1 dimension, this is called a Bragg reflector. For reflections in multiple directions, we call it a photonic crystal. An example of other kinds of reflection would be in metals where electric fields are simply excluded by the high concentration of 'free' charges.

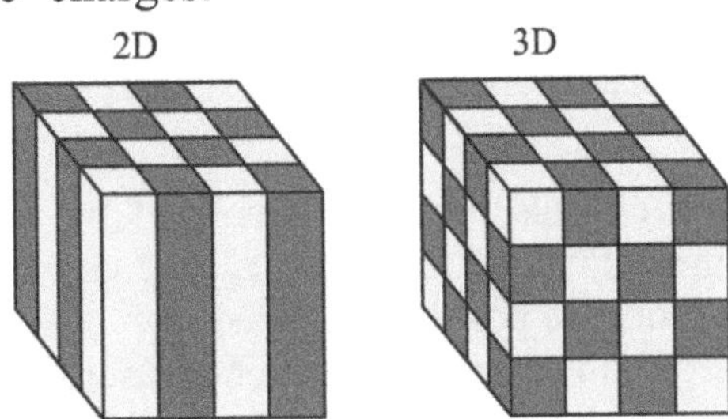

Figure 11.19 Potential geometries for 2D and 3D photonic crystals

It can be seen in Figure 11.19 Photonic crystals have periodic arrangements of dielectric materials so as to reflect light in multiple directions .An important feature of photonic crystals is that they have photonic bands, and these bands are separated by band gaps in which there is no propagation of light. These bands are similar to the semiconductor conduction and valence bands for electrons, which are separated by a band gap. One difference between the two is that photons are not bound to atoms, so they do not come with an energy level structure which is the origin of the band structure in solids. Instead, the band gaps in photonic crystals result from multiple reflections interfering constructively. That is, in a semiconductor, the atomic energy levels and the gaps between them result in bands and band gaps. In a photonic crystal, interference effects result in band gaps. One example of a photonic crystal would be silica for the red material and air for the yellow. Some other examples include titania and air, titania and silica, and aluminum arsenide and air. In Figure 11.20, the dispersion relation, i.e. a plot of energy, ", vs. wave vector, k, for a simplified photonic crystal is shown.

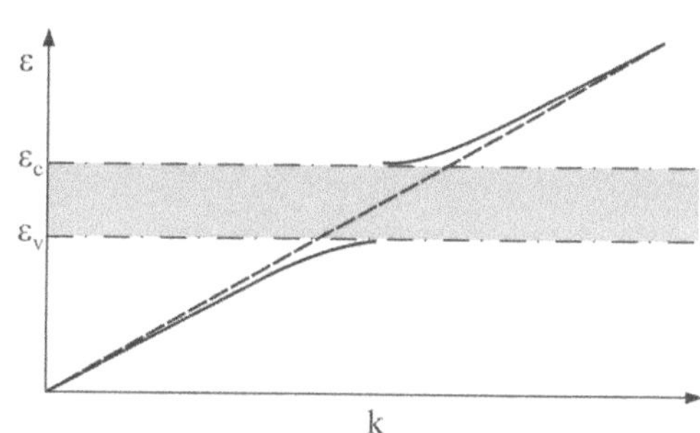

Figure 11.20 Dispersion relation for a simple photonic structure in 1 dimension, showing a single valence and conduction band. Note the band gap where photons cannot propagate

As can be seen from Figure 11.20 near the band edge there is a region where the effective index of refraction is large, since the slope here is small. In other words, the speed of light in this material is slow, since the slope of the energy vs. wave vector curve is small. This means that the effective index of refraction for that frequency is large. In real photonic crystals the dispersion relation tends to be highly directional (51).

Given that the defining characteristic of a photonic crystal is the existence of a photonic band gap, the defining parameters for an extant photonic crystal tend to be the size, location, and directionality of the band gap in question (51-53). The gap-to-midgap ratio, for example, is often used, and it is defined as:

$$r_{gm} = \frac{2(\varepsilon_c - \varepsilon_v)}{\varepsilon_c + \varepsilon_v}$$

where ε_C is the energy at the upper edge of the band gap and ε_V is the energy at its lower edge. These letters are chosen to emphasize the similarities with semiconductors, where conduction and valence bands exist.

The properties of photonic crystals can be understood by considering an analogy with semiconductors. In general, in a semiconductor, the valence or conduction electrons can move in a periodic potential arising from the positively charged ion cores (the nucleus). The potential is characterized by two areas of allowed energy, which are separated by a region of forbidden energy, called the energy gap, which means that there are certain wavelengths that will not propagate in the lattice (i.e. no electrons will be found in the band gap region). This is true for a perfect silicon crystal. However, in real materials, the situation is different: electrons can have energy within the band gap if the periodicity of the lattice is broken. This can be the result of a missing silicon atom, or the presence of an impurity atom occupying a silicon site, or if the material contains interstitial impurities (additional atoms located at non-lattice sites). Doping of semiconductors is intentionally done in microelectronics. Now consider photons moving through a block of transparent dielectric material containing a number of tiny air holes arranged in a lattice pattern. The photons will pass through regions of high refractive index — the dielectric — alternating with regions of low refractive index — the air holes. This is analogous to the periodic potential that an electron experiences when travelling through a silicon crystal. If there is a large difference in refractive index between the two regions, then most of the light will be confined either within the dielectric material or the air holes. This confinement results in the formation of allowed energy regions separated by a forbidden region — the so-called photonic band gap. Since the wavelength of the photons is inversely proportional to their energy, the patterned dielectric material will block light with wavelengths in the photonic band gap, while allowing other wavelengths to pass freely.

One way to increase the size of a band gap is to increase the refractive index contrast between the two materials used. It is worthwhile to have large band gaps, since the larger they are, the more control one would have in adjusting the behaviour of light within an arrangement of photonic crystals.The simplest way to increase refractive index contrast is to have one high index material, such as silica, and let everything else be air or vacuum, which do not absorb significantly. Metals typically have a real part of their refractive index which drops below one, but this always accompanies some absorption (54).

11.2.3 Theory of Nano Photonics

Nanophotonics is the field of nanotechnology concerned with discovering and developing nanomaterials that can control the flow of light and in some cases localize or confine it within a volume. Intuitively, we view light as rays, which either propagate in a single direction, being absorbed or reflected to some extent by any object on which it impinges. However, the propagation of light through a material is itself a quantum effect, involving the excitation and relaxation of electrons in the material. Therefore, creating a material with structural and compositional features on a length scale comparable to the wavelength of light i.e. 300 - 750 nm for the visible light enables us to guide light in any direction of our desire. There are two major challenges to reach the molecular scale and develop mass production techniques. In molecular photonics the aim is to achieve sufficient control over light-matter interactions through nanoscale physical, chemical and structural modification interactions using a few or single molecules

To understand the mechanism of nano photonics the basic equations describing propagation of photons in dielectrics that has some similarities to the propagation of electrons in crystals (Figure 11.19) need to be examined. There are similarities between photons and electrons.

Wavelength of light (photon) is given by

$$\lambda = \frac{h}{p} = \frac{g}{mv}$$

Wave length of electron is given by

$$\lambda = \frac{h}{p} = \frac{c}{v}$$

The Maxwell's electromagnetic equations for the light (photon) describes the allowed frequencies of light and in a similar way Schrodinger's equation for the electrons describes allowed energies of electrons. Free space solution of the wave equation for both the photon and the electron are plane wave. Interaction potential in a medium for propagation of light affected by the refractive index of the dielectric medium and the propagation of electrons are affected by the Coulomb potential. In this case, both electrons and photons have propagation through classically forbidden zones. Photon tunnels through classically forbidden zones, the electric, and the magnetic fields decay exponentially that results k-vector is imaginary. Electron wave function decays exponentially in forbidden zones and has finite tunneling probability as shown in the Figure 11.21.

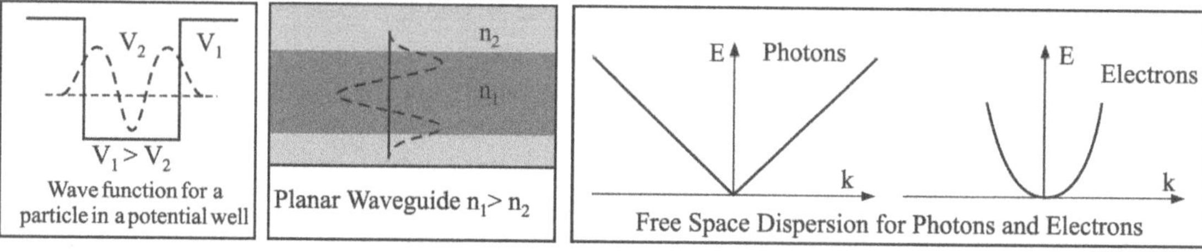

Figure 11.21 (a) Free space dispersion relations for photons (and electrons (b)Tunneling of Photon (c) Tunneling of Electron from classically forbidden regions

As discussed earlier Quantum confinement (Figure 11.22) causes manifestations of new optical effects due to the change in the nature of density of states from continuous to discrete structures as the dimension is reduced from three to zero. It is the density of states that guides optical and electrons correlation effects in quantum confined nanostructures.

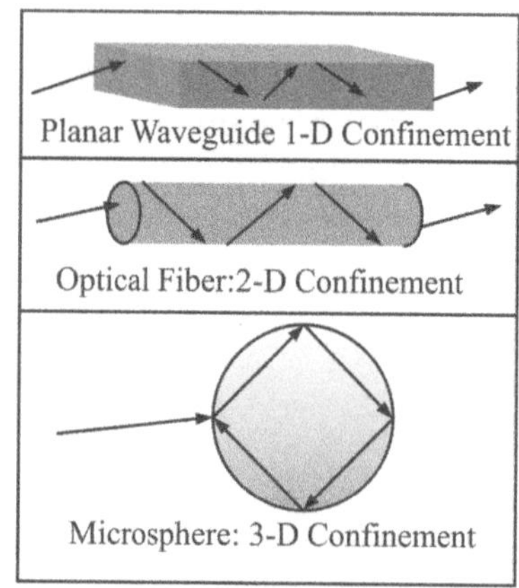

Nano Scale confinement of Photons

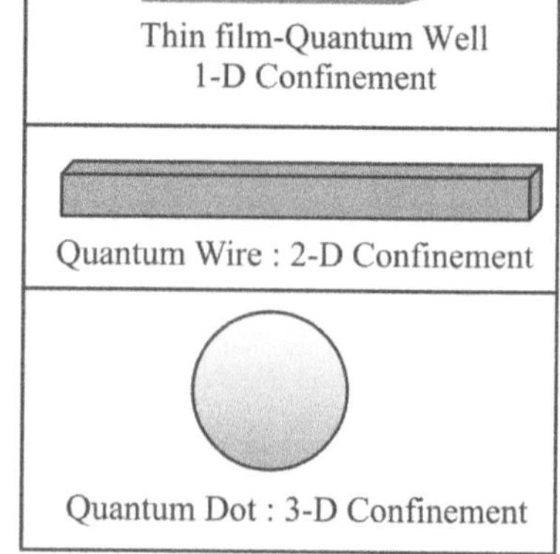

Nano Scale confinement of electrons

Figure 11.22 Nano Scale confinement of photons and electrons in different spatial dimensions

Size dependence of optical properties due to confinement produces a blue shift of the band-gap in semiconductors, and increases the oscillator strengths and in turn increases the optical transition probability. Location of discrete energy levels depends on the size and nature of the confinement. The oscillator strength increases with increasing confinement from bulk to quantum well to quantum wire to quantum dot. Confinement produces sub-bands within the conduction and valence bands, enabling intra-band optical transitions, which are not allowed in the bulk structures. These infrared transitions have applications in making new quantum cascade lasers and detectors. The oscillator strengths increase as the width of the quantum well decreases. The most important optical feature of these structures is that absorption/emission spectra shift to shorter wavelengths as the size become smaller. In case of metallic nanostructure, somewhat spectacular things happen due to which they show strong luminescence and plasmon absorption bands.

11.2.4 Fabrication of Photonic Crystals

There are different approaches to building a photonic crystal. The first researchers to report this idea were Yablonivich and John in 1987, who fabricated a crystal at microwave wavelengths.

The fabrication consisted of covering a block of dielectric material with a mask consisting of an ordered array of holes and drilling through these holes in the block on three perpendicular facets. The material, which became known as 'Yablonovite', prevented microwaves from propagating in any direction — in other words it exhibited a 3D photonic band gap.

Despite this early success, it has taken over a decade to fabricate photonic crystals that work in the near-infrared and visible regions of the spectrum. For example, for visible light, a lattice dimension of approximately 500 nm would be required, which means fabricating a material with holes separated by 500 nm. This is an extremely difficult task which requires suitable materials and processing techniques.

One way to produce a photonic crystal with a band gap in 3 dimensions is to arrange spheres periodically in a container, fill the empty spaces with a background material, and dissolve the spheres away (58).This leaves air spheres with a background material acting as a photonic crystal. Another technique involves a photosensitive material which is exposed to several lasers that interfere constructively in order to cross the intensity threshold for exposure, after which either the ex-posed or unexposed portion is dissolved (59). The geometry in that reference is also different, in that their crystal follows the 'log cabin' approach. This is where the high-index material is arranged in rods in a perpendicular cross-hatch, where one layer has lines parallel to the x-axis, and the next layer has lines parallel to the y-axis, as seen in

Figure 11.23 A log cabin geometry for a photonic crystal. 4 layers in different colours for a photonic crystal. Each layer consists of rods of the same material

Figure 11.23 If the low index of refraction material is air or a vacuum, then that leaves a lot of empty space in which objects smaller than 100 nm or so (such as quantum dots) may be placed.

Instead of drilling holes on a surface, another approach to creating a photonic crystal is to build the lattice out of isolated dielectric materials that are not in contact, as illustrated in Figure 11.24.

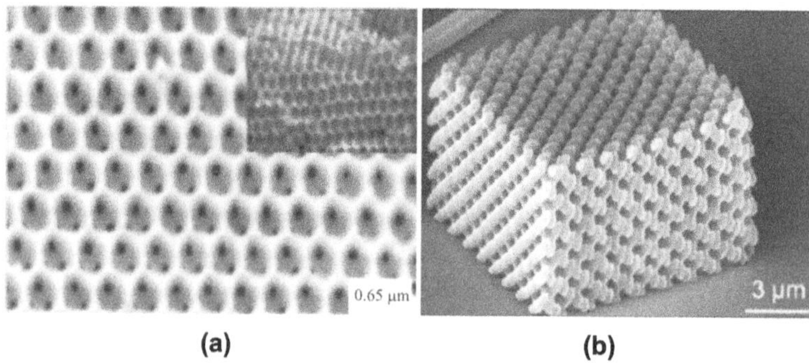

Figure 11.24 (a) An inverse opal photonic crystal structure (inset shows dielectric and air holes). (b) A photonic bandgap structure for channeling light scaled in microns

If an entire line defect is introduced in the lattice as in Figure 11.25. By removing an entire row of rods, the regions where the rods have been removed act like a waveguide, and an allowed energy level is created in the photon gap. A waveguide is like a pipe that confines electromagnetic energy, enabling it to flow in only one direction.

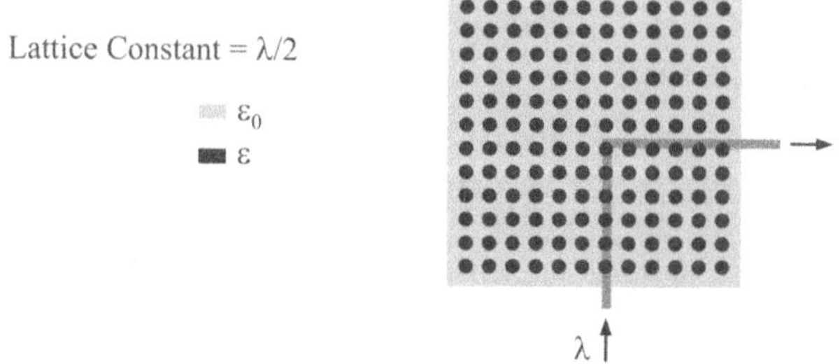

Figure 11.25 Wave guide by photon localization PC

A gap (defect) in the crystal traps light in the photonic crystal while a line defect acts as a wave guide which has great potential for applications in optical integrated circuits

One interesting characteristic of such a waveguide is that light passing though it can turn very sharp corners, which is not true in an optical fibre. If only one rod is eliminated in the lattice in Figure 11.25 (or if its diameter is changed), a resonant cavity is formed, which also puts an energy level into the gap. As already mentioned, this energy gap depends on the radius of the rod, which provides an opportunity to tune the frequency of the cavity. This ability to tune the light and concentrate it in small spaces makes photonic crystals useful for filters and couplers in lasers.

11.2.5 Applications of Nano Photonics

Periodic Arrays in Photonic Crystals: When light travels through a media, if the periodicity of the electromagnetic wave matches the periodicity of the refractive index, then the photonic crystal would rather emit or scatter that light instead of transmitting light at that wavelength.

The position of this stop-band wavelength, λ_C, can be tailored by adjusting the size or the refractive index of periodic structures. Refractive index in a nanocrystal-polymer composite can be adjusted by controlling the quantum dot concentration. When spheres are made with tightly controlled diameters, they can be self assembled into regular arrays with special optical properties.

Artificial structures or meta-materials with the optical equivalent of the energy gap in semiconductors promise a wealth of new devices that could satisfy the demand for ever-faster computers and optical communications (55).

Presently, the availability of personal computers that operate at 1 GHz (109 Hz), are really very impressive. However the likelihood of a 100 GHz desktop computer appearing on the market at affordable price is difficult to imagine while with the current understanding of semiconductor technology, even producing a 10 GHz personal computer would seem to be difficult.

However, by transmitting signals with light rather than electrons (a domain of optical computing), it might be possible to build a computer that operates at hundreds of terahertz (1012 Hz) which is of the order of the processing speed of the fastest earth simulator computer. Researchers now believe that such an awesome processing engine could be built from optical components made from so-called photonic crystals and quasi-crystals. These materials possess structures with periodically modulated refractive indices that can be designed to control and manipulate the propagation of light. They are attracting more and more attention not only in the field of photonics but also in the fields of chemistry, physics, and microelectronics. This is because the photonic crystals enable us to control the flow of photons by means of photonic band gaps, which is a photonic analogue of energy band gaps in semiconductors. In recent years, much of the developments on the photonic band gap are taking shape.

In theory, photonic crystal applications reach across the entire electromagnetic spectrum, from UV to radio waves. In practice, the challenge is the nanofabrication of these devices. Novel nanofabrication methods, including self-assembly approaches, will play an essential role in the development of this field.

Nano Photonic crystals can efficiently manipulate the direction in which light travels. Photonic crystals are usually viewed as an optical analog of semiconductors that modify the properties of light similar to a microscopic atomic lattice that creates a semiconductor band-gap for electrons. It is therefore expected that by replacing relatively slow electrons with photons as the carriers of information, the speed and bandwidth of advanced communication systems will be dramatically increased.

However the integration of photonic components such as lasers, detectors, couplers, and wave-guides is still at a very primitive stage, compared with that of microelectronics. This is due to difficulties with implementing integrated optical components smaller than certain sizes.

For instance, small bends or curves of wave-guides lead to the leakages of optical signals, so the bends have to be bigger than certain critical length. Photonic crystals can then be used as they offer a potential to provide solutions to the challenge of photonic component integration by means of its nature to confine photons within their structures. Moreover, intriguing phenomena in photonic crystals can be exploited, which are not achievable by conventional isotropic media. Photonic crystals can be a breakthrough in photonic technology in the near future (56).

The absence of allowed propagating electromagnetic modes inside the structures in a range of wavelengths called a photonic band gap, gives rise to distinct optical phenomena such as inhibition of spontaneous emission, high-reflecting omni-directional mirrors and low-loss-wave guiding amongst others (53-57). Since the basic physical phenomenon is based on diffraction, the periodicity of the photonic crystal structure has to be in the same length-scale as half the wavelength of the electromagnetic waves ~300 nm for photonic crystals operating in the visible part of the spectrum. The periodicity, whose length scale is proportional to the wavelength of light in the band gap, is the electromagnetic analogue of a crystalline atomic lattice. This periodicity acts on the electron wave function to produce the familiar band gaps as in the case of semiconductors in condensed matter physics. Based on these concepts a number of applications of nano photonics being explored in the following areas

1. Nano Scale confinement of light: Ex: Near field microscopy, SHG microscopy, nano cavities
2. Nanoscale confinement of photoprocesses Ëx: Nano scale optical memory, nanoscale conformational dynamics
3. Nanoscale confinement of matter: Ex: nano particles, nano composites

Photonic crystals and photonic band gap materials have a myriad of uses which range from computing to telecommunication and from lasers to circuits that can also be found in almost anything digital. Light-emitting diodes play a key role in optical-communication systems. These devices are made from so-called photo emissive materials that emit photons once they have been excited electrically or optically. These photons are typically emitted in many different directions and have a range of wavelengths, which is not ideal for communications applications. A type of light emitting diode can be created that only emits light in the forward direction by placing a reflector behind the photo emissive layer. However, the efficiency of such a device is limited by the efficiency of the reflector.

References

1. Zhou, M. and Ghosh, I. (2007). Quantum Dots and Peptides: A Bright Future Together. Biopolymers, 88, 325-39.
2. De Mel A., Oh J. T., Ramesh B. and Seifalian A. M. (2012). Biofunctionalized, Quantum Dots For Live Monitoring of Stem Cells: Applications in Regenerative, Medicine. Regen Med, 7, 335-47.
3. Dubertret B., Calame M. and Libchaber A. J. (2001). Single-Mismatch Detection Using Gold-Quenched Fluorescent Oligonucleotides. Nat Biotechnol, 19, 365-70.
4. Bentolila L.A., Michalet X., Pinaud F.F., Tsay J.M., Doose S., Li J.J., Sundaresan G., Wu A.M., Gambhir S.S., and Weiss S. (2005). Quantum Dots for Molecular Imaging and Cancer Medicine. Discov Med, 5, 213- 8.

5. Green M. and Howman E. (2005). Semiconductor Quantum Dots and Free Radical Induced DNA Nicking. Chem Commun (Camb), 121-3.
6. Jamieson T., Bakhshi R., Petrova D., Pocock R., Imani M. and Seifalian A. M. (2007). Biological Applications of Quantum Dots. Biomaterials, 28, 4717-32.
7. Kidane A. G., Burriesci G., Edirisinghe M., Ghanbari H., Bonhoeffer P. and Seifalian A. M. (2009). A Novel Nanocomposite Polymer for Development of Synthetic Heart Valve Leaflets. Acta Biomater, 5, 2409-17.
8. Arya H., Kaul Z., Wadhwa R., Taira K., Hirano T. and Kaul S. C. (2005). Quantum Dots in Bio-Imaging: Revolution by the Small. Biochem Biophys Res Commun, 329, 1173-7.
9. Alivisatos P. (2004). The Use of Nanocrystals in Biological Detection. Nature Biotechnology, 22, 7-52.
10. Brus Louis (1986). Electronic Wave Functions in Semiconductor Clusters: Experiment and Theory. The Journal of Physical Chemistry 90: 2555-2560.
11. Juzenas P., Chen W., Sun Y. P., Coelho M. A., Generalov R., Generalova N. and Christensen I. L. (2008). Quantum Dots and Nanoparticles for Photodynamic and Radiation Therapies of Cancer. Adv Drug Deliv Rev, 60, 1600-14.
12. Smith A. M., Duan H., Mohs A. M. and Nie S. (2008a). Bioconjugated Quantum Dots for *in vivo* Molecular and Cellular Imaging. Adv Drug Deliv Rev, 60, 1226- 40.
13. Reiss Peter., Myriam Protière, and Liang Li. (2009). Core/Shell Semiconductor Nanocrystals, Small 5: 154-168.
14. Rogach Andrey L. (2008). Semiconductor Nanocrystal Quantum Dots: Synthesis, Assembly, Spectroscopy and Applications. Wien: Springer.
15. Drbohlavova J, Adam V, Kizek R, and Hubalek J (2009). Quantum dots-characterization, preparation and usage in biological systems. Int J Mol Sci 10(2):656–673.
16. Murray, C. B., C. R. Kagan, and M. G. Bawendi. (2000). Synthesis and Characterization of Monodisperse Nanocrystals and Close-Packed Nanocrystal Assemblies. Annual Review of Materials Science 30: 545-610.
17. Yu William W. and Xiaogang Peng (2002). Formation of High-Quality Cds and Other II-VI Semiconductor Nanocrystals in Noncoordinating Solvents: Tunable Reactivity of Monomers. Angewandte Chemie (International Ed. In English) 41: 2368-2371.
18. Chomoucka J., et al. (2009). Synthesis of Glutathione-coated Quantum Dots. 2009 32nd International Spring Seminar on Electronics Technology. New York: IEEE, pp. 653-657.
19. Wang H.Q., et al. (2008). Solubilization and bioconjugation of QDs and their application in cell imaging, Journal of Biomedical Materials Research Part A, 86A, 3, 833-841.
20. Koole R., et al. (2008). On the incorporation mechanism of hydrophobic quantum dots in silica spheres by a reverse microemulsion method, Chemistry of Materials, 20, 7, 2503-2512.
21. Cdse/Zns Core Shell Evidots in Toluene. Evident Technologies.Web
22. Grabolle Markus, Jan Ziegler, Alexei Merkulov, Thomas Nann, and Ute Resch-Genger (2008). Stability And Fluorescence Quantum Yield of Cdse-Zns Quantum Dots-Influence of the Thickness of the Zns Shell. Annals of the New York Academy of Sciences 1130: 235-241.

23. R. Ober, K. F. Jensen, and M. G. Bawendi (1997). (Cdse) Zns Core–Shell Quantum Dots: Synthesis and Characterization of a Size Series of Highly Luminescent Nanocrystallites. The Journal of Physical Chemistry B 101: 9463-9475.
24. Hines Margaret A. and Philippe Guyot-Sionnest (1996). Synthesis and Characterization of Strongly Luminescing Zns-Capped Cdse Nanocrystals. The Journal of Physical Chemistry 100: 468-471.
25. Quantum Light Optic. QD Vision Inc. - Quantum Dot Technology for Lighting and Displays. QD Vision Inc. Web.
26. Coe-Sullivan, Seth. Hybrid Organic/Quantum Dot Thin Film Structures and Devices. Phd Thesis, Massachusetts Institute of Technology, 2005.
27. Kamat Prashant V (2008). Quantum Dot Solar Cells. Semiconductor Nanocrystals as Light Harvesters. The Journal of Physical Chemistry C 112: 18737-18753.
28. Savage Richard N., Hans Mayer, Matthew Lewis, and Dan M. Marrujo (2008). Utilizing Quantum Dots to Enhance Solar Spectrum Conversion Efficiencies for Photovoltaics. MRS Online Proceedings Library 1120.
29. Klarreich, E. (2001). Biologists Join the Dots. Nature, 413, 450-2.
30. Rizvi S. B., Ghaderi S., Keshtgar M. and Seifalian A. M. (2010). Semiconductor Quantum Dots as Fluorescent Probes for in vitro and in vivo Bio Molecular and Cellular Imaging. Nano Rev, 1, 5161-76.
31. Gao J. H and XU B (2009). Applications of nano materials inside cells, Nano Today, 4, 1, 37-51.
32. Smith A. M., Duan H., Rhyner M. N., Ruan G. and Nie S. (2006). A Systematic Examination of Surface Coatings on the Optical and Chemical Properties of Semiconductor Quantum Dots. Phys Chem Chem Phys, 8, 3895-903.
33. Jin T., Yoshioka Y., Fujii F., Komai Y., Seki J. and Seiyama A. (2008). Gd3+-Functionalized Near-Infrared Quantum Dots for in vivo Dual Modal (Fluorescence/Magnetic Resonance) Imaging. Chem Commun (Camb), 5764-6.
34. Bharali D. J., Klejbor I., Stachowiak E. K., Dutta P., Roy I., Kaur N., Bergey E. J., Prasad P. N. and Stachowiak M. K. (2005). Organically Modified Silica Nanoparticles: A Nonviral Vector for *in vivo* Gene Delivery and Expression in the Brain. Proc Natl Acad Sci U S A, 102, 11539-44.
35. Maysinger D., Lovric J., Eisenberg A. and Savic R. (2007). Fate of Micelles and Quantum Dots in Cells. Eur J Pharm Biopharm, 65, 270-81.
36. Schroeder J. E., Shweky I., Shmeeda H., Banin U. and Gabizon A. (2007). Folate-Mediated Tumor Cell Uptake of Quantum Dots Entrapped in Lipid Nanoparticles. J Control Release, 124, 28-34.
37. Wang X. S., Dykstra T. E., Salvador M. R., Manners I., Scholes G. D. and Winnik M. A. (2004). Surface Passivation of Luminescent Colloidal Quantum Dots with Poly (Dimethylaminoethyl Methacrylate) Through a Ligand Exchange Process. J Am Chem Soc, 126.
38. Nann T. (2005). Phase-Transfer of Cdse@Zns Quantum Dots using Amphiphilic Hyperbranched Polyethylenimine. Chem Communication, 1735-1736.
39. Pinaud F., Michalet X., Bentolila L.A., Tsay J.M., Doose S., Li J.J., Iyer G. and Weiss S. (2006). Advances in Fluorescence Imaging With Quantum Dot Bio-Probes. Biomaterials, 27, 1679-87.
40. Alivisatos A. P., Gu W. and Larabell C. (2005). Quantum Dots as Cellular Probes. Annu Re Biomed Eng, 7, 55-76.

41. Alivisatos P. (2004). The Use of Nanocrystals in Biological Detection. Nature Biotechnology, 22, 47-52.

42. Benoit Dubertret P. S., David J. Norris, Vincent Noireaux, Ali H. Brivanlou, and Albert Libchaber (2002). *In vivo* Imaging of Quantum Dots Encapsulated in Phospholipid Micelles. Science 298, 1759-1762.

43. Kim S., Lim Y.T., Soltesz E.G., De Grand A. M., Lee J., Nakayama A., Parker J. A., Mihaljevic T., Laurence R. G., Dor D. M., Cohn L.H., Bawendi M.G. and Frangioni J.V. (2004). Near-Infrared Fluorescent Type II Quantum Dots for Sentinel Lymph Node Mapping. Nat Biotechnol, 22, 93-7. Dots Encapsulated in Phospholipid Micelles. Sciencemeg, 298.

44. Xing Y., Xia Z. and Rao J. (2009). Semiconductor Quantum Dots for Biosensing and in Vivo Imaging. IEEE Trans Nanobioscience, 8, 4-12.

45. Lakowicz J.R., Gryczynski Gryczynski Z., Nowaczyk K. and Murphy C.J. (2000). Time-Resolved Spectral observations of Cadmium- Enriched Cadmium Sulfide Nanoparticles and the Effects Of DNA Oligomer Binding. Anal Biochem, 280, 128-36.

46. Clapp A.R., Goldman E.R. and Mattoussi H. (2006). Capping of Cdse-Zns Quantum Dots with DHLA and Subsequent Conjugation with Proteins. Nat Protoc, 1, 1258-66.

47. Weng J. and Ren J. (2006). Luminescent Quantum Dots: A Very Attractive and Promising Tool in Biomedicine. Curr Med Chem, 13, 897-909.

48. Sinha R., Kim G. J., Nie S. and Shin D.M. (2006). Nanotechnology in Cancer Therapeutics: Bioconjugated Nanoparticles for Drug Delivery. Mol Cancer Ther, 5, 1909-17.

49. Cubukcu, Aydin, Foteinopolou, Soukolis, Ozbay et al (2003). Physical Review Letters, 2140.

50. Miao X.; Wilson B.K.; Pun S.H. and Lin L.Y. (2008). Optical manipulation of micron/submicron sized particles and biomolecules through plasmonics. Opt. Express, 16(18), 13517-13525, ISSN 1094-4087.

51. E. Yablonovitch and T. J. Gmitter (1989). Photonic band structure: The face-centered-cubic case, Phys. Rev. Lett., vol. 63, 1950.

52. S. Imagawa, K. Edagawa, K. Morita, T. Niino, Y. Kagawa, and M. Notomi (2010). Photonic band-gap formation, light di usion, and localization in photonic amorphous diamond structures, Phys. Rev. B, vol. 82, 115116.

53. P. N. Dyachenko, N. D. Kundikova, and Y. V. Miklyaev (2009). Band structure of a photonic crystal with the clathrate Si-34 lattice, Phys. Rev. B, vol. 79, 233102.

54. A. Hatef and M. R. Singh (2010). The study of quantum interference in metallic photonic crystals doped with four-level quantum dots, Nanoscale Res. Lett., vol. 5, 464.

55. J.E.G.J. Wijnhoven, L. Bechger, and W. L. Vos (2001). Fabrication and characterization of large macroporous photonic crystals in titania, Chem. Mater., vol. 13, 4486.

56. S. Noda, N. Yamamoto, H. Kobayashi, M. Okano, and K. Tomoda (1999). Optical properties of three-dimensional photonic crystals based on III-V semiconductors at infrared to near-infrared wavelengths, Appl. Phys. Lett., vol. 75, 905.

57. Rayleigh J.W.S. (1888). On the remarkable phenomenon of crystalline reflexion described by Prof. Stokes. Phil. Mag., 26, 256–265, ISSN 1478-6443.

12

Thin Films

12.1 Introduction

What is a Thin film?

A thin film is a layer of material ranging from fractions of a nanometer (monolayer) to several micrometers in thickness. Thin films are created through a process called "deposition." Deposition is a thin film coating process, which is achieved by modifying the four states of matter, solid, liquid, vapor and plasma. This can be accomplished through a variety of vacuum processes. Conventional categories of vacuum deposition processes include chemical vapor deposition (CVD), physical vapor deposition (PVD), evaporation through vacuum sublimation, or occasionally, some combination of these methods.

In the field of nanotechnology-based thin films and coatings, new approaches using nanoscale effects can be used to design, create or model nano coating systems with significantly optimized or enhanced properties of high interest to electronic, food, health and biomedical industry.

Polymers in Thin films

In many cases the films can be of complex composition with different types of polymers with complex architecture and other components such as nanoparticles. Polymers in thin films and nano composite structures can exhibit unusual physical properties due to the geometric constraints imposed by the presence of surfaces and interfaces. Furthermore, polymers and nanoparticles can self-assemble into a variety of nanostructures and nano patterns at interfaces offering alternative ways of facile and inexpensive nanofabrication routes which have great potential for many applications ranging from micro/nano electronics to biomedical implants.

Technological properties of thin films

The performance of thin film optical coatings (e.g., antireflective, or AR, coatings) is typically enhanced when the thin film coating consists of multiple layers having varying thicknesses and refractive indices. Similarly, a periodic structure of alternating thin films of different materials may collectively form a super lattice, which exploits the phenomenon of quantum confinement by restricting electronic phenomena to two dimensions. Thin film materials are the key elements of continued technological advances made in the fields of optoelectronic, photonic, and magnetic devices. The processing of materials into thin films allows easy integration into various types of devices. The properties of material significantly differ when analysed in the form of thin films. Most of the functional materials are rather applied in thin film form due to their specific electrical, magnetic, optical properties or wear resistance. Thin film technologies make use of the fact that the properties can particularly be controlled by the thickness parameter.

12.2 Applications of Thin Films

Thin films find extensive applications in wireless communications, telecommunications, integrated circuits, rectifiers, transistors, solar cells, light-emitting diodes, photoconductors, light crystal displays, magneto-optic memories, audio and video systems, compact discs, electro-optic coatings, memories, multilayer capacitors, flat-panel displays, smart windows, computer chips, magneto-optic discs, lithography, micro electromechanical systems (MEMS), and multifunctional emerging coatings, as well as other emerging cutting technologies. Other main stream applications include optical coatings, photovoltaic solar cells, and thin film batteries. Thin polymer films have numerous technological applications in various industrial and biomedical sectors related to protective and functional coatings, non-fouling biosurfaces, biocompatibility of medical implants, separations, advanced membranes, microfluidics, sensors & devices, adhesion, lubrication and friction modification(1,2).

Some major areas of application of thin films are discussed in detail.

12.2.1 Optical Coatings

An optical coating is one or more thin layers of material deposited on an optical component such as a lens or mirror, which alters the way in which the optic reflects and transmits light. Optical thin film coatings improve the performance of optical systems and most modern optical systems could not function without them. Examples of these coatings are mirrors, anti-reflection coatings, beam splitters, and filters. They are used in a diverse range of applications including telecommunications, displays and lighting, architectural glass and automotive glass.

One type of optical coating is an antireflection coating, which reduces unwanted reflections from surfaces, and is commonly used on spectacle and photographic lenses. Another type is the high-reflector coating which can be used to produce mirrors which reflect greater than 99.99% of the light which falls on them. More complex optical coatings exhibit high reflection over some range of wavelengths, and anti-reflection over another range, allowing the production of dichroic thin-film optical filters.

12.2.2 Photovoltaic

Thin film solar cells are the new generation solar cells that contain multiple thin film layers of photo voltaic materials. The thin film solar cells (TFSC) are also known as Thin Film Photo Voltaic cell (TFPV). The thicknesses of thin film layers are very less as (few nano meters) compared to traditional P-N junction solar cells. According to the type of photo voltaic material used, the thin film solar cells are classified into four types. They are

1. Amorphous silicon (a-Si) and other thin-film silicon (TF-Si)
2. Cadmium Telluride (CdTe)

3. Copper indium gallium deselenide (CIS or CIGS)
4. Dye-sensitized solar cell (DSC) and other organic solar cells

12.2.2.1 Structure of Thin Film Solar Cell

The structure of thin film solar cell (Figure 12.1) is almost same as that of normal silicon wafer cells. The only difference is in the thin flexible arrangement of the different layers and the basic solar substance used. The thin flexible arrangement of the layers helps to produce very thin form of cells (Figure 12.2) that is much more efficient than the conventional silicon wafer cells.

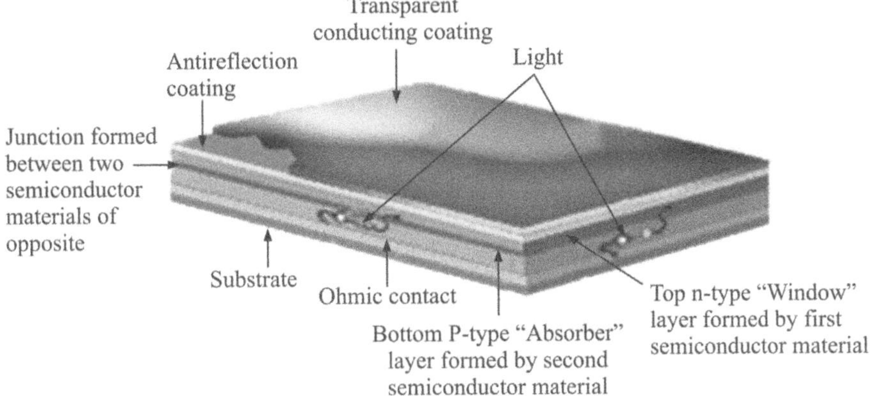

Figure 12.1 Structure of thin film solar cell

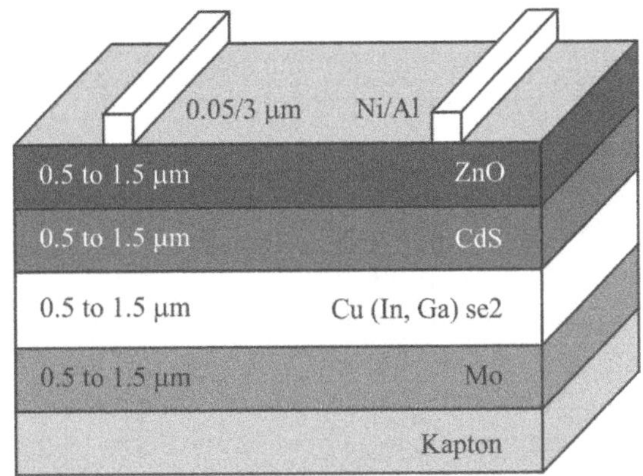

Figure 12.2 Layer arrangement of thin film solar cells

The basic substance of a photovoltaic cell is semiconductor which is doped with phosphorus to provide an excess of free electrons (usually called N type material) and a semiconductor doped with boron, gallium or indium develop a vacancy (called holes) and this doped materials are known as P type materials. These n type and p type materials combine (join) to form a Photo voltaic cell. During the absence of light, a very small amount of atoms are excited and move across the junction. This causes a small voltage drop across the junction. In the presence of light, more atoms are excited and flow through the junction and cause a large current at the output. This current can be stored in a rechargeable battery and used for several applications based on the requirement.

The old solar panel technology use silicon semiconductor for the production of p-type and n-type layers and has several disadvantages. But in the case of Thin Film Layer technology, the silicon semiconductor materials is replaced by either cadmium telluride (CdTe) or copper indium gallium deselenide (CIGS).

Advantages

1. Easy to handle
2. More flexible than conventional solar cells
3. Available as thin wafer sheets
4. Cheaper than traditional panels

Disadvantages

1. Less efficiency (20 to 30% of light converted into electricity)
2. Complex structure
3. Need to be very careful in handling
4. Can't be used in astronomical devices

12.2.3 Thin films in Semiconductors

Historically, the semiconductor industry has relied on flat, two-dimensional chips upon which to grow and etch the thin films of material that become electronic circuits for computers and other electronic devices. This thin layer (only a couple of hundred nanometers thick) can be transferred to glass, plastic or other flexible materials, opening a wide range of possibilities for flexible electronics. In addition, the semiconductor film can be flipped as it is transferred to its new substrate, making its other side available for more components. This doubles the possible number of devices that can be placed on the film. By repeating the process, layers of double-sided, thin-film semiconductors can be stacked together, creating powerful, low-power, three-dimensional electronic devices. It's important to note that these are single-crystal films of strained silicon or silicon germanium. The strain is introduced in the way we form the membrane. Introducing strain changes the arrangement of atoms in the crystal such that we can achieve much faster device speed while consuming less power.

For non-computer applications, flexible electronics are beginning to have significant impact. Solar cells, smart cards, radio frequency identification (RFID) tags, medical applications, and active-matrix flat panel displays can use thin film semiconductors for achieving high efficiency. The techniques could allow flexible semiconductors to be embedded in fabric to create wearable electronics or computer monitors that roll up like a window shade.

The ability to create fast, low-power, multilayer electronics has many exciting applications. Silicon germanium membranes are particularly interesting. Germanium has a much higher adsorption for light than silicon. By including the germanium without destroying the quality of the material, we can achieve devices with two to three orders of magnitude more sensitivity." That increased sensitivity could be applied to create superior low-light cameras, or smaller cameras with greater resolution.

12.2.4 Photo Electrochemical Cells (PEC)

In photo electrochemical experiments, irradiation of an electrode with light that is absorbed by the electrode material causes the production of a current (a photocurrent). The dependence of the photocurrent on wavelength, electrode potential, and solution composition provides information about the nature of the photoprocess, its energetics, and its kinetics. Photocurrents at electrodes can also arise because of photolytic processes occurring in the solution near the electrode surface. Photoelectrochemical studies are frequently carried out to obtain a better understanding of the nature of the electrode-solution interface. Photoelectrochemistry and Electrogenerated Chemiluminescence photocurrent can represent the conversion of light energy to electrical and chemical energy; such processes are also investigated for their potential practical applications.

12.2.5 Optoelectronics

An optoelectronic thin-film chip, comprising at least one radiation-emitting region in an active zone of a thin-film layer and a lens disposed downstream of the radiation-emitting region, said lens being formed by at least one partial region of the thin-film layer, the lateral extent of the lens being greater than the lateral extent of the radiation-emitting region. The thin-film layer is provided for example by a layer sequence which is deposited epitaxially on a growth substrate and from which the growth substrate is at least partly removed. That is to say that the thickness of the substrate is reduced. In other words, the substrate is thinned. It is furthermore possible for the entire growth substrate to be removed from the thin-film layer. The thin-film layer has at least one active zone suitable for generating electromagnetic radiation. The active zone may be provided for example by a layer or layer sequence which has a pn junction, a double heterostructure, a single quantum well structure or a multiple quantum well structure. Particularly preferably, the active zone has at least one radiation-emitting region. In this case, the radiation-emitting region is formed for example by a partial region of the active zone. Electromagnetic radiation is generated in said partial region of the active zone during operation of the optoelectronic thin-film chip.

12.2.6 Flat Panel Displays

The Flat Panel Display (FPD) fabrication environment is among the world's most competitive and technologically complex. Device designers and manufacturers continually strive to satisfy the worldwide consumer demand for larger displays, greater pixel resolution and feature-rich performance – all at a lower cost than the previous generation of technology. The need to control contamination in air, gas and liquid process streams is now a paramount focus of process engineers and designers.

12.2.7 Data Storage

As the data storage density in cutting edge microelectronic devices continues to increase, the super paramagnetic effect poses a problem for magnetic data storage media. One strategy for overcoming this obstacle is the use of thermo mechanical data storage technology. In this approach, data is written by a nanoscale mechanical probe as an indentation on a surface, read by a transducer built into the probe, and then erased by the application of heat. An example of such a device is the IBM millipede, which uses a polymer thin film as the data storage medium. It is also possible, however, to use other kinds of media for thermo mechanical data storage, Nanometer-scale indentations made in martensite phase Ni-Ti SMA thin films recover substantially upon heating. Issues such as repeated thermo mechanical cycling of indentations, indent proximity, and film thickness impact the practicability of this technique. While there are still problems to be solved, the experimental evidence and theoretical predictions show SMA thin films are an appropriate medium for thermo mechanical data storage.

12.2.8 Super Capacitor

Since first patents were in the 1950, the idea of storing charge in the electric double layer that founds at the interface between a solid and an electrolyte has presented itself as an attractive notation while was considerable development was carried out by the slander oil company, Cleveland, Ohio (SOHIO) in the 1960 a lack of sales lead them to license the technology to Nippon electric company. The first high power double layer capacitor were developed for military applications by the pinnacle research institute in 1982.

Conventional capacitor stores energy electrostatically on two electrode separated by a dielectric, the capacitance of which is C= €A/d, where € being the dielectric constant, A surface area and d the dielectric thickness. In a double layer capacitor, charges accumulate at the boundary between electrode and electrolyte to form two charge layers with separation of several Angstroms. Super capacitors are electrochemical energy storage device. The away that super capacitor store energy is based in principal on two types capacitive behaviour the electrical double layer capacitance from the pure electrostatic charge accumulation at the electrode interface and the pseudo-capacitance developed from fast and reversible surface redox process at characteristic potential. Super capacitor integrated in to hybrid vehicle, power source, power control, multi source supply, fuel save etc.

12.2.9 Gas Sensors

Advancements in micro technology and the evolution of new nonmaterial and devices have been playing a key role in the development of very accurate and reliable sensors. The technology of sensors has developed tremendously in the last few years owning many scientific achievements from various experiments, offering newer challenges and opportunity to the quest for every smaller devices capable of molecular level imaging and monitoring of pathological samples and the macromolecules has lately gained the focus of attention of the scientific community, particularly for remote monitoring due to the increasing need for environmental safety and health monitoring.

Gas sensors generally operate different principles and various gas sensing elements have been developed past years, out of which resistive metal oxide sensors comprise a significant part. However these sensing elements typically operate at an elevated temperature for maximum performance. This makes higher power consumption which is not suitable for the inflation. Over the last decayed carbon nanotube has attracted considerable attention, and great efforts have been made to exploit their unusual electronic and mechanical properties. These properties make them potential candidates for building blocks of active materials in nanoelectronics, field emission devices, gas storage and gas sensors. Among these the room temperature gas sensing property is very attractive for many applications.

12.3 Thin Film Deposition Processes

Thin film deposition is a process in which the physical characteristics of a surface are modified by applying a very thin coating, often just a few millionths of a millimeter thick. Thin films can be used to modify the optical characteristics of a surface (the substrate), its electrical conductivity, its hardness or lubricity, to provide corrosion resistance or chemical inertness, or simply for decoration. Every-day examples of thin films include the reflective coating on CDs and DVDs, anti-reflective coatings on spectacles and thermally efficient double glazing (optical films), the tiny electrical pathways within a micro-processor (electrical films), hardened machine tool bits (hardness), to food and drug packaging (decoration and passivation).

In order to produce such thin films, the deposit must be applied in a very controlled manner, sometimes even atomic monolayer by atomic monolayer, under extremely clean, particulate free conditions. Individual depositing particles must be able to travel from some deposition source to the substrate to be coated with little or no collision with gas molecules on the way (a long 'mean free path') - thin film deposition is a high vacuum process. Thin film deposition processes broadly fall into two main categories viz; Physical Vapor Deposition (PVD) and Chemical Vapor Deposition (CVD) methods (3-5).

The widely practiced PVD and CVD methods which are capable of producing thin films, i.e., 1 μm or less and thick films, i.e., 1 μm or more are summarized in Table 12.1.

Table 12.1 Classification of thin film deposition techniques (6-8)

Thin Film Deposition Techniques			
Physical		Chemical	
Sputtering	**Evaporation**	**Gas Phase**	**Liquid Phase**
Glow discharge DC sputtering	Vacuum Evaporation	Chemical vapour Deposition	Electro-deposition
Triode sputtering	Resistive heating Evaporation	Laser Chemical vapour deposition	Chemical bath deposition (CBD)/Arrested Precipitation Technique (APT)
Getter sputtering	Flash Evaporation	Photo-chemical vapour deposition	Electro less deposition

Table 12.1 ***Contd...***

Sputtering	Evaporation	Gas Phase	Liquid Phase
Radio Frequency sputtering	Electron beam Evaporation	Plasma	Anodisation
			Liquid phase Epitaxy
Magnetron sputtering	Laser Evaporation	Metal-Organo Chemical vapour Deposition (MO-CVD)	Sol-gel Spin Coating
Ion Beam Sputtering			Spray-pyrolysis technique (SPT)
A.C. Sputtering	Arc		Ultrasonic (SPT)
	R.F Heating		Polymer assisted deposition (PAD)

12.3.1 Physical Vapour Deposition (PVD)

The advantages of physical methods are due to dry processing, high purity and cleanliness, compatibility with semiconductor integrated circuit processing and epitaxial film growth can be achieved. However, there are certain disadvantages such as slow deposition rates, difficult stoichiometry control, high temperature post deposition annealing often required for crystallization and high capital expenditure.

In general, there are three steps in any physical vacuum deposition (PVD) process:

(i) creation of an evaporant from the source material,

(ii) transport of the evaporant from the source to the substrate (item to be coated), and

(iii) condensation of the evaporant on to the substrate to form the thin film deposit.

There are two reasons why this process is best conducted under vacuum:

1. The process of evaporation involves significant amounts of heat, if oxygen were present, any reactive metal would form oxides;
2. Collisions with gas molecules during the transport of evaporant from source to substrate would impact net deposition rate significantly, and could also prevent growth of dense films.

12.3.1.1 Film

Film deposition and growth: Upon arrival at the substrate, evaporated material condenses on the substrate in a complex sequence of events that determine many of the physical properties of the deposited film.

The steps in the growth of thin films are generally referred to as nucleation and growth.

In nucleation, the atoms and molecules which are arriving (called ad atoms) at the surface lose thermal energy of the surface, and the surface absorbs that energy. Depending on the amount of thermal energy of the ad atoms and of the surface have, the ad atoms move about on the surface until they lose the thermal energy required to move about the surface (referred to as Adam mobility). As nuclei continue to form, the film grows into a continuous sheet covering the substrate. Chemical interactions between the ad atoms and the surface determine the strength of the bond between the film and substrate. Gold, for example, does not form a chemical bond with silicon dioxide, and therefore, the adherence of gold films on glass are very weak.

Improvement of this adhesion may be made by first depositing a thin (500Å thick) "Binder" layer of chromium or niobium, then depositing the gold over the binder layer. Chromium and niobium do form chemical bonds with the silicon dioxide in glass, and also form metallic bonds with the following gold layer. Once a few monolayers of evaporant have condensed on the substrate, the film continues to grow in thickness as if the entire substrate were made of the material being deposited. During film growth the microstructure of the deposited film will be developed. This microstructure may be described in terms of grain size, orientation, porosity, impurity content, and entrained gases.

12.3.1.2 Desired Properties of Thin Films

Normally, vacuum deposition processes are selected over other processes (electrochemical deposition, flame spraying, etc.) to achieve the following desirable properties (6):

1. High chemical purity.
2. Good adhesion between the thin film and substrate.
3. Control over mechanical stress in the film.
4. Deposition of very thin layers, and multiple layers of different materials.
5. Low gas entrapment.

12.3.1.2.1 Control Parameters for Thin Film Characteristics

For each of the vacuum deposition process the ultimate goal is to provide a means for depositing a thin film having the required physical and chemical properties. The parameters one can control to achieve the specified goals are:

1. Kinetic energy of the ad atoms.
2. Substrate temperature.
3. Deposition rate of the thin film.
4. Augmented energy applied to the film during growth.
5. Gas scattering during transport of the evaporant.

By varying these parameters one can generate thin films of a given material that have different mechanical strength, adhesion, optical reflectivity, electrical resistivity, magnetic properties and density.

12.3.1.3 Evaporation

Evaporation or sublimation techniques are widely used for the preparation of thin layers. A very large number of materials can be evaporated and, if the evaporation is undertaken in vacuum system, the evaporation temperature will be very considerably reduced, the amount of impurities in the growing layer will be minimized.

In order to evaporate materials in a vacuum, a vapour source is required that will support the evaporant and supply the heat of vaporisation while allowing the charge of evaporant to reach a temperature sufficiently high to produce the desired vapour pressure, and hence rate of evaporation, without reacting chemically with the evaporant. To avoid contamination of the evaporant and hence of growing film, the support material itself must have a negligible vapour pressure and dissociation temperature of the operating temperature. Laser beam evaporation has also come in to use recently. The laser source is situated outside the evaporation system and the beam penetrates through a window and is focused on to the evaporate material, which is usually fine powder form.

12.3.1.3.1 Thermally Induced Evaporation

In this process, heat is input into the source material (often called the charge) to create a plume of vapor which travels in straight-line paths to the substrate. Upon arrival at the substrate, the atoms, molecules, and clusters of molecules condense from the vapor phase to form a solid film. The heat of condensation is absorbed by the substrate. On a microscopic scale the localized heating from this process can be enormous. It is common, in the development of metal coating techniques for thin cross-section plastic parts, to melt substrates during the initial deposition runs.

With experience, one can select source-to-substrate distances and deposition rates which will allow coating of temperature sensitive substrates without melting.

There are several methods by which heat can be delivered to the charge to cause vaporization:

(a) Electric resistance heating

(b) Induction heating

(c) Electron beam heating

(d) Laser ablation

(e) Cold & hot cathodic arc

(f) Sputtering

(g) Ion platting

12.3.1.3.1.1 Electrical Resistance Heating

Evaporation of material by electrical resistance evaporation is very likely the easiest of the thermal evaporation techniques. Quite simply, in a vacuum environment the charge (which may be an elemental metal, an alloy, a mixture or a compound) is heated to become a vapor. Low voltage, high current (typically 10 to 40 VDC, 1 to 10 amps) power is brought into the vacuum vessel using electrical power feed throughs. The electrical power is passed through a filament which is in intimate contact with the charge. Filaments are often heated to 1000 to 2000 °C (Table 12.2). A materials requirement for efficient thermal evaporation is that the charge have an appreciable vapor pressure at the operating temperature of the filament.

Table 12.2 Temperatures for efficient evaporation of selected metals.

Material	Evaporation temperature, °C	Comments
Zinc	325	High vapor pressure at RT
Aluminum	1390	
Copper	1516	
Chromium	1612	
Lead	1680	Toxic
Iron	1829	
Nickel	1848	

All materials evaporate, even at room temperature. The addition of heat simply accelerates the process. At a specified temperature the pressure of the vapor emitted by a material is called the equilibrium vapor pressure.

Thermal evaporation is typically conducted under high to ultra-high vacuum conditions. Bell jar type vacuum vessels are often selected to perform this process on a laboratory scale, as they offer a great deal of design flexibility as well as the advantage of direct observation of the process at a very reasonable cost. Filaments are usually made of refractory metals such as Tungsten, Tantalum, or their alloys. Some of the requirements for a good filament material are:

1. High melting point.
2. Low solubility for the charge materials.
3. Filament should be wettable by the charge materials.
4. Filament should withstand thermal shocks well.

There exists a great variety of filament configurations ranging from straight and coiled wires to "boats" and boxes (Figure 12.3). Each type of filament is designed for a unique application. Coil filaments made of refractory metal strands are loaded with the charge by hanging small sections of wire made of the charge material on the coil. Upon heating, the charge melts, and wets the coil. Further heating causes the evaporation of the charge from the coil filament. Very rapid heating of a filament with hanging charge material may cause the charge to melt locally and fall off the filament.

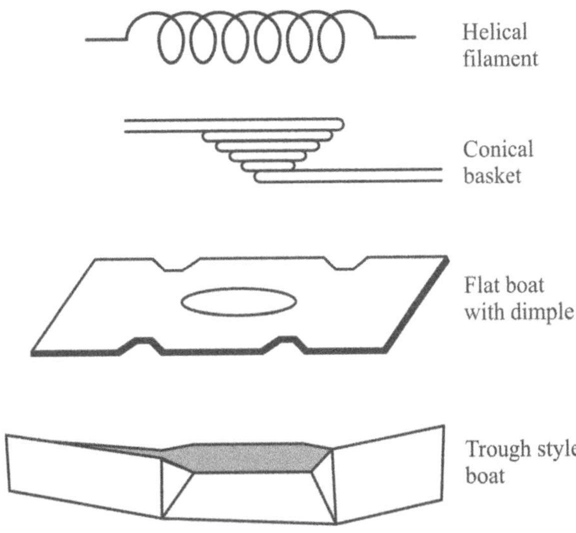

Figure 12.3 Configurations for filaments used in electrical resistance heated thermal evaporation

12.3.1.3.1.2 *Molecular Beam Epitaxy (MBE)*

Molecular beam epitaxy (MBE) is a modern application of electric resistance heated thermal evaporation. This technique is used to accurately deposit ultra-high purity semiconductor materials with specified crystallographic orientations between the layers of different materials. The vacuum environment for MBE is typically in the extreme ultrahigh vacuum range (10-9 to 10-11 Torr). Evaporation of materials in an MBE vacuum system is conducted using a special thermal evaporation source called a Knudsen cell.

There are some inherent disadvantages of resistance heated thermal evaporation that should be kept in mind when selecting a deposition technique:

1. The source may generate impurities which may co-deposit in the condensing thin film.
2. Accurate control of the deposition rate is difficult.
3. The composition of alloy thin films deposited may differ from that of the charge material (especially if the elements in the alloy have markedly different vapor pressures).
4. The amount of material which may be evaporated per run is limited.
5. The substrate will experience heating due to radiant energy from the source.

12.3.1.3.1.3 *Induction Heated Thermal Evaporation*

In this technique an electric current is induced to flow through an electrically conductive charge material by the application of radio-frequency (RF) alternating current. The RF current is generated by a power supply which may range in output from 1 to 50 kilowatts, depending on the size of the charge. The AC current is flowed through the copper coil which surrounds a refractory ceramic crucible Figure 12.4.

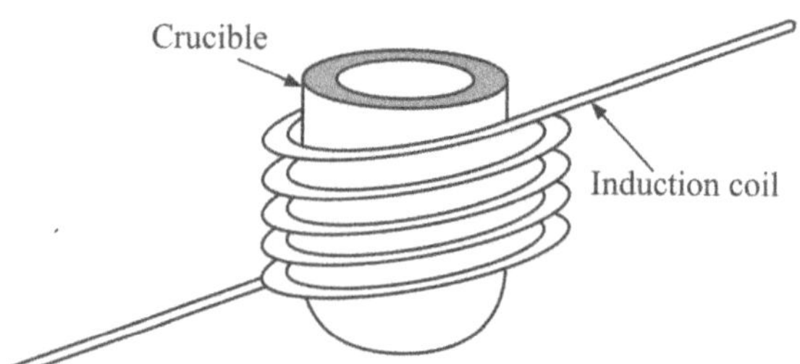

Figure 12.4 A crucible and coil used for induction heating for thermal evaporation

For all forms of thermal evaporation which employ ceramic crucibles, selection of the appropriate material for the crucible is vitally important. If the incorrect selection is made, the charge may chemically react with the crucible, ruining both and possibly harming other components of the vacuum deposition system.

Advantages of induction-heated thermal evaporation as compared to electrical resistance evaporation include:

1. Low contamination of the deposited thin films.
2. Improved control of deposition rate.
3. Larger charges can be loaded per deposition run.

Disadvantages of induction heated thermal evaporation include:

1. The charge must be electrically conductive.
2. RF power supplies and matching networks can be expensive and quite large.
3. Chemical interaction between the charge and crucible can occur.

12.3.1.3.1.4 Electron Beam Evaporation

In this thermal evaporation process, a beam of energetic electrons generated from a heated filament supplies the thermal energy to evaporate of the charge. There are several variants of the electron beam evaporation process. All electron beam thermal evaporation systems have an anode (biased positively) and a cathode (either grounded, or biased negatively with respect to the anode). There are two types beam evaporators

(i) self- accelerated and

(ii) work-accelerated electron beam evaporators.

In the work-accelerated type electrons emitted from a heated tungsten filament are attracted to the charge material by an applied high voltage bias (10- 40 kV DC). A focusing aperture aids in minimizing spurious heating of the vessel interior by off-axis electrons (Figure 12.5).

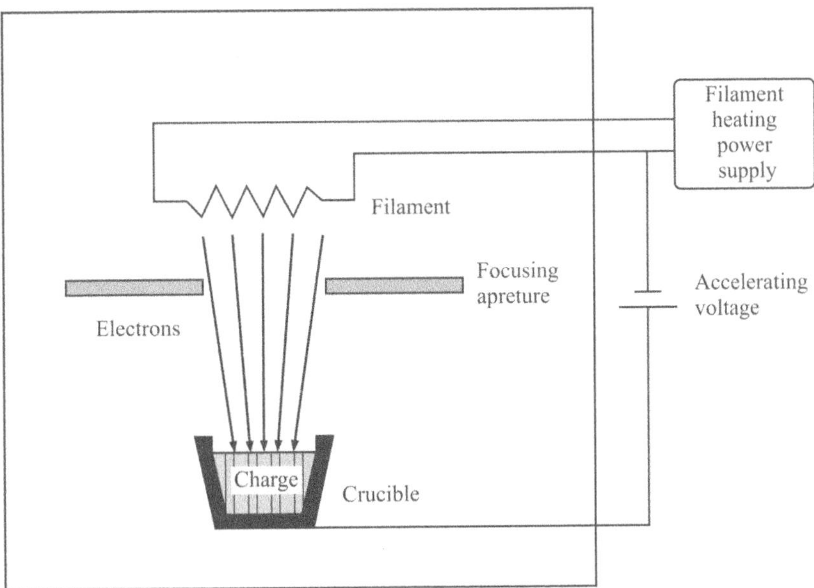

Figure 12.5 A work-accelerated electron beam evaporator

Several configurations of the work-accelerated electron beam evaporator have been designed and used (7,8). In Figure 12.6 electrons emitted from the heated cathode are caused to travel in an oval shaped path by the electrostatic negative field applied to the cylindrical focusing electrode. Water cooling is provided via concentric tubes connected thermally to the bottom of the charge. This design has a distinct advantage over that shown in Figure 12.5: the evaporant may be directed at a substrate placed above the source without interference by a focusing aperture or filament. Additionally, the focusing aperture and filament do not become heavily overcoated with evaporant.

Some work-accelerated electron beam evaporators use electromagnetic coils to steer and focus the electron beam as shown in Figure 12.7. This system provides several advantages: source utilization can be controlled and maximized by rastering the electron beam; and the effective source size can be made large by rastering the beam, improving the thickness uniformity and coverage of the substrate. Note that in this design the electrons emitted from the filament impact the backside of the cathode, heating it so that it will in turn emit electrons. The cathode area which emits electrons is hemispherical, which improves the three-dimensional uniformity of the electron beam emitted from it.

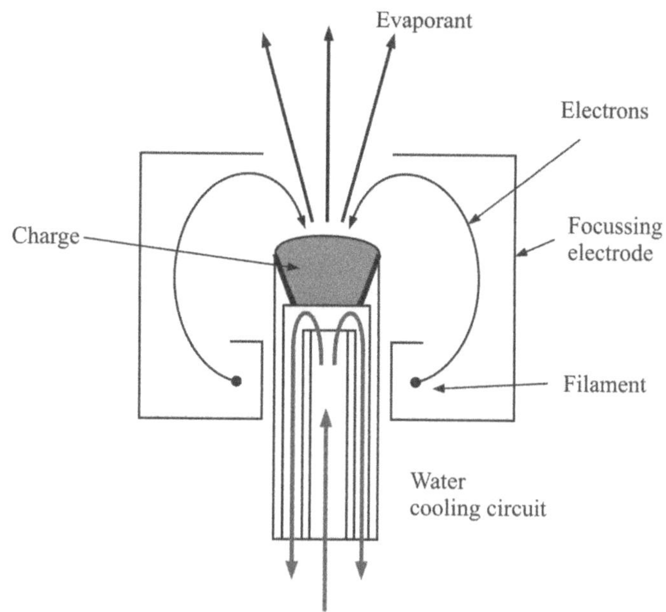

Figure 12.6 Work-accelerated electron beam evaporator with a cylindrical electron focusing electrode

Self-accelerated electron beam evaporators use an auxiliary anode to impart kinetic energy to the electrons emitted by the filament (cathode). This allows the evaporation of charge materials which are not electrically conductive, such as calcium fluoride, which is used to coat camera lenses for improved optical performance. A self-accelerated electron beam gun configuration is presented in Figure 12.8.

As was the case with the work accelerated electron beam evaporator in Figure 12.7 the self-accelerated gun has a set of electromagnets which may be used to scan the electron beam during evaporation. Automated scan controls for both types of guns are commercially available. These scan controls vary the current in the electromagnets so as to sweep the beam in a "Lissajous" pattern, which is sinusoidal in two dimensions.

The majority of commercial electron beam evaporators are of the transverse design, as shown in Figure 12.9. These guns use a permanent magnet to steer the electrons emitted from the cathode around 270°, and a set of electromagnets to raster the beam on the charge material. Modern electron beam guns of the transverse design are available with a rotating multiple pocket hearth, which allows deposition of up to five different materials without venting the vacuum vessel. Some of these units also have integral shutters which allow the gun to achieve a stable operating temperature before deposition of the substrate is begun.

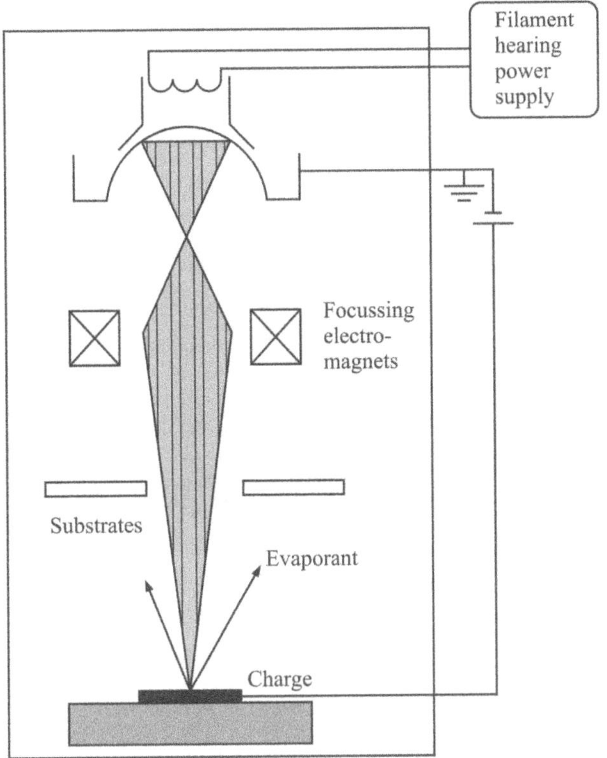

Figure 12.7 Work-accelerated electron beam evaporator electromagnets to focus and steer the electron beam on the charge material

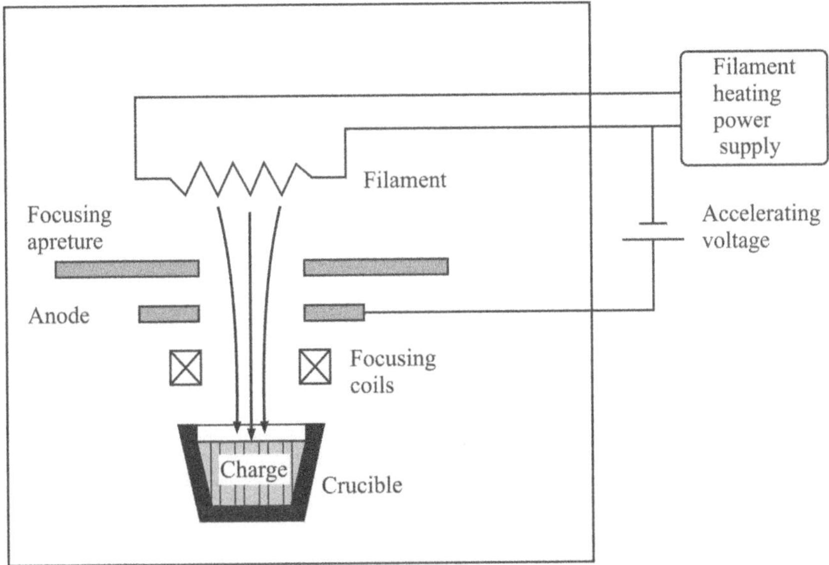

Figure 12.8 Self-accelerated electron beam evaporation

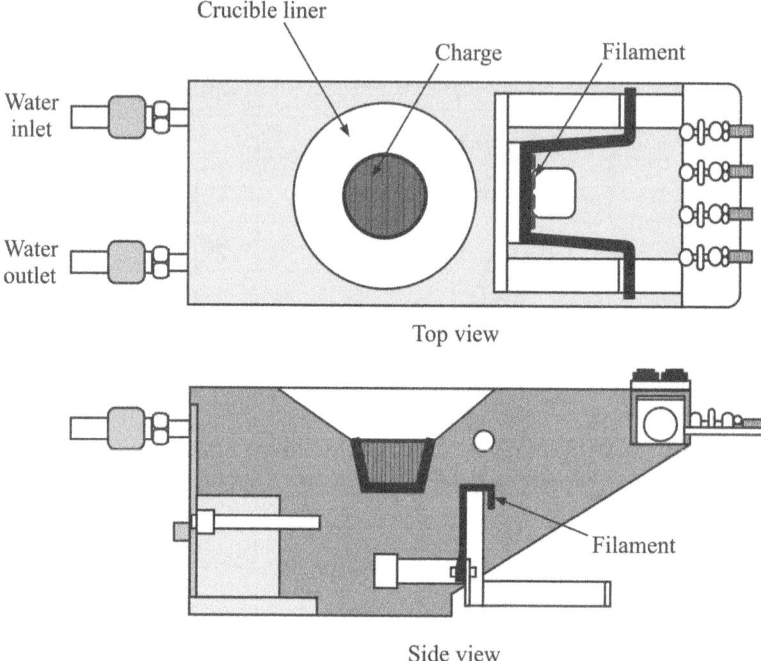

Figure 12.9 Transverse design electron beam evaporation gun

12.3.1.3.2 Laser Ablation

High energy density pulsed laser beams have been used to deposit thin films of a variety of elements, alloys and compounds. In this process, a laser source, external to the vacuum vessel generates a beam which is focused, passed through a viewport and impinges on a target within the vacuum vessel (Figure 12.10). Sufficient energy is generated to blast (ablate) material from the surface of the target. This ablated material consists of neutral atoms, ions, clusters of atoms and macro particles. The amount of material deposited per laser pulse is very consistent, allowing one to accurately deposit films of a specified thickness. The deposition rate is low compared to other techniques (electron beam evaporation and sputtering, for example). The range of commonly used operating parameters is given below:

laser power	100 to 500 mJ
spot size at target	0.1 to 0.25 cm^2
power density at target	400 to 5000 mJ/cm^2
pulse duration	5 to 50 ns
wavelength	250 to 308 nm
pulse frequency	1 to 20 Hz

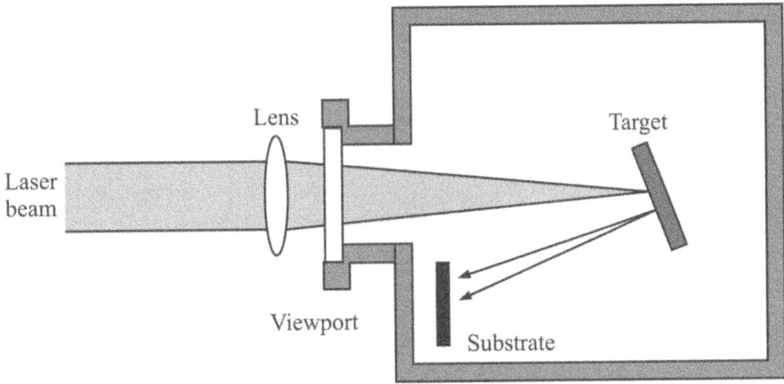

Figure 12.10 Deposition of a thin film by laser ablation

Laser ablation, as a deposition technique is currently limited to research and development laboratories due to the low deposition rate, the additional safety issues involved with the use of UV lasers and the expense of the equipment. Some of the thin films that have been deposited using laser ablation include super conducting thin films, ceramic coatings, and amorphous metallic layers.

12.3.1.3.3 Cold Cathode and Hot Cathode Arc as Plasma Electron Beam Deposition Sources

Instead of using a heated filament to generate the electrons for evaporation of a material, plasma guns utilize the electrons which exist in high temperature gases called plasmas. Two types of plasma sources exist: cold cathode and hot cathode.

In the cold cathode design (Figure 12.11), the cathode is biased negatively from –5 to –20 kV, the rest of the source is at ground potential. Following evacuation of the vacuum system to its base pressure, process gas, such as helium is injected into the system. A partial pressure from 1 to 100 mTorr is dynamically maintained using flow controllers at the process gas inlet, and adjustable apertures at the high vacuum stack inlet. Electrons emitted by the cathode can ionize process gas atoms, creating positive ions. These ions are accelerated to the cathode, and upon impact, generate a cascade of secondary electrons. Many of these secondary electrons escape the source and are accelerated and focused on the work piece as described in the section on work accelerated electron beam guns.

Following evacuation to the vessel's base pressure, a process gas, such as Argon is flowed in a controlled manner through the cylindrical hollow cathode. Radio frequency AC electric current is supplied to the cathode of the electron beam source from a low voltage, high current power supply. Ionization of the process gas occurs as a result of the applied electrical power. The cathode operating temperature is very high. For this reason, the cathode is made of a refractory metal, such as tungsten, and must be actively water cooled. Electrons emitted from the hot cathode and from the plasma may be steered and accelerated as described for other electron sources.

Very high deposition rates may be achieved using the hot hollow cathode electron beam source to perform thermal evaporation of materials. The electrons emitted from this source may be work or self-accelerated, and may be magnetically steered.

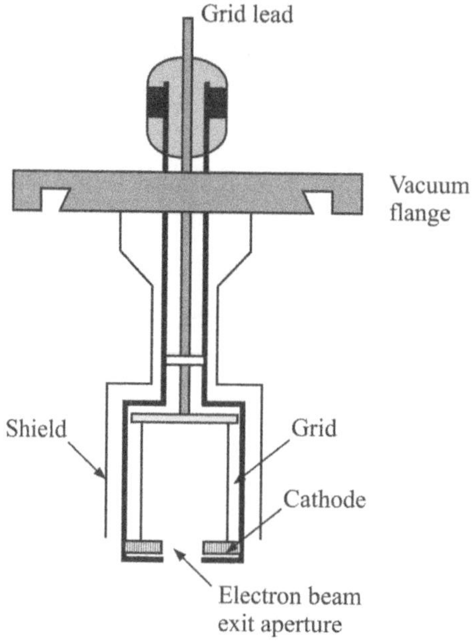

Figure 12.11 Cold cathode electron beam source

12.3.1.3.3.1 Safety Issues for Electron Beam Evaporation Systems

There are several potential dangers associated with the use of high energy electron beams that operators should be aware of:

1. *Radiation:* A significant amount of X-rays are generated whenever high energy electrons impact materials. The maximum energy of the X-rays is limited by the electrical potential applied to accelerate the electrons. A radiation survey of the exterior of the vacuum vessel, especially viewports, should be conducted to insure a safe working environment.
2. If the beam control systems fail, high energy electrons may be directed to the vacuum vessel walls or to internal fixturing, including water lines. The electron beam sweep pattern should be monitored carefully at the beginning of a deposition run, and periodically during the run. Leaving an electron beam system to run unattended is not recommended.
3. High voltage and high current are used in the process. The vacuum vessel and power supplies should be electrically grounded.

12.3.1.4 Sputter Deposition of Thin Films

If a surface of target material is bombarded with energetic particles, it is possible to cause ejection of the surface atom: this is the process known as sputtering. The ejected atoms can be condensed on to a substrate to form a thin film. This method has various advantages over normal evaporation techniques in which no container contamination will occur. It is also possible to deposit alloy films which retain the composition of the parent target material. DC sputtering, radio frequency sputtering and magnetron sputtering methods are the oldest types of sputtering used (5). High pressure oxygen sputtering and facing target sputtering are the two new methods introduced for deposition of thin films for applications in superconducting and magnetic films.

Unlike thermal evaporation, in which a material is heated to increase its equilibrium vapor pressure, in sputter deposition, material is made to go into the vapor phase by the physical interaction of particles impacting the source material (often referred to as the "target"). Though a number of sputtering techniques that are currently used to deposit thin films for use in magnetic storage media (cassette tapes, computer discs), optical thin films and microcircuits two main types are discussed in the following pages i.e., (a) diode sputtering: magnetron sputtering, RF sputtering, and ion beam sputtering.

12.3.1.4.1 Diode Sputtering

A simple diode sputtering process system is presented in Figure 12.12. Following evacuation of the vessel to its base pressure to reduce contamination of the film by atmospheric gases or water vapor, a process gas (most often Argon) is admitted into the vessel. The pressure of this process gas is maintained at a user-selectable pressure between 1-100 mTorr using a set of upstream mass flow controllers to regulate the process gas mix, as well as a downstream throttling gate valve. Such a valve, situated right at the turbpump, is equipped with a controller that can perform fast, real time pressure control using a capacitance manometer as a sensor element. An electric bias of from 500 to 5000 V DC is applied to the target. Electrons emitted by the target strike process gas molecules in the vicinity of the target, and may cause the gas to become ionized. The positive ions thus created are accelerated towards the cathode by the applied negative bias. When the positive ions collide with the cathode, the kinetic energy transferred is sufficient to eject atoms of the cathode material. Secondary electrons, ions, and light (IR, visible, UV and X-rays) are also emitted during this collision.

The ejected (sputtered) material travels towards surfaces in the vacuum vessel where it condenses to form films. Since the process gas pressure is in the range of from 1 to 100 mTorr, a significant amount of scattering of the sputtered material by the process gas occurs. In this scattering, sputtered material loses its directional identity, may become neutral, if it was ionized, and also loses kinetic energy. The visible glow that surrounds the cathode during sputtering is called the "glow discharge" or plasma. Visible light of the plasma has a color which is characteristic of the process gas and the material being sputtered. This visible light may be used to monitor the chemical composition of the plasma using a visible light spectrometer (details of this and other deposition monitoring techniques will be covered later in this). One requirement of the diode sputtering technique is that the cathode be electrically conductive. Elements and compounds which are insulators can be sputtered by other techniques

Typical operating parameters for DC diode sputtering are:

Current density at the cathode	1 mA/Cm^2
Process gas pressure (Argon)	30 to 80 mTorr
Cathode bias	–3000 VDC
Working distance	2 to 8 cm

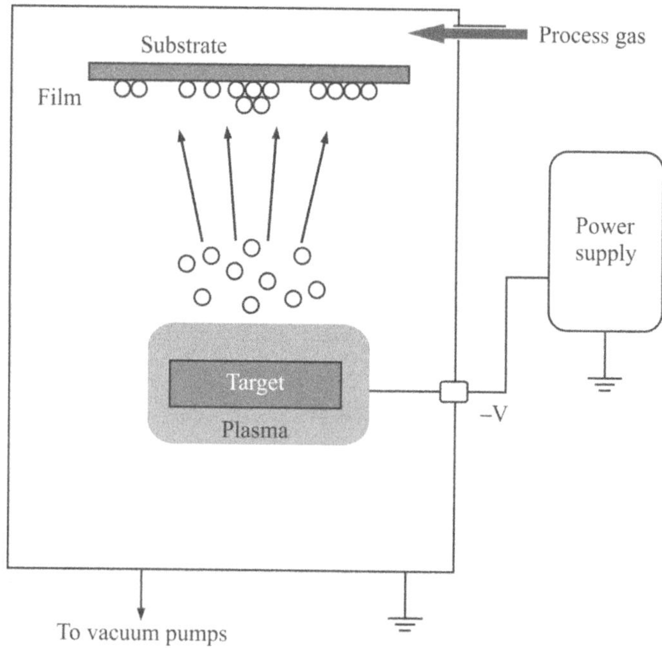

Figure 12.12 Simplified representation of a diode sputter deposition process

Depending on the composition of the cathode, this set of parameters should yield a deposition rate from 60 to 400 Å/minute. It should be noted that the rate of erosion of the cathode is expected to be greater than the rate of deposition at the substrate. Some of the sputtered material condenses on the inner surfaces of the vacuum vessel (a good reason for removable deposition shields), and some may react chemically with residual gases to form volatile by-products which are pumped away.

Now that the concept of diode sputtering has been introduced, a few of the relevant concepts that apply to all sputtering processes will be covered.

12.3.1.4.2 Sputtering Yield

The number of target atoms which are ejected from the cathode for every incident ion is called the "Sputtering yield". The magnitude of the sputtering yield is a function of the composition of the process gas and target material, the energy of the incident ions, and the angle of incidence of the ions on the target. In general, the sputter yield is greatest for the following set of conditions (5):

1. High atomic weight process gas.
2. Low atomic weight cathode material.
3. Low concentration of reactive gas species in the vessel.

The noble gas Argon is the most commonly employed process gas for sputter deposition processes, as it has a high sputter yield for most metals, is chemically inert and non-toxic, and is relatively inexpensive (compared with the other noble gases Krypton and Xenon).

Table 12.3 Sputter yields for several elements at 200, 600 and 1000 V DC.

Target Material	200 V	600 V	1000 V
Silver	1.6	3.4	No data
Aluminium	0.35	1.2	No data
Gold	1.1	2.8	3.6
Cobalt	0.6	1.4	No data
Molybdenum	0.40	0.9	1.1
Silicon	0.2	0.5	0.6

12.3.1.4.3 Ion Plating

In this atomistic, essentially sputter-deposition process the substrate is subjected to a flux of high energy ions, sufficient to cause appreciable sputtering before and during film deposition.

The advantages of physical methods are laid in dry processing, high purity and cleanliness, compatibility with semiconductor integrated circuit processing and epitaxial film growth. However, there are certain disadvantages such as slow deposition rates, difficult stoichiometry control, high temperature post deposition annealing often required for crystallization and high capital expenditure (6).

12.3.2 Chemical and Electrochemical Methods for Thin Film Deposition

Among chemical and electrochemical methods the most important are chemical vapour deposition, cathode electrolytic deposition, anodic oxidation and Chemical Bath Deposition.

12.3.2.1 Chemical Vapour Deposition

Chemical vapour deposition can be defined as a material synthesis method in which the constituents of vapour phase react together to form a solid film at surface. The chemical reaction is an essential characteristic of this method; therefore, besides the control of the usual deposition process variables, the reactions of the reactants must be well understood. Various types of chemical reactions are utilized in CVD for the formation of solids are pyrolysis, reduction, oxidation, hydrolysis, synthetic chemical transport reaction etc.

12.3.2.1.1 Chemical Bath Deposition (Deposition by Chemical Reactions)

Among the chemical methods of thin film depositions, Chemical Bath Deposition (CBD) is probably the most simplest method available for this purpose and it is also known as Chemical Solution Deposition (CSD) or Chemical Deposition (CD). It is not a new technique as early as in 1835 Liebig reported the first deposition of silver, the silver mirror deposition using a chemical solution technique.

Chemical deposition is the deposition of films on a solid substrate from a reaction occurring in a solution using the prototypical MS (metal sulphide) as an example, a M salt in solution can be converted to MS by adding sulphide ions (e.g., H_2S), MS immediately precipitates unless the solution is very dilute- a few milli molar or less in which case MS often forms as a colloidal solution. Another pathway for MS formation, one that does not require free sulphide ions, is decomposition of a MS-thiocomplex. In MS, the trick is to control the rate of these reactions so that they occur slowly enough to allow the MS either to form gradually on the substrate itself (at the early stages of deposition) or to the growing film, rather than aggregate in to larger particles in solution and precipitate out.

This rate control can be accomplished by generating the sulphide slowly in the deposition solution. The rate of generation of sulphide, and therefore reaction, can be controlled through a number of parameters, in particular the concentration of sulphide forming precursor, solution temperature, and pH. Although CBD can be carried out both acidic and alkaline solutions, most of the CBD reactions have been carried out in alkaline solutions.

12.3.2.1.2 Arrested Precipitation Technique

Arrested precipitation technique (APT) is modified chemical bath deposition process. The arrested precipitation technique based on Ostwald ripening law is simple and inexpensive method used for deposition of wide variety of metal chalcogenide thin films.

Arrested precipitation technique can be distinguished from other conventional techniques in the following aspects.

(i) It is ideally suited for large area thin film depositions; substrate surfaces of both accessible and non-accessible nature could easily be deposited.

(ii) It is simple, inexpensive and does not require sophisticated instrumentation.

(iii) The deposition is usually at low temperature and avoids oxidation or corrosion of the metallic substrates.

(iv) Stoichiometry of the deposits can be maintained since the basic building blocks are ions rather than atoms.

(v) Slow film formation process facilitates better orientation of the crystallites with improved grain structures over the substrate surface. Doped and mixed films could be obtained by merely adding the mixant/dopant solution directly into the reaction bath.

Electrical conductivity of the substrate material is not an important criterion. An intimate contact between reacting species and the substrate material permits pinhole free and uniform deposits on the substrates of complex shapes and sizes. Wide varieties of conducting/nonconducting substrate materials can be used. Dissociation rate of organometallic complex to release free metal ions for reaction is well control by maintaining the pH of reacting solution.

12.3.2.1.3 Anodic Oxidation

Anode oxidation (anode electrolytic deposition) is used mainly in the formation of films of the oxides of certain metals, e.g. ,Al, Ta, Nb, Ti, and Zr. The oxidized metal is an anode dipped in the electrolyte from which it attracts the oxygen ions. The ions pass through the already formed oxide film by diffusion forced by a strong electric field and combine with metallic atoms to form molecules of the oxide. For anode oxidation it is possible to use either the constant current or constant voltage method. Solutions or melts of various salts or in some cases acids are used as electrolytes. This is an electrolytic method for producing oxide films on the surface of metal.

12.3.2.1.4 Cathodic Deposition

This is a standard method of electroplating. Two metal electrodes are dipped into an electrolyte solution and on application of an external field across the electrodes; metal ions from the solution are deposited on cathode as a film. Deposition of the films is mainly controlled by the electrical parameters such as, electrode potential and current density. The mass of the substance deposited is proportional to the amount of electrical charge. The proportionality constant is the electrochemical equivalent of the given substance, which is for example, is 2.04×10^{-3} g./C, for Au and 0.09 × 10-3 g/C for Al. By this method it is possible to deposit films only on conducting substrates.

12.3.3 Electro Chemical Deposition for Thin Films

12.3.3.1 Advantages of Electro Chemical Deposition

Electro deposition is a long established way to deposit metal layers on a conducting substrate. Among the various chemical deposition techniques, electrodeposition technique is more popular and extensively used for synthesis of binary transition metal dichalcogenides. Ions in solution are deposited onto the negatively charge cathode carrying charge at a rate that is measured as a current in the external circuit.

The electro deposition is the simplest of the chemical methods, and it has many advantages like

1. Structurally and compositionally modulated alloys and compounds can be deposited which are not possible with other deposition techniques.
2. In most of the cases the deposition can be carried out at room temperature enabling to form the semiconductor junctions without inter diffusion.
3. Deposition on complex shapes is possible.
4. Toxic gaseous precursors need not to be used (unlike gas phase methods).
5. The deposition process can be controlled more accurately and easily
6. The process is relatively cheap and fast and allows complex shapes.
7. The layer thickness simply depends on current density and the time for which the current flows.
8. The deposit can be detached if the substrate is chosen to be soluble by dissolving it away

During last two decades, electro deposition has become a tool of materials technology for obtaining films of wide variety of materials including binary-ternary semiconductors, high T_c superconductors, polymer films etc.

12.3.3.2 Basics of Electrochemical Processing

Electrodeposition, also known as electrochemical deposition or electro crystallization, it is one of the most useful techniques for preparing thin films on the surface of conducting substrate. Besides advantages such as low temperature and pressure, the technique is directly related to many academic challenges in materials chemistry and physics.

Electrodeposition of metallic films has long been known and used for preparing metallic mirrors and corrosion resistant surfaces, etc.

12.3.3.2.1 Mechanism of Electrodeposition

The principle of electro deposition is inducing chemical reactions in an aqueous electrolyte solution with the help of applied voltage. This is the process of using electrical current to coat an electrically conductive object with a relatively thin layer of metal. This method is relevant to deposition of nano structured materials include metal oxides and chalcogenides. The schematic diagram explaining the electro deposition is shown in Figure 12.13.

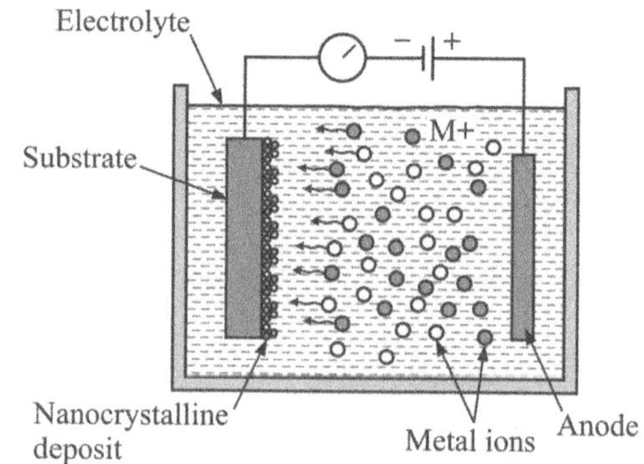

Figure 12.13 Schematic of Electro deposition

The typical electrodeposition set up consists of following, 1. Electrolyte 2. Cathode and anode 3. Source of electricity.

When direct current is passed through cathode and anode, immersed in electrolyte containing the metal(s) ions, the metal ions get attracted towards the cathode, neutralized electrically by receiving electrons and get deposited on cathode. The deposition is controlled by monitoring the amount and the rate of charge passing through the electrolyte. Thus the electrical energy is used to cause chemical change. Electro deposition process can be written as

$$M_n^+ \text{ (Aq)} + ne^- \rightarrow M \text{ (s)} \qquad .....(12.1)$$

On the other hand, if the electrolyte contains more than one ionic species that can be simultaneously deposited then the electrodeposition process for, say two types of ionic species can be written as

$$M_n^+ \text{ (aq)} + ne^- \rightarrow M \text{ (s)} \qquad .....(12.2)$$

$$N^{++} e^- \rightarrow N \qquad \text{Or} \qquad M^+ + N^+ + 2e^- \rightarrow MN \qquad .....(12.3)$$

Accordingly, one can deposit a compound or an alloy of a multicomponent system.

12.3.3.2.2 Faraday's Laws of Electrolysis

Michael Faraday (1834) established the relationship between the electricity passed through the electrolyte and the chemical change produced in terms of solid material liberated/deposited at the electrode.

Faraday's First Law

"*The amount of substance liberated or deposited on the electrode is proportional to the quantity of electricity passed.*"

Mathematically, W Q where, W is the amount of substance liberated in grams, and Q is the quantity of electricity passed through electrolyte, in coulombs. If current strength I is passed for t seconds, then the quantity of electricity.

Quantity of electricity = Current strength × time

$$Q = c\,t \qquad \text{.....(12.4)}$$

$$W \;\; c\,t \text{ or } W = z\,c\,t \qquad \text{.....(12.5)}$$

Here z is the proportionality constant, known as 'Electrochemical equivalent'. It can be defined as, "The amount of substance liberated (in gm) on the electrode on passing a current 1 A for 1 Sec or passing 1 coulomb of electricity".

Faraday's Second Law

"If same quantity of electricity is passed through different electrolytes, then the amounts of substance liberated on the respective electrodes are in the ratio of their equivalent weights."

An important implication of the Faraday's second law is that the ratio of the mass of electrodeposit to its gramequivalent weight is a constant equivalent to 1 Faraday or 96,500 Coulombs (c) or 26.3 ampere-hour (Ah).

12.3.3.2.3 Electrode-Electrolyte Interface

The electrode-electrolyte interface plays an important role in electrodeposition process. The electrode used in electrodeposition should necessarily be a good conductor. When an electrode is immersed in the liquid electrolyte, the anisotropic forces developed at the interface, results in a new arrangement of solvent dipole and ions of the electrolyte. At the beginning, the charge comes in two phases (one electrode and other electrolyte), start accumulating at the phase boundary and giving rise to the interface, acting as a barrier to flow of charge in either direction. As the built-up grows on the two sides, the electrical forces overpower the barrier, resulting in the flow of charges. The charge flow stops when the electrochemical potential on the two sides of the interface become equal. The interfacial region thus acquires a potential gradient that acts as a barrier to the further flow of charges. The whole process occurs at about 1000 Å region at the phase boundary. The layer of charges is developed on both sides of the boundary and hence termed as an "electrical double layer."

The electrode-electrolyte interface according to Stern's model can be divided into two parts:

1. A compact double layer, known as Helmholtz double layer adjacent to the electrode, and
2. A diffuse layer known as Gouy-Chapman layer.

12.3.3.2.3.1 The Helmholtz Layer

The Helmholtz double layer is a dense layer of ions stuck to the electrode. In this region the potential varies linearly with distance as shown in Figure 12.14. This dense layer is divided into inner and outer Helmholtz planes. The inner Helmholtz Plane (IHP) is adjacent to the electrode surface and consists of completely oriented solvent dipoles and specifically adsorbed (or contact-adsorbed) ions. The orientation of solvent dipoles depends on the specific interaction with the electrode surface as well as the electric field. Large ions with negative free energy of contact adsorption are expected to be contact adsorbed. In aqueous system, the cations react rather strongly with the water molecule, and their inner hydration sphere is retained. This limits their closest distance of approach to the electrode. They are thus separated from the electrode by approximately one or two electrolyte molecules. On the other hand, anions interact weakly with

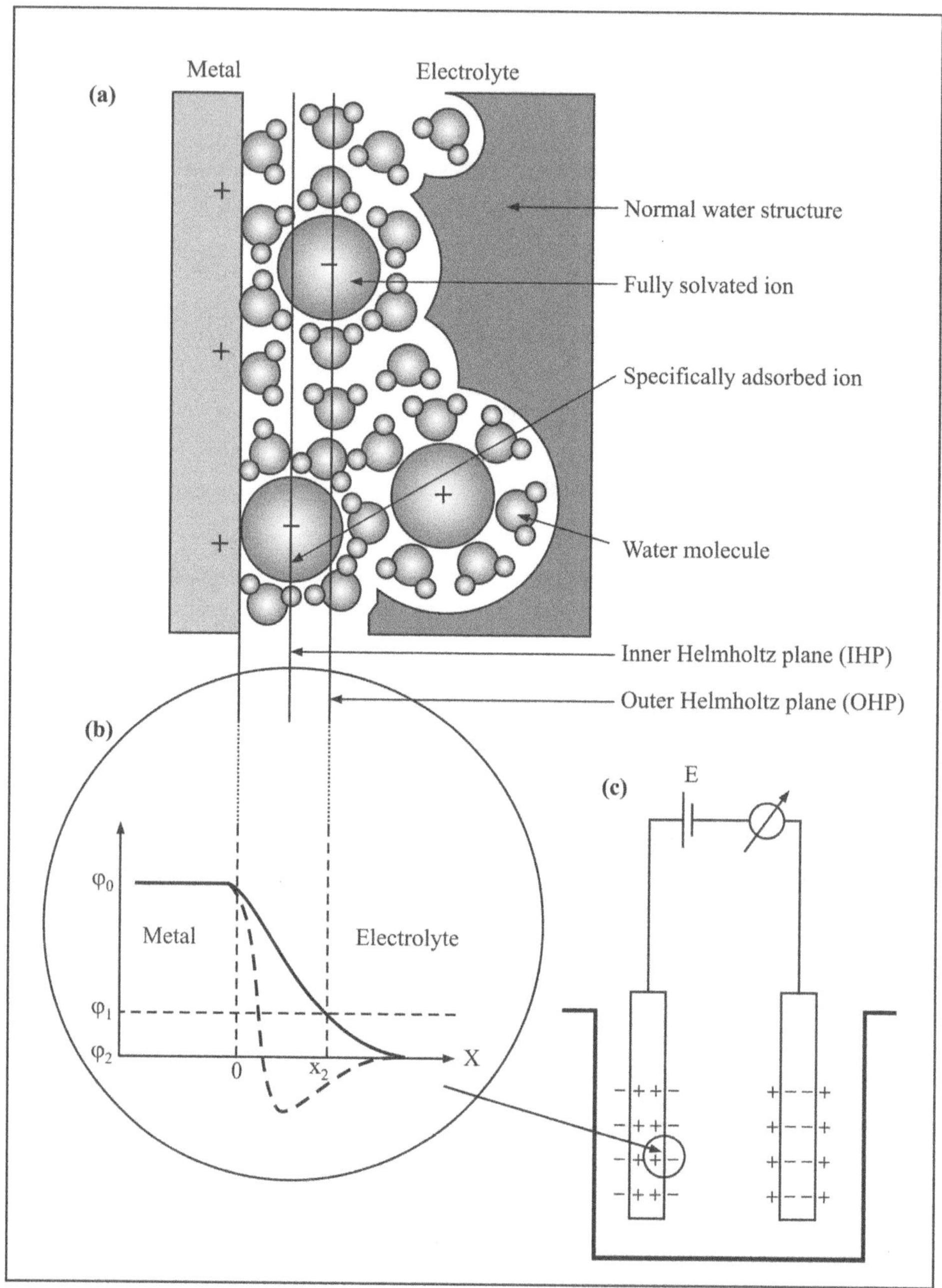

Figure 12.14 (a) Electrode electrolyte interface, (b) Potential distribution across the double layer & (c) The schematic of electrode immersed in electrolyte

electrolyte; hence hydration sheath is not covered on them. Thus the closest distance of approach could correspond to direct contact; they can be a part of IHP. The Outer Helmholtz Plane (OHP) consists of solvated ions (usually cations) at the closest distance of approach from the electrode surface. The OHP thus consists of partly ionized electrolyte molecule dipole layers.

Bockris et al. (9) developed a model of electrode-electrolyte interface and is shown in Figure 12.14(a). In presence of contact adsorbed ions, it can be seen from Figure 12.14(b) that the potential distribution changes across the double layer. The schematic of electrode immersed in electrolyte is shown in Figure 12.14(c). OHP is the site at which non-specifically adsorbed ions arrive to take part in the charge transfer process.

12.3.3.2.3.2 Gouy-Chapman Layer or Diffuse Layer

The size of the ions forming the outer Helmholtz plane (OHP) is such that the sufficient number of them cannot neutralize the charge on the electrode. Therefore, the remaining charges are held with increasing disorder as the distance from the electrode surface increases and the electrostatic forces become weaker and dispersion by thermal motion is more effective. These less ordered charges forcing opposite to that on the electrode constitutes the diffuse part of double layer. Thus, all the charges, which neutralize on the electrode, are held in a region between OHP and the bulk of electrolyte. The additional charges required to neutralize the total charge on the electrode forms the Gouy-Chapman layer or diffuse layer.

12.3.3.2.4 Steps Involved in Electrodeposition Process

Electrodeposition of ionic species from the electrolyte occur in following successive steps (Figure 12.15).

1. Ionic transport
2. Discharge
3. Breaking of ion-ligand bond (if the bath is complexed)

Incorporation of a atoms on the substrate surface followed by nucleation and growth.

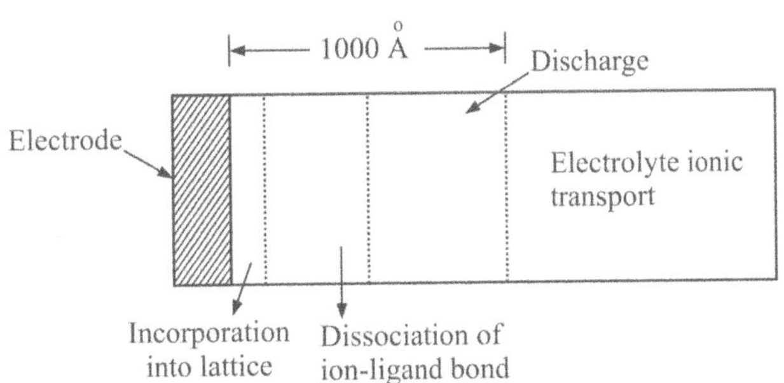

Figure 12.15 Approximate region in which various stages of ion transport occur leading to various steps involved in electrodeposition process

All above steps occur within 1–1000 Å from the substrate; however each has its own region of operation. These various processes can be classified with respect to distance from the electrolyte as:

(a) In the electrolyte
(b) Near the electrode
(c) At the electrode

12.3.3.2.5 Process in the Electrolyte

The ions in the electrolyte can move towards the electrode under the influence of;

1. Potential gradient leading to ion drift, d/dx
2. Concentration gradient leading to diffusion of ions, dc/dx
3. A density convective current, d/dx due to consumption of ions at the electrode.

The potential where a reduction reaction occurs can be estimated from the Nernst equation, (12.6) for the reaction (12.1)

$$E = E^0 - RT/ZF \ln\{a\,(M)/a(M^{Z+})\} \qquad \text{.....(12.6)}$$

Where β is the reduction potential, E^0 is the standard potential for the reaction

R the gas constant, T the temperature, F the Faraday constant, z the number of electrons, and a(M) and a (M^{z+}) are the activities of M and M^{z+}. Activities of the ions are usually approximated to be the same as the concentrations and the activity of a pure element is one. The calculated potential is valid only under the equilibrium conditions, which are hardly met in practice. The reduction needs a certain overvoltage, which is the irreversible excess potential required for a reaction to occur. The origin of the over potential may be for example adsorption reactions, concentration profiles, diffusion rates, nucleation, or reaction rates. Therefore it is more useful to measure the actual reduction potential under the experimental conditions, for example by cyclic voltammetry.

In order to achieve a suitable potential range for the deposition process, the precursors can be complexed which shifts their reduction potentials. When the metal ion M^{z+} is complexed with a ligand L^{P-}

Where $$M^{Z+} + qL^{P-} \leftrightarrow (ML_q)^{Z-pq} \qquad \text{......(12.7)}$$

it's reduction potential is shifted according to the equation 12.6

$$E = E^0_M{}^{Z+} - RT/zF\ln\beta + RT/zF\ln\{(a\,ML_q{}^{Z-pq}/a(L^{P-})^q\,a(M)\} \qquad \text{.....(12.8)}$$

where β is the stability constant of the complex. The use of the complexing agent may also be necessary to prevent unwanted precipitation in the precursor solution. Complexing agents are widely used in the industrial metal deposition processes as they stabilize the deposition solution, dissolve metals and improve the quality of the deposit.

A monolayer of an element may also deposit by under potential deposition (UPD), i.e., at the potential more positive than the equilibrium, bulk deposition potential. UPD occurs if the bonding energy between two different metal atoms is greater than the bonding energy between the deposit atoms and it is common for example for lead on noble metals.

12.3.3.2.6 Processes Near the Electrode but Within Electrolyte

Ionic species are normally surrounded by hydration sheath or by other complex forming ion or legand present in the electrolyte. They move together as a single entity and arrive near the electrode surface where the ion-ligand system either accepts electrons from the cathode (or donates electrons to the anode). This ionic discharge occurs between 10 to 1000 Å from the electrode.

12.3.3.2.7 Processes that Occur on the Electrode Surface

The discharged ions arrive near the electrode, where step by step they lead to the formation of a new solid phase or the growth of an electrodeposit. The atoms deposited have a tendency to form either an ordered crystalline phase or a disordered amorphous phase.

The electrodeposit formation steps of transport, discharge, nucleation, and growth are interlinked.

12.3.3.2.8 Pathways for the Growth of an Electrodeposition

The entire pathway for the growth can be divided into following steps (Figure 12.16).

1. Transport of ions in the electrolyte bulk towards the interface.
2. Discharge of ions reaching the electrode surface, giving rise to generation of adatoms.
3. Nucleation and growth, where again alternative routes are possible
   (a) Growth assisted by surface diffusion.
   (b) Growth assisted by formation of clusters and critical nuclei.
   (c) Formation of monolayer and final growth of electrodeposit.

Surface defects such as steps, kinks and dislocations generally control the growth kinetics. The kinks sites and screw dislocations together sustain the growth of the electrodeposit.

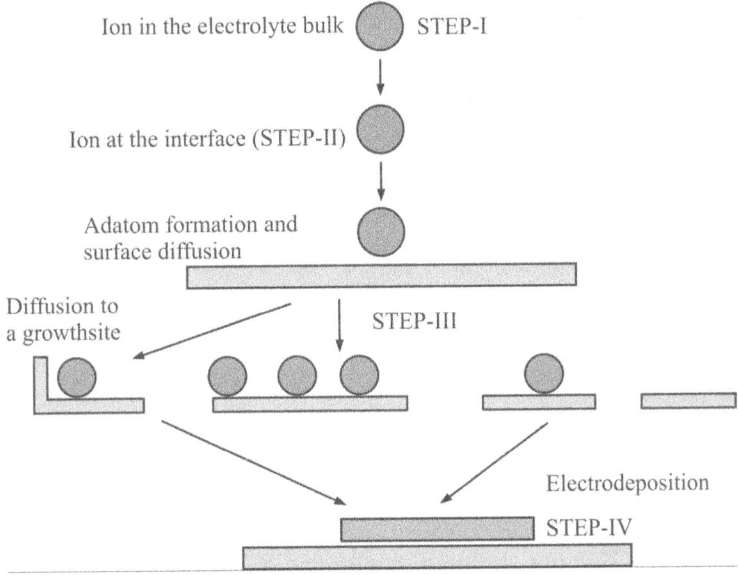

Figure 12.16 Pathways for the growth of an electrodeposition

12.3.3.2.9 Factors Governing Electrodeposition

The preparative parameters directly affect the structural, morphological and optical properties of the electrodeposits. The various preparative parameters like substrate, applied field and current density, bath temperature, complexant and pH of the bath etc, should be controlled to obtain uniform, smooth and stoichiometric electrodeposits. Some preparative parameters are discussed below.

(i) **Substrate:** Substrates play an important role in electrodeposition. Besides providing mechanical support to the electrodeposition, influences on the morphological characteristics of the growing layer and on the electronic and optical properties of the electrodeposition. For the choice of suitable substrate following criteria should be applied for their selection (9).

It should have good conductivity because it is one of the electrodes in electrodeposition. A good conductivity is beneficial in improving carrier collection efficiency.

(a) It should have good conductivity because it is one of the electrodes in electrodeposition. A good conductivity is beneficial in improving carrier collection efficiency.

(b) The thermal expansion coefficient should match with that of electrodeposition. A mismatch may cause cracking or peeling of the film.

(c) It should have good mechanical strength.

(d) It should be stable in electrolyte bath.

(e) It should be smooth. Uneven, porous, voids and other irregularities influence the local current distribution.

Metals have been widely used as substrates because of their good conductivity, easy availability, lower cost and relative ease of handling.

(ii) **Bath Temperature:** The rise in bath temperature enhances the rate of diffusion and increases ionic mobility, hence the increase in conductivity of the bath. The increase in temperature increases the rate of crystalline growth favoring the coarse deposits. This increase in crystal size corresponds to decrease in polarization. At higher temperature, current densities increase, which increases the rate of nucleation, hence fine-grained, smooth deposits can be obtained. The rise in bath temperature decreases the hydrogen over voltage so facilitates the evolution of gas, as well as precipitation of basic salts. The opposing effects make it difficult to predict the choice of bath temperature, however, it can be optimized by performing actual experiments.

(iii) **Current Density:** At lower current densities (or over-potentials) the discharge of ions occurs slowly, so the growth rate decreases but increases the crystallinity forming closely packed structures. As the current density is raised, the nuclei formation rate will increase and the deposits become fine grained. At higher current densities, the rate of discharge of ions becomes greater compared to rate of supply of ions and there is duplicity of ions near the cathode, which favours the growth perpendicular to the substrate surface. Usually spongy, dendritic growth can be observed under this condition. Secondly, at very higher current densities, hydrogen evolution occurs at faster rate, which interferes the crystal growth and spongy, porous deposits may be obtained. This can also favour the precipitation of hydrous oxides or basic salts due to increase in local pH.

(iv) **Metal Ion Concentration:** The plating bath is always an aqueous solution containing compound of metal to be deposited. It is always advantageous to use higher concentration of metal components in the bath solution. A high current density can be employed in high metal bearing bath. An increase in metal concentration, under given condition, decreases the cathode polarization and increases the crystallite size.

(v) **Hydrogen Ion Concentration (pH):** In order to operate a bath with optimum efficiency and maintain the desired physical properties of the deposit, control of pH of plating bath is necessary. Besides too low pH may lead accumulation of hydroxide ions in the vicinity of the cathode and consequent precipitation of basic salts, which may get included in electrodeposition, thereby altering deposit properties. All aqueous solutions contain H^+ ions, infact in every deposit from an aqueous bath, there is a possibility of the hydrogen gas evolution at the cathode due to H^+ ions. It takes place, the efficiency of metal deposition is lowered. As this efficiency and hydrogen discharge potential partly depends upon hydrogen ion concentration, at low pH, the bath permits the use of higher current density to produce a sound deposit with relatively high efficiency.

(vi) **Addition Agents:** In almost all cases of electrodeposition of metals, it is observed that addition of small quantities of certain substances often result in production of smooth, fine grained and nanocrystalline deposits. Such substances are known as addition agents. Addition agent such as brightening agents, surfactants, complexants etc., are often added to the bath to obtain smoother, brighter deposits, controllable reaction rates, better adhesion and better texture. The adsorbed additives influence the rate of deposition by (i) changing Helmholtz layer potential, (ii) acting as a bridge for mediating electron transfer reactions between electrode and discharging species (iii) forming complexes between the adsorbed additive and the ionic species to be plated.

(vii) **Complexing Agent:** The unstable metal ions are capable of combining chemically with neutral molecules and with ions of opposite sign to form stable complex ion. The combination is through the covalent bond, when neutral molecules interact with positively charged metal ions to yield negatively charged complex ions. Complex compound in a plating bath serves two purposes. Firstly they make possible to maintain a high metal concentration but low metal ion concentration. The complex ions of the complex compound serve as reserves and continuously supply of the simple ions necessary for the discharge at the cathode occurs. A low metal ion concentration enables the production of deposits with small grains and improves the throwing power. Secondly, complex formation enables us to enhance appreciably the solubility of slightly soluble salts.

12.3.3.3 Electrodeposition Applications

Electrodeposition is relatively cheap and can be performed at low temperatures which will minimize interdiffusion of materials in the case of multilayered thin film preparation.

The film thickness can be controlled by monitoring the amount of charge delivered whereas the deposition rate can be followed by variation of the current with time.

The composition and defect chemistry can be controlled by the magnitude of applied potential which can be used to deposit non equilibrium phases. Pulsing or cycling the applied current or potential in a solution containing a mixture of precursors allows the production of a multilayered material.

The potential during the pulse will determine the species deposited while thickness of individual layers is determined by the charge passed. Alternatively the substrate can be transferred periodically from one electrolytic cell to another. The final films can range in thickness from few nanometers to tens of microns and can be deposited on to a large specimen areas of complex shape making the process highly suitable for industrial use.

Electrodeposition is being exploited now to make complex 3D electrical inter connects in computer chips. The key concept is that electrode deposited materials grow from the conductive substrate outward and the geometry of the deposit can be controlled using an insulated mask.

The insulating mask need not have a straight line of sight path between substrate and the electrolyte, even tortuous masks can be filled with materials so long as the whole path through the mask is wetted with ED electrolyte. The resulting deposit is a high fidelity negative replica of mask itself. Through mask ED has been used extensively to pattern metals semiconductors, polymers on conductive substrates.

As a water based process it is often more environmentally friendly than deposition methods that require hazardous solvents and reactive precursors, chemicals like organo metallic compounds etc.

Metal chalcogenide thin films preparation by Electro Chemical methods are currently attracting considerable attention as it is relatively inexpensive, simple. Using these methods, thin films of group II–VI, V–VI, III–VI etc., have been deposited. Solar selective coatings, solar control, photoconductors, solid state and photo-electrochemical solar cells, optical imaging, hologram recording, optical mass memories etc., are some of the applications of metal chalcogenide films deposited by electro deposition.

Electrochemical Thin Film Deposition finds extensive applications in Corrosion resistant films (Zn, Ni, etc.)– Conductive films, e.g., Cu interconnect wiring in semiconductor devices.

Also, Au, Ag, Ni. – Thin film photovoltaic materials (CdTe, CIGS, CZTS)– Sensor electrodes (Ir, Au, Pt)– Decorative coatings (Au, Ag. Thin film photovoltaic materials (CdTe, CIGS) by electro chemical deposition is less costly than thick film photovoltaics (Si). First Solar (FSLR), other companies sell thin film CdTe solar cells manufactured by electro deposition. Coating of large-area substrates can be done economically by electrodeposition (e.g., Cu on Si wafers).

Biological Fabrication: Proteins are responsible for nucleation, growth, composition and shape of functional biological structures like bones, teeth and shells. Using proteins to control the growth of ED materials is truly frontier area where biology meets nano technology. One way that proteins are being used in electrochemical nano technology is as masks for masks for electrodeposition. Proteins can self organize into complex structures representing all possible two dimensional (2D) space groups built from chiral molecules. Moreover they are readily engineered through molecular biology providing an attractive foundation for nano technology.

Nanometer-Scale cuprous oxide (Red coloured) (Figure 12.17) can be electrodeposited through the openings in the hexogonally packed intermediate layer protein (White regions) from the bacterium Deinococcus radiodurans.Purofied crystalline protein sheets are first adsorbed to conductive substrate and then electrodeposition is carriedout to fill the nano meter scale pores in the protein.

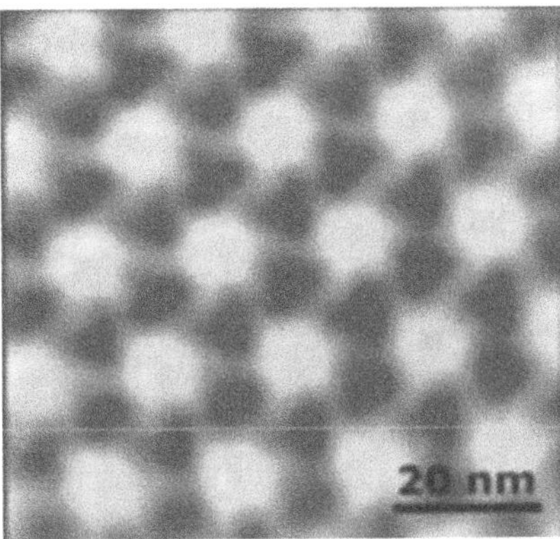

Figure 12.17 Electro deposition nano scale cuprous oxide (Red coloured) (*Source:* 10)

12.4 Plasma Deposition/Plasma Enhanced Surface Modification (11-14)

In general plasma is referred to as a partially ionized gas that contains positively and negatively charged particles. The plasma state is more highly activated than the solid liquid, or gas state and is often called the fourth state of materials. Plasma deposition or polymerisation can be defined as the formation of materials under influence of plasma. Solid deposits from organic compounds from a plasma generated by some kind of electric discharge. Advantage plasma deposition is that thin films can be deposited on most substrates using relatively simple one step coating and considered as valuable technique to modify the surfaces of polymers or of other materials. Wide range of compounds can be chosen as a monomer for plasma polymerization. Even saturated hydrocarbons provide great diversity of possible surface modifications. These advantages have resulted in the rapid development of plasma technologies during past decades for applications ranging from adhesion to composite materials, protective coatings, printing, membranes, bio medical applications and so on. However plasma polymerization remains a very complex process that is not well understood. Recently interest has shifted towards the synthesis of films that contain high concentration of specific functional groups at the surface.

12.4.1 Polymer Surface Modification Techniques

Many polymers posses good bulk mechanical properties which make them very useful for a number of applications. In fields such as protective coatings, adhesion and biomaterials however their surface properties are equally important for their success. Since polymers often do not posses the required surface properties for these or similar applications intensive research efforts are made recently to develop surface treatment methods that alter and improve the chemical and physical properties of polymers of surfaces. Improving adhesion, characters increasing hydrophobicity, introducing special functional groups at a surface or modifying the surface morphology are examples for the surface treatment.

12.4.2 Advantages of Plasma Surface Modification or Polymerization

1. The modification is limited to the top surface layer and does not affect the bulk properties of the polymer.
2. The modification by the plasma process is largely independent on the structure or chemical reactivity of the substrate.
3. A broad range of functional groups can be introduced at the surface by variation of the gas that is used
4. In general the modification is fairly uniform over the whole substrate. The plasma treatment or polymerization is a simple one step procedure and is all dry process.

12.4.3 Mechanism

Plasma surface modification involves the interaction of the plasma-generated excited species with a solid interface. The plasma process results in a physical and/ or chemical modification of the first few molecular layers of the surface, while maintaining the properties of the bulk. Typical materials employed in the microelectronics and optoelectronics industries include ceramics, glass, polymers and metals— such as gold, copper, aluminum, nickel, palladium, tungsten, and silver. The effectiveness of the plasma on these complex interfaces is determined by the plasma source gases, the configuration of the plasma system and the plasma operating parameters.

Plasma enhanced surface treatment of polymers: The most common application of plasma treatments are surface cleaning or etching to increase the wettability and adhesion of polymers to reduce friction and to improve cell attachment in tissue culture studies or to alter surface protein interactions. Functional groups and cross links are introduced at the surface of the polymer by reaction of gas phase species and surface species.

12.4.4 Classification of Plasma Enhanced Surface Modification Process

Plasma enhanced Surface modification processes can be classified into four categories:

(i) Contamination Removal
(ii) Surface Activation
(iii) Etch
(iv) Cross Linking

12.4.4.1 Contamination Removal

Surface contamination removal involves the use of the plasma's physical and/or chemical energy to remove micron-level contamination.

This process employs ablation, where the positive ions bombard the surface. The ablation process can dislodge contamination from the surface, and can roughen the surface on an atomic scale, as revealed by atomic force microscopy. The chemical process is widely employed to remove residual materials, typically less than a few microns, such as organic films, and oxidation.

In cases where oxidation is of concern, a physical process can be employed to prepare the bond pad surfaces. An argon plasma treatment of PBGA strips has been shown to improve the wire bond pull strength by as much as 24.3 percent.

Metal oxidation can act as a physical barrier for both wire bonding and solder reflow. A combined physical and chemical process using argon and hydrogen can reduce the metal oxides. For example, reduction of copper oxide to copper is achieved in a hydrogen plasma via the reaction of the hydrogen radicals with the metal oxide.

$$CuO + 2H\bullet \rightarrow Cu + H_2O \qquad \text{......(12.9)}$$

Even in the absence of a contamination source, ablation will roughen the surface and provide a larger surface area for wire-bonding, resulting in improved wire-bond uniformity bond to bond.

12.4.4.2 Surface Activation

Plasma surface activation employs gases, such as oxygen, nitrogen, hydrogen, and ammonia, which, when exposed to the plasma, will dissociate and react with the surface, creating different chemical functional groups on the surface.

The different functional groups modify the chemical activity of the surface. The new functional groups have strong chemical bonds with the bulk material and have the capability to further bond with adhesives to promote better adhesion. The functional groups also increase the surface area available for the adhesive and thus will distribute the load over a larger area resulting in improved adhesive strength. Gas selection and surface type determine the functional group that will be substituted on the surface. In microelectronic applications, plasma surface activation prior to die attach provides better contact, improved heat transfer and minimal voiding.

The purpose of the mold/encapsulant material for semiconductor applications is to provide adequate mechanical strength, adhesion to various package components, good corrosion and chemical resistance, matched CTE to the materials it interfaces with, high thermal conductivity and high moisture resistance in the temperature range used.

The ability to form good adhesion with package components and to remain bonded is paramount importance, since delamination along the interfaces is a major reliability issue for plastic-encapsulation.

Plasma treatment has been demonstrated to improve the bond strength at the plastic encapsulant, gold-plated copper leadframe interface via an enhanced chemical compatibility with the molding compound.

12.4.4.3 Etch

Plasma etch is characterized by the chemical reactivity of the discharge. The etching process utilizes source gases that dissociate within the plasma, creating a mixture of highly reactive species. The advantage of this chemical plasma is its chemical selectivity. The process chemistry can be optimized so that one material can be selectively etched in the presence of other materials. For example, the dissociation of carbon tetrafluoride (CF_4) and oxygen in the appropriate concentrations produces highly reactive oxy, oxyfluoro, and fluoro radicals that rapidly break carbon-carbon (Table 12.4).

Table 12.4 Process applications for plasma surface enhancements

Process Applications for Plasma Surface Enhancements		
Plasma source gas	**Surface Modification Processes**	**Advanced Technology Application**
Argon (Ar)	Contamination Removal-Abiation Cross Linking	Wirebond Die Attach Substrate Polymer-Metal Adhesion
Oxygen (O_2)	Contamination Removal-Chemical Oxidation Process (Organic Removal) Surface Activation Etch	Wirebond Die Attach Mold and Encapsulant Adhesion Photoresist Removal
Nitrogen (H_2)	Surface Activation	Mold and Encapsulant Adhesion
Hydrogen (H_2)	Contamination Removal-Chemical Reduction Process (Metal Oxide Removal)	Wirebond Eutectic Die Attach
Carbon Tetrafluoride (CF_4) and Oxygen (O_2) or Sulphur Hexafluoride (SF_6) and Oxygen (O_2)	Etch	Polymer Etch-Fiber Stripping Photoresist Removal Thin Film Etch-Oxides, Nitrides

The process chemistry can be optimized so that one material can be selectively etched in the presence of other materials. For example, the dissociation of carbon tetrafluoride (CF4) and oxygen in the appropriate concentrations produces highly reactive oxy, oxyfluoro, and fluoro radicals that rapidly break carbon-carbon.

12.4.4.4 Cross Linking

Plasma-induced cross linking employs inert gases such as argon or helium to remove some atomic species from the surface, and generates reactive surface radicals. These radicals react within the surface forming chemical bonds, which results in cross-linked surface. This approach is employed on polymeric substrates, such as those used for PBGA packages.

Argon plasma effectively sputters nano-meters of material from the sample surface, literally roughening the surface on the nano-meter scale. The cross linking that results improves the adhesion of metal layers to the plasma-treated polymer laminate.

Selection of a specific process is determined by the physical and chemical composition of the material to be processed, as well as by the ensuing process required. Surface modification is often sensitive to time and environmental exposure, where the surface may lose its plasma-induced physical and chemical properties.

12.5 Nano Structured Thin Films by Colloidal Phase Synthesis

Thin films with nano porous, nano layered, nano no patterned structures with porous size or layer thickness or surface roughness at nano meter scale are termed as nano structured thin films (Figure 12.18).

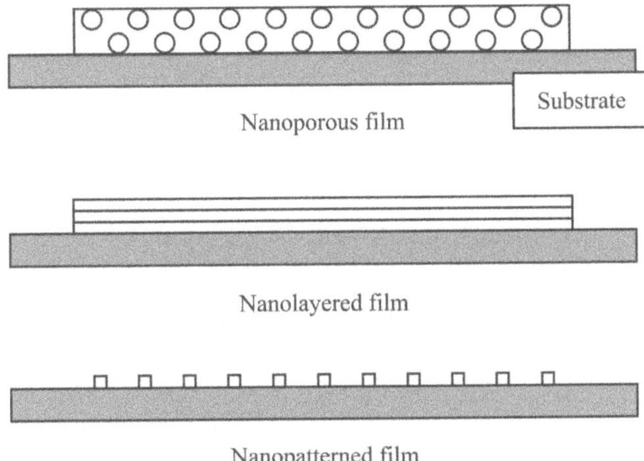

Figure 12.18 Structures of nano porous, nano layered and nano patterned films

12.5.1 Colloidal Phase Methods for Thin Films

Nanostructured films are two-dimensional objects which can be produced with the support of any solid substrates either flat, curved spherical as long as they have good affinity with the incoming building units or precursors. However the way those building units or precursors interact with the solid surface differ in different methods.

The following are some common colloidal based methods for production nano thin films.

(a) Layer-by-Layer assembly

(b) Self Assembled Monolayers (SAMs) and

(c) Sol-Gel Method

12.5.1.1 Layer-by-Layer Assembly

This is a process for preparing nano-layered films mainly by controlling the electrostatic interactions between the building units and the solid substrate. Physio sorption of the building units or precursors on the solid surfaces is the main driving force in layer-by-layer deposition.

Typical primary building units are polyelectrolytes with charged functional groups, which include poly (ethylene imine), poly (allylamine hydrochloride), poly (sodium styrene sulfonate). If the solid surface is positively charged the anionic polyelectrolyte from the solution will be easily layered as a monolayer of the polyelectrolyte on the surface via electrostatic attraction. Then, the polyelectrolyte with cationic charge is employed on this surface, which will be layered on top of the first layer, also via electrostatic attraction.

Successive layering can be achieved by alternatively employing cationic and anionic polyelectrolytes, which will produce multilayers of polyelectrolyte films. This layer-by-layer assembly is one facile method to prepare nano-layered films. It does not require serious equipment. Also, the building units are not limited to polyelectrolytes. As long as the conditions (both building unit and solvent) for the electrostatic interaction are properly set, a wide variety of organic, inorganic, colloidal, and biological materials can be employed as its building units.

12.5.1.2 Self Assembly Monolayers (SAMs)

The main driving force in SAM process are Chemisorption or possible subsequent surface reactions like surface oxidation/reduction, surface dissociation, surface alloying, or surface metallization.

Self assembled monolayers of long chain thiols on noble metals such as gold and silver are the common systems used for various micro electronic applications.

Similarly silylation of long chain silanes on properly functionalized surfaces of hydroxylated silicon and complexation of long chain carboxlic acid on metal or metal oxides are some common SAM systems in use.

SAMs find wide application in nanotechnology in areas like biosensors, selective binding of functional entity, bio-comparable substrate for cell engineering, circuit-boards for the creation of nano-electronics and nano-devices, friction control such as lubricants etc.

The ability to control the structural and functional features of SAMs both on molecular and nanometer levels makes this technique adoptable for various applications (15).

12.5.1.3 Sol-gel Deposition of Thin Films

In general, the sol-gel process involves the transition of a solution system from a liquid "sol" (mostly colloidal) into a solid "gel" phase. Utilizing the sol-gel process, it is possible to fabricate advanced materials in a wide variety of forms: nano structured thin films and extremely porous composites.

The details of this process is discussed in detail in Chapter 7.

Figure 12.19 presents the general scheme for Sol-Gel process.

For the nano porous thin films the inorganic precursors are hydrolyzed first, and then these hydrolyzed forms co-self-assemble with organic cations to form microstructured composites. Alkoxides of silicon, aluminum, and titanium are among common inorganic precursors and tetra-alkyl quaternary ammonium cations are typical organic components.

Final zeolites can be obtained as single components such as pure siliceous or multicomponents such as titanosilica, aluminosilica, and so on. Removal of the water and organics by caking produces the highly nano porous composites as thin films.

The pre-self-assembled precursor sol also can be deposited on pre patterned solid substrate. By controlling the structural or chemical characters of the patterned solid surface such as size/thickness of the patterned area or the types of surface micelles preassembled on a solid surface, the sol can be selectively introduced as the guide of the pre patterned solid. This can be applied as one of the direct deposition nano composite or hybrid film or nanoporous film .

12.5.1.3.1 ZnO Thin Films Produced by the Sol-gel Method

ZnO thin films have been extensively studied of late, due to their potential applications as metal oxide semi-conductors in optoelectronic devices, piezoelectric devices and as transparent electrodes in displays. Zinc Oxide has a hexagonal wurtzite structure. Films with preferred orientation along the c-axis can operate as surface acoustic wave devices because of their large piezoelectric constant. The sol-gel method is of interest, as it enables film deposition in conventional lab environments without the need for expensive equipment.

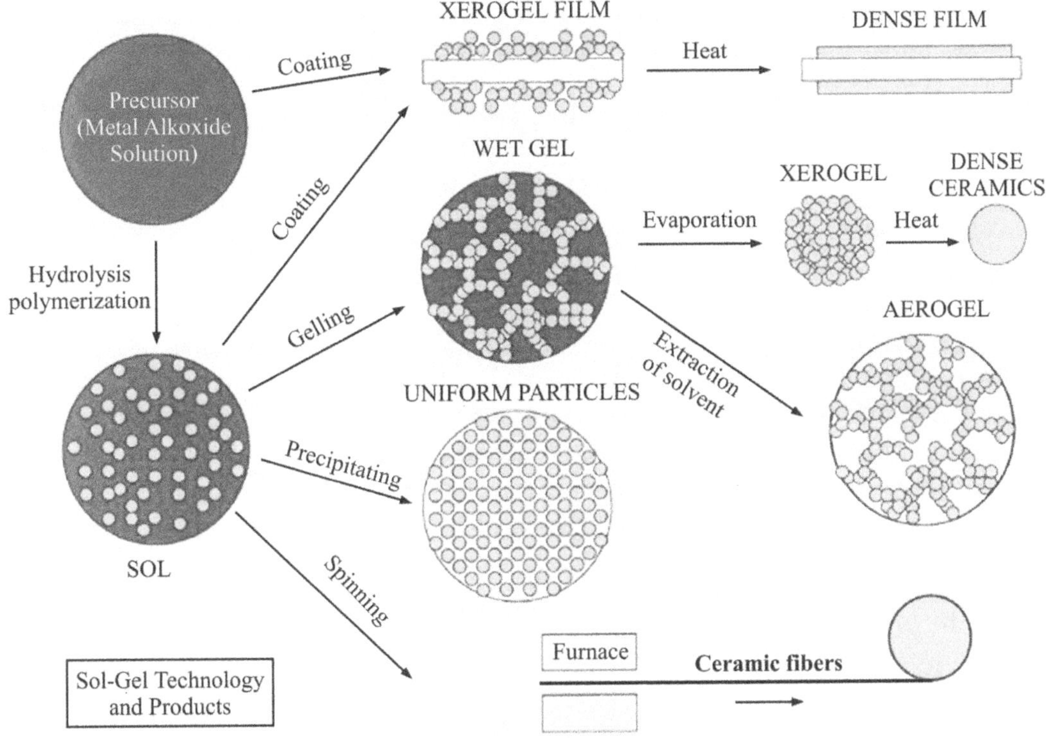

Figure 12.19 General scheme of sol gel process

12.6 Characterization of Thin Film

There are two types of thin film characterization techniques

(i) *in-situ* techniques which provide information while the film is being deposited. These can be used for process control and QC.

(ii) *Ex-situ* techniques which provide information after deposition process.

12.6.1 *In-situ* Characterization Techniques-Quartz Crystal Microbalance

In this technique a small quartz crystal is caused to oscillate by the application of an electric field. This crystal is placed inside a vacuum vessel, and during deposition of a film, the crystal gains mass due to the material condensing on it. The frequency of this oscillation is reduced as the mass of the film on the crystal increases. Through cross-calibration using a technique such profilometry, the quartz microbalance can be used to accurately monitor film thickness as the film is deposited, and to provide feedback information to control the process (close shutters, increase or decrease power to the deposition source) (16,17).

12.6.2 *Ex-situ* Thin Film Characterization Techniques

Care must be taken to prevent altering the thin film coating in the performance of *ex-situ* characterization techniques. In general, it is good practice to allow a coated substrate to cool to room temperature before venting the vacuum vessel. This minimizes the possibility of oxidation of the part. Thermal shock, due to rapid cooling caused by premature venting may cause films to tear or delaminate from the substrate, making them virtually useless for physical characterization purposes. Contamination of the coated substrate due to handling and storage should be kept to a minimum for best characterization results.

Contact Profilometry

In this *ex-situ* characterization technique, the sample is prepared by masking off a portion of a smooth substrate, such as float glass, then depositing a thin film onto the sample. This creates coated and uncoated areas (a so-called "step slide") on the smooth substrate which can be used to determine film thickness. After the part is removed from the deposition system, a ruby or diamond tipped miniature stylus is scrolled across the sample.

Contact profilometry is frequently used to establish deposition rates, and as a means of calibration for *in-situ* rate monitoring techniques, such as the quartz crystal microbalance. Films having thicknesses of from 50 to 100,000Å have been accurately measured using contact profilometry.

12.6.3 Sheet Resistivity

Some thin films are deposited with the goal of attaining a specified electrical characteristic, such as resistivity. These films may be characterized using a commercial instrument called a four-point resistivity probe. Current is passed between each of the points of the probe through the thin film, and is measured. The data is output as sheet resistivity expressed in ohms per square centimeter (Ω/cm^2). Other electrical properties of interest in thin film coatings are the dielectric strength (the ability of a film to prevent high voltage from dissipating to ground through the film), and the critical current density (current density below which a material is super conducting at a given temperature).

12.6.4 Optical Transmission

Ex-situ optical transmission measurements often are more widely used methods a single wavelength, or narrow band of wavelengths are often used, along with low resolution spectrometers for the *in-situ* optical measurements. Optics is often carefully inspected following deposition of a thin film coating to establish the transmission over a broad range of wavelengths. For the determination of thin films thickness with optical microscope e.g., metallographic cross sections are used in the Transmission Electron Microscope(TEM) (18).

In the Scanning Electron Microscope (SEM) the step itself or a fracture surface of the film can be imaged (19, 20). The resolution and metric range depends on the instrument and on the magnification. In the optical microscope film thickness up to some mm with a resolution of 0.1um can be determined. In the SEM a resolution of 5 nm can be achieved and in the high resolution TEM (HRTEM) 0.1 nm are possible.

The details of these instruments and their working function are discussed in detail in Chapter 5.

12.6.5 Compositional Analysis

There are a variety of techniques used to determine the chemical composition of thin film coatings following a deposition run. These techniques may be grouped into two broad categories: destructive and non-destructive. Wet chemical analyses, in which the film is dissolved, the solution being analyzed using techniques such as atomic emission spectroscopy are obviously harmful to the coating. If destructive analytical techniques are to be used, it is possible to perform them on "dummy parts" that were coated along with the part of interest. Non-destructive analytical techniques include x-ray techniques (x-ray florescence, x-ray diffraction), electron spectroscopy for chemical analysis (ESCA) and other surface science techniques. Most of these non-destructive analytical methods are conducted under vacuum in relatively small chambers, which places size limitations on the parts to be analyzed (21, 22, 23).

12.6.6 Mechanical Testing

Frequently the mechanical property of a thin film coating which is of most interest is its adhesion to a substrate. Simple tests of adhesion include the "scotch tape test" in which a piece of scotch tape is pressed firmly to the coated side of a part, then peeled away. Other more sophisticated tests of adhesion include the Sebastian pull test (Figure 12.20) and the ring shear test. In the Sebastian test, a metal stump is glued to the thin film deposited onto a substrate. The force required to pull the stump away from the substrate is recorded, and plotted as a graph of stress (force per unit area) versus strain (deformation).

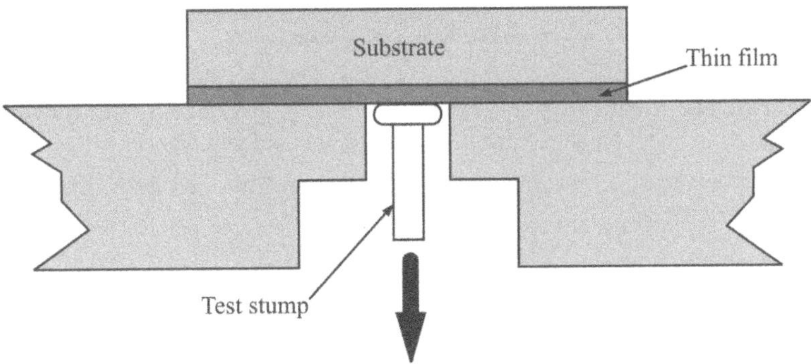

Figure 12.20 Sebastian pull test for adhesion of thin films

References

1. A.R. West (2003). Solid State Chemistry, John willey & Sons, Singapore.
2. K. L. Chopra (1969). Thin Film Phenomena, McGraw Hill, New York.
3. Y. Gao, H. Niu, C. Q. Chen (2003). Chem. Phys. Lett., 367:(1-2), 141.
4. Zeng Acosta R.E., Romankiw L.T., and VonGutfeld R. J (1982). Thin Sol. Films 95,131.

5. Behrisch R (1981). Sputtering by Particle Bombardment, Springer, Berlin.
6. Deposition Technologies for Films and Coatings, Bunshah, Roitan et.al; Noyes Publications, Park Ridge, New Jersey,1982, 170-237.
7. Thin film Processes, Vossen, John L. Krn, Werner Academic Press, Inc. Orlando Florida.1978, 12-62,115-170.
8. Electron Beam Technology, Schiller, Sgfried, et.al, Wiley Interscince, New York, New York 1982.
9. The structure of water in the double layer Joh OM Bockris et.al., (Tetrahedron letters 2503 (1977)).
10. Daniel Schwartz, and Walther Schwarzacher (2006). The Electrochemical Society Interface Spring, 34.
11. F.D. Egitto and L.J. Matienzo (1994). Plasma Modification of Polymer Surfaces for Adhesion Improvement, IBM Journal of Research and Dcvclopment, 423.
12. S. Yi, J. Kim et al. (2000). Bonding Strengths at Plastic Encapsulant-Gold-Plated Copper Leadframe Interface, Microelectronics Reliability, 1212.
13. M. White (1982). The Removal of Die Bond/Epoxy Bleed Material by Oxygen Plasma, Proceedings 32nd IEEE Electronic Components Conference, 262.
14. L. Wood, C. Fairfield et al (2000). Plasma Cleaning of Chip Scale Packages for Improvement of Wire Bond Strength, Chip Scale Package Seminar.
15. Julie N.L. Albert and Thomas H Epps III (2010). Materials Today, Vol 13.
16. O. Bohnke and G. Robert (1982). Solid State Ionics, 6, 115.
17. Y. Hiruta, M. Kitao and S.Yamada (1984). Journal of Appl. Phys., 23, 1624.
18. S. Banerjee (2004). Proc. of National Workshop on Advanced Methods for Materials Characterization (NWMMC-2004), Mat. Res. Soc. India.
19. Oatley, D. McMullan and K. C. A. Smith (1985). The Development of the Scanning Electron Microscope, in The Beginnings of Electron Microscopy, ed P. W. Hawkes. Advances in Electronics and Electron 1Physics Suppl. 16 (London: Academic Press) 443.
20. Oliver C. Wells, Scanning Electron Microscopy, McGraw Hill, Inc. USA. EDAX.
21. Taylor (1942). X-ray Metallography, Wiley, New York.
22. Guinier (1963), X-ray Diffraction, Freeman, San Francisco.
23. B.D. Cullity (1978). Elements of X-rays 2nd edition Addiso Wesley, London.

13

Nano-Bio-Info-Cogno (NBIC) Convergent Technologies

13.1 Introduction

13.1.1 What are Convergent Technologies?

"Convergence" is a broad term applicable in many ways to science and technology and refers to merging of concepts from different systems of knowledge. Physical chemistry, Polymer chemistry, mechatronics, electronics, Biochemistry, molecular biology, evolutionary medicine, computational linguistics, cognitive psychology, etc., are all result of convergence of previously separate disciplines and domains. In the area of information technology, "convergence" is usually used to designate multi-functionality as the telephone, display screen, computer, internet access, and video camera all merge into a single device.

Nanotechnology by itself a convergence of a number of disciplines such as biomedicine, information technology, chemistry, photonics, electronics, robotics, and materials science into a single engineering paradigm where all the phenomenon are re configured at molecular level.

Social and cognitive sciences have also developed considerable knowledge about social interactions, effective communication, etc., which can be implemented in engineered systems to enable far-reaching technological developments.

An important step in the history of convergent technologies (CTs) is the realization that apart from nanotechnology, there are other enabling technologies and knowledge systems that are open to new R&D challenges and ready to enable one another.

Convergent Technologies (CTs) represent the synergistic combination of four major Nano-Bio-Info-Cogno (NBIC) disciplines of science and technology for improving human performance and quality of life. The integration of emerging scientific areas (a) nanoscience and nanotechnology; (b) biotechnology and biomedicine, including genetic engineering; (c) information technology, including advanced computing and communications; (d) cognitive science, including cognitive neuroscience which are currently progressing at a rapid pace is based on material unity at the nanoscale and on technology integration at that scale.

Convergent Technologies (CTs) facilitate improvement of human performance through integration of technologies as shown in (Figure 13.1).

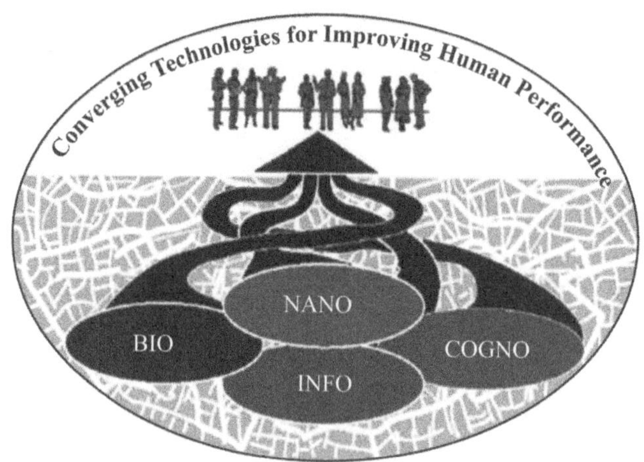

Figure 13.1 Convergent technologies for improving human performance (Source R.E.Horn 2002) (1)

The building blocks of matter that are fundamental to all sciences originate at the nanoscale. Revolutionary advances at the interfaces between previously separate fields of science and technology are ready to create key transforming tools for NBIC technologies. Developments in systems approaches, mathematics, and computation in conjunction with NBIC allow us for the first time to understand the natural world, human society, and scientific research as closely coupled complex, hierarchical systems. Improved scientific and social understanding with accelerated progress can be achieved by combining research methods and results across Nano-Bio-Info-Cogno (NBIC) disciplines of science and technology. Figure 13.2 shows the "NBIC tetrahedron," which symbolizes this convergence. Each field is represented by a vertex, each pair of fields by a line, each set of three fields by a surface, and the entire union of all four fields by the volume of the tetrahedron.

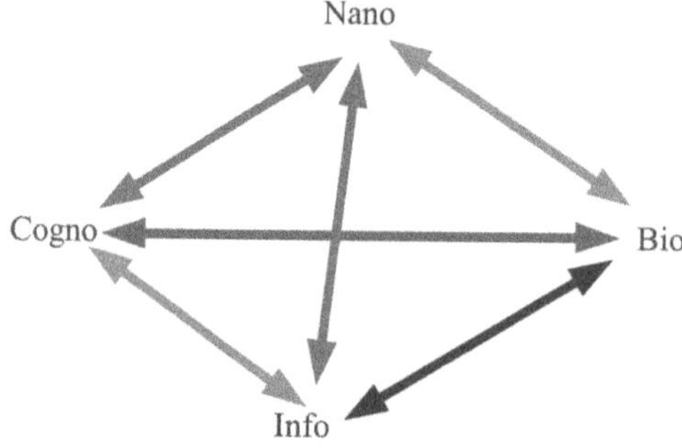

Figure 13.2 NBIC tetrahedron

Examples of areas where CT may play important role include improving work efficiency, learning, enhancing individual sensory and cognitive capabilities. CTs also help greatly revolutionary changes in healthcare, improving both individual and group creativity. Highly effective communication techniques including brain-to-brain interaction, perfecting human-machine interfaces including neuromorphic engineering, sustainable and "intelligent" environments including neuro-ergonomics and enhancing human capabilities for defense purposes are some recent areas where CT technologies have emerged with tremendous progress. NBIC tools are also finding wide applications in sustainable development with proper attention to ethical issues and societal needs, converging technologies could achieve a tremendous improvement in human abilities, societal outcomes, the nation's productivity, and the quality of life. This technology will be a broad, cross-cutting, emerging and timely opportunity of interest to individuals, society and humanity in the long term.

13.2 Synergic Interplay of Intersecting Technologies in CT

Many technology trends have been enabled by the contributions of two or more intersecting technologies. Consider, for example, MEMS-based molecular diagnostics, biomaterials, biological-based computing, and biomimetic robotics. Various technologies have combined in the past to enable applications, but there has been an increase in multidisciplinary areas to assess system challenges and examine approaches in a unified way rather than through a hierarchical relationship.

Materials scientists, for example, are working increasingly with computer scientists and application engineers to develop biomedical materials for artificial tissues or to develop reactive materials to facilitate active system control surfaces. Materials are also being developed and adapted as embedded sensors and actuators for smart structures.

13.3 The Synergistic Interplay of Technologies in NBIC Convergent Technologies

The combination of materials technology with biology and the life sciences has at the same time provided knowledge and materials obtained from living organisms to enable a further integration with pervasive effects from design (biomimetics) to end product (bionics). It can be noted that these examples show different combinations of technologies resulting in different advances. The entire set of technologies provides a rich mix of contributions to the overall technology revolution.

In addition to their technology and artifacts, different fields also tend to produce different views and approaches to the world. Combining these views also enriches the scientific toolbox used on a problem, resulting in advances that combine the best of each world and enabling applications that would not be possible otherwise. For example, engineers increasingly turn to biologists to understand how living organisms solve problems in the natural environment. Rather than blindly copying nature, such "biomimetic" endeavors often combine the best solutions from nature with artificially engineered components to develop a system that is better for the particular environment than any existing organism.

Many significant new trends result in evolving emerging technologies from multiple areas. Figure 13.3 shows examples of the interplay between biotechnology, nanotechnology, materials technology, and information technology areas. Smart materials are contributing both function and processing capabilities. This figure depicts how smart systems are enabled by progress in materials (sensor/actuator) and information technology capabilities together with microsystem trends. Smart systems in turn can be engineered to provide living organisms with bionic capabilities. Living organisms are informing new ways to construct systems (biomimetics) as well as manufacturing components themselves.

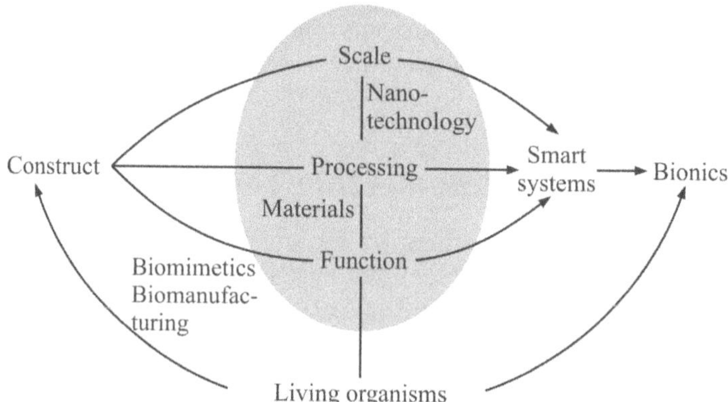

Figure 13.3 Illustrates examples of how nanotechnology (scale), information technology (processing), materials (processing and function), and living organisms interrelate to produce new systems and concepts. Materials provide function, and the emergence of nanotechnology has enabled construction at a scale that integrates function with processing (smart materials)

13.4 Definition and Scope of Various Convergent Technologies as per International Scientific Organizations

The meaning of "converging technologies" (CT) was established in December 2001 workshop organised by the US National Science Foundation and Department of Commerce (2).

(i) The workshop report suggests that Converging technologies - Nanotechnology, Biotechnology, Information Technology and Cognitive enable each other in the pursuit of a common goal for improving human performance and quality of life

(ii) A science and technology foresight report for the Canadian National Research Council also defined the scope of converging technologies for bio-health, eco and food system integrity and disease mitigation – nanotechnology, ecological science, biotechnology, information technology and cognitive sciences (3).

(iii) Norwegian researchers reported "Converging technologies for salmon-productive aquatic environments – bioinformatics, environmental science, systems theory, salmon genomics, production biology, and economics (4).

(iv) Expert group of NSF Workshop also deliberated on the "Converging technologies for natural language processing – information and nanotechnology, linguistics, cognitive and social science and "Converging technologies for the treatment of obesity," and "Converging technologies for intelligent dwelling."

The experts group of European R&D council defined Converging technologies as enabling technologies and knowledge systems (CTEKS) that enable each other in the pursuit of a common goal. This definition captures the scientific and technical potential of CTs and suggests opportunities for European R&D in the larger societal context.

All these CTs definitions provide a list of enabling technologies and technology-enabling sciences, stating that these are converging technologies for the achievement of a more or less general goal to improve human performance and quality of life.

13.4.1 Enabling Technologies

"Enabling technologies" prepare the ground for a wide variety of technical solutions. Because they unlock vast potential and open the door to radically novel technological developments, they are also referred to as "key technologies." Nanotechnology is a prominent enabling technology. Biotechnology and information technology are also enabling, as is the knowledge base of cognitive, social, and other sciences

Thus nanotechnology, biotechnology, and information technology which act as enable technologies need to enable each other as discussed below.

(i) Nanotechnology enables other technologies by providing a common framework for all hardware-level engineering problems. Everything that consists of molecules can, in principle, be integrated with each other. An understanding of properties at the nanoscale allows for the realisation of desirable architectures at the micro - and macroscale. Instrumentally, nanotechnology enables biotechnology by developing new imaging techniques, probes and sensors. It contributes to the miniaturisation demands of information technology. Also, nano-chips and nano-sensors are set to enable advances in the new world of bioinformatics.

(ii) Biotechnology enables other technologies by identifying chemical-physical processes and algorithmic structures in living systems that are traced to their material basis in cellular and genetic organisation. Instrumentally, biotechnology enables nanotechnology by providing mechanisms of cellular recognition and targeted transport. It promises to enable information technology by developing, for example, the foundations for DNA-based computing. Also, biomimetics and the investigation of cellular motors can enable nano-info R&D in nano-robotics.

(iii) Information technology enables other technologies through its ability to represent ever more physical states as information and model processes with a variety of computational methods. Instrumentally, information technology provides the computing power which is essential to the research process in all technical disciplines. It enables nanotechnology through precision control of patterning and intervention. It enables biotechnology by providing the means to model complex processes and thereby solve difficult research problems. Also, simulation software can enable nano-bio R&D in environmental monitoring.

(iv) Social sciences and humanities can enable science and technology in a variety of ways. Familiar examples include development of theoretical strategies for maximising benefit and minimising cost, models for the representation of economic and other forms of exchange, patterns of gestalt recognition in human perception or by machine intelligence, semiotics as a general theory of humanly and naturally produced signs, etc.

Instrumentally, they offer techniques of probabilistic reasoning and statistical inference, methodologies for qualitative research, or an understanding of the social dynamics of the creation and diffusion of technological innovation. Economics and the law enable technology R&D by shaping the incentive structure for its support and diffusion. Philosophy, cultural studies, and ethics provide orientation where new technologies disrupt traditional ways of life. And of course, without an understanding of society, it is all too easy for a technology to be launched improperly and subsequently rejected by society.

Thus the list of enabling technologies and knowledge systems that enable each other can be expanded by including cognitive science, environmental science, systems theory, and social science, including philosophy, economics and the law.

13.5 Major Thrust Areas and Overall Potential of Converging Technologies

US NSF December 2001 workshop, evaluated the potential impact of NBIC convergent technologies on improving human capabilities at the microscopic, individual, group, and societal levels and identified vast potential in six different thrust areas of relevance as presented below

(i) **Synergistic development of nano-, bio-, information- and cognition-based technologies:** These converging technologies provide an outstanding opportunity at the interface and frontier of sciences and engineering in the following decades.

(ii) **Expanding human cognition and communication**: "The Human Cognome research programs" provide potential multidisciplinary areas to understand the structure, functions, and potential enhancement of the human mind. Development of Personal sensory device interfaces; enriched community through humanized technology; learning how to learn; and enhanced tools for creativity are also possible areas of great impact of Convergent Technologies in this area.

(iii) **Improving human health and physical capabilities:** Six priority areas have been identified which are expected to have great potential for development of convergent technologies in this area:

- (a) nano-bio processors for research and development of treatments including those resulting from bioinformatics,
- (b) genomics and proteomics;
- (c) nanotechnology-based implants and regenerative biosystems as replacements for human organs or for monitoring of physiological well-being;
- (d) nanoscale machines and comparable unobtrusive tools for medical intervention;
- (e) multi-modality platforms for increasing sensorial capabilities, particularly for visual and hearing impaired people;
- (f) brain-to-brain and brain-to-machine interfaces; and virtual environments for training, design, and forms of work unlimited by distance or the physical scale on which it is performed.

(iv) **Enhancing group and societal outcomes:** An NBIC system called "The Communicator" would remove barriers to communication caused by physical disabilities, language differences, geographic distance, and variations in knowledge, thus greatly enhancing the effectiveness of cooperation in schools, corporations, government agencies, and across the world. Other areas of focus are in enhancing group creativity and productivity, cognitive engineering and developments related to networked society. Revolutionary new products and services based on the integration of the four technologies from the nanoscale will play important role.

(v) **National security:** Given the radically changing nature of conflict in this new century, several opportunities emerge to strengthen national defense offered by technological convergence in the following areas: data linkage and threat anticipation; uninhabited combat vehicles; war fighter education and training; responses to chemical, biological, radiological and explosive threats; war fighter systems; non-drug treatments to enhance human performance; and applications of human-machine interfaces.

(vi) **Unifying science and education:** To meet the coming challenges, scientific education needs radical transformation from elementary school through post-graduate training. Convergence technologies which are based on previously separate scientific disciplines and fields of engineering cannot take place without the emergence of new kinds of people who understand multiple fields in depth and can intelligently work to integrate them. New curricula, new concepts to provide intellectual coherence, and new forms of educational institutions based on the adaptation of CTs will be necessary.

13.6 Future Visions for Unique & Significant Applications of Convergent Technologies for improvement of Human Performance and Quality of Life

US NSF sponsored workshop 2001 on Convergent technologies provided the future visions in the following typical areas where CTs likely to play significant role in improving human performance and quality of life in the coming 10-20 years.

(i) **Engineering for the mind and a healthy body:** As opposed to the "engineering of the mind"-approach, CTEKS prioritize engineering for the mind. It favors CT "software" over ambitions to engineer the mind by implanting electronic enhancement devices, creating brain-machine interfaces, etc. Information technology, in particular, offers a vast resource of designs that can support and improve social interaction and social decision-making in diverse ageing societies. Many of these draw on enabling knowledge systems such as social science, cognitive psychology, economics, or the law. For their physical implementation, many rely on nanotechnological miniaturization trends. CTs for development of Fast, broadband interfaces directly between the human brain and machines which will transform work in factories, control automobiles, ensure military superiority, and enable new sports, art forms and modes of interaction between people.

(ii) **Converging technologies for the treatment of obesity:** Obesity is simultaneously a problem of individual and public health, it concerns genetics, physiology, nutrition research, exercise science, the psychology of eating disorders, food producers and sellers, social and cultural studies of gender, ethnicity, class, and the representation of the body in advertisement and popular culture. All these areas of research must converge in the treatment of obesity. They may be aided by nano-tagging of food products, integrated models of consumption patterns, and information technology assistants for self-monitoring of food-intake.

(iii) **Converging technologies for intelligent dwelling:** Buildings can generate and distribute energy, they can process and recycle waste, they can collect and purify water. These capabilities should be improved through bio-mimetic technologies, by integrating photo-voltaics in smart materials, by incorporating environmental sensors in information and regulation systems. In any city of the future, each building contributes to the maintenance of neighboring buildings and urban infrastructure at large by exploiting its own specific capabilities that are determined by its location, structure, and material make-up. Buildings will generate and share energy among each other, invisibly small renewable energy collectors may be integrated in devices in a manner that emulates nature, new energy carriers and forms of energy storage can be explored. The goal of this research could be that within a century nearly all energy needs can be met by way of locally generated renewable energy.

(iv) Comfortable, wearable sensors and computers will enhance every person's awareness of his or her health condition, environment, chemical pollutants, potential hazards, and information of interest about local businesses, natural resources, etc.

(v) Robots and software agents will be far more useful for human beings, because they will operate on principles compatible with human goals, awareness, and personality.

(vi) People from all backgrounds and of all ranges of ability will learn valuable new knowledge and skills more reliably and quickly, whether in school, on the job, or at home.

(vii) Individuals and teams will be able to communicate and cooperate profitably across traditional barriers of culture, language, distance, and professional specialization, thus greatly increasing the effectiveness of groups, organizations, and multinational partnerships. The human body will be more durable, healthier, more energetic, easier to repair, and more resistant to many kinds of stress, biological threats, and aging processes.

(viii) Machines and structures of all kinds, from homes to aircraft, will be constructed of materials that have exactly the desired properties, including the ability to adapt to changing situations, high energy efficiency, and environmental friendliness.

(ix) A combination of technologies and treatments will compensate for many physical and mental disabilities and will eradicate altogether some handicaps that have plagued the lives of millions of people.

(x) National security will be greatly strengthened by lightweight, information-rich war fighting systems, capable uninhabited combat vehicles, adaptable smart materials, invulnerable data networks, superior intelligence-gathering systems, and effective measures against biological, chemical, radiological, and nuclear attacks.

(xi) Anywhere in the world, an individual will have instantaneous access to needed information, whether practical or scientific in nature, in a form tailored for most effective use by the particular individual.

(xii) Engineers, artists, architects, and designers will experience tremendously expanded creative abilities, both with a variety of new tools and through improved understanding of the wellsprings of human creativity.

(xiii) The ability to control the genetics of humans, animals, and agricultural plants will greatly benefit human welfare; widespread consensus about ethical, legal, and moral issues will be built in the process.

(xiv) The vast promise of outer space will finally be realized by means of efficient launch vehicles, robotic construction of extraterrestrial bases, and profitable exploitation of the resources of the Moon, Mars, or near-Earth approaching asteroids.

(xv) New organizational structures and management principles based on fast, reliable communication of needed information will vastly increase the effectiveness of administrators in business, education, and government.

(xvi) Average persons, as well as policymakers, will have a vastly improved awareness of the cognitive, social, and biological forces operating their lives, enabling far better adjustment, creativity, and daily decision making.

(xvii) Factories of tomorrow will be organized around converging technologies and increased human-machine capabilities as "intelligent environments" that achieve the maximum benefits of both mass production and custom design.

(xviii) Agriculture and the food industry will greatly increase yields and reduce spoilage through networks of cheap, smart sensors that constantly monitor the condition and needs of plants, animals, and farm products.

(xix) Transportation will be safe, cheap, and fast, due to ubiquitous real time information systems, extremely high-efficiency vehicle designs, and the use of synthetic materials and machines fabricated from the nanoscale for optimum performance.

(xx) The work of scientists will be revolutionized by importing approaches pioneered in other sciences, for example, genetic research employing principles from natural language processing and cultural research employing principles from genetics.

(xxi) Formal education will be transformed by a unified but diverse curriculum based on a comprehensive, hierarchical intellectual paradigm for understanding the architecture of the physical world from the nanoscale through the cosmic scale.

If Governments make the correct decisions and investments today, any of these visions could be achieved within 20 years' time. Moving forward simultaneously along many of these paths could achieve a golden age that would be a turning point for human productivity and quality of life. Technological convergence could become the framework for human convergence (5). The twenty-first century could end in world peace, universal prosperity, and evolution to a higher level of compassion and accomplishment.

13.7 Limits of Convergent Technologies

Though CTs which are enabling technologies and knowledge systems that enable each other indicate that their potential is unlimited, they have number of limitations which are discussed below.

(i) In future CTs using Lab on Chip technologies will enable individuals with physical as well as psychological problems to be quickly identified, and courses of treatment tailored to relate to their genome and severity of the illness. However self-monitoring may also open a negative feedback- loop that amplifies anxieties and obsessions. Since anxieties can have measurable effects on physical variables, these may be reinforced for individuals who constantly monitor themselves for changes in their state of health.

(ii) CTs create a new challenge for the allocation of R & D resources. Research administrations will have to dedicate substantial resources to research initiatives that cannot yet be expressed in the standard form of work programs. Only an understanding of the medium- and long-term feasibility and limitations of CTs that promise great social benefits and high marketability can justify the allocation of public funds.

(iii) CTs can be used to advance the goal of sustainable development. However it might be thought that consumption and harmful emissions may not have to be curbed if CTs can reduce overall environment stresses. Since complacency induced by the fix-all potential of technology could be dangerous in the extreme, sound environmental policy requires an awareness especially of present and medium-term technical limits. The limits of CT research need to be assessed in order to frame perceptions of risk.

(iv) Nanotechnology's promise to harness physical self-organisation and to make materials with novel properties has led at least one reinsurance company to question the insurability, for example of nanoparticles (6). To the extent that CT researchers really set out to produce unpredictabilities, their research must be monitored and constrained accordingly. Here, philosophers of science are called upon to study the limits of the CT research process. They can investigate to what extent scientists and engineers actually draw on theories of complexity and self-organisation and whether these theories can be technically useful for creating genuine novelty.

    (i) CT visionaries and their critics tend to assume not only that brain implants to enhance mental capacities are technically feasible. They also make assumptions about their diffusion. However the brain-implants fail as consumer products due to varying human behavioural problems and complex cognitive processes.

    (ii) While it is possible to invent, manufacture, and fit an artificial leg with little understanding of anatomy, physiology and the dynamics of walking, this would obviously amount to a dramatic technical and economic failure.

    (iii) Similarly, some visions for NBIC convergence imagine cognitive enhancements while underestimating the complexity of cognitive processes.

(iv) To the extent that CT engineering is done from the bottom up in the nano-, micro- or bio-domains it is presupposed that the problems can be resolved at their physical basis. This approach is limited by problems that have no physical properties. Arguably, cognition, questions of meaning, patterns of social interaction and emotional response derive from socialisation rather than physical facts. Thus, the true extent to which illness and disease have social, perhaps psychosomatic origins may only become apparent when complete medical self-monitoring CTs give the patient a clean bill of physical health and thereby make him or her feel more poorly still.

(v) The broadly transformative potential of CTs sets limits to their public acceptance. The pace of the diffusion of new technologies is constrained by the pace in which societies accept and, if so, accommodate them. Here the social sciences and humanities are needed to inform and accompany CT research and to serve as intermediaries. They should create settings within which science and technology researchers on the one hand and their various publics on the other, can learn from each other.

(vi) Innovation studies and the history of technology have demonstrated that even in a global economy the absorption of a new technology and its economic exploitation vary according to the particular society into which it is introduced. We know, for example, that cell-phone technology has developed very differently in Portugal, Germany, Finland, Italy, and Poland. At each location, the diffusion of technology and its integration into every-day life are shaped by economic policies, cultural resistance and support, complementarity with established technologies, and many other factors.

Thus nanotechnology and cognitive science, information technology and philosophy, biotechnology and the social sciences need to work together when the prospects, including near- and medium-term limits of proposed CT research programs are determined. CT research must therefore include a study of current limits. In order to avoid bad public investments, cognitive science research is needed for an assessment of how it can most usefully enable specific nano-, micro- or bio- technologies

13.8 Strategies for Effective Realization of Benefits of Convergence Technologies for the Improvement Human Performance and Quality of Life

The NSF workshop identified the following general strategies for achieving convergence:

(i) We should prepare key organizations and social activities for the envisioned changes made possible by converging technologies. This requires establishing long-term goals for major organizations and modeling them to be most effective in the new setting.

(ii) Activities must be enhanced that accelerate convergence of technologies for improving human performance, including focused research, development, and design; increasing synergy from the nanoscale; developing interfaces among sciences and technologies;

and taking a holistic approach to monitor the resultant societal evolution. The aim is to offer individuals and groups an increased range of attractive choices while preserving fundamental values such as privacy, safety, and moral responsibility. A research and development program for exploring the long-term potential is needed.

(iii) Education and training at all levels should use converging technologies as well as prepare people to take advantage of them. Interdisciplinary education programs, especially in graduate school, can create a new generation of scientists and engineers who are comfortable working across fields and collaborating with colleagues from a variety of specialties. Essential to this effort is the integration of research and education that combines theoretical training with experience gained in the laboratory and industry.

(iv) Experimentation with innovative ideas is needed to focus and motivate needed multidisciplinary developments. Professional societies could set performance targets and establish criteria for measuring progress toward them.

(v) Concentrated multidisciplinary research thrusts could achieve crucially important results. Among the most promising of such proposed endeavors are the Human Cognome Project to understand the nature of the human mind, the development of a "Communicator" system to optimize human teams and organizations, and the drive to enhance human physiology and physical performance. Such efforts probably require the establishment of networks of research centers dedicated to each goal, funded by coalitions of government agencies and operated by consortia of universities and corporations.

(vi) Flourishing communities of NBIC scientists and engineers will need a variety of multiuser, multiuse research and information facilities. Among these will be data infrastructure archives, that employ advanced digital technology to serve a wide range of clients, including government agencies, industrial designers, and university laboratories.

(vii) Integration of the sciences will require establishment of a shared culture that spans across existing fields. Interdisciplinary journals, periodic new conferences, and formal partnerships between professional organizations must be established..

(viii) We must find ways to address ethical, legal, and moral concerns, throughout the process of research, development, and deployment of convergent technologies. This will require new mechanisms to ensure representation of the public interest in all major NBIC projects, to incorporate ethical and social-scientific education in the training of scientists and engineers, and to ensure that policy makers are thoroughly aware of the scientific and engineering implications of the issues they face.

(ix) It is necessary to accelerate developments in medical technology and healthcare in order to obtain maximum benefit from converging technologies, including molecular medicine and nano-engineered medication delivery systems, assistive devices to alleviate mental and emotional disabilities, rapid sensing and preventive measures to block the spread of infectious and environmental diseases, continuous detection and correction of abnormal individual health indications, and integration of genetic therapy and genome-aware treatment into daily medical practice. To accomplish this, research laboratories, pharmaceutical companies, hospitals and health maintenance organizations, and medical schools will need to expand greatly their institutional partnerships and technical scope.

Thus, in order to attain the maximum benefit from scientific progress, the goal can be nothing less than a fundamental transformation of science and engineering.

Fundamental research will also be essential in engineering, including computer engineering, because engineers must be ready in the future to take on entirely new tasks from those they have traditionally handled. The traditional tool kit of engineering methods will be of limited utility in some of the most important areas of technological convergence, so new tools will have to be created. This has already begun to happen in nanotechnology, but much work remains to be done developing engineering solutions to the problems raised by biology, information, and the human mind.

References

1. Horn R. (2002). Visual languge, Stanford U. http://www.stanford.edu/~ rhorn/index.html
2. Converging Technologies for Improving Human Performance, Nanotechnology, Biotechnology, Information Technology And Cognitive Science (NSF/DOC-sponsored report), Edited by Mihail C. Roco and William Sims Bainbridge, National Science Foundation, June 2002, Arlington, Virginia.
3. Raymond Bouchard, Bio-Systemics Synthesis; Science and Technology Foresight Pilot Project, Canadian National Research Council, June 2003, 8.
4. In his presentation at the Norwegian Research Council, May 5 2004, StigOmholt (Agricultural University of Norway) proposed a somewhat more general CT research agenda: Converging technologies for a new production biology to revert world-wide habitat fragmentation and loss of biodiversity on land and at sea.
5. Ostrom E., T. Dietz., P.C. Stern., S. Stonich, and E.U. Weber. (2002). The drama of the commons. Washington D.C.: National Academy Press.
6. Swiss Reinsurance Company, Nanotechnology - Small Matter, Many Unknowns (2004).

14

Opportunities and Challenges in Newly Emerging Nano-Technologies and Strategies for their Utilization for Sustainable Development

14.1 Opportunities and Issues/ Concerns in the Application of Nano Technologies

14.1.1 Opportunities

Nano Technologies are enabling a number of present technologies for greatly improving quality of life and human performance. Advances in bio/nano/materials/info technologies are for their integration in devices and systems which help greatly in monitoring now facilitaties individual and public health and safety; economic, social and political systems; and business and commerce.

The emerging nano technology revolution, together with the ongoing process of globalization enabled by the information technology and continued improvements in transportation are providing potential possibilities for increased life span, economic prosperity, and quality of life (1). Enormous benefits are expected in diverse areas like health care, drug development, information and communication technology (ICT), water decontamination, and production of stronger and lighter materials due to which governments and businesses are rapidly investing huge amounts of money in nanotechnologies.

14.1.2 Concerns/Issues

While it is true that nanotechnology has the potential to revolutionize electronics, aerospace, food, healthcare, environment, energy, water and medicine, as with other technologies, its misuse also has the potential to harm humanity.

Nanotechnology can be used to make weapons with enhanced properties, such as improved body armour and enhanced battlefield communication systems (2), which can be misused.

Regulated and beneficial drugs could augment healthcare in rich nations but become unavailable to the poor. Unregulated nanoengineered recreational drugs could be produced and distributed with criminal intent.

Nanotechnology promises smaller, cheaper, and more ubiquitous sensing devices (3) that could be linked over networks to provide greater safety, security, and better healthcare. However, these networks could also be used to limit individual or group privacy through covert surveillance, to collect and distribute personal information without consent, and to concentrate access to such information to enable discriminatory policing, profiling, and social sorting (4).

To assess the impact of a nanomaterial, it will be critically important to determine how nanomaterials interact with their environment as a result of intentional or unintentional release. Determining their distribution, fate, and transformation processes will be vital. It will be especially critical to know the biopersistence of nanomaterial, how much of it there is, where it is, and in what chemical form. Currently, very little is known about how nanoscale materials move in the environment.

In view of the above concerns it has been recognized by many scientific agencies that extensive applications of emerging nanotechnologies (due to their complex and multi disciplinary nature) face new challenges in the education, inter-disciplinary training, research infrastructure, safety, regulatory and ethical areas which need to be analyzed and assessed before for implementing at large scale for the benefit of societies.

14.2 Challenges in the Application of Nano Technologies

14.2.1 Grand Challenges

The following are some "grand challenges" identified by National Science Foundation (NSF) USA which need to be addressed for rapid growth of nanoscale science and engineering at commercial level.

(a) "How do we perfect cheap, efficient and scalable ways for the synthesis and fabrication of nanomaterials and nanostructures with tailored properties?"

(b) "How do we characterize and control nanoscale materials and phenomena?"

(c) How do we study and evaluate the environmental, health, safety, economic and societal impacts of nanomaterials and nanotechnology?

(d) How can we predict and design new properties of nanomaterials?

(e) How do we foster the safe and ethical development of nanomaterials and technologies?

These grand challenges in nanotechnology lead to more fundamental questions. For example to increase the efficiency of nano synthesis and nanofabrication methods, can we develop methods with precise enough control to facilitate the assembly of elaborate multifunctional architectures, atom by atom, from the bottom up or can we engineer functional systems at the atomic and molecular scale? The attempt to answer these questions provides the basis of nanotechnology (5,6).

14.2.2 Emerging Nanotechnologies – Specific Challenges/Barriers/ Solutions in their Application in Various Fields

Many scientific agencies have identified that there are major research challenges for completely understanding the implications of nanotechnology on health and the environment as they being highly interdisciplinary by nature and need to be thoroughly examined and addressed before taking up large scale application.

To assess the impact of new nano materials/systems with the surrounding ecosystems and human population and their ecosystem compatibility, it is necessary to examine various emerging nano technologies for use in different fields, new materials likely to be used, existing barriers/challenges and nano-solutions to assess and plan for future skilled work force and infrastructure to be developed. These issues are briefly discussed below.

14.3 Nanotechnology-Sustainable Processes-Materials and Resources

Materials and process technology sustainability has become a topic of great interest in nanotechnology (7). Materials scientists have had a great impact on the improved quality of released effluent and exhaust plumes, the use of catalysts to avoid unwanted by products in chemical processes, and the treatment of waste. They also have led the way in moving from the use of solvents in industrial processes to the use of biodegradable materials and the generation of cleaner energy.

Designers of materials for such purposes, however, have not always considered the environmental impact as well as material functionality, and "the costs and benefits of synthesizing and processing" (7). For example, efforts at greening vehicles, such as battery-powered vehicles, fall short of impacting the environment in a positive way. The vehicles appear to be a cleaner choice when compared with gasoline-powered vehicles because there are no tailpipe emissions produced during operation. However, large quantities of toxic and hazardous materials are used to make batteries, and this must be considered in the assessment of environmental benefits of this technology (8). Thus, applications must address societal needs with respect to sustainability over their entire lifecycle. To develop these applications, it will be necessary for nano-scientists to collaborate with scientists and engineers from other disciplines.

Issues: Life Cycle Design: For achieving sustainable processes, materials and resources need a life cycle design approach for which technology specific R & D is necessary.

Considering the photovoltaic example, durable functionality, recyclability, and use of benign materials in developing the products must be considered. In the area of nutrient optimization, green manufacturing and global nitrogen/phosphorous cycles need to be addressed.

Thus Nanotechnology processes are expected to have significant potential to impact conventional and future manufacturing processes. However, these processes may also carry environmental and social impacts as well, so it is necessary to try to foresee and prevent negative impacts as new processes are developed. Integrated processes and bottom-up assembly can serve human needs and are compatible with the surrounding ecosystems and human population.

14.3.1 Manufacturing using Integrated Nanodevices

The following operations integrated using nano material usage will help in achieving environmental compatibility

1. sensors, actuators, and multifunctional devices (e.g., a catalytic reactor that also performs separations);
2. transformation of unit operations (e.g., building nanosize reactors, pumps, mixers, and separators) moving from micro to nanofluidics and constructing "factories on particles"; and
3. self-assembly, bottom-up manufacturing, directed assembly.

14.3.2 Designing and Manufacturing Nano-Materials

Challenges: Recent developments have suggested a new paradigm for ecofriendly manufacturing process (Figure 14.1) need to have the following objectives:

1. eliminate material waste;
2. reduce resources used in manufacturing processes;
3. minimize energy use; and
4. assess the safety, environmental, & ethical aspects of nanoscale manufacturing.

For incorporating these aspects in nano production process we need to address following challenges:

1. large-scale production of nanoscale building blocks (e.g., nanotubes, diamondoids, quantum dots);
2. design and production of complex, nanostructured materials (e.g., nanocomposites, multifunctional catalysts, electrolytes); and
3. the possible creation of a Govt funded user facility in bottom-up processing for pilot-scale synthesis and production of nanomaterials.

14.3.3 Optimization of Nano Materials Engineering Process – Concept to Product /Application

Developments in materials science and engineering result from interdisciplinary materials research. Figure 14.1 provides an example of the development process (the schematic description of the materials engineering process from concept to product/application) in the area of electroactive polymers for robotic devices and artificial muscles.

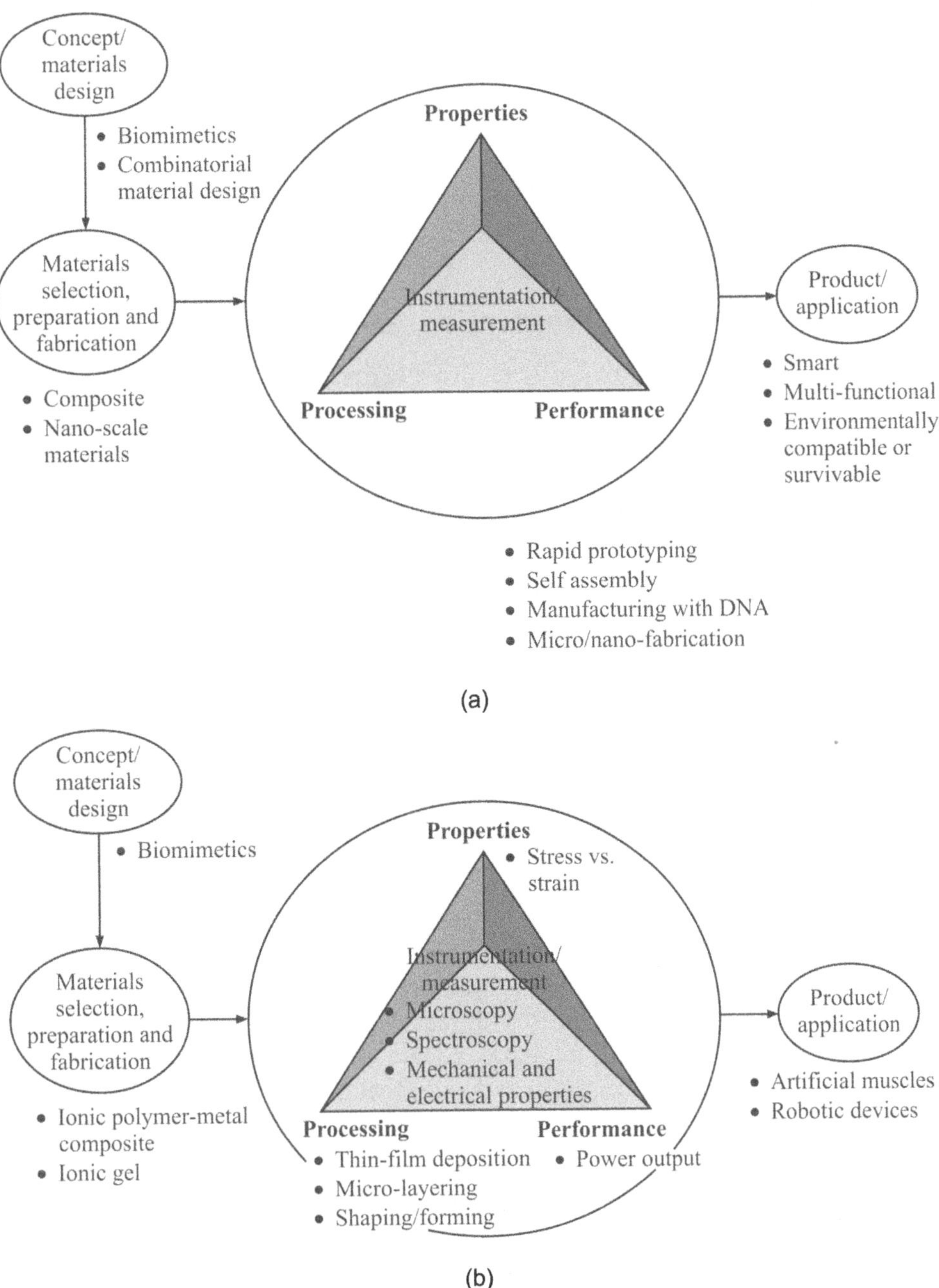

Figure 14.1 (a) General materials engineering process, (b) Engineering Process for electro active polymers (Source National Research Council, 1989 [123], p.29)

Nano materials with properties that can be controlled at submicrometer ($<10^{-6}$ m) or nanometer (10^{-9} m) level, are an increasingly active area of research because properties in these size regimes are often fundamentally different from those of ordinary materials. Examples include carbon nanotubes, quantum dots, and biological molecules. These materials can be prepared either by purification methods or by tailored fabrication methods.

However these process need to be examined with reference their impact on environment, cost and process optimization for desired properties of products.

Combinatorial materials design uses computing power (sometimes together with massive parallel experimentation) to screen many different materials possibilities to optimize properties for specific applications (e.g., catalysts, drugs, optical materials).

14.3.4 Materials Selection, Preparation, and Fabrication

Composites are combinations of metals, ceramics, polymers, and biological materials that allow multi-functional behavior. One common practice is reinforcing polymers or ceramics with ceramic fibers to increase strength while retaining light weight and avoiding the brittleness of the monolithic ceramic. Materials used in the body often combine biological and structural functions (e.g., the encapsulation of drugs).

14.3.5 Processing, Properties, and Performance

These areas are inextricably linked to each other: Processing determines properties that in turn determine performance. Moreover, the sensitivity of instrumentation and measurement capability is often the enabling factor in optimizing processing, for example, as for nanotechnology and micro-electromechanical systems (MEMS).

14.3.6 Product/Application

The trends described above will likely work in concert to provide materials engineers with the capability to design and produce advanced materials that will be:

- *Smart*—Reactive materials combining sensors and actuators, perhaps together with computers, to enable response to environmental conditions and changes thereof. (Note, however, that limitations include the sensitivity of sensors, the performance of actuators, and the availability of power sources with required magnitude compatible with the desired size of the system.) An example might be robots that mimic insects or birds for applications such as space exploration, hazardous materials location and treatment, and unmanned aerial vehicles (UAVs).
- *Multi-functional*—MEMS and the "lab-on-a-chip" are excellent examples of systems that combine several functions. Another example is a drug delivery system using a hydrogel with hydrophilic exterior and hydrophobic interior. Consider also aircraft skins fabricated from radar-absorbing materials that incorporate avionic links and the ability to modify shape in response to airflow.
- *Environmentally compatible or survivable* —The development of composite materials and the ability to tailor materials at the atomic level will likely provide opportunities to make materials more compatible with the environments in which they will be used. Examples might include prosthetic devices that serve as templates for the growth of natural tissue and

structural materials that strengthen during service (e.g., through temperature- or stress-induced phase changes).

14.4 Possible Nano based Solutions for Addressing the Challenges

Developing theories, models, and experimental data on nanoscale materials and processes in nanoscale manufacturing processes will greatly help in addressing the challenges for sustainable production processes (7-10).

Research programs on

1. thermo/kinetics/transport fundamental studies at the nanoscale level;
2. linking macro/micro/nano/atomic regimes using quantum, molecular, and continuum modeling; and
3. surface properties (intermolecular forces, surface area, surface charge, surface chemistry) will help in solving the many process optimization issues.

Development of safety and environmental metrics and ethical principles appropriate for nanotechnology, such as modifying existing indicators and metrics for use in nanoscale manufacturing (e.g., research decision tools, considering particle aspect ratio, surface area, and reactivity) and adapting current ethical principles from professional societies to the needs of nanotechnology (e.g., what determines responsible use of nanotechnology) are some important challenges. Finally, it also will be challenging to incorporate the concept of sustainability into current and future educational activities related to nanotechnology.

If the challenge of designing and manufacturing nanomaterials through sustainable approach can be achieved, it will be possible to produce bulk quantities of other nanoparticles such as carbon nanotubes, diamon-doids, and fullerenes. At present MEMS/NEMS are produced using photolithography, a material-and energy-intensive process. However, bottom-up techniques that use self-assembly and directed assembly could produce molecular electronic devices that are smaller and use the quantum effects inherent in the nanoparticles (7). Thus, the architectures would be as novel as the particle properties. DNA functionality could be used to drive assembly.

Current efforts in pharmaceutical delivery have included drug delivery and laboratory-on-a-chip techniques. One paradigm shift envisions a "factory on a particle," in which small-scale pharmaceutical production occurs inside the body (11). The nanoscale factory that produces the drug could be implanted, and benign raw material could be injected. Sensors in the "factory" would determine when to turn it on and when to turn it off. Transportation, production, and legal costs could be greatly reduced. The factories would be distributed spatially, as each patient would have a personal factory

14.5 Impact of Nano Technologies on Various Industries and Challenges

14.5.1 Nano Informatics

The use of massively parallel arrays brings with it a need for the development of new computational methods for understanding signal transduction in nanoscale systems and analyzing large amounts of data from nanoscale systems in real time. This may lead to a unique field of "nano-informatics," analogous to the explosion of interest in bioinformatics. There are several aspects of nano-informatics applicable to nanotechnology and the environment. New challenges in understanding signal transduction and data analysis in nanoscale arrays and networks, systems issues for deployment, communication networking.

Challenges: The fabrication of thousands or millions of "perfect" sensing elements will be extremely difficult or impossible. The need for perfection in the sensor "hardware" can be avoided by using advanced computer training and measurement methods.

The development of nanoscale sensor arrays ultimately requires the ability to fabricate, integrate, and interface them with the macroscopic world. Furthermore, this must be achieved in a cost effective manner. Recent advances in the ability to prepare and chemically/biochemically modify nanoscale "components" need to be leveraged into a more general set of nanoscale assembly methods that would constitute a type of "nano-toolbox" for fabricating more complex structures. One of the fundamental problems in using nanoscale materials is that "bottom-up" fabrication processes, in which complex nanoscale assemblies are made in a linear, sequential manner, are extremely susceptible to failure because of the large number of steps involved.

Solutions: Recent experiments have shown that direct current and alternating current electric fields can be used to translate and align nanotubes and nanowires; magnetic fields can be used in a similar way to manipulate nanoscale materials such as nickel nanowires and quantum dots (5, 6). These results suggest that it should be possible to use electric and magnetic fields, together with fluid flow and other methods, to control the assembly of nanoscale objects into more complex two-and three-dimensional assemblies. However the implementation these new schemes need well trained work force and infrastructure.

14.5.2 Telecom and Datacom

The ability to engineer and integrate both optical and electrical nanomaterials will bring about tremendous advances in the applications in telecommunication and data communication sectors by providing low power, high speed, interference-free devices such as electro optic and all-optical switches integrated on computer chips. These developments will lead to the further integration of related components such as the massive storage of data on compact storage devices with high-performance computing features. This in turn will enable new high speed communications and high performance information processing that will impact several sectors (Information Technology, Energy, Health, Medicine, Environment).

14.5.3 Impact of Nanotechnolgies on Energy Technologies (12-32)

Nanoscience and technology will significantly impact the renewable energy sector. For instance, the incorporation of low-cost, earth abundant nanomaterials in energy conversion and storage device architectures will improve the efficiency, energy and power densities of fuel cells, capacitors, and batteries for portable applications, transportation, and large scale grid storage. These energy storage and conversion devices are important to many areas of technology, including portable electronics, medical devices, power tools, transportation and the storage of electricity produced by intermittent renewable sources (wind and solar).

Advances in this area will accelerate the design and implementation of on-site energy generation and decentralized energy generation systems. These systems will change how we generate, deliver and use electricity. The development of decentralized generation networks enables collection of energy from many sources which offer the promise of reducing environmental impacts and improving our security of supply.

***Energy and transportation*:** In the near term, fossil fuels will be used as the main source of energy. Improvement in the performance of both gas and diesel engines is needed. To enable the production of more super ultra low emission vehicles, higher quality fuels are needed. This will require advances in catalyst technology to: improve catalyst reactivity, selectivity, and yield; optimize and reduce active species loading levels; improve catalyst durability and stability under exposure to the operating environment; reduce reliance on precious metal-based and corrosive catalysts; and produce lower cost, less energy-intensive, and more environmentally friendly catalysts. In essence, catalytic processes are nanoscale because reactions take place on the surface.

***Challenges*:** To realize the above improvements, continued research and development are needed to advance understanding of molecular and particle behavior in catalytic reactions. Basic research to develop methods and approaches for optimizing control of stability at all scales and conditions of use will be critical to achieve these advances. Basic research in synthesis, assembly, and processing will be required to develop fabrication techniques for controlling surface area, cluster and particle structure, component dispersion, and other defining characteristics (12). Advances in nanoscale science that enable sequestering of CO_2 and separations applications also will also be beneficial.

***Solutions*:** Alternative energy transduction, storage, and transmission materials can be improved in the near term. Ultracapacitors and batteries will benefit from moving to the nanoscale. Breakthroughs in the performance of thermos -electrics have already occurred due to advancements at the nanoscale. Methods for assembling nanoparticles need to be explored in more depth, as does nanoscale wiring to move electrons when designing new energy-conversion devices. Photovoltaics can be greatly improved by implementing photonic crystals concepts, which improve light collection and emissivity. Quantum effects made possible by nanotechnology might lead to increases in efficiency. Bridging the molecular scale and ordered bulk will be particularly important for space-efficient energy storage. For any country to realize the benefits of the nano technological innovations necessary work force and infrastructure need to be created with appropriate funding.

14.5.4 Impact of Nano Technologies on Environment (33-39)

Nanotechnology will play a heavy role in addressing critical environmental problems for water purification and remediation. For instance, new nanomembrane technologies will enable efficient water purification processes, and facilitate the development of improved remediation strategies for clean-up of hazardous waste streams. Nanomaterials in various shapes/morphologies (particles, tubes, wires, fibers) will be made to function as adsorbents and catalysts for the detection and removal of toxic gases (SO_2, CO, NO_x, H_2S), inorganics contaminants (arsenic, nitrate, heavy metals), organic pollutants (polyaromatic hydrocarbons, Bisphenol A) and biological substances (spores, viruses, bacteria). Nanotechnology will also create new 'green' synthesis and processing technologies that can minimize the generation of undesirable by-products in effluents and emissions streams.

***Nanotechnology applications for measurement of environmental parameters*:** The unique properties of nanoscale materials will enable the development of a new generation of environmental sensing systems. Examples include: (1) distributed arrays of smart sensor networks that could be used to quantify the spatial, chemical, and biological dynamics of an eco-system in real time; (2) small, inexpensive, low-power, multifunctional sensor arrays, distributed in public places, homes, or on individuals, that could be used to warn of pollutants and other environmental hazards; and (3) improved characterization of the life cycle of natural and anthropogenic nanoparticles in the environment. Measurement science and technology will enable the development of a comprehensive understanding of the interaction and fate of natural and anthropogenic nanoscale and nanostructured materials in the environment. Examples of uses of these measurements include identifying conditions for safe manufacturing, use, and disposal of nanotubes, nanoparticles, and other nanoscale materials; and understanding how the morphology, size, composition, surface reactivity, and transport of combustion-generated particles affect human health and global climate change.

***Challenges*:** The biotechnology revolution has led to rapid advances in the detection of biomolecules. Nanotechnology provides a way to strongly leverage these advances toward the development of environmental sensing systems that can operate on a continuous mode. The important role of microbes in controlling a large number of environmental processes has been known for many years; however, only recently has it been possible to use genomic analysis to understand biological diversity in environmental systems (40).

From the standpoint of environmental measurement, problems exist in measuring anthropogenic and natural nanoparticles that are present in the soil, air, and water. Particles in liquid phases present unique measurement challenges. Little is known about the diversity of chemical composition at the nanoparticle level and the transformations that occur.

A major challenge in using nanotechnology for environmental analysis is the need for generic nanoscale assembly methods. The usefulness of rapid advances in the ability to fabricate and functionalize individual nanoparticles and nanotubes will be realized when these materials are integrated into more complex assemblies (i.e., dense arrays) and combined with macroscopic materials (i.e., silicon-based microelectronics) in a cost-effective manner. Nanoscale materials such as carbon nanotubes can be fabricated on silicon, but *in situ* chemical/biochemical modification of these structures in an array format has not yet been demonstrated.

Other challenges exist in collecting spectroscopic data for analysis at smaller spatial scales in a timely manner. Existing methods for analyzing nanoparticles, for example, are prohibitively slow and generally non quantitative. In addition, the analytical cost of reducing the interaction volume of the spectroscopic probe (e.g., electrons) to achieve smaller scale measurement is a decrease in the signal-to-noise ratio. Thus, advances in spectroscopic instrument technologies must allow rapid detection of adequate signal strength, while probing increasingly smaller volumes of the nanoparticulate sample.

Nanoscale systems also can be expected to achieve higher sensitivity due to intrinsic quantum effects. For example, there has been a great deal of attention paid to single-molecule or single-particle detection and single-electron transistors. In an analogous manner, a chemically-modified quantum dot or other nanoscale structure should be able to achieve single-molecule sensitivity. A single molecule interacting with the quantum dot, in turn, can relay the sensing signal in terms of an optical property or electrical conductivity.

14.5.5 Impact of Nanotechnologies on Water Purification and Global Drinking Water Supply

New purification schemes, such as microfiltration, reverse osmosis, and photocatalysis, are needed to replace current water purification techniques. New highly distributed and smart water monitoring schemes are needed, as are new membrane technologies (self-assembling pores, adaptable, smart, photoactive, reporter functionality) and new models for nanofiltration/nano separations. New composite materials, water purity sensors, e.g., for chemical speciation, and adaptive multi-functional materials will help to achieve this goal. Use of new water purity sensors will greatly enhance purification schemes.

14.5.6 Role of Nano Technologies on Global Sustainable Agriculture - Optimization of Production and Distribution of Agriculture Products

Current technology requires use of fertilizers or crop rotation to provide nutrient fixation to agricultural crops (35). Nutrient optimization is needed to avoid these excess usage fertilizers. Nano-particles for direct nitrogen fixation might prove revolutionary, or there may be opportunities to engineer soil for fertilization improvement. Excessive use of pesticides, herbicides, and rodenticides is another area that may benefit greatly from R&D at the nano-scale. It will be a great challenge to eliminate or optimize delivery and use of these polluting materials for pest-management applications. Nanotechnology may add benefit to the development of photo-catalysts to facilitate the breakdown of biocides and smart dust to identify and locate biocides. Development of a knowledge base that relates structure and function at the nanoscale and design of new materials and architectures with tailored multi-functionality will aid in the development of agricultural applications.

Challenges: Maintaining and improving soil, water, and air quality represent some of the most formidable challenges facing global society in the 21st century. The earth's population is expected to reach between 10 and 11 billion by 2050 (35). The materials and resources required to sustain such a population are staggering. Furthermore, increasing amounts of pollutants from a variety of sources enter the atmosphere and hydrosphere on a daily basis. For example, at the current rate of fuel consumption, global energy consumption is expected to double by 2050 to nearly 30 TW, and atmospheric levels of carbon dioxide are expected to double by the end of the 21st century (35). It has been suggested that materials and resource use be reinvented. An interdisciplinary and life cycle design approach will be required to achieve the goals of developing and deploying sustainable materials.

14.5.7 Nano Technologies in Health Care, Medicine, Bio-Diagnostics

There will be rapid advancement of the development of biological nano sensors for fast and accurate monitoring of health and biological functions at the point of care level. Nanotechnology is now facilitating the creation of 'lab on a chip' technologies involving the assembly of artificial organs to enable organ-based screens for drug development and treatment.

Advances in nano-based environmental monitoring will lead to real-time, rapid, multimedia measurements of thousands of pollutants. The ability to "mine" this immense database will provide epidemiologists with an unprecedented ability to associate adverse health effects associated with exposures to complex mixtures of pollutants. This application of nano-based monitoring can be broadened to include applications to monitor medical health conditions.

Challenges: Information describing the relative health and environmental risk assessment of nanotechnology and associated nano materials is severely lacking. Due to this knowledge gap, it is uncertain whether one can extrapolate nano material health and environmental risk assessments from information derived from natural or pollution-derived nanoparticles. Only recently some studies have assessed the relative toxicities of nanomaterials (36-39).

Understanding the health and environmental impacts of nanomaterials will be essential given the ever-increasing applications of nanotechnology in society. Several research challenges need to be addressed for the applications of nanotechnology to proceed in a safe and "environmentally friendly" manner.

For nanotechnology to develop in a rapid and responsible manner, it will be critical to generate the appropriate health, safety, and environmental guidelines based on sound science for handling and employing nanomaterials.

14.5.8 Risk Assessment for Nano Materials

14.5.8.1 Characterization of Nanomaterials for Risk Assessment

To conduct risk assessment research, it will be necessary for investigators to have access to well-characterized nanomaterials. Due to the size and physiochemical characteristics of nanomaterials, new monitoring methods and instrumentation will be needed to perform nanomaterial exposure assessments. Nanomaterial exposure assessments in the medical, occupational, and environmental areas using instrumentation that can accurately detect nanomaterials in each of these areas are needed.

14.5.9 Exposure Assessment of Nanomaterials

Nanotechnology will be pervasive and incorporated into all aspects of daily life. Therefore, information regarding the exposure to nanomaterials resulting from medical, occupational, environmental, and accidental release of nanomaterials is critically needed for nanotechnology risk assessment. Information with regard to the concentration of nanomaterials as well as what form(s) they may assume upon release into the environment is needed. Nanotechnology exposure risk assessment will provide critical information on the routes of exposure to nanomaterials.

14.5.10 Risk Management of Carbon Nanotubes

CNTs are molecular scale manufactured three dimensional forms of carbon, falling into two general groups: a) single walled (SWCNTs); and b) multiwalled (MWCNTs). CNTs may be present as long, straight fibres or tangled bundles. CNTs can differ in terms of chemical composition; they may be pure carbon or contain metals or other materials, by design, through contamination or as a result of residual catalyst. They are sixty times stronger than steel, yet six times lighter. CNTs have chemical, physical and bioactive characteristics of considerable research and commercial interest.

Occupational exposure to CNTs can occur:

(a) during manufacture;
(b) through incorporation in other materials, e.g. polymer composites, medical applications and electronics; and;
(c) generating nanoparticles in non-enclosed systems;
(d) during research into their properties and uses;
(e) cleaning of dust collection systems used to capture nanoparticles;
(f) as a result of incorrect disposal and;
(g) as a result of accidental spillage

Emerging data indicates that when CNTs are breathed in they can cause lung inflammation and fibrosis. The type of CNT, its physical form and presence of impurities and surface modifications may influence the severity of the response but at present there is not enough information to identify which factors are of greatest concern. It is also not clear if inhaled CNT have a role in the development of adverse health effects at other sites in the body.

There is an increasing body of evidence to suggest that CNTs and other nanomaterials with a long, thin and straight shape (referred to as high aspect ratio nanomaterials or HARN) may be particularly hazardous. However, there are insufficient data to confirm the health consequences of long-term repeated exposure. There is some evidence to suggest that CNTs may be able to provoke inflammatory reactions in the skin but more information is required to properly understand the conditions of exposure that are required to produce such effects. In view of the evidence for lung damage and lack of information on the effects of long-term repeated exposure a high level of control is warranted for CNTs (37-40).

Legal duty: The occupational use of nanomaterials is regulated under the US - Control of Substances Hazardous to Health Regulations (COSHH) 2002 (as amended).The principle of risk assessment is embedded in COSHH and applies even though all the necessary information on nanomaterials is not available. Since there is uncertainty about the risks of being exposed to CNTs, the regulatory and safe response is to take a precautionary approach. The toxicity of CNTs has not yet been fully investigated. However it is clear that Safety Data Sheets for CNTs that are based on conventional graphite or graphene will not provide suitable adequate information to assess the risk from CNTs. Furthermore as measuring potential airborne exposure levels of CNTs is not a simple task, it is therefore difficult to carry out a rigorous COSHH risk assessment. Thus it is important that an assessment is done for all work involving CNTs and suitable and sufficient risk management measures put in place. It is important that everyone potentially exposed to CNTs receives a high standard of information, instruction and training, particularly on controlling exposure and maintaining that control. This guidance should help with these processes.

Supply of CNT materials: When supplying CNTs to other companies or university departments, health and safety information must be provided with the material. This information should include a warning that the material contains CNTs, with an indication of the CNT percentage or concentration. It is good practice to label the material.

Caution: It is also important to bring this guidance to the attention of the person receiving the CNTs informing that substance not yet fully tested.

Risk management advice: CNTs are substances of high concern and unless, or until, sound evidence is available on the hazards from inhalation a precautionary approach should be taken to the risk management of all CNTs. If the use of CNTs cannot be avoided then the implementation of a risk management program in the workplaces where exposure to CNTs exists can help to minimize the potential for exposure to CNTs.

14.5.11 Nanotechnology Implications in Natural and Global Processes

It is necessary to understand and quantify nanoparticles in earth system processes in order to anticipate their impacts to optimize and integrate environmental sustainability and nanotechnology. The intentional and unintentional release of nanoparticles and nanomaterials into the environment poses challenges and potential dangers. Nanoparticles and nanoscale processes impact such long-term phenomena as climate change, ore and petroleum deposition, paleo magnetism, and groundwater composition (41-44). Environmental science and technology link the concentrations of both anthropogenic and natural constituents to their effects on the well-being of humans and other organisms. Quantifying these links is especially needed when assessing the effects of nanoparticles and the uses of nanotechnology. The concepts of nanoscale science and the tools of nanotechnology offer unique opportunities for understanding and monitoring these links. Environmental and biological sciences at all levels of organization must be closely integrated because human survival requires robust and diverse ecosystems.

Current scientific and technological advancements: Advances to date have been identified from several different perspectives, such as technology for detection and characterization of nanoparticles, how molecular and nanoscale events trigger processes at all scales important to the environment, and discoveries of new and different properties inherent at the nanoscale. However, experts still do not know how to anticipate the fate of nanoparticles in most environmental systems (soils, groundwater, bed sediments, lakes, etc).

Technology that led to the understanding of particles in the atmosphere (mobility measurements) has allowed size resolution of particles from a polydispersed system and quantification of particles within a size class, allowing for rapid determination of particle size distributions in real time (45-52). Recent advances in the physical characterization of atmospheric nanoparticles now allow for the detection of these nascent particles when they are only hours old (corresponding to diameters of approximately 3 nm), and thus, correlate local metrology and gas-phase atmospheric chemical composition to new particle production (53). This capability has enabled direct observation of homogeneous nucleation events over boreal forests and tidal zones, in the free troposphere, and even in the polluted urban atmosphere, as well as the detection of nanoparticles in the on-highway emissions from diesel engines. Such data have stimulated advances in modeling the dynamics of atmospheric nanoparticles. These physical characterization techniques already are undergoing rapid advances that are decreasing the detectable particle size, even to the molecular scale.

Thermodynamically stable crystal structures are different at the nanoscale than at macroscales. For example, transport of aluminum in aqueous solution occurs through a series of labile clusters that aggregate to form larger particles, eventually defining the reactivity of the bulk phase (54-57). This phenomenon occurs in many environments, such as water treatment plants where aluminum sulfate is used in coagulation, a critical step for efficient removal of particles, pollutants, and pathogens affecting human health.

Understanding the distributions of chemical species in aquatic systems among dissolved, colloidal, and larger particle phases has been a major success in the field of geochemistry. An improved knowledge of nanoscale species has explained periodic failures of filtration- based water treatment. The discovery that small nanoparticles and their chemical burdens pass through the filters led to improved water treatment processes (58). Further, recognition of colloidal phases enabled great improvements in knowledge of solid-solution partitioning (59). Likewise, this improved understanding has allowed experts to anticipate how subsurface transport of contaminants (e.g., radionuclides) can be facilitated by suspended nanoparticles (60-61).

Research needs/challenges: Future advances as a result of changing technology will help to develop predictive models of the spatial and temporal distribution, transport, transformations of nanoparticles in the environment, and impacts on organisms. The ultimate goal of these models is to predict the effects of nano-particles on earth system processes, local-to-global ecology, and human health. One key product will be predictions of the likely sources, types, and concentrations of nanoparticles released to the environment.

The interaction between nanotechnology and ecosystem biology is insufficiently developed. Biological research at all scales of organization is essential to appropriately address the environmental impact of nanoscale science and technology. Goals are to identify, quantify, and predict effects on individual, population, community, and ecosystem phenomena. Then, it will be possible to distinguish between adverse and beneficial perturbations on both short and long time scales.

14.6 Global Challenges in Nano Technologies Application

14.6.1 Global Collaboration in Nanotechnology Applications for Meeting Development Challenges

For Addressing Global Challenges using Nanotechnology (62-63) some top ten nanotechnology applications need to be studied with global collaboration by construction of a plan of action for mobilizing the international community. One way to accelerate the application of nanotechnology to meet the needs of developed and developing countries consists in involving the world's best scientific minds to address grand challenges to channel their efforts towards a specific scientific or technological breakthrough that will overcome one or more significant development challenges.

14.6.2 International Competition for Developing Cutting Edge Nano Technologies

For developing cutting-edge nano technologies, a range of possibilities exists in each area. These possibilities range from a nationally competitive system in which both technology investments and technology products need to be identified with respect to national boundaries, to a situation in which they are highly fluid across national and regional boundaries. The actual direction will depend on a number of factors, including future regional economic arrangements (e.g., the European Union), international intellectual property rights and protections, the character of future multi-national corporations, and the role and amount of public sector research and development investments. Currently, there are moves toward competition among regional (as opposed to national) economic alliances, increased support for a global intellectual property protection regime, more globalization, and a division of responsibilities for R&D funding (public sector research funding with private sector development funding).

14.6.3 Global Governance to Promote Nanotechnology for Development

A global strategy is needed to assess and promote nanotechnology for development. While traditional governance models tend to focus on risks and restrictions, a more holistic model should include a focus on innovation and the enormous potential of nanotechnology to address global challenges. Among the issues to be addressed are: how can nanotechnology be used responsibly? Which applications ought to be given priority? What policies are needed to develop capacity and ensure that the benefits of nanotechnology reach those in greatest need?

How ought regulatory regimes to be developed and implemented? Recently several organizations have shown an increased interest in answering these questions (64). The most salient initiatives include those spearheaded by the Academy of Sciences for the Developing World (formerly the Third World Academy of Sciences; see http://www.twas.org) and its president, Mohamed Hassan; the Meridian Institute's Global Dialogue on Nanotechnology and the Poor (GNDP; http://www.meridian-nano.org/gdnp.php); and the Foresight Institute's 'Foresight Nanotechnology Challenges' (1986).

Thought should be given to what governance model is most appropriate for mobilizing international action and institutionalizing cooperation in nanotechnology. International treaties and conventions may not be well suited to promoting the global benefits of a rapidly advancing field of S&T. They are often slow to be negotiated and ratified, and may still suffer from poor adherence once they come into force.

A flexible global network of representatives from government, academia, industry and civil society could serve as a forum for international dialogue between the public and policy-makers to shape institutional decisions and discuss policy options.

Equality of participation through engagement of developing country voices and the inclusion of diverse sectors like civil society and industry would facilitate collaborative strategies for managing risks in order to promote the global benefits of nanotechnology. Existing networks, public-private partnerships and coalitions working to establish worldwide capacity in S&T, including the Inter Academy Council (www.interacademycouncil.net) need to be further strengthened.

14.6.4 Challenges for Developing Countries in Promotion of Nanotechnologies for Development

Some people expect that nanotechnology will upset developing countries export markets which rely heavily on raw materials, such as rubber, cotton, and agricultural products; the demand for these primary commodities will decrease as nanotechnology produces cheaper laboratory-created substitutes. It is contended that nanotechnology will largely displace poor mine, factory and agricultural labourers, among others (65).

However advances in S&T have inevitably brought about the automation of manual tasks resulting decrease in manual labour which will be a natural consequence of any technology, not just nanotechnology. In both developed and developing countries, the workforce necessarily need to be adapted to the changes derived from successive waves of technology. Nanotechnology will create more opportunities for local scientists and researchers and will help to accelerate the transition of innovating developing countries towards robust knowledge-based economies. It is important to bear in mind that the economic impact of nanotechnology will vary depending on whether we are considering the more advanced developing countries, such as India and China, or the less developed.

Earlier studies on the impact of the health biotechnology innovation systems on seven developing countries identified private-sector involvement as a key factor in moving from knowledge to product.

14.7 Important Strategies to be Adopted for Realizing the Benefits of Newly Emerging Nano Technologies and Addressing the Challenges in their Implementation for Sustainable Growth

Nanotechnology will have an impact on all sectors of the economy. Those individuals, industries or nations who adapt to the changes brought about by nanotechnology and promote R&D to innovate in this field, will enjoy the benefits of this new technology while those who do not keep up with the changes will find it difficult to survive in a rapidly changing environment (66).

The following are some key strategies to be adopted by any country for developing nano technologies for realizing their benefits for betterment of society.

14.7.1 Funding by Government and Private Industries and Global Collaboration in Nanotechnology Applications for Meeting Development Challenges

At present the major sources of funding for nanotechnology R&D include government agencies, large corporations, venture capital, and universities.

Continued Governments funding in basic research and in small and intermediate scale instrumentation for characterization and study of nano materials is a critical need.

The continued support for larger-scale advanced instrumentation facilities and networks is crucial, as well as, increased investments of multi-disciplinary research activities that integrate both theorists and experimentalists.

The US government is continuously investing in nanotechnology (from \$116 million in 1997 to \$774 million in 2003 and increasing to \$2.18 billion in 2011). Other nations, e.g., Japan, the countries of the EU, China, Australia, Taiwan and Israel, are also investing substantially in nanotechnology (67) while the total global investment in 2011 was \$67.5 billion (68). Worldwide spending on nanotechnology R&D is projected to be around \$100 billion by the end of 2014 (69). These investments are depicted through Figure 14.2.

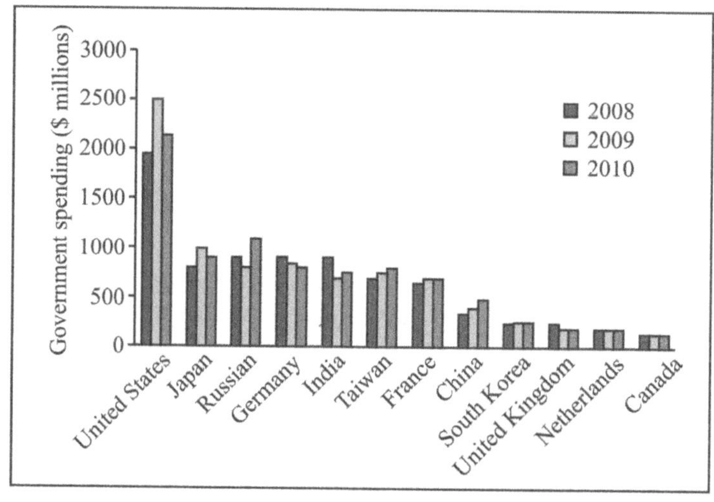

Figure 14.2 Government investment in nanotechnology

Support for nanotechnology will greatly help in advancing new research initiatives. Existing nanotechnology facilities should be leveraged to attract grant funding in other emerging areas. New investments for the training of a nanotechnology workforce (undergraduates, graduate students, postdoctoral fellows, and early stage scientists) are vital for maintaining a competitive scientific and technological advantage. While the most recent focus on nanotechnology funding includes support of environmental, health and safety (EHS) areas, solar energy conversion, sustainable nanomanufacturing, and nanoelectronics, the support for work in newly emerging areas (nanophotonics, nanobiotechnology, energy, health and medicine) should be increased. The recent Govt. focus in advanced countries on the materials genome and mesoscale science initiatives are highly integrated with advances in nanoscience and nanotechnology.

Establishment of an extensive nanotechnology infrastructure for nanofabrication and characterization capabilities, increased and continued investments from various governmental agencies are essential for evolving nanotechnology discoveries from the research laboratory into innovative technologies. Both new facilities and networks should be established and the existing capabilities and user facilities should be supported as they broadly support R&D on many levels including nano-scale characterization, synthesis, simulation and nano manufacturing.

14.7.2 Private/Industry Participation in Nano Technology Growth

The invention of tools to image and manipulate molecular species and nanoparticles, such as the scanning tunnelling microscope (STM) and atomic force microscope (AFM) helped nanotechnology R&D greatly resulting explosive growth in publications and patents (71-72) due to substantial support from governments, universities and investors (69-73).

The two major pathways for nanotechnology commercialization have been through start-up firms funded mainly by government grants and venture capital (with examples of Carbon Nanotechnologies Inc., Molecular Electronics Inc., InMat Inc. and Zyvex), or through large corporations, e.g., IBM, Hewlett-Packard, Motorola and Texas Instruments. Most start-ups and spin-off firms were nurtured and formed in a university environment, typically founded by professors, graduate students and research scientists. The success of early stage commercialization depends on the future marketplace and the entrepreneurial expertise of the researchers and entrepreneurs who conceive products. The nanotechnology marketplace is promising and includes, for example, the automotive, aerospace, electronics, energy, food, medical, and pharmaceutical industries. Existing uses of nanotechnology typically focus on performance improvement.

The U.S. National Science and Technology Council estimates revenue from nanotechnology products and services to be about $1 trillion (72). This includes $300 billion for electronics and communication, $340 for materials and material processing, $180 billion for pharmaceuticals, $100 billion for chemical manufacturing, $70 billion for aerospace, $45 billion for sustainability, $22 billion for sensors and instrumentation, and $30 billion for improved healthcare.

As per one report more than 2500 companies in over 48 countries are involved in nanotechnology research, manufacturing or applications (74). The U.S., with around 1200 companies, is the major hub for nanotechnology firms. Figure 14.3 shows the distribution of nanotechnology firms in different countries (75). Over 270 of these companies are involved in the manufacture of raw materials such as nanoparticles, nanofibers and nanowires, carbon nanotubes, and quantum dots. More than 340 companies are active in the life sciences and pharmaceuticals (Figure 14.3). The vast majority with well over half of all companies manufacture instruments, devices, or advanced materials and components.

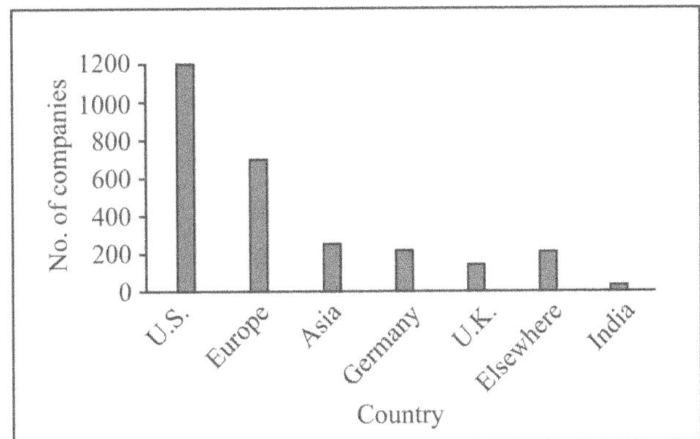

Figure 14.3 Number of companies involved in nanotechnology in different countries (74,75).

Large companies are directly or indirectly involved in nanotechnology R & D investment. Some major companies in the U.S. include Hewlett Packard, IBM, Intel, Motorola, Texas Instrument, Dow Chemical and Raytheon. Their Japanese counterparts are NEC

Corporation, Toshiba, Nihon Shinku, Gijustu, NTT Docomo, Fujitsu and Sony. Venture capitalists have spent almost $900M on nanotechnology start-up firms since 1999. Universities have also funded nanotechnology research and development and start-up firms.

14.8 Educational Challenges and Research Funding in Promotion of Nano Technologies

The enormous challenges presented by nanotechnology and its environmental implications for education at all levels, requires state-of-the-art education and research funding management. There is a general need to develop an educational system that links the biological sciences, physical sciences, engineering, and computer sciences. This need exists at all levels, from high school to graduate level through expert faculty, where fundamental barriers in terminology and approach can inhibit vital cross-disciplinary interaction. Undergraduate and graduate curricula are needed to provide students with the broad knowledge base required for interdisciplinary research, while including an introduction the unique properties of nano structured materials. The natural interest of young people in nanotechnology and environmental science should be cultivated and used as a motivation for basic science learning and the development of knowledge-based interaction with the world of technology. An opportunity also exists to incorporate the concept of sustainability into current and future educational activities related to nanotechnology.

Curricula that emphasize the unique properties of materials at nanometer length scales are need to be developed. One recommendation is to develop a summer research program for High School students and teachers that focuses on nanotechnology and environmental sciences. Funding mechanisms to link high school students with industries should be developed, and support is needed to create outreach programs to enable students to visit universities and other sites that are actively engaged in nanotechnology research. In addition, support is needed for the development of a more complete set of instructional materials and hands-on demonstration tools that teachers could use to teach students about nanotechnology. This could be accomplished with a 1-day course in nanotechnology, involving in-house and outside experts. Local media could be invited to broaden the outreach effort beyond the participants.

14.8.1 Development of an Inter-agency Group to Foster Research, Curricula, and Evaluation

Creation of an interagency group that could effectively support inter institutional, interdisciplinary research, curricula design, and evaluation will greatly help nano technology education on uniform format. Such a group could develop, design, and advocate for the diverse set of programs essential for creation of a technologically and environmentally literate population and a highly functional nanotechnology workforce. Examples of activities that this group might oversee include:

1. a summer research program for High School students and teachers focusing on nanotechnology and environmental sciences;
2. a funding mechanism that would promote linkage between school students and industry, academic organizations, and others where nanotechnology research is being pursued,

including programs that would bring the students and teachers to the research sites;

3. the development of a more complete set of instructional materials and hands-on demonstration tools that high school teachers would be able to use to teach nanotechnology; and
4. a variety of 1 - or 2 - day programs for continuing teacher education that would involve government laboratory and industry experts as well as local media.

14.8.2 Creation of Scientific and Technological Infrastructure for Advanced Nano Technology Research

Long-term research support (4-5 years) for interdisciplinary projects is critical as is increased support for tightly knit, small collaborative groups (3–4 investigators) in focused areas of research. Additionally, it is recommended that a mechanism be developed for providing mid-sized instrumentation grants ($100K–$1M) to small groups. In addition, research on the development of novel manufacturing techniques on the nanoscale for sustainable processes should be encouraged. A key component for any sustainable nano technology development planning needs building an infrastructure for training scientists, engineers, and researchers from other subject areas to enhance their ability to work together in an interdisciplinary manner. Another infrastructure need is incorporating the concept of sustainability into current and future educational activities related to nanotechnology.

14.8.3 Interdisciplinary Training on Nano Technologies

A key feature for achieving the goals of sustainable development of nano technologies will be the ability of scientists, engineers, and researchers from diverse subject areas to collaborate in an interdisciplinary manner. In general, research in engineering as well as synthesis, assembly, and processing will be required to scale-up production of the new materials. Collaboration with other fields will be necessary to develop specific applications. Solar energy/photovoltaic applications for example, will require that chemists work with materials scientists as well as architects and urban planners to optimize and integrate photovoltaic technology into existing and future physical infrastructure. Work on creating virtual transportation, however, will require a different set of collaborators, such as computer scientists and engineers, manufacturing engineers, and urban planners.

14.9 Safety and Regulatory Issues Need to be Addressed on the Application of Nano Technologies

While there is clear evidence that nanotechnology can contribute to human welfare, particularly in developing countries, it is also recognized that they also cause some adverse effects on environment and human health and not all of the effects of nanomaterials in the environment are currently known. It is possible that nanoparticles could accumulate in different organisms, become incorporated in living tissue, and move up the food chain in a process known as bioaccumulation. For clear understanding of these aspects extensive research programs need to be initiated by each country.

Regulatory issues: It is critical to develop regulatory frameworks, either by building upon existing regimes or by creating entirely new ones that take into account the novel properties of matter at the nanoscale. Given that regulatory reform is necessary, governments in both developed and developing countries ought to ensure that new regulatory frameworks are logical, efficient, transparent and readily adaptable to fast-paced technological change. At the international level, clear standards for workplace exposure to nanomaterials must be set. Additionally, international definitions and standards for measurement of the concentration and toxicity of nanomaterials in the environment must be developed. This should be carried out by independent bodies that remain under public scrutiny. Specific attention also needs to be directed towards recognizing individual responsibility of scientists and engineers who are responsible for generating advances.

14.9.1 Public Engagement on the Use of Nanotechnology

Public involvement in crucial decisions about the use of nanotechnology is not optional but essential. However, the public cannot make informed decisions if it knows very little about nanotechnology. Thus public education about nanotechnology is absolutely necessary. Governments should encourage candid discussion of the risks and benefits which is receptive to public concerns. Public debate should include discussions on control of nanotechnology, access to trustworthy information, the terms according to which nanotechnology is introduced into society, the benefits of nanotechnology, the risks of nanotechnology, the groups that will either benefit or be at risk, and the definition of the individual(s) who will take responsibility for problems that arise through research, development, commercialization and use of nanotechnology applications.

Social science research in nanotechnology need to be funded which can be a good way to lend transparency to the debates on nanotechnology, since social scientists can add richness to the discussion on this topic. Engaging youth in discussions of emerging technologies is vital to developing an informed future voting population.

14.9.2 Communication Efforts on Health, Safety, and Environmental Guidelines

There are needs to generate appropriate health, safety, and environmental guidelines based on sound science for handling and employing nano-materials and creating an increased awareness of the importance of interactions with the media for scientists in nanotechnology through programs that assist or train scientists to interact with the news media. There is also a need to create an awareness of the importance of the news media in clearly presenting the risks and benefits of nanotechnology in layman terms.

14.10 Policy Issues on Nanotechnology

The best way to prevent the misuse or abuse of nanotechnology may involve bringing together large groups of diverse stakeholders so that the areas where controversial issues may arise can be identified at the earliest possible stage. A role for benefit-sharing between developed and developing countries in the commercialization of nanotechnology products should be established. Consideration must be given to the best ways to harness intellectual property (IP) for the development of nanotechnology, bearing in mind that the inappropriate use of patents from universities, companies and the military might inhibit R&D efforts in less industrialized countries. There is an inherent tension with IP rights between providing incentives and rewards for innovation on the one hand, and promoting wide and easy access to the resultant technologies on the other hand, especially if the technologies are important to save lives or are needed by the majority. While such issues are complex, it seems prudent to avoid undue concentration of resources in a few hands, for this leads to 'patent thickets' which may actually impede R&D (76).

In summary, both developed and developing countries must do their part to ensure that advances in nanotechnology lead to solutions for humanity's most significant hurdles. The key feature consists in addressing sustainable development issues by focusing on S&T innovation. Both developed and developing nations must leverage their national nanotechnology assets into foreign policy and, simultaneously, marshal both public and private funds to assist the developing world to address sustainable development challenges.

14.11 India's Research Initiatives in Nanotechnologies and Specific Challenges

India is one of the developing countries which has initiated a large number of programs in nanotechnology covering a wide spectrum of areas related to energy, agriculture, water, health and the environment (77). A National Nantechnology program is initiated by DST with a funding of $10 M for three years which greatly helped many institutes to take-up R&D programs in nanotechnologies.

India involving five Govt Agencies has also created a Nanomaterials Science and Technology Initiative and a National Program on Smart Materials; with a funding of $15 million over five years. This program mainly supports research centers on the development of smart materials that respond quickly to environmental stimuli. The Ministry of Defense is also supporting programs on the development of nanostructured magnetic materials, thin films, magnetic sensors, nanomaterials and semiconductor materials. India has also formed a joint nanotechnology initiative with the European Union (EU).

Application of Nano Technologies for Health care systems has been supported by DST, DBT, ICMR, CSIR and DRDO. MNRE is supporting the development of nanomaterials for photovoltaics, hydrogen energy and fuel cells. DBT supports plant and animal biotechnology, bio pesticides, genetics, and protein engineering. However greater coordination among these agencies in their support of nanotechnology R&D will help to optimize resource distribution.

Several academic institutions are pursuing nanotechnology research and development, among them the Institute of Smart Materials Structures and Systems of the Indian Institute of Science; the Indian Institute of Technology (IIT); Indian Institute of Chemical Technology, Hyderabad (IICT); JNTU(H) the Saha Institute of Nuclear Physics; and the Universities of Delhi, Pune and Hyderabad. The Council for Scientific and Industrial Research, India's premier research and development body, holds numerous nanotechnology-related patents, including novel drug delivery systems, production of nanosized chemicals, and high temperature synthesis of nanosized titanium carbide.

Nanotechnology-based interventions in solar energy systems are also being pursued with an aim to improve the overall SPV process efficiency, reduce the cost of solar cells and search for alternate semiconductors due to a shortage of silicon.

However, industry-university interactions in India are not as strong as in other countries due to which product development and characterisation is often ignored so that significant contributions are not coming out (77).

There is also lack of regulatory framework that brings the stakeholders together in developing the technology. No concerns' regarding risk is shown in India. Lately some initiatives have been taken for addressing risk issues by Nano Mission and key scientific agencies. National Institute of Pharmaceutical Education and Research (NIPER) is developing regulatory approval guidelines for nanotechnology based drugs and standards for toxicological tests in nano-based drug delivery systems. In 2010, DST appointed a task force which has been asked to advice Nano Mission Council to develop a regulatory body for nanotechnology in India. However, there is only a few academic debates but not much progress in establishing regulation. Firms involved in nanotechnology based product development primarily products addressing water, textile, drug delivery have undertaken Life Cycle Analysis (LCA) partnering with research institutes/universities. Standardization remains an area of concern. India, has only taken initial first steps in addressing standardization issue. There is lack of private sector involvement human resource base is restricted to selected institutes and limited number of research groups.

There should be a policy focus on infrastructure, education, standards and workforce preparation. Nanotechnology commercialization in India can be improved by enhancing IP protection, which should attract investors and facilitate commercialization. Government agencies should coordinate their efforts.

Public discussions of technologies such as nuclear energy, agricultural biotechnology and embryonic stem cells illustrate that some nanotechnologies could also raise significant social and ethical concerns (78), which need to be critically examined with reference to Indian conditions

In India, there is very little research with respect to risks associated with nanotechnology although nanotechnology applications might conceivably pose risks to the public. A few toxicological studies have been supported by various agencies (79-80). Therefore, larger initiatives are to be made on this front.

14.12 Careers Opportunity in Nanotechnology

14.12.1 Major Categories of Nanotechnologies for Career Opportunities

Nanotechnology is expected to offer exciting applications in bioscience, medical science, environment, electronics, cosmetics, security and variety of other fields. There are two major categories of nanotechnology which can provide large career opportunities in future i.e., (i) nanoscale technology and (ii) molecular manufacturing.

The nanoscale technology covers small structures and can be used for introducing stronger materials, better medicines, faster computers and so on.

Molecular manufacturing is an attempt at building mechanical and chemical manufacturing systems that join molecules together.

14.12.2 Scope of Applications of Nano Technologies

Nanotechnology continuously generates new capabilities, innovative products and new markets. It has varied applications viz. drug delivery to treat cancer tumour (without using radiotherapy & chemotherapy), solar energy, batteries, display technologies, optoelectronic devices, semiconductor devices, biosensors, luminous paints, and many others. Thus as nanoscale science and technology come to have an increasing impact on many aspects for our daily lives, the opportunities for careers are expanding rapidly. A major challenge nanotechnology now faces is the education and training of a new generation of skilled workers. In areas as diverse as designing medical diagnostic devices to building better batteries, from creating cosmetics to enhancing energy efficient windows, from auto and plane manufacturing to researching the nature of matter itself, knowledge of nanoscale science and technology will be increasingly important during upcoming years and decades. The trained persons in nanotechnology get job opportunities in health industry; pharmaceutical industry; agriculture industry; environment industry; food and beverage industry as well in government and private research institutes.

14.12.3 What are Major Career Areas/Fields in India?

Current applications of nano scale science and technology and areas which can provide career opportunities, exist in areas such as

- Electronics/semiconductor industry
- Materials science including textiles, polymers, packaging, among others. Auto and aerospace industries
- Sporting goods
- Biotechnology
- Medical fields and pharmaceuticals
- Environmental monitoring, control, and remediation
- Food science including quality control and packaging
- Forensics
- University and Govt. lab research
- Military and national security
- Energy capture and storage and many more areas

14.12.4 Great Potential for Future Career Opportunities in Material Processing Informatics and Quantum Computation

Material Informatics: Materials informatics can be broadly divided into two main parts – data management and knowledge discovery. The term "informatics" has long been used to describe the entire field of computer science from computer engineering to information systems and related fields. Modern materials science and engineering research produces a large amount of heterogeneous data. As a result computational methods that are used to organize, manage, interpret and analyze these data and are now becoming essential tools for materials science and engineering community.

14.12.5 Current Career Opportunities in Nano Technologies

14.12.5.1 US Scenario

The National Science Foundation's estimated job projection for the U.S. is nearly one million workers by 2015.

The U.S. market value of products using nanotechnology is estimated to be $1 trillion or 5% of the GDP by 2020.

(i) What type of education is needed for a career in nanotechnology?

Not everyone working in nanotechnology will require a doctorate degree in science or engineering. A skilled workforce trained at a variety of levels is needed to meet the projected workforce challenge. Education levels include:

Technical program (a high school degree plus additional training).

- Associate's degree (2 years of college)
- Bachelor's degree (about 4 years of college total)
- Master's degree (about 6 years of college total)
- Doctorate (about 9 years of college total)

More information can be obtained on these websites:

- National Nanotechnology Initiative's Schools and Training Programs - http://www.nano.gov/education-training/workforce-training-center
- Nanotechnology Applications and Career Knowledge Center - http://nano4me.org/

(ii) What are the Expected Salaries?

- Two Year Associates - $35,000 - $50,000
- Four Year Bachelors - $40,000 - $65,000
- Six Year Masters - $50,000 - $80,000
- Eight Year Doctorate - $75,000 - $100,000

U.S. salaries for nanotechnology engineering technicians range from $30,000 to $94,000 according to Recruiter.com (http://www.recruiter.com/salaries/nanotechnology-engineering-technicians) with the average at $79,000 (http://www.simplyhired.com/a/salary/search/q-Nanotechnology). You can go online to see the average for your state or particular positions.

Some additional sites to explore:

- http://www.nano.gov/education-training
- http://www.tinytechjobs.com
- http://www.nanowerk.com/nanocareer/homepage.php
- http://www.nanovip.com
- http://www.foresight.org/cms/resources/111

14.12.6 Indian Scenario

The scope and application of nanotechnology is growing tremendously in India. It is one of the hottest career options available to Indian Engineering and Science graduates. Nanotechnology has applications in medicine, pharmaceuticals, information technology, electronic, optoelectronics, energy, chemicals, advanced materials, textiles and in many other fields. A lot of job opportunities and research career are being generated in the above areas and in particular nano device, nano packaging, nano wires, nano tools, nano biotechnology, nano crystalline materials, nano photonics and nano porous materials to name a few. It is estimated that around two million nanotechnology skilled workforce will be required world wide by 2020. Many government institutes and Indian industries have focused on nanomaterials. It is also estimated nanotechnology will create another five million jobs worldwide in support fields and industries. A professional in the field of nanotechnology can easily find lucrative jobs in most of fields.

14.12.7 What Education is needed and in which Fields?

Nanoscale phenomena underlie many of the properties and interactions of matter, and thus the sciences of physics, chemistry, and biology. However, nanoscale science is an interdisciplinary field – one where science and engineering intersect. Studying science or engineering and paying attention to the developments in nanoscience that advance these fields can provide you with a solid foundation for any broad range of careers.

14.12.8 Remuneration

The salary is based on qualification, skills acquired, experience, nature of the job and the employer. The nanotechnologist can have monthly salary of Rs. 20,000 to Rs. 1,20,000. In US nanotechnologist draws very high salary.

14.12.9 Eligibility

M.Tech/M.Sc. in Nanotechnology is the eligibility criteria in this field. B.Tech with Material science/Mechanical /Biomedical/Chemical/Biotechnology/Electronics /Computer science offers many career opportunities. Those who have completed M.Sc. in Physics/ Chemistry/ Material Science/ Biotechnology/ Computer Science can also join for M.Tech Course in nanotechnology.

References

1. Friedman Thomas L. (2000). The Lexus and the Olive Tree, Anchor Books, New York.
2. Oriakhi C (2004). Commercialization of nanotechnologies. M.S. Thesis, Massachusetts Institute of Technology, Sloan School of Management, Management of Technology Program.
3. Department of Biotechnology (2009). Annual report 2008-2009.
4. The Royal Society & The Royal Academy of Engineering (2004). Nanoscience and nanotechnologies: Opportunities and uncertainties. Cardiff, Clyvedon Press. Social and ethical issues: 51-6.
5. Chen X.Q.; Saito T.; Yamada H; and Matsushige K (2001). Aligning single-wall carbon nanotubes with an alternating current electric field. Appl. Phys. Lett. 78, 3714-3716.
6. Garmestani H.; Al-Haik M.S.; Dahmen K.; Tannenbaum R.; Li D.S.; Sablin S.S.; Hussaini M.Y (2003). Polymer-mediated alignment of carbon nanotubes under high magnetic fields. Adv. Mat. 15, 1918-1921.
7. Campbell P (2002). Opinion: materials for sustainability. Nature, 419, 543.
8. Lave L.; MacLean H.; Hendrickson C.; and Lankey R (2000). Life cycle analysis of alternative automobile fuel/propulsion technologies. Environ. Sci. Technol. 34(17), 3598-3605.
9. Chusid O.; Gofer Y.; Gizbar H.; Vestfrid Y.; Levi E.; Aurbach D.; and Riech I (2003). Solid-state rechargeable magnesium batteries. Adv. Mater. 15, 627-630.
10. Aurbach D.; Lu Z.; Schechter A.; Gofer Y.; Gizbar H.; Turgeman Y.; Cohen Y.; Moshkovich M.; and Levi E (2003). Prototype systems for rechargeable magnesium batteries. Nature, 407, 724-727.
11. Mirkin C. A (2000). A DNA-based methodology for preparing nanocluster circuits, arrays, and diagnostic materials, MRS Bulletin, Vol. 25, No. 1, 43-54.
12. Ying J.Y.; and Sun T (1997). Research needs assessment on nanostructured catalysts. J. Electroceramics, 1(3), 219-238.
13. Kumar M.S.; Lee S.H.; Kim T.Y.; Kim T.H.; Song S.M.; Yang J.W.; Nahm K.S.; and Suh E.K (2003). DC electric field assisted alignment of carbon nanotubes on metal electrodes. Solid-State Electronics, 47, 2075-2080.
14. Huang S.; Maynor B.; Cai X.; and Liu J (2003). Ultra-long, well-aligned single-walled carbon nanotube architectures on surfaces. Adv. Mater. 15, 1651-1655.
15. Ural A.; Li Y.M.; and Dai H.J (2002). Electric field-aligned growth of single-walled carbon nanotubes on surfaces. Appl. Phys. Lett. 81, 3464-3466.
16. Huang S.M.; Cai X.Y.; and Liu J (2003). Growth of millimeter-long and horizontally aligned single-walled carbon nanotubes on flat substrates. J. Am. Chem. Soc. 125, 5636-5637.
17. Joselevich E.; and Lieber C.M 2002. Vectorial growth of metallic and semiconducting single-wall carbon nanotubes. Nano. Lett. 2, 1137-1141.
18. Zhong Z.H.; Wang D.L.; Cui Y.; Bockrath M.W.; and Lieber C.M (2003). Nanowire crossbar arrays as address decoders for integrated nanosystems. Science, 302, 1377-1379.
19. Huang Y.; Duan X.F.; Cui Y.; Lauhon L.J.; Kim K.H.; and Lieber C.M (2001). Logic gates and computation from assembled nanowire building blocks. Science, 294, 1313-1317.
20. Cui Y.; and Lieber C.M (2001). Functional nanoscale electronic devices assembled using silicon nanowire building blocks. Science, 291, 851-853.

21. Huang Y.; Duan X.F.; Wei Q.Q.; and Lieber C.M (2001). Directed assembly of one-dimensional nanostructures into functional networks. Science, 291, 630-633.
22. Messer B.; Song J.H.; and Yang P.D (2000). Microchannel networks for nanowire patterning. J. Am. Chem. Soc. 122, 10232-10233.
23. Nam J.M.; Park S.J.; and Mirkin C.A (2002). Bio-barcodes based on oligonucleotide-modified nanoparticles. J. Am. Chem. Soc. 124, 3820-3821.
24. Chan W.C.W.; and Nie S (1998). Quantum dot bioconjugates for ultrasensitive nonisotopic detection. Science, 281, 2016-2018.
25. Goldman E.R.; Anderson G.P.; Tran P.T.; Mattoussi H.; Charles P.T.; and Mauro J.M (2002). Conjugation of luminescent quantum dots with antibodies using an engineered adaptor protein to provide new reagents for fluoroimmunoassays. Anal. Chem. 74, 841-847.
26. Thomas K.G.; and Kamat P.V (2003). Chromophore functionalized gold nanoparticles. Acc. Chem. Res. 36, 888-898.
27. Brongersma M.L 2003. Nanoshells: gifts in a gold wrapper. Nature Materials, 2(5), 296-297.
28. Wohlstadter J.N.; Wilbur J.L.; Sigal G.B.; Biebuyck H.A.; Billadeau M.A.; Dong L.W.; Fischer A.B.; Gudibande S.R.; Jamieson S.H.; Kenten J.H.; Leginus J.; Leland J.K.; Massey R.J.; and Wohlstadter S.J (2003). Carbon nanotube-based biosensor. Adv. Mat. 15, 1184-1187.
29. Li J.; Lu Y.J,; Ye Q.; Cinke M.; Han J.; and Meyyappan M (2003). Carbon nanotube sensors for gas and organic vapor detection. Nano. Lett. 3, 929-933.
30. Pengfei Q.F.; Vermesh O.; Grecu M.; Javey A.; Wang O.; Dai H.J.; Peng S.; and Cho K.J (2003). Toward large arrays of multiplex functionalized carbon nanotube sensors for highly sensitive and selective molecular detection. Nano. Lett. 3, 347-351.
31. Shim M.; Kam N.W.S.; Chen R.J.; Li Y.M.; and Dai H.J (2002). Functionalization of carbon nanotubes for biocompatibility and biomolecular recognition. Nano. Lett. 2, 285-288.
32. Cui Y.; Wei Q.Q.; Park H.K.; and Lieber C.M (2001). Nanowire nanosensors for highly sensitive and selective detection of biological and chemical species. Science, 293, 1289-1292.
33. Feldman R.A.; and Harris D.W (2000). Beyond the human genome: high-throughput, fine scale, molecular dissection of earth's microbial diversity. J. Clin. Ligand Assay, 23(4), 256-261.
34. Background Report on Fertilizer Use, Contaminants and Regulations. U.S. Environmental Protection Agency Report EPA 747-R-98-003, 1999, 6.

    http://www.epa.gov/opptintr/fertilizer.pdf.
35. Dagani R (2003). Nanomaterials: safe or unsafe? Chem. Engineering News, 81(17), 30-33.
36. Lam C.W.; James J.T.; McCluskey R.; and Hunter R.L (2004). Pulmonary toxicity of carbon nanotubes in mice 7 and 90 days after intratracheal instillation. Toxicol. Sci. 77, 126-134.
37. Warheit D.B.; Laurence B.R.; Reed K.L.; Roach D.H.; Reynolds G.A.M.; and Webb T.R (2004). Comparative pulmonary toxicity assessment of single-wall carbon nanotubes in rats. Toxicol. Sci. 77, 117-125.
38. Dreher K (2004). Toxicological highlight: toxicological assessment of manufactured nanoparticles. Toxicol. Sci. 77, 3-5.
39. ISO/TR 12885:2008(E) Nanotechnologies – Health and Safety practices in occupational settings relevant to nanotechnologies.

40. Banfield J.F.; and Zhang H (2001). Nanoparticles in the Environment. Rev. Mineral. Geochem. 44, 1-58.

41. Ryan J.N.; and Elimelech M (1996). Colloid mobilization and transport in groundwater. Colloid Surface, A107, 1-56.

42. Sanudo-Wilhelmy S.A.; Rossi F.K.; Bokuniewicz H.; and Paulsen R.J (2002). Trace metal levels in uncontaminated groundwater of a coastal watershed: importance of colloidal forms. Environ. Sci. Technol. 36(7), 1435-1441.

43. Hellerich L.A.; Oates P.M.; Johnson C.R.; Nikolaidis N.P.; Harvey C.F.; and Gschwend P.M (2003). Bromide transport before, during, and after colloid mobilization in push-pull tests and the implications for changes in aquifer properties. Water Resourc. Res. 39(1), Art. No. 1301.

44. Flagan R.C (1998). History of electrical aerosol measurements. Aerosol Sci. Technol. 28(4), 301-380.

45. Gard E.; Mayer J.E.; Morrical B.D.; Dienes T.; Fergenson D.P.; and Prather K.A (1997). Rcal-time analysis of individual atmospheric aerosol particles: design and performance of a portable ATOFMS. Anal. Chem. 69(20), 4083-4091.

46. Hughes L.S.; Allen J.O.; Kleeman M.J.; Johnson R.J.; Cass G.R.; Gross D.S.; Gard E.E.; Gaelli M.E.; Morrical B.D.; Fergenson D.P.; Dienes T.; Noble C.A.; Liu D.-Y.; Silva P.J.; and Prather K.A (1999). Size and composition distribution of atmospheric particles in Southern California. Environ. Sci. Technol. 33(20), 3506-3515.

47. Higgins K.J.; Jung H.; Kittelson D.B.; Roberts J.T.; and Zachariah M.R (2002). Size-selected nanoparticle chemistry: kinetics of soot oxidation. J. Phys. Chem. A, 106, 96-103.

48. Tobias H.J.; Beving D.E.; Ziemann P.J.; Sakurai H.; Zuk M.; McMurry P.H.; Zarling D.; Waytulonis R.; and Kittelson D.B (2001). Chemical analysis of diesel engine nanoparticles using a nano-DMA/thermal desorption particle beam mass spectrometer. Environ. Sci. Technol. 35, 2233-2243.

49. Liu D.-Y.; Wenzel R.J.; and Prather K.A (2003). Aerosol time-of-flight mass spectrometry during the Atlanta Supersite Experiment: 1. Measurements. J. Geophys. Res. D: Atmos. 108, SOS 14/11-SOS 14/16.

50. Middlebrook A.M.; Murphy D.M.; Lee S.-H.; Thomson D.S.; Prather K.A.; Wenzel R.J.; Liu D.-Y.; Phares D.J.; Rhoads K.P.; Wexler A.S.; Johnston M.V.; Jimenez J.L.; Jayne J.T.; Worsnop D.R.; Yourshaw I.; Seinfeld J.H.; and Flagan R.C (2003). A comparison of particle mass spectrometers during the 1999 Atlanta Supersite Project. J. Geophys. Res. D: Atmos. 108, SOS 12/11-SOS 12/13.

51. Wenzel R.J.; Liu D.-Y.; Edgerton E.S.; and Prather K.A (2003). Aerosol time-of-flight mass spectrometry during the Atlanta Supersite experiment: 2. Scaling procedures. J. Geophys. Res. D: Atmos. 108, SOS 15/11-SOS 15/18.

52. Marti J.J.; Weber R.J.; McMurry P.H.; Eisele F.; Tanner D.; and Jefferson A (1997). New particle formation at a remote continental site: assessing the contributions of SO_2 and organic precursors. J. Geophys. Res. 102(D5), 6331-6339.

53. Kulmala M.; Vehkamaki H.; Petaja T.; dal Maso M.; Lauri A.; Kerminen V.-M.; Birmili W.; and McMurry P.H (2004). Formation and growth rates of ultrafine atmospheric particles: a review of observations. J. Aerosol Sci. 35, 143-175.

54. Urrer G.; Phillips B.L.; Ulrich K.-U.; Pothig R.; and Casey W.H (2002). The origin of aluminium flocs in polluted streams. Science, 297, 2245-2247.

55. Casey W.H.; and Swaddle T.W (2003). Why Small? The use of small inorganic clusters to understand mineral surface and dissolution reactions in geochemistry. Rev. Geophys. 41(2), 1-20.
56. Casey W.H.; Phillips B.L.; and Furrer G (2001). Aqueous aluminum polynuclear complexes and nano-clusters: a review. Rev. Mineral Geochem. 44, 167-190.
57. Howe K.J.; and Clark M.M (2002). Fouling of microfiltration and ultrafiltration membranes by natural waters. Environ. Sci. Technol. 36(16), 3571-3576.
58. Gustafsson Ö.; and Gschwend P.M (1997). Aquatic colloids: concepts, definitions, and current challenges. Limnol. Oceanogr. 42, 519-528.
59. Honeyman B.D (1999). Geochemistry: colloidal culprits in contamination. Nature, 397, 23-24.
60. Kersting A.B.; Efurd D.W.; Finnegan D.L.; Rokop D.J.; Smith D.K.; and Thompson J.L (1999). Migration of plutonium in ground water at the Nevada Test Site. Nature, 397, 56-59.
61. Salamanca-Buentello F., Persad D. L., Court E. B., Martin D. K., Daar A. S. and Singer P. A. (2005). Nanotechnology and the developing world. PLoS Medicine, Vol. 2, No. 4, e97,
62. Singer P.A., Salamanca-Buentello F. and Daar A. S (2005). Harnessing nanotechnology to improve global equity. Issues in Science and Technology, 9 July, pp. 57-64.
63. Dowdeswell E., Daar A. S., Acharya T. and Singer P. A (2006). Realizing the promise of genomics: Exploring governance. International Journal of Technology and Globalization.
64. ETC Group. (2004). Down on the Farm: The Impact of Nano-scale Technologies on Food and Agriculture. Winnipeg, ETC Group.
65. Enriquez J (2001). As the Future Catches You. New York, Three Rivers Press.
66. Roco MC (2003). Broader societal issues of nanotechnology. J Nanoparticle Res 5: 181-9.
67. Sargen JF (2013). USET: The national nanotechnology initiative (operating report).
68. Paull R, Wolfe J, Hébert P, and Sinkula M (2003). Investing in nanotechnology. Nat Biotechnol 21: 1144-7.
69. Xue A, and Hwang D (2011). Nanotechnology funding: Corporations grab the reins. Lux Research.
70. Ramón C, and Angela H (2002). Forecasting the development of nanotechnology with the help of science and technology indicators. Nanotechnology 13: 243.
71. Dora M, and Michael (2003). Nanotechnology strength indicators: International rankings based on US patents. Nanotechnology 14: R1.
72. Fishbine G (2002). The investor's guide to nanotechnology & micromachines. Newyork, John Wiley & Sons, Inc.
73. Berger M (2010). Well over 2000 companies worldwide already involved in nanotechnology.
74. Wolfe J, and Hebert P (2003). The Nanotech report 2003. New York, Lux Capital.
75. Davies J. C (2006). Managing the Effects of Nanotechnology. Washington, DC, Project on Emerging Nanotechnologies.
76. Sabety T (2004). Nanotechnology innovation and the patent thicket: Which IP policies promote growth? Abstract of paper presented at the First Conference on Advanced Nanotechnology: Research, Applications, and Policy, 21-24.
77. The Energy and Resources Institute (TERI) (2009). Nanotechnology developments in India – a status report TERI project: Capability, governance, and nanotechnology developments - a focus on India New Delhi: The Energy and Resources Institute. New Delhi.

78. The Royal Society & The Royal Academy of Engineering (2004). Nanoscience and nanotechnologies: Opportunities and uncertainties. Cardiff, Clyvedon Press. Social and ethical issues: 51-56.

79. Sharma V, Shukla RK, and Saxena N (2009). DNA damaging potential of zinc oxide nanoparticles in human epidermal cells. Toxicology Letters 185: 211-8.

80. Singh S, D'Britto V, Prabhune AA, Ramana CV, Dhawan A, et al. (2010). Cytotoxic and genotoxic assessment of glycolipid-reduced and –cappevf rd gold and silver nanoparticles. New J Chem 34: 294.

15

Laboratory Exercises

15.1 Introduction

Nanotechnology impact is growing in magnitude in every activity of our growth. New industrial revolution based on emerging Nano technologies is expected to drastically change the present production processes in automobiles, computers, light emitting diodes (LEDs), shape memory alloys, amorphous metals, ferrofluids, catalysts, nano medicines etc.

Due to the importance of this emerging technology for industrial development and sustainable growth, skilled man power in these specialized and multi disciplinary areas are required. This demands more focus of nano technologies in graduate and undergraduate curricula in science and engineering. Students need an interdisciplinary education in the basic sciences, the engineering sciences, and the information sciences, as well as an understanding of the relationships of these fields to nanotechnology.

The Laboratory Exercises presented in this Chapter are aimed to explain the students various nano Science/Technology concepts and their Emerging applications in various fields

15.2 Experiment -1 Testing of Natural Materials for Nano-Structures

15.2.1 Natural Nano Materials to be Tested

The chemical identity and properties of a substance depend on its molecular structure. The nanostructure of a biological material is due to its supramolecular organisation and the arrangement of tens to hundreds of molecules into shapes and forms in the nanoscale range. The interaction of light, water and other materials with these nanostructures gives the natural materials remarkable properties that can be appreciated at the macroscale

In this experiment, the natural nanomaterials that will be analysed are **gelatin and milk which are colloids and composed of nano structured particles (10-300 nm size) dispersed.**

The nano structures in these molecules are small enough to be dispersed evenly and maintain a homogenous appearance, but large enough to **scatter light**. The particles in a colloid can be so well dispersed that they have the appearance of a solution (e.g. transparent).

A simple way to test if a mixture is a solution or a colloid is to **pass a laser beam through the mixture;** the light will be scattered only by the colloid.

In this experiment students will:

1. Prepare gelatin and test it with a laser pen to confirm its colloidal nature.
2. Confirm that milk is a colloid and treat it with acid to induce its aggregation.

This experiment will enlighten students

(i) without these nanostructures, common materials like milk lose their appearance and function.

(ii) with practical evidence of the link between structure and function, and how manipulation of the molecular organization of a material, such as milk, leads to materials with different colour, odour and taste.

Experiment 1.1: Tests with Gelatin

Gelatin is a tasteless solid substance, derived from the collagen inside the skin and bones of animals. It is used as a gelling agent in food products (cakes etc.), in pharmaceuticals (*e.g.*, gelatin capsules), in cosmetic products and photography.

Gelatin is a protein produced by partial hydrolysis of collagen found in the bones, connective tissues, organs and some intestines of mammalian animals such as pigs. However, gelatin from fish is also becoming a common source.

During the partial hydrolysis of collagen, some molecular bonds between individual collagen strands are broken down into a form that rearranges more easily (gelatin). For this reason, gelatin chemical composition is, in many respects, very similar to that of its parent collagen.

Gelatin is often found in the form of powder. When mixed with water, it forms a solution of high viscosity, which sets to a gel on cooling, forming a colloid gel. Gelatin gel melts to a liquid when heated and solidifies when cooled again. Therefore, its existence as a gel is limited to a specific temperature.

The fact that gelatin is a colloid rather than a solution can easily be seen by using a laser pen and passing light through the gel. A path of scattered light is clearly visible (Tyndall effect). This effect is due to the scattering of light by the nanoparticles inside the colloid and students will test this effect.

Nano Structures of Gelatin

Recent studies with the Atomic Force Microscope (AFM) have shown that gelatin is, indeed, formed by numerous nanostructures which have various shapes depending on the type of gelatin being analysed. For example, AFM analysis of gelatin extracted from catfish (*Ictalurus punctatus*) skin has revealed the presence of annular pores with diameters averaging 118 nm and spherical nanoagregates with diameters around 260 nm. It is hypothesised that these structures are formed during the penetration of water inside the collagen molecules during hydrolysis. (Labs equipped with AFM facility can give a lab experiment to students to examine the nano structures of gelatin).

Procedure

***Step* 1:** Prepare a 10 mg/mL gelatin sample by mixing 0.5 mg gelatin powder with 50 mL cold water. Place on the hotplate and heat the water and gelatin mixture. Stir with the spatula as the mixture heats up. Bring close to the boil (check the temperature with the thermometer), then turn off the hotplate and let the gelatin cool down.

***Step* 2:** Once the gel is formed, test it with a laser pen. Place a piece of white paper on the far side of the beaker. Pass the laser beam through the gelatin sample and ask students to record their observations.

***Step* 3:** Repeat the laser test using a beaker of plain water. Does plain water scatter light?

The student from this experiment establish the presence nanostructures in gelatin which makes it a colloid and explains its light scattering behavior.

Experiment 1.2: Testing Milk for Nano Structures

Bovine milk contains a number of biomolecules, such as lipids and proteins, which are dispersed in water. The amount of proteins is between 2.5 and 3.5% depending on the animal breed, of which about 80 % are caseins (the rest being whey or serum proteins). Four proteins comprise the casein group and are: αs1-casein, αs 2-casein, β-casein and k-casein. The caseins are characterized by the fact that they are phosphoro proteins that precipitate at pH 4.6 (isoelectric point, IP), at this pH whey proteins remain soluble. Another property of caseins is they exist as casein micelles which are 50-300 nm in size. Micelles contain the caseins combined with calcium, phosphate and small amount of citrate. As such, milk is a colloid (a mixture of nanoparticles evenly dispersed but only suspended in a liquid medium). The presence of these micelles (together with other biomolecules like lipids) determines the white colour of milk due to their light scattering. Although milk is a colloid, it is not transparent: therefore, if you pass a light through milk in a glass, the Tyndall effect is not visible. However, if milk is diluted (1% milk in water) the effect is seen. Students will see this effect.

From Structure to Function: The fine molecular self-organisation of proteins and minerals in milk is fundamental to realise its natural function of transporting calcium from the mother to the off spring. Numerous studies have revealed that this organisation results in nanostructures which have precise functions casein micelles. Without this fine organisation, calcium would not be 'trapped' inside the milk micelles and the biological function of milk would not be realised.

Milk Processing: Processing of milk with various treatments is widely used in the dairy industry. For instance, yogurt is a fermented milk product obtained by the controlled growth of specific microorganisms, mainly bacteria that convert lactose (milk sugar) into lactic acid. By lowering the pH of milk, its consistency and taste change. In cheese making enzymes are used to induce the aggregation and precipitation of caseins. In all milk processing methods, the molecular organisation of caseins is altered, which leads to thickening, precipitation and other effects. The appearance, taste and other 'macro' properties of milk are deeply connected to its supra-molecular (nano) structure suspended in a liquid medium.

In this experiment Students will use vinegar and heat to alter the properties of milk.

Caseins: Caseins are one of the types of proteins found in milk. Casein in milk (which has a pH near to neutral, about 6.7) are negatively charged (IP = 4.6). All caseins, except k-casein, possess the ability to **bind to Ca^{2+} which occurs mainly through their phosphate residues.** The binding of Ca2+ is fundamental for milk to fulfil its function (i.e., to transport calcium (and other nutrients) from the mother to the offspring). Each casein is composed of a different peptide sequence and, therefore, has a different secondary and tertiary structure.

The most current model of casein tertiary structure is based on considering **caseins as block copolymers**. In the case of α-casein, the protein has **two hydrophobic regions separated by a hydrophilic core**. It is predicted that this protein self-assembles into a 'train-loop-train structure' as illustrated in **Figure 15.1.**

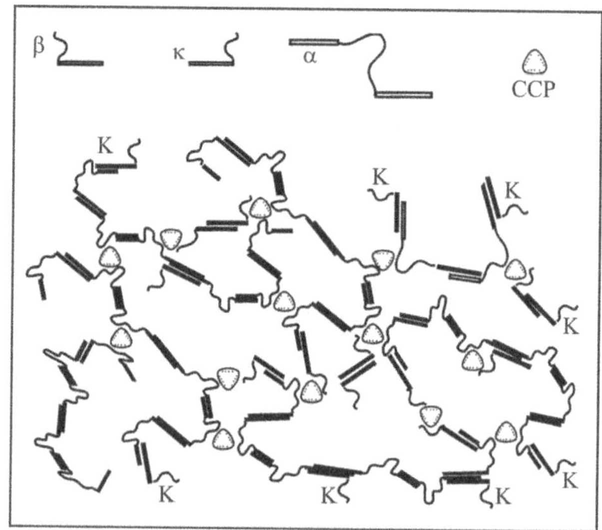

Figure 15.1 Structures of caseins and their polymers
(*Source:* D. S. Horne, International Dairy Journal. (1998 8(3), 171-177)

Casein Micelles Dissociation and Aggregation

As already discussed, casein micelles are believed to have an intricate structure which is an interplay of hydrophobic and electrostatic interactions. Maintenance of micellar integrity is a balancing act and numerous methods exist to disrupt this balance. These methods are widely used in the dairy industry to make cheese and fermented products such as yogurt.

Increasing the pH (to about 8) leads to casein micelle dissociation, and the effect is that heated milk becomes more translucent. The reason for this is that increasing the pH from the natural neutral point converts the phosphoseryl groups from singly to doubly charged units which are no longer capable of linking the calcium phosphate nanoclusters.

The increased negative charge of the micelle induces electrostatic repulsion, and the micelles dissociate. Thus decreasing the pH to the isoelectric point (4.6) induces dissociation of the casein micelles. The reason for this is that calcium micelles only exist because of the presence of calcium phosphate; therefore, its dissolution necessarily causes changes in the stability of the micelle. The consequence of lowering the pH is the titration of the phosophoseryl and carboxyl groups in the proteins. Without their negative charge these groups cannot link to the colloidal calcium phosphate nanoclusters, so these are released from the micelle. It should be noted that this does not necessarily cause the caseins to dissociate from the micelles and lead to precipitation. At temperatures below 25 °C, increasing dissociation occurs, but otherwise the caseins remain in the micelles. This effect will be tested in this experiment by adding vinegar (a source of acid) to cold milk. The reason lies in the fact that the stability of casein micelles is not entirely connected to elctoelectrostatic interaction, but also to hydrophobic interactions. The latter are extremely temperature dependent: hydrophobic interactions are stronger at higher temperatures. Therefore, hydrophobic interactions maintain the stability of casein micelles in cold milk even when its pH has been lowered to the isoelectric point. On the other hand, if acidification occurs after milk has been warmed (to about 60 °C), micelles are dissociated (calcium phosphate is released from the micelle) and will aggregate due to increased electro-static forces and increased hydrophobic interaction. This will be tested in this exercise by adding vinegar to warm or cold milk.

Attack by chymosin leads to micelle precipitation and formation of a curd. This process is employed in cheese-making. Chymosin is a proteolytic enzyme which is the active ingredient in rennet, the extract of calves' stomachs used in cheese-making. Chymosin specifically attacks a single bond in the k-casein, breaking the molecule into two peptides: one remains attached to the micelle, while the other diffuses in solution. As already mentioned, the presence of k-caseins is fundamental for the overall stability of the casein micelle; therefore, its disruption leads the micelle to lose stability, aggregate and eventually form a curd. The controlled addition of lactic acid bacteria (bacteria that produce lactic acid such as Lactobacillus, Lactococcus and Leuconostoc) under specific processing conditions leads to fermented milk products such as yogurt. This process differs from simple acidification as milk is heat-treated and whey proteins are also incorporated. The coagulation is induced by the acidification but does not lead to the formation of a curd but to a product which is more viscous than plain milk.

The phenomenon of aggregation of milk can easily be seen in milk that is old and has long passed its sell-by date. In this case, it is lactic acid bacteria that are responsible for the acidification of milk and consequent aggregation to form acid-smelling lumps. In this experiment, students will use vinegar (a source of acid) and heat to alter the properties of milk.

What can this Experiment Teach about Nanotechnology?

Through this exercise, students will learn about two fundamental concepts.

(i) Structure means appearance: Materials in the 'real' natural world, such as milk, appear as they do because of the fine nanostructures they possess. Milk is white because it contains colloidal nanoparticles (micelles). If the structure of these micelles is altered, some of the 'macro' properties of milk such as the colour and odour will be changed.

(ii) Structure means function: Natural materials have very specific functions which are dictated by the fine supra-organisation of their molecules (nanostructures). If this structure is altred, a material with a new function can be produced. In cheese production, altering the casein micelles through specific processes (*e.g.*, chymosin treatment or lactic acid bacteria) fermentation leads to different products (cheese, yogurt, etc.).

Thus the student can understand the concept of nanotechnologies: to engineer new materials with new functions from the manipulation of their molecular organisation.

Experimental Procedure

Materials

- white vinegar
- laser pen
- hotplate,

Testing Milk

Step 1: Milk is a natural colloid but unlike gelatin it is not transparent. Pour 400 ml of milk into a beaker. Try to test it with a laser pen as you did with gelatin, can you see a path of scattered light? No, because the light is reflected back (back scattering).

Now ask the students to dilute milk. Take 150 mL of distilled water in a beaker or glass, and add 1-2 droplets of milk (using a pipette). Mix and let the solution stand for a couple of minutes (so the air bubbles can dissipate). The solution will look pale grey. Now ask the students to test it again using the laser pen. They will see a path of scattered light.

Step 2: With the use of a pH paper/pH meter record the pH of skimmed milk. Now place the same beaker containing milk used in STEP 1 on a hotplate, turn the hotplate on and warm the milk up to about 60 °C.

Step 3: To the hot milk (about 60 °C) add two table spoons of white vinegar stirring well. It will be seen that milk will immediately aggregate and form a 'milk ball'. Ask the students to record the pH of the liquid. Students should write down their observations.

Caution: Aggregated acid-milk should not be tasted by the students!

Step 4: Repeat the test but without heating the milk. Give each pair of students another 400 mL cold milk, and ask them to add two tablespoons of white vinegar. The milk will not agglomerate but only become a bit thicker. The student to record the pH of the liquid.

What will student learn from this experiment

- Protein self-assembly into nanostructures
- Relationship between the 'macro' properties (colour, smell, taste, consistency) of milk and its molecular structure and how these can be manipulated to obtain different products (cheese, yogurt, etc.)

The following are reasons why the milk only becomes a bit thicker but does not agglomerate

In normal milk (pH close to neutral), the casein micelles are formed of caseins (charged negatively, Ionization Potential = 4.6), which are neutralized by calcium clusters resulting in a stable micelle. In addition to electrostatic interaction, hydrophobic interactions among the caseins stabilize the micelle. By adding vinegar, the pH of milk is lowered enough to induce dissociation of the casein micelles because calcium phosphate is released from the milk micelles. The reason for this is that lowering the pH to the isoelectric point leads to the titration of the phosophoseryl and carboxyl groups in the proteins. Without their negative charge, these groups cannot link to the colloidal calcium phosphate nano clusters, so these are released from the micelle.

When vinegar is added to warm milk, a curd is formed because the proteins that make up the micelles are no longer charged and aggregate due to hydrophobic interactions. This aggregation leads to precipitation because the milk is warm. In fact, if the vinegar is added to cold milk, an increase in viscosity is noted (sign of a coagulation of the micelles) but a curd is not formed. This is because at temperatures below 25 °C, acidification of milk leads to an increase in micelle dissociation but otherwise the caseins remain in the micelles. The reason lies in the fact that the stability of casein micelles is not entirely connected to electrostatic interactions but also to hydrophobic interactions. The latter are extremely temperature-dependent: hydrophobic interactions are stronger at higher temperatures.

Therefore, hydrophobic interactions maintain the stability of casein micelles in cold milk even when its pH has been lowered to the isoelectric point. On the other hand, if acidification occurs after milk has been warmed (to about 60 °C), micelles are dissociated (calcium phosphate is released from the micelle) and will aggregate due to increased electrostatic forces and increased hydrophobic interaction.

15.3 Experiment 2: Testing Self Assembly in Liquid Crystals and Effect of Temperature

Theory: Liquid crystals (LCs) are an example of self-assembled molecules that are sensitive to external factors, such as temperature, and that change their assembly as a consequence of these variations. The effect in some types of liquid crystals is a change of colour. Liquid crystals have properties that are directly connected to the way their long molecules self-assemble into nanostructures. Although these nano structures cannot be seen with the naked eye, macroscale changes in the optical properties (colour) of the liquid crystal as its temperature is changed can be seen. When the temperature of a liquid crystal changes, its molecules self-assemble into a different nanostructure: this influences the way light is reflected by the liquid crystal, and so that a different colour is observed. Therefore, this colour change is a direct consequence of a change in the self-organisation of the liquid crystal molecules

The molecules in a liquid crystal are often shaped like rods or plates or some other forms that encourage them to **align collectively along a certain direction (Figure** 6.8) **A liquid crystal is formed by the self-assembly of molecules into ordered structures**, called phases. An external disturbance, such as a change in temperature or magnetic field, even very small, can induce the liquid crystals to assume a different phase. The molecules in liquid crystal displays, for instance, are reoriented by relatively weak electrical fields. Different phases can be distinguished by their different optical properties.

Lyotropic liquid-crystalline phases are **abundant in living systems**, such as biological membranes, cell membranes, many proteins (*e.g.*, the protein solution extruded by a spider to generate silk), as well as tobacco mosaic virus. Soap is another well-known material which is in fact a lyotropic liquid crystal. The thickness of the soap bubbles determines the colours of light they reflect.

Experiment 2.1 Testing Thermotropic Liquid Crystals

In this experiment, students will study the properties of a **thermotropic liquid crystal** meaning that its properties change with changes in temperature.

This experiment will show students two fundamental concepts:

(a) that the way a material behaves at the macroscale depends on its structure and

(b) at nanoscale liquid crystals are self-assembled molecules that organise themselves into nanostructures which have specific optical properties.

Liquid crystals that change colour when temperature is changed

Liquid crystal phases that respond to certain temperature ranges are called **thermotropic liquid crystals**. Many thermotropic liquid crystals exhibit a variety of phases as temperature is changed,. The ordering inside a thermotropic liquid crystal exists in a specific temperature range (Figure 15.2).

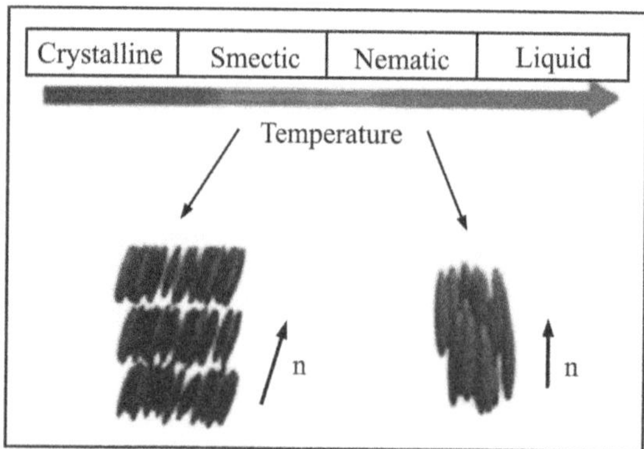

Figure 15.2 Structure transition of an thermotropic LC from smectic to nematic phase as the temperature is increased

(*Source*: Share alike .3.0 IPSE Education Resources, Liquid Cryastals,University of Wisconsin-Maddison)

At high temperatures, the thermal motion will destroy the delicate cooperative ordering of the liquid crystal phase, pushing the material into a conventional isotropic liquid phase. At too low a temperature, most liquid crystal materials will form a conventional (though anisotropic) crystal. These intermediate phases have some level of order, which is progressively lost as the temperature rises.

In the smectic phases, which are found at lower temperatures than the nematic, molecules form well- defined layers that can slide over one another. The smectics are thus positionally ordered in one direction. In the nematic phase, molecules have no positional order but they have long-range orientational order. This means that the molecules move quite randomly but they all point in the same direction (within each domain).

What is important to note is that some specific structural and optical properties are associated with each liquid crystal phase: **as the temperature is changed, the colour of the liquid crystal changes**. Thus, changing the temperature of the material allows the material to move from one phase to the other, so that the material exhibits these different colours. This type of liquid crystal is used in thermometers and in miniaturised temperature sensors (*e.g.*, to locate short circuits on circuit boards).

A particular type of liquid crystal phase is the chiral nematic phase. The chiral nematic phase exhibits chirality (handedness). This phase is often called the cholesteric phase because it was first observed in cholesterol derivatives.

In this experiment, students will analyse this type of liquid crystal. Only chiral molecules (i.e., those that lack inversion symmetry) can give rise to such a phase. In this phase, the molecules are stacked in rotating layers, like a spiral staircase (helix). On each 'step' of the staircase, the molecules are arranged in a specific order, but there is a finite angle between each 'step' and the next. The chiral pitch, p, refers to the distance over which the liquid crystal molecules undergo a full 360° twist. As the temperature of the liquid crystal changes, the pitch changes, which leads to tighter or looser helixes.

What gives the Colour to the Liquid Crystal?

When light strikes a liquid crystal, some of the light is reflected. What is seen is the reflected light.

The colour (i.e., the wavelength) of the reflected light depends on how tightly the helix is twisted. If the pitch in the liquid crystal is of the same order as the **wavelength of visible light (400-700 nm)**, then interesting optical interference effects can be observed. The colour of the light reflected depends on the pitch in the liquid crystal (i.e., **how tightly the helix is twisted**). When the helix is tightly twisted, the pitch is smaller, so it reflects smaller wavelengths (blue end of the spectrum); when the liquid crystal is less twisted, it has a larger pitch, so it reflects larger wavelengths (red end of the spectrum). An increase in temperature leads to a decrease in the pitch. By increasing the temperature of the liquid crystal, a colour change should be expected from the red end of the spectrum to the blue end of the spectrum, so from orange, to yellow, green, blue and violet.

The chiral twisting that occurs in chiral liquid crystal phases also makes the system respond differently in right and left-handed circularly polarised light. These materials can thus be used as polarisation filters.

What can This Experiment Teach about Nanotechnology?

The properties of materials at the macroscale are affected by the structure of the material at the nanoscale. Changes in a material's molecular structures are often too small to see directly, but sometimes changes can be seen in the properties of the material. Liquid crystals are an excellent example, in particular the type used in this experiment, since their optical properties (colour) change visibly as the temperature of the liquid crystal is changed. In nanotechnology, scientists take advantage of the peculiar properties of materials at the nanoscale to engineer new materials and devices.

Liquid crystals are an example of self-assembled molecules which change their spatial organization depending on external such as temperature. Self-assembly is another fundamental concept in nanoscience.

Experimental Materials

- cholesteryl oleyl carbonate (Sigma-Aldrich No 151157),
- cholesteryl pelargonate (Sigma-Aldrich No C78801),
- cholesteryl benzoate (Sigma - Aldrich No. C75802),

Procedure

Samples to be prepared

Table 15.1 indicates the quantities of materials needed to make four samples of liquid crystals each sensitive to a different temperature interval. Depending on the class level, students can synthesise four different liquid crystal mixtures, or the instructor could prepare them in advance for the class.

Table 15.1 Liquid crystal samples

Liquid Crystal	Cholesteryl oleyl carbonate	Cholesteryl pelargonate	Cholesteryl benzoate	Temperature (°C)
Type 1	0.65	0.25	0.10	17-23
Type 2	0.45	0.45	0.10	26.5-30.5
Type 3	0.40	0.50	0.10	32-35
Type 4	0.30	0.60	0.10	37-40

1. ***Synthesis of four different liquid crystal mixtures*:** Measure the weights of solids in three different petridishes and mix the three solids in the large glass vial using a plastic funnel. Secure the funnel on the mouth of the glass vial with clear wrapping tape to make sure the funnel stays in place when the students add the solids. Note that the solids (particularly cholesteryl oleyl carbonate) are quite sticky so be sure to gently push the solids down the funnel and remove as much solid as possible from the funnel walls. If a lot of solid remains on the spatula or funnel walls, keep both of these in place as you heat the glass vial, so the solid can melt and is not lost. Heat the glass vial using a hot plate. The hot plate should be set at 185 °C. While the sample is still liquid, gently move the vial around at an angle so that the liquid crystal spreads around the vial walls. Clearly mark the vial with a number corresponding to the type of liquid crystal you have made ('1' for Type 1 etc.).

   Students should observe the colour change as the liquid cools down.

   Prepare four different vials and then let the students test them.

2. **Preparation of four liquid crystal sheets and testing in a water bath:** *Prepare four different crystal liquid sheets*: Cut two pieces of transparent contact paper (about 10 × 10 cm), peel off the backing paper and place on the laboratory bench. With aspatula, place some liquid crystal type 1 in the centre of the sheet. If the liquid crystal is very cold and has turned solid, heat the vial first with a hairdryer (it should be the consistency of honey). Two to three spatulas of material will be needed. Place the second piece of contact paper on top of the first, so that the two sticky parts attach to each other. Gently press the centre where the liquid crystal is to distribute it evenly. A thin layer of liquid crystal about 4 × 4 cm is needed. Do not press too hard or the material will come out from the edges. Cut the sheet at the end around the edges. With a permanent marker, mark the number corresponding to the type of liquid crystal ('1' for Type 1 etc.) in the corner of the sheet. (Figure 15.3).

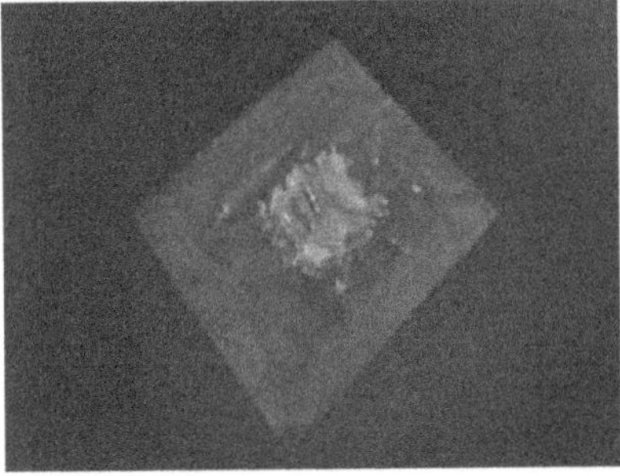 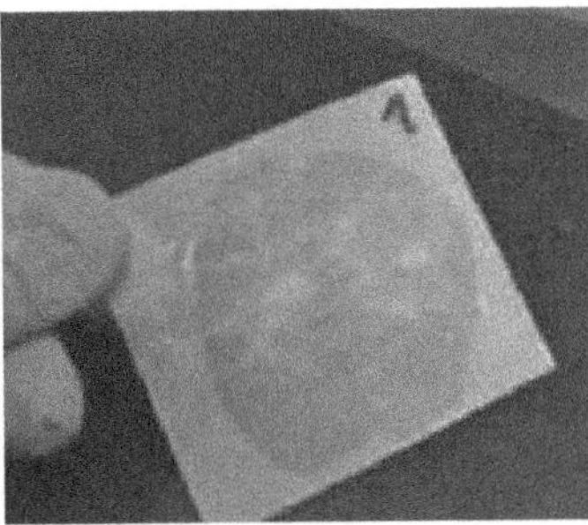 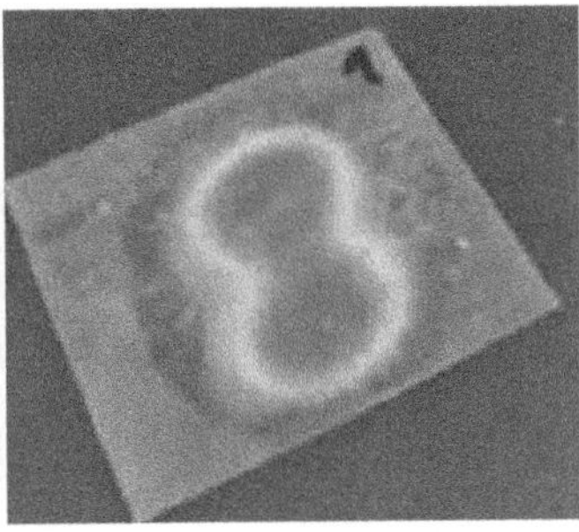

Figure 15.3 The liquid crystal sheet before and after it has been touched
(Images: *courtesy*: Nanoyou project)

Press the liquid crystal sheets with your fingers: Place the four liquid crystal sheets you have just made on a piece of **white A4 paper**. Wait few seconds. What do you see? Now press against each of the liquid crystal sheets with a finger (wear gloves). Repeat the experiment putting the sheets on the **black A4 cardboard**. In order to compare the different sheets, the finger should be kept on each sheet for the same time. Students should record their observations.

Test each liquid crystal sheet in a water bath: The temperature of the water bath should initially be 15 °C, and then gradually increased to reach the working temperature of each liquid crystal mixture.

To confirm the sensitivity of liquid crystals to temperature changes the students should test sheet 1 (Type 1) at higher temperatures and/or sheets 2 to 4 at lower temperatures (compared to their working temperature) and record data in the following tables.

	White Paper	Black Paper
Sheet 1 (mixture 1)		
Sheet 2 (mixture 2)		
Sheet 3 (mixture 3)		
Sheet 4 (mixture 4)		

Liquid crystal sheet number 1		
Temperature (°C)	Colour	Comments
16		
17		
18		
19		
20		
21		
22		
23		

Liquid Crystal Water	Temperature range (°C)
Type 1	15-23
Type 2	23-30
Type 3	30-35
Type 4	35-40

Students should observe the colour change of each liquid crystal as the water temperature increases (it is best to test the samples one at the time, so students can monitor closely what happens in each sample.

15.4 Experiment 3: Sensor Applications of Localized Surface Plasmon Resonances (LSPR) in Colloid Nano Particles

Experiment 3.1 Application of Colloidal Gold Nano Particles

Distinct properties of Nano Gold

One of the distinguishing properties of metal nanoparticles in general is their optical properties, which are different to those of their bulk counterpart. This is due to an effect called localised surface plasmon resonance (LSPR), which is described in detail in Chapter 1.

Metallic NPs with sizes smaller than the wavelength of light show strong dipolar excitations in the form of localized surface plasmon resonances (LSPR). LSPRs are non propagating excitations of the conduction electrons of metallic NPs coupled to the electromagnetic field.

Thus Surface Plasmon resonance (SPR) can be described as the resonance collective oscillation of valence electrons in a solid stimulated by beam of light incident (electromagnetic field) (Figure 15.4). The resonance condition is established when the frequency of light matches the natural frequency of valence electrons oscillating against from this restoring force. SPR in nanometer-sized structures is called LSPR. For gold and silver NPs, the resonance falls into the visible region of the electromagnetic spectrum. A striking consequence of this is the bright colors exhibited by particles both in transmitted and reflected light, due to resonantly enhanced absorption and scattering.

The resonance frequency of the oscillation, i.e., the surface plasmon (SP) energy, it is essentially determined by the dielectric properties of the metal and the surrounding medium, and by the particle size and shape. The collective charge oscillation causes a large resonant enhancement of the local field inside and near the NP. This field enhancement is used in surface-enhanced Raman scattering (SERS) (5).

One of the consequences of the LSPR effect in metal nanoparticles is that they have very **strong visible absorption** due to the resonant coherent oscillation of the plasmons. As a result, colloids of metal nanoparticles such as gold or silver can display colours which are not found in their bulk form, such as red, purple or orange, depending on the shape, size and surrounding media of the nanoparticles. The energy of LSPR is sensitive to the **dielectric function** of the material and the surroundings and to the shape and size of the nanoparticle. This means that if a ligand, such as a protein, attaches to the surface of the metal nanoparticle, its LSPR energy changes. Similarly, the LSPR effect is sensitive to other variations such as the **distance between the nanoparticles**, which can be changed by the presence of surfactants or ions. The fact that LSPR depends on the dielectric environment means that the refractive index can be used as the sensing parameter: changes in the local dielectric environment, induced by the sensing process, are used to detect the binding of molecules in the particle nano-environment.

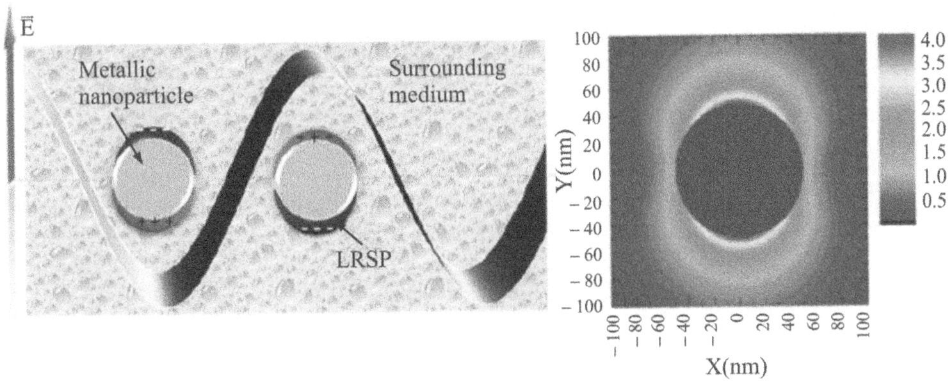

Figure 15.4 Left: Schematic of the electron charge displacement (valence electrons) in a metallic NP (LSPR) interacting with an incident plane wave, with electric field polarized E into a host matrix. **Right:** The corresponding electric field strength pattern for a 100 nm silver sphere, irradiated at a wavelength of 514 nm

Colorimetric Gold Nanosensor

Aim: Plasmonic colorimetric nanosensors are used in nanomedicine because they allow the precise detection of specific analytes with high sensitivity and without the need for a label. In a colloidal plasmonic biosensor (*e.g.*, a sensor made of gold nanoparticles), the sensing event results in a change of aggregation in the colloid, which can determine a colour change of the colloid.

The aim of this experiment is to synthesise and test a plasmonic colorimetric sensor. First, the students will synthesise a colloid of gold nanoparticles and learn that the colour of the gold colloid is ruby-red, rather than golden as for bulk gold, so they will have direct evidence that the size of a material at the nanoscale is fundamentally important in determining its properties.

The gold colloid prepared is then tested as a chemical colorimetric sensor: by adding a salt to the colloid, its agglomeration changes, which results in a change in colour (from ruby-red to blue). In the experiment, students test other solutions, such as sugar, and observe the effect on the colloid. The last test involves the use of a protein and shows how these can coat the nanoparticles to a point they can no longer aggregate. In this experiment, students use a plasmonic colorimetric sensor made of nanoparticles: the gold colloid (containing gold nanoparticles) is the sensing material, and the addition of an analyte (such as salt or sugar) induces a change in the aggregation of the nanoparticles in the colloid, which is reflected in a change in its colour.

Expected Outcomes: Understanding of the effect of size on – Properties of colloids; optical properties of gold nano particles

Understanding of the use of nano metal colloids for sensing and in particular medical diagnostics.

Plasmonic Colorimetric Gold Colloid Sensor

In a plasmonic biosensor, the nanoparticles can be dispersed in a medium (in which case, the biosensor is a **colloidal plasmonic biosensor**) or supported on a surface (**surface plasmonic biosensor**). Both types of sensors exploit the fact that the sensing event changes the LSPR of the metal nanoparticles, but they use different read-out (measurement) strategies.

In a **colloidal plasmonic biosensor** (*e.g.*, made of gold nanoparticles), the sensing event results in a change of aggregation among the nanoparticles that form the colloid, which can determine a colour change of the colloid. Absorption spectroscopy can be used to quantify the biosensing event.

In the case of gold colloid, which is normally red, the sensing event can result in the colloid becoming blue. Thus, metal colloids can be used as **plasmonic colorimetric biosensors**. In nanomedicine, this effect is used, for example, in **genetic screening**, where scientists look for a specific gene sequence in a sample which can be indicative of a specific disease.

In gene screening first, the sequence of bases in the target DNA is identified. Then two sets of gold particles are prepared one has DNA attached that binds to one end of the target DNA, and the second set carries DNA that binds to the other end (**Figure 15.5**).

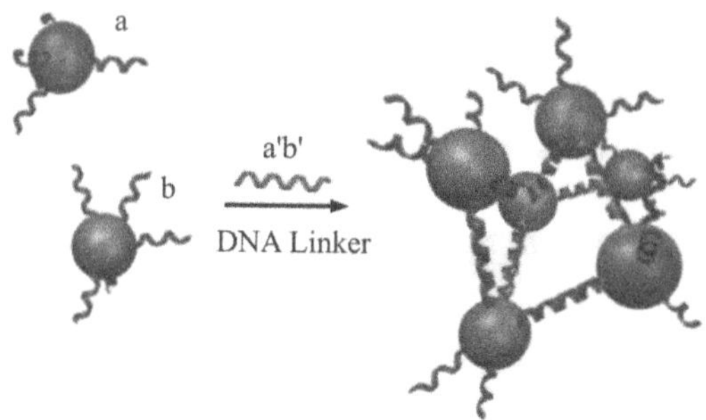

Figure 15.5 Plasmonic colloidal nano sensor
(*Source*: Jin. etal..Journal of American Chemical Society 2003,125, (6)1643-54)

The nano particles are dispersed in water. When the target DNA is added, it binds both types of nano particle together, linking them together to form an aggregate. The formation of this aggregate causes a shift in the light-scattering spectrum from the solution (i.e., a colour change in the solution that can easily be detected). The example is illustrated in Figure 15.6. Since the colloid changes colour as a result of the sensing event, this is called colorimetric sensing. Gold colloids can be used for colorimetric sensing specifically because they have different colours depending on the environment that surrounds them.

The method is not limited to sensing DNA strands and can sense also other complementary biomolecules such as antigen-antibody systems; it is used to study the stability of proteins.

Here we will only use a qualitative approach: the outcome of the sensing process is a change in colour of the solution which is visible with the naked eye (visible absorption).

Gold Colloid Synthesis

In this experiment, students will prepare a colloid of gold nanoparticles in water starting from a solution of gold chloride hydrate and a solution of sodium citrate. This is the simplest reaction to synthesise gold nanoparticles and was pioneered by J. Turkevich et al. in 1951 and refined by G. Frens in the 1970s. The method described in this reaction is adapted from 'Color my nanoworld', *Journal of Chemical Education*, 2004, 81(4).

This synthesis method generally produces nanoparticles 10-20 nm in size, so the end-product is a gold colloid. In the reaction, the citrate acts as a weak reducing agent (reducing $AuCl_4$-to Au) and as a stabiliser. A layer of citrate anions adsorbs around each nanoparticle and prevents these from aggregating: the electrostatic repulsion of the anions keeps the nanoparticles separated. (Figure 15.6).

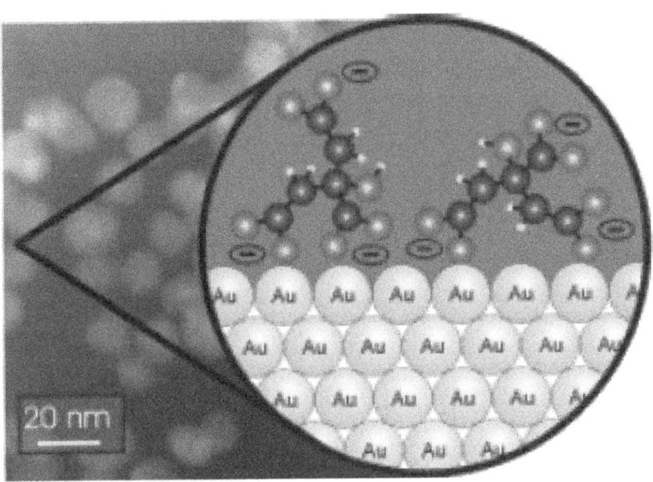

Figure 15.6 Left: Au nano particles 13 nm diameter **Right:** Au nano particle surface (each nano particle is made up of more than 50,000 Au atoms) Citrate ions cover the surface of nano particle

In this state, the colloid appears ruby-red owing to the absorption of light by the free electron oscillations (the surface plasmon) (Figure 15.7).

The position of the surface plasmon λmax for colloid gold is between 500 and 600 nm depending on the particles' size and shape, as well as the solvent refractive index and inter-particle distance. In this experiment, λmax is around 520 nm (green) and the solution appears red. A Colorimeter or a Spectro photometer can be used to observe colour change and its absorbance.

If the anion layer is removed, the nano particles start to approach and agglomerate.

Figure 15.7 Colloidal nano gold ruby red (the surface plasmon λ_{max} for colloid gold is between 500 and 600 nm)

This effect can be used to sense a certain chemicals (Figure 15.8). If a strong electrolyte is added, such as NaCl, the ions of the salt shield the negative charges on the particles, allowing them to approach and aggregate into larger and larger clumps. The formation of agglomerates is reflected in a change of the optical spectrum and the appearance of a second absorbance peak around 650-750 nm, causing the solution to **turn deep blue**. If a high concentration of salt is added, the nano particles aggregate to a point at which they precipitate, and the solution becomes clear.

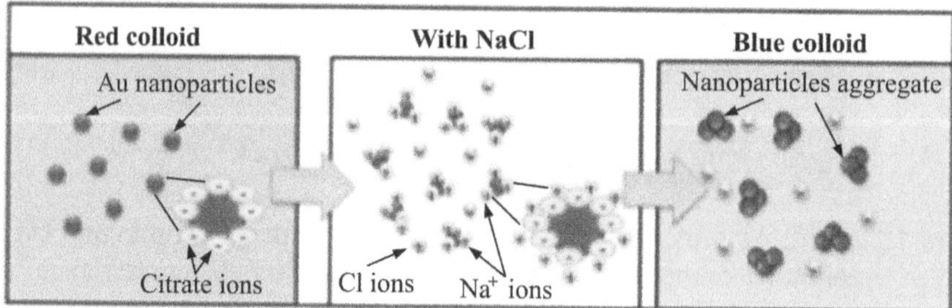

Figure 15.8 Schematic representation of the optical changes of a colloidal gold as a solution of salt is added (***Image:*** L. Filipponi, iNANO, Aarhus University, Creative Commons Attribution ShareAlike 3.)

If a stabiliser of high molecular weight, such as a protein or polyethylene glycol, is added, it adsorbs to the surface of the nanoparticles with the effect of inhibiting aggregation, even at high salt concentration (Figure 15.9). In this exercise, egg white is used as a very economic source of protein (mainly ovoalbumin). If a weak or non-electrolyte is added (*e.g.*, sugar), the electrostatic repulsion between the gold and the citrate ions is not disrupted and the solution remains. A small change in colour (*e.g.*, from ruby-red to slightly pink) can be observed, the consequence of a very small agglomeration that occurs as the protein is added (Figure 15. 10).

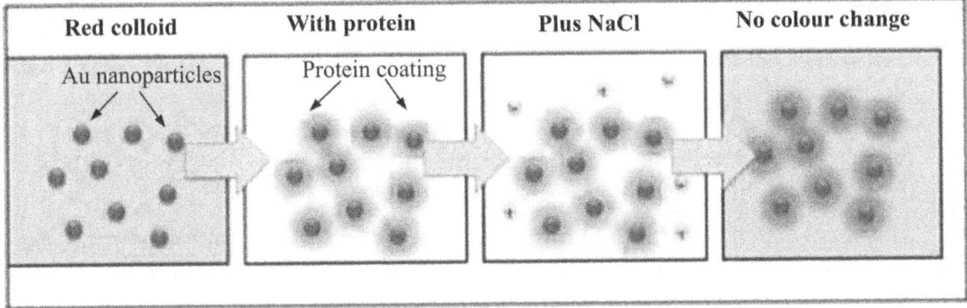

Figure 15.9 Schematic of colour changes in gold colloids when protein is added

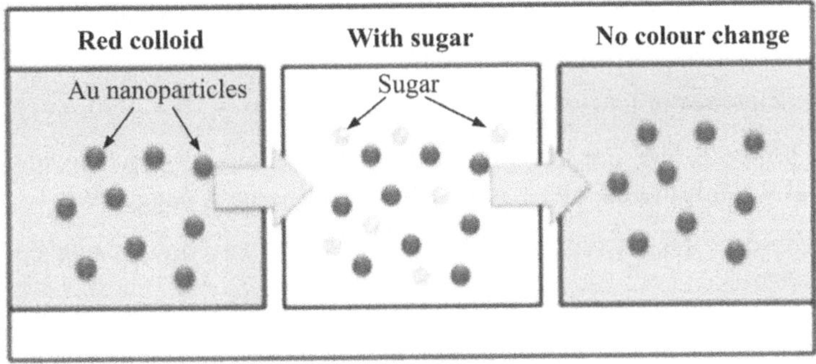

Figure 15.10 Schematic of optical changes of colloidal gold when a solution of sugar is added

Experimental

Synthesis of a Gold Colloid

Materials

Chemicals

- 0.1 g $HAuCl_4*3H_2O$ (Sigma-Aldrich No G4022),
- 0.5 g $Na_3C_6H_5O_7*2H_2O$ (Sigma-Aldrich No S4641),
- 1 litre distilled water

Procedure

1. **Preparation of stock solutions**

   1.0 mM hydrogen tetrachloroaurate: Dissolve 0.1 g $HAuCl_4*3H_2O$ (orange solid) in 500 mL distilled water. Solution appears light yellow. This stock solution of gold (III) ions can be prepared in advance if stored in a brown bottle or in a bottle well covered (including cap) in aluminium foil. Stored in this way, the solution will last a couple of months.

   1% trisodium citrate: Dissolve 0.5 g $Na_3C_6H_5O_7*2H_2O$ in 50 mL distilled water.

   Approximately 1 M NaCl: Dissolve 0.5 g NaCl (or table salt) in 10 mL distilled water.

   Approximately 1 M sugar solution: Dissolve 2 g sugar in 10 mL distilled water.

2. **Synthesis procedure for the gold colloid**
   (a) Add 20 ml 1.0 mM $HAuCl_4$ stock solution to a 50 mL Erlenmeyer flask on a stirring hotplate (turned off). The solution appears light yellow. Add a magnetic stir bar to the flask. Turn the stirring and the temperature on. Bring the solution to the boil. After the temperature is turned on, some bubbles are seen forming at the bottom of the flask, then they become more and more evident.

   This is when the citrate should be added, even if full bubbling is not seen. The reaction should ideally be performed under reflux conditions, as evaporation occurs which leads to a volume loss.

   (b) To the boiling solution, add (all at once) 2 mL of a 1% solution of trisodium citrate dihydrate, ($Na_3C_6H_5O_7*2H_2O$). The gold colloid gradually forms as the citrate reduces the gold (III). Students should observe the colour change as the citrate is added.

Gold Colloid as a Plasmonic Colorimetric Sensor

In nanomedicine, gold colloids are now being investigated as colorimetric sensors, for example to detect specific DNA fragments. Here, a simpler experiment is suggested.

The colloid should be tested with different solutions to see how it can be used for sensing certain species.

- Strong ionic solutions induce a colour change from ruby-red to blue (e.g. NaCl). Therefore, it can be used to detect the presence of **electrolytes.**
- Weak or non-ionic solution does not induce this change (e.g. sugar).
- Surfactants or proteins coat the colloid to such a point that it is no longer affected by strong ionic solutions (*e.g.* adding a white egg solution followed by NaCl does not induce a colour change).
- Gold colloid in other solutions, such as vinegar (which is a mild acid) can be used for sensing

Materials

Chemicals

- the gold colloid (prepared as described earlier) about 15 mL can be used
- 0.5 g NaCl fine kitchen salt can be used – 2 g sugar
- A fresh egg
- 1 litre distilled water

Method

1. Divide the ruby-red gold colloid into five vials each containing about 3 mL.
2. Use one vial as control, and the other four to perform different plasmonic colorimetric tests.

Test 1: Add 6 droplets NaCl solution; the colloid becomes blue (if it is general kitchen salt, it looks more grey-blue).

In the colloid citrate, ions act as stabilisers: they cover the surface of the nanoparticles and prevent their aggregation.

When a strong ionic solution (NaCl) is added, its ions shield the charges on the particles, allowing them to approach and aggregate.

The formation of agglomerates is reflected in a change of the optical spectrum causing the solution to turn deep blue.

Test 2: Add 15-20 droplets NaCl: the colloid becomes dark grey, and a black precipitate starts to appear. If the sample is left for some time (30 minutes), the solution becomes clear and much black precipitate forms.

Explanation: If a high concentration of ionic solution (NaCl) is added, the nano particles aggregate to a point they precipitate, the solution becomes clear and a black precipitate is clearly seen forming. If the vial is left to rest for about 30 minutes, aggregation continues until a big precipitate forms, and the solution appears clear.

Test 3: Add 10 drops sugar solution: the colloid doesn't change colour and it remains ruby-red.

Explanation: Sugar is anionic; therefore, the citrate ions covering the gold nanoparticles remain undisturbed.

Test 4: Crack open a fresh egg and with a Pasteur extract some egg white (about 1 mL or 2-3 full Pasteurs), place it in an empty glass vial and add one Pasteur distilled water. Mix gently together: it will form a foam so let it stand for a minute to reduce the foam, then take the solution from the very bottom to avoid the foam/bubbles. Add this water-egg white mixture to the gold colloid: the colloid turns dark pink and an 'emulsion' is seen.

Explanation: As the protein is added to the gold colloid, the nanoparticles become coated with a layer of protein, which acts as a stabiliser. Nevertheless, during the addition of the protein, a little aggregation might occur, which is reflected in a colour change (dark pink instead of ruby-red). More sophisticated addition methods (*e.g.*, using stirring) might prevent this from happening.

Test 5: If NaCl (6 drops) are added to the above solution (**Test 4**), the colour does not change; it does not turn blue. If more NaCl is added, the colour still doesn't change.

Explanaion: The protein (mainly ovo albumin) in the egg white acts as a surfactant, shielding the gold nanoparticles from the ions, in a sense 'protecting' the nanoparticles. The NaCl ions don't have access to the gold nanoparticle surface and therefore do not induce a change in their aggregation.

Experiment 3.2 Gold Colloid as a Medical Diagnostic Tool

Students should reflect on how a gold colloid can be used as a diagnostic colorimetric tool. The general idea is that any diagnostic process that induces a change in aggregation can (potentially) induce a visible colour change. To quantify this change, a spectrophotometer would be needed to measure the absorption of the solution. To be relevant in diagnosis, the process must be specific, based on a biomolecular recognition system, which can be the complementary of DNA strands, or other processes such as antigen-antibody recognition. Compared to other diagnostic tools, gold colloid has the advantage of not requiring a label, since the detection reaction is visible. This is an advantage because labeling is normally a long process; therefore, a label-free detection method is faster and requires fewer consumables, thus lowering the cost of the process.

Experiment 3.3 Gold Nanoparticles (GNP) for Bacterial Detection *(Neisseria Meningitidis Bacteria)*

Theory: *Neisseria meningitidis* bacteria: This bacterium is responsible for bacterial meningitis. This disease is contracted through body fluids and is very contagious. The outbreaks commonly affect areas with young populations, such as colleges. Symptoms include headache, fever, vomiting, confusion, and sensitivity to light.

Cerebral spinal fluid (CSF) can also be tested by a variety of means for bacterial meningitis antigen (one method is an ELISA (Enzyme Linked Immuno Sorbant Assay) Test. The present experiment is based on gold nanoparticle assay test.

Assay Test: An assay test is a quantitative *in vitro* test for a molecule such as an antibody or antigen in which the test material is absorbed on a surface and exposed either to a complex of an enzyme linked to an antibody specific for the antigen. Upon GNPs bonding with target molecule a color change will occur.

Antigen: A foreign molecule that does not belong to the host organism and causes an immune response.

For this assay test the GNP's surface need to be modified to be specifically attach to the bacterial meningitis antigen thus reacting with and only with the antigen. One way in which GNP surfaces can be modified to be specific to a molecule (ex specific strain of DNA, RNA or protein) is by coating the gold surface with alternating layers of cationic and anionic polyelectrolytes. Synthesizing DNA or peptide linkers between GNPs confer these building blocks with programmable spatial coordinates. This can give an exquisite level of control, as the polymerase chain reaction can be performed directly on the surface of GNPs for this kind of organization.

In this experiment the outbreak of bacterial meningitis will be simulated swapping test tube fluids.

Experimental

Producing GNPs

1. Prepare 250 ml of 1.0 mM $HAuCl_4$ solution by dissolving 0.1 g of $HAuCl_4$ in 250 mL of distilled water. Equally distribute this solution into 8 vials that are large enough to hold 20 mL of solution. This solution is unstable and keeps only a few days, so make it fresh for the lab.
2. Prepare 50 mL of 38.8 mM sodium citrate solution by dissolving 0.5 g of $Na_3C_6H_5O_7$ in 50 mL of distilled water.

Add 0.5 mL (approx. 15 drops) auric chloride solution to 28 mL distilled water. Heat the solution to 100 °C on the stirrer or hot plate. Once the solution reaches 100 °C and starts to bubble, add 1.5 mL citrate solution as quickly as possible, stirring vigorously. The redox reaction of tetrachloraurate (also known as tetrachlorauric acid or tetra-chlorauric (III) acid trihydrate) takes place in which gold ions are reduced to atomic gold clusters as shown in Figure 15.11. The reductant sodium citrate (also called trinatrium citrate dihydrate) not only reduces the gold but also acts as a dispersion medium to stabilise the gold clusters that are created. By adding the reductant, the atomic coagulation of the metal ions is halted and the result is a colloidal cluster (A cluster, or nanoparticle, is a collection of 3 to 50 000 atoms) enclosed by a ligand case. The diameter of the gold nanoparticles will be generally between 12-18 nm. If the clusters are spatially distributed in another physical medium, the entire system is known as a colloid.

$$3\,HO-C(COOH)(CH_2COOH)_2 \xrightarrow{+\,2(AuCl_2)^-} 3O{=}C(CH_2COOH)_2 + 2Au\,(O)$$

$$3O{=}C(CH_2COOH)_2 \xrightarrow{+\,16\,(AuCl2)^-} 15\,CO_2 + 16\,Au\,(O)$$

Figure 15.11 Chemical reactions involved in Gold Nano Particles (GNP) formation

The initial red colour of the auric chloride solution intensifies until it becomes a deep red. At temperatures between 85 and 90 °C, it will take approximately 5 minutes until the colour changes; at 100 °C, the reaction is even faster. Depending on the size of the particles formed, you may get a violet colour instead of red.

3. Making NaCl solution for GNP Assay Test Simulation. Prepare a 1M solution of NaCl by dissolving 0.5 g of NaCl in 10 mL of distilled water to be used as antigen

For a group of 10-15 students give each student an eye dropper and at a test tube filled with 10 ml of distilled water EXCEPT for one student who you select to receive the meningitis sample (NaCl solution). Instruct students to squirt 2 mL of their test tube solution into a partner's test tube and vice versa. Be sure students don't touch other test tubes or submerge in other test tubes to avoid contamination. Ask students to add 1.5 mL of their GNP from day 1 to mimic an assay test. A color change will signal a positive result for meningitis (positive results should change blue). No color change or a slight brightening of the pink red signals a negative result for meningitis.

For laboratories having microbiological facilities *Neisseria meningitidis infected* bacterial fluids can be used for real test. *However we need use surface modified GNP's to be specifically attach to the bacterial meningitis antigen need to be used.*

Experiment 3.4 Gold Nano Particles for Early Pregnancy Test

Theory: The light absorbing properties of gold nano particles can be used to detect biomolecules This allows antibodies to be tagged by coupling with gold nano particles. When light is passed through these GNP they exhibit different colours due to LSPR effect discussed earlier. This can be used for early pregnancy test in which nano gold particles conjugated with the antibody are deposited in finely distributed form on a test strip. The attached antibody recognizes the peptide harmone in the urine with which the early pregnancy can be detected. A positive test result is indicated by a red stripe in the window. This pregnancy test is a monophase sandwich immune test for rapid detection for early pregnancy.

Procedure: The UltiMed® pregnancy test relies on the principle of detection of choriongonadotropin (hCG) which is a peptide hormone (gonadotropin) formed in the human placenta during pregnancy and is released early in pregnancy by the fertilised egg and the lining of the uterus. hCG consists of two subunits: a and b. On the test strip, a-subunits of hCG will be immobilised, forming a line that will turn red to indicate a pregnancy. Elsewhere in the strip, colloidal gold particles are tagged with monoclonal antibodies specific to the b-subunit of hCG. When the strip is dipped in urine, the liquid allows the tagged gold particles to move through the strip by capillary action. If the urine contains hCG (i.e., if the woman is pregnant), b-subunits of hCG bind to the tagged gold particles. When the b-subunits bound to the gold reach the immobilised a-subunits, the a- and b-sub-units bind together, forming a gold-hCG complex. If the concentration of hCG is high enough, the complex is visible as a red line, indicating that the woman is pregnant. Further gold particles bind to a second line, indicating that the test (whether positive or negative) was correctly performed. If there is no or only a little hCG is present in the urine sample the hormone and gold particle is not formed or an insufficient amount produced to generate a visible colouring in the area of the test line (Figure 15.12).

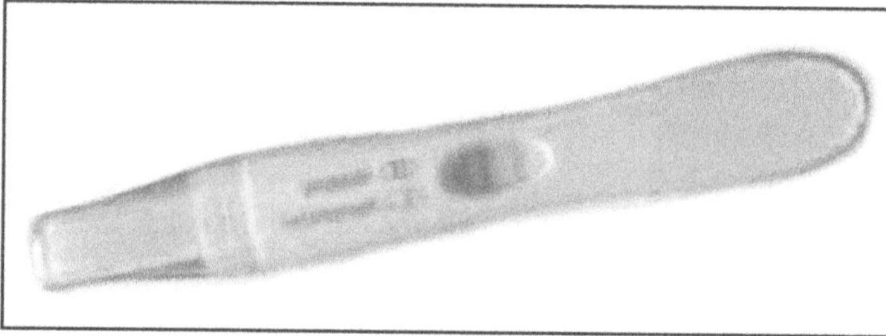

Figure 15.12 UltiMed® pregnancy test strip

Resources for Further Reading

- J Pharm Bioallied Sci. "Introduction to metallic nanoparticles" (July, 2013) http://www.ncbi.nlm.nih.gov/pmc/articles/PMC2996072/
- National Cancer Institute. "Video Journey into Nanotechnology (July, 2013) http://nano.cancer.gov/learn/understanding/video_journey.asp
- National Cancer Institute. Nano devices (July, 2013) http://www.cancer.gov/cancertopics/understandingcancer/nanodevices.
- BioAssay Works. "Ultra-sensitive gold technologies" (July, 2013) http://www.bioassayworks.com/
- Center for Disease Control and Prevention. "Meningococcal Diseases" http://www.cdc.gov/meningococcal/about/symptoms.html

Experiment 3.5 Application of Silver Nanoparticles Synthesis, Spectroscopy and Antibacterial activity

Introduction: In this experiment the students will learn the reduction synthesis of silver nanoparticles differences in optical properties at the nanoscale as compared to the same materials at the macro-scale and antibacterial activity of silver nano particles.

Nanoscale silver has been used as an antibacterial agent, and can even kill harmful strains of bacteria which are resistant to antibiotics. Silver nanoparticles have been used in the dressing of wounds, surgical masks, food packaging, water treatment, and even socks which prevent the growth of bacteria that cause foul odor. Because of the widespread use of nanosilver, it is important for students to understand its applications as well as potential problems.

Silver nanoparticles (AgNPs), or colloidal silver, can be synthesized in the presence of starch according to the following redox reaction:

$$2AgNO_3\,(aq) + C_6H_{12}O_6(aq) + H_2O(l) \longrightarrow 2Ag(s) + 2HNO_3(aq) + C6H_{12}O_7(aq)$$

In this reaction, glucose ($C_6H_{12}O_6$) reduces the silver cations from the silver nitrate. As the silver metal forms, starch coats the outsides of the particles, preventing them from aggregating and forming larger particles. Nano-sized materials often have different properties than the bulk materials; for example, silver nanoparticles appear yellow. A visible spectrum of AgNP solution (Figure 15.13) can be used to characterize the optical properties of silver nano particles and the difference between nano and macro particles using a spectrophotometer.

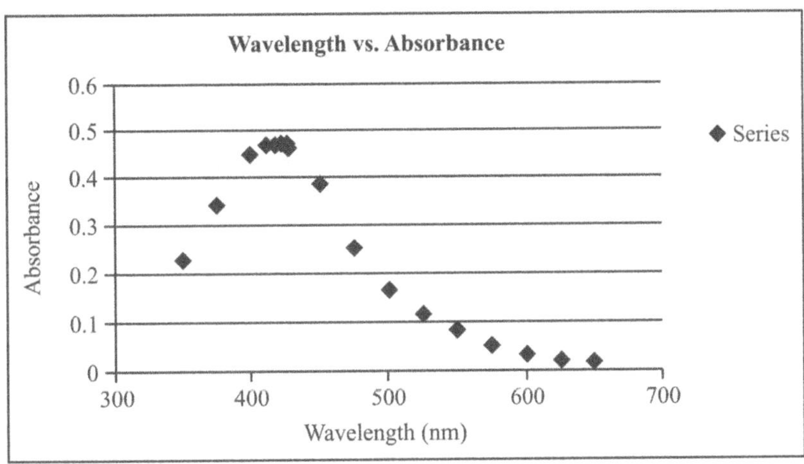

Figure 15.13 Absorption spectra of silver nano particles

Materials

- 0.1M $AgNO_3$ ($AgNO_3$)
- 0.1M α-D-glucose (α-D-glucose)
- 0.2% wt. soluble starch (soluble starch)

For Spectroscopy: Spectrophotometer

Procedure

Synthesis of Ag Nanoparticles

1. Place 200 μL of 0.1M $AgNO_3$ into a small Erlenmeyer flask or beaker.
2. Add 500 μL of 0.1M glucose, making sure that it comes into contact with the $AgNO_3$.
3. Invert the starch solution several times. Add 10 mL of the starch solution.
4. Heat the solution on a hot plate on a high setting until it is boiling vigorously. **Do not stir the solution!**
5. Boil the solution for 10 minutes. The solution should turn yellow.
6. Remove the sample from the hot plate, and let it cool.

Spectrophotometric Analysis of Ag Nanoparticles

1. Turn on the spectrophotometer and allow it to warm up for the recommended length of time.
2. Put the instrument in Transmittance mode. With the sample compartment empty, set the dark current to zero.
3. Invert the starch solution several times. Fill a clean cuvette with the starch solution. Wipe the outside with a tissue paper, then insert it into the sample compartment.
4. Set the wavelength at 350nm and zero the absorbance.
5. Insert a clean cuvette filled with the AgNP solution, and record the absorbance
6. Repeat steps 4 and 5 for different wavelengths; increasing the wavelength by 25 nm each time until a reading is obtained for 650 nm. Remember to zero the machine each time you change the wavelength. Record the data.
7. Identify the wavelength with the highest absorbance. Use the spectrophotometer to collect additional absorbance readings by varying the wavelengths in that specific area of the spectrum to determine which gives the wavelength of maximum absorbance (λ_{max}), then test the absorbance of the AgNPs at 490, 495, 500, 505, and 510 nm, looking for the wavelength that gives the maximum absorbance).
8. Divide the absorbance at the λ_{max} by 2 as absorbance at ½ max.

Calculations

1. To determine the Peak Width at Half Max (PWHM), first subtract the λ_{max} from the $\lambda_{1/2max}$. The difference is half the peak width. Multiply by 2 to get the PWHM.

$$PWHM = (\lambda_{1/2max} - \lambda_{max}) \times 2$$

$$PWHM = (480nm - 425nm) \times 2 \quad PWHM = 110 \text{ nm}$$

2. How many silver atoms are in one nanoparticle?

Determine the average number of atoms per nanoparticle using the following formula:

$$N = \frac{\pi \rho D^3}{6M} N_A \qquad .....(15.1)$$

N = number of atoms per nanoparticle, $\pi = 3.14$

ρ = density of face centered cubic (fcc) silver = 10.5 g/cm^3 (convert to g/nm3)

D = average diameter of nanoparticles 50 nm = 50 × 10^{-7} cm

M = atomic mass of silver (107.8688)

NA = number of atoms per mole (Avogadro's No = 6.023 × 10^{23})

$$N = \left[\frac{\pi \rho D3}{(6M)}\right] \times N_A = \frac{3.14 \times 10.5 \times \left(50.0 \times 10^{-7}\right)^3 \times 6.023 \times 10^{23}}{6.0 \times 107.868} = 383723.003$$

N = [(π (1.05 × 10–20) (11)3) / (6(108))] × 6.022 × 1023 N = 4.08 × 104 atoms/NP

3. What is the concentration of your silver nanoparticle solution?

Determine the molar concentration of the Nanoparticle solution using the following formula: (2)

C = NT/NVNA

C = molar concentration of nanoparticle solution

NT = Total number of silver atoms added on $AgNO_3$=1 M

N = number of atoms per nanoparticle (determined in Calculation 2)

V = volume of the reaction solution in L

NA = Avogadro number of nanoparticles per mole (= 6.023 × 10^{23})

$$C = \frac{1.6023 \times 10^{23}}{3837234 \times 6.023} = 2.606 \times 10^{-11} M / L = 260 \text{ Nm}$$

Note: This equation also assumes that the reduction of Ag^+ to Ag^o was complete (2).

4. What effect might a large variance in particle size have on the width of the absorbance peak in the visible spectrum?

Different sized nanoparticles will absorb different wavelengths of light. A large variance in particle size will result in a wider absorbance peak. A Peak Width at Half Max (PWHM) of 100 or less would indicate a fairly uniform distribution of particle sizes.

5. Name a physical property of silver that changes at the nanoscale.

The color of silver changes; this procedure produces yellow silver nanoparticles (AgNPs). The melting point of silver also drops at the nanoscale.

Experiment 3.6 Testing Antibacterial Activity of Silver Nanoparticles

Introduction: This experiment is aimed to demonstrate how effective silver nanoparticles are in inhibiting bacterial growth using the silver nanoparticles synthesized earlier.

Silver nanoparticles are used to control spoilage of foods, to fight bacterial infections, and to limit bacterial growth in clothing. In minute concentrations, silver is nontoxic to human cells. The high surface area to volume ratio of the nanoparticles allows for increased contact with the bacteria with minimum exposure to silver.

Materials

Earlier Synthesized silver nanoparticles

- 100 uL micropipetter with tips
- *E. coli* in nutrient broth, at an optical density of 0.03 at 600 nm
- Table disinfectant
- Marking pens

Bunsen burner

- Incubating oven (optional)

Needed for zone of inhibition procedure:

- Sterilized filter paper dots
- Ethanol with bent glass elbow
- Forceps
- Dilute hand soap solution
- 8 sterile nutrient agar Petri dishes
- Sterile deionized water
- Metric ruler with mm calibration

Needed for colony counting procedure:

- 14 sterile nutrient agar Petri dishes
- Ethanol with bent glass elbow

Needed for growth in nutrient broth:

- Spectrophotometer
- 13 Sterile cuvettes containing 5 mL sterile nutrient broth – set aside one cuvette of sterile broth as a blank to zero the spectrophotomer
- Sterile synthesized AgNPs

Safety

Use aseptic technique when handling the *E. coli.* Wear nitrile/latex gloves when working with *E. coli*

Procedure

Zone of Inhibition Procedure

Disinfect table

Use marking pens to label the bottoms of Petri dishes – two control, two silver nanoparticle, two hand soap, and two disinfectant dishes.

Aseptically pipette 100 uL of the *E. coli* stock solution to a control Petri dish. Use a sterile bent glass elbow to aseptically spread the bacteria over the surface of the agar. Close the dish. Using the same aseptic techniques, pipette 100 uL of *E. coli* to each of the remaining seven dishes.

Dip tips of forceps in ethanol and flame to sterilize. Use sterile forceps to pick up a sterile paper dot. Soak the dot in sterile deionized water. Place the paper dot in the center of the control inoculated Petri dish. Repeat this procedure to place the appropriate paper dots in the Petri dishes.

Place lids on petri dishes and tape closed.

Wash hands with soap and water.

Incubate Petri dishes upside down at room temperature or place in an oven at 37° C for 24 hours.

Observe Petri dishes through the plastic. Do not open the Petri dishes. Using the ruler, measure the zone of inhibition around the paper dots and record in the data table.

Colony Counting Procedure

Disinfect table

1. Use marking pens to label the bottoms of Petri dishes with the varying volumes of silver nanoparticles solutions to be used – 0 uL AgNPs, 10 uL AgNPs, 25 uL AgNPs, 50 uL AgNPs, uL AgNPs, 100, 150, 200 uL AgNPs, and 300 uL AgNPs. Label two Petri dishes for each volume of silver nanoparticles, for a total of 14 Petri dishes.
2. Aseptically pipette 10 uL AgNPs to the appropriately labeled Petri dish. Use a sterile bent glass elbow to aseptically spread the nanoparticles over the surface of the agar. Be careful to spread the nanoparticles over the entire surface. Repeat this procedure (with aseptic pipette), pipetting the appropriate volumes of nanoparticles to each of the Petri dishes.
3. Aseptically pipette 100 uL of *E. coli* stock solution to the first Petri dish, 0 uL AgNPs. Use the sterile bent glass elbow to spread the *E. coli* over the surface of the agar. Be careful to spread the bacteria over the entire surface. Repeat this procedure to aseptically inoculate and spread uL *E. coli* onto each of the remaining Petri dishes.
4. Cover Petri dishes with lids and tape closed. Allow Petri dishes to sit for 10 minutes undisturbed.
5. Wash hands with soap and water.
6. Incubate the Petri dishes upside down at room temperature or in an oven at 37° C for 24 hours.
7. Observe colonies through the plastic – do NOT open the Petri dishes. Count the colonies on each Petri dish and record in the data table.
8. *Growth in Nutrient Broth Procedure*
9. Disinfect table.
10. Zero the spectrophotometer at 600 nm using a sterile broth cuvette as the blank.
11. Use small pieces of marking tape to label the cuvettes of sterile nutrient broth with the varying volumes of silver nanoparticles to be used – 0 uL AgNPs, 10 uL AgNPs, 25 uL AgNps, uL AgNPs, 75 uL AgNPs, and 100 uL AgNPs. Label two cuvettes for each volume of nanoparticles. Put each piece of labeled tape near the top of the cuvette.

12. Aseptically pipette 10 uL of sterile AgNPs to the appropriate sterile cuvette of broth. Repeat this procedure for the varying volumes of the AgNPs solutions, aseptically pipetting the sterile AgNPs to the sterile broth cuvettes (12 total).
13. Aseptically pipette 100 uL *E.coli* to a cuvette with 0 uL AgNPs and gently finger vortex to mix. Record the absorbance of this solution at 600 nm in the data table at time 0 hours. Repeat this procedure, aseptically inoculating *E. coli* into each of the cuvettes and recording the time zero absorbances.
14. Wash your hands with soap and water.
15. Incubate in an oven at 37° C for 24 hr. Gently finger vortex & take absorbance readings at 24 hr. (If possible, take readings at recorded time intervals during the 24 hr. Before each reading, gently finger vortex.)

Analysis

Table 15.2 Zone of Inhibition

Plate	Test Reagent	Diameter of Zone of Inhibition (mm)	Average Zone (mm)
1	Control		
2	Control		
3	AgNPs		
4	AgNPs		
5	Soap		
6	Soap		
7	Disinfectant		
8	Disinfectant		

Table 15.3 Colony Counting

Plate	Amount of AgNPs (μL)	Colony Count	Average Colony Count
1	0		
2	0		
3	10		
4	25		
5	25		
6	50		
7	50		
8	100		
9	100		
10	200		
11	200		
12	300		
13	300		

Table 15.4

Tube	Amount of AgNPs (μL)	Absorbance at 600 nm							
		0 hrs	Avg. 0 hrs	2 hrs	Avg. 2 hrs	12 hrs	Avg. 12 hrs	24 hrs	Avg. 24 hrs
1	0								
2	0								
3	10								
4	10								
5	25								
6	25								
7	50								
8	50								
9	75								
10	75								
11	100								
12	100								

References

1. Kalaugher, L. (2004, January 12). Green Technique Makes Silver Nanoparticles. Retrieved July 3, 2007, from Nanotechweb.org: http://nanotechweb.org/articles/news/3/1/1/1.
2. Liu, X., Atwater, M., Wang, J., & Huo, Q. (2006). Extinction coefficient of gold nanoparticles with different sizes and different capping ligands. Colloids and Surfaces B: Biointerfaces.
3. Solomon, S. D., Bahadory, M., Jeyarajasingam, A. V., Rutkowsky, S. A., Boritz, C., & Mulfinger, L. (2007). Synthesis and Study of Silver Nanoparticles. Journal of Chemical Education, 322-325.
4. Journal of chemical Education 81(4) 2004.
5. K. Kneipp, H. Kneipp, I. Itzkan, R. R. Dasari, and M. S. Feld. J. Phys.: Condens. Matter 2002; 14:597–624.

Resources for Further Reading

1. Silver as an antimicrobial agent:
   http://microbewiki.kenyon.edu/index.php/Silver_as_an_ Antimicrobial_Agent# Current_uses
   Silver nanoparticles and the environment:
   http://pubs.acs.org/doi/abs/10.1021/es2001248

15.5 Experiment 4: Testing Super Hydrophobic Materials

Experiment 4.1 Measurement of water drop -hydrophobic/hydrophylic surface contact angle

Aim: The aim of this experiment is to show students some innovative materials that are **highly water repellent, stainless and require less cleaning due to their surface nano engineering**. These materials have been developed using as inspiration, since some plant leaves (such as the lotus leaf or nasturtium (*Tropaeolum*)) leaf have the exceptional property of being super hyrophobic due to nanostructures present in their surface composition. (as discussed in chapter 1) Water simply rolls off these leaves. The effect is due to the interplay of surface chemistry and surface topography at the micro and nano-level.

The students will:

(a) analyse different materials, including leaves, polymers, glass, etc., and analyse their wetting properties;

(b) see and test functional nano materials that have been engineered at the nanoscale to be super hydrophobic.

Expected Outcomes

- Understanding of the effect of surface topography on contact angle, and the insurgence of super hydro phobicity
- Knowledge of novel advanced materials that have been engineered at the nanoscale to be super hydrophobic and their applications

Background Information

1. *Fundamentals of surface properties*: The surface properties of a material are largely related to chemical species that are present at the surface. A very important surface property is its wetting behaviour (i.e., how water interacts with the surface). This property is related to the terminal groups of the molecules at the surface interface, which can be either hydrophilic (water-loving) or hydrophobic (water-hating).

One of the methods to quantify the wetting behaviour of a surface is to measure its contact angle (CA). The contact angle is the angle at which a liquid/vapour interface meets the solid surface as illustrated in **Figure 15.14**. The contact angle is the angle formed by the liquid and the three phase boundaries. The shape of the droplet is controlled by the three forces of interfacial tension, as shown in **Figure 15.14**. The contact angle provides **information on the interaction energy between the surface and the liquid.**

Measurement of Contact angle goniometer: The contact angle θ can be measured using an instrument called a contact angle **goniometer**. This gives a static measurement of contact angles. A droplet of water will be deposited over the surface under investigation and the angle θ measured either manually or, in modern instruments, digitally, by capturing a digital image and using dedicated software.

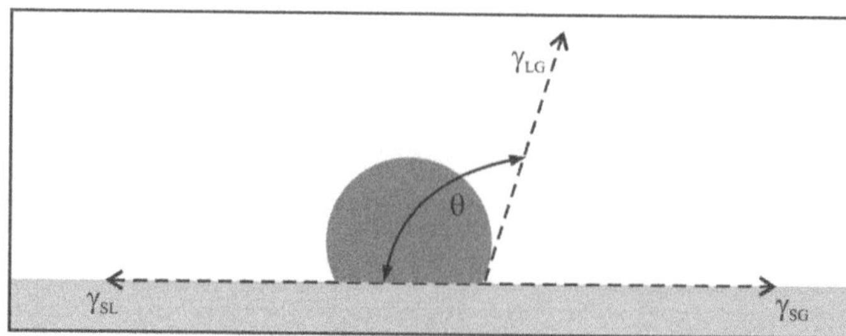

Figure 15.14 Static contact angle measurement of a droplet of water on a flat solid surface

Synthetic Surfaces with nanostructures tend to have very high contact angles, which can reach the super hydrophobic level: In principle, water resistance is always related to a micro or nanostructure of the surface. The physical reason for water resistance is the relationship of the interfacial surface tensions of water to air, of water to solid and of solid to air. The respective conformation of the relationships to each other determines the contact angle of the drop to the solid. According to the physical definition, a surface is hydrophilic (water friendly) if the contact angle to the drop is below 90 degrees. If it lies above 90 degrees, the surface is hydrophobic (water repellent, i.e., water resistant). In the case of full wet ability, the contact angle then stands at 0 degrees - the water forms a monomolecular film on the surface of the solid and a drop of water will, at the most, spread over the leaf. At a contact angle of 180 degrees (which is not present in nature) a drop would theoretically come into contact with the surface of the solid at only one point. The contact angle for conventional water-repellent surfaces (e.g. Teflon) is maximum 120 degrees. Surfaces can be classified depending on their contact angle as illustrated in the Table 15.5.

Table 15.5 Classification of surfaces based on contact angle

Contact angle value	Type of surface	Example
~0	Superhydrophilic	UV irradiated TiO_2
< 30	Hydrophilic	Glass
30-90	Intermediate (*)	Aluminium
90-140	Hydrophobic Superhydrophobic	Plastic > 140 Lotus leaf

(*) If the contact angle is towards 30, the material is defined as hydrophilic; if the contact angle is towards 90, the material is defined hydrophobic

The **superhydrophobic effect found in lotus leaves** has been extensively researched. The lotus plant (*Nelumbo nucifera*) is a native Asian plant which has the distinct property of rendering its leaves particularly clean even if its natural habitat is muddy. The leaves of the lotus plant have the outstanding characteristic of totally repelling water because they are superhydrophobic. This can be understood by imagining that in lotus plant leaf a surface with nano-roughness is formed of a series of very small pillars. (Figure 15.15) When a droplet rests

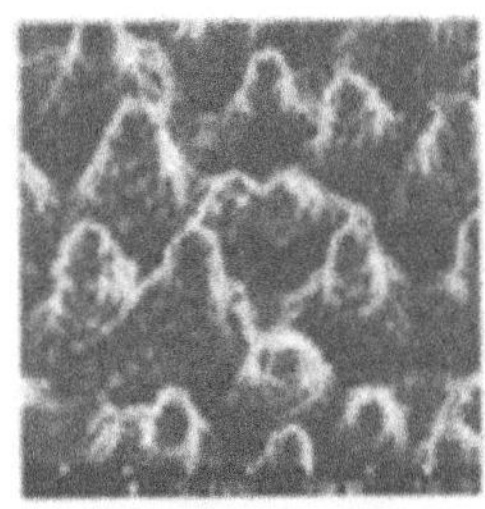

Composition of a lotus leaf viewed under a scanning probe microscope

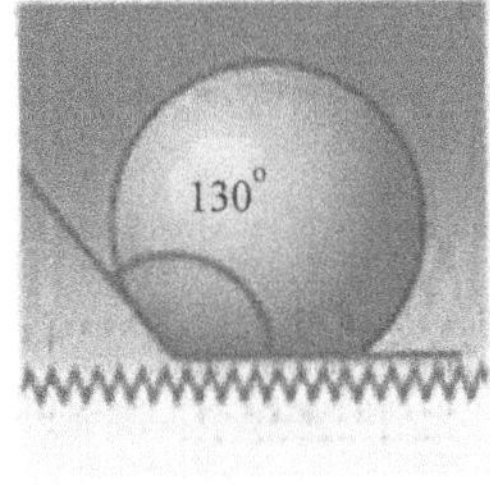

Contact angle of 130°

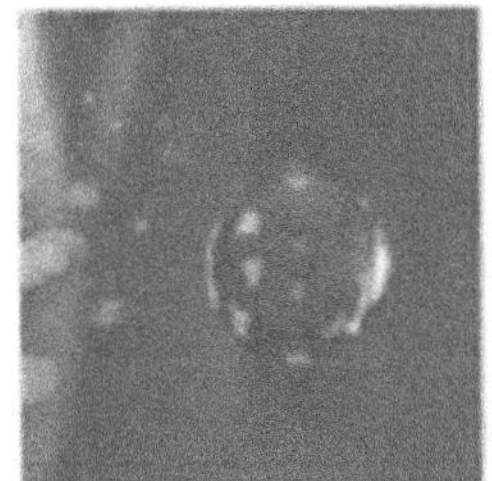

Drops on a superhydrophobic surface (photo: Nano-X GmbH, Saarbrücken)

Figure 15.15 SEM picture of lotus super hydrophobic surface and contact angle of water drop on its surface

on this 'mat of pillars', it is in contact with a large fraction of air. Consider the ideal case of a single droplet of water in air, it will have a totally spherical shape (contact angle = 180°). For a droplet of water on a surface with a large air fraction, the larger this fraction, the closer we get to this 'ideal' situation.

The same effect is found in other leaves such as those of nasturtium (*Tropaeolum*) and some *Cannas*

Some artificial hydrophobic surfaces can reach a contact angle of 170 degrees. These surfaces are known as super-hydrophobic. A drop of water is not able to spread over this microrough water-repellent surface. Due to its own surface tension it draws together into a ball. The adhesion (bonding force), which keeps the drop on the surface of the leaf, is reduced to a minimum of 130 °C. This enables the drop to roll away over the leaf at the slightest inclination of the surface.

Experiment 4.2 Determination of diameter of water drop with ruler/scale for qualitative assessment of hydrophobic/hydrophilic nature of surfaces and assessment of their self cleaning(Lotus) effect.

Description of Experiment: The aim of this experiment is to determine whether certain materials possess a hydrophobic surface by qualitatively measuring water drop diameter using micro ruler on different surfaces. For laboratories which are not equipped with gonio-meter , this qualitative method will be suitable. In order to do this one need to collect different sorts of paper (e.g. high-gloss paper from the printer, type writing paper, coffee-filter paper) and parts of plants, such as the leaves of grass, nasturtiums, kohlrabi, rubber plants, grapes, salad leaves or a lotus leaf and measure the contact angle of water drop on these surfaces

Materials: pipettes, different types of paper, plant leaves, a millimeter-scale ruler, fine dust (finely ground garden clay), fine ash, (e.g., from cigarettes) or soot.

Procedure: Collect different types of papers and plant leaves. Then allow water drops of the same volume (use the pipettes) to drip onto the paper and leaves of different structures. One can also do this with other surfaces, such as glass, wood or plastic.

Measure the diameter of the drops or spots of water with the micro ruler/scale. The more hydrophobic the surface, the smaller the diameter of the drop and, consequently, the greater its curvature.

Filter paper completely absorbs the drop of water, which means that it is not a hydrophobic (but hydrophilic) surface. The diameter of the drop on the high-gloss paper is very small, but the curvature very high. Find plant leaves which clearly demonstrate a drip-off effect.

Now put dust or ash on the leaves showing a drip-off effect. Then add some drops of water with a pipette to the contaminated surface and tilt the leaf so that the drops can roll away. The clay particles are absorbed by the drops of water. The drop leaves a clean track on the dirtied leaf.

Now rub the leaf with your thumb (not so hard that the If, at this stage, you drip water onto the leaf again with the pipette, you will see that the drip-off effect has clearly diminished or even been completely destroyed.

Observations

1. The hydrophilic pulp fibres of some of the leaf surfaces exercise adhesive forces on the water, which increase the greater their surface (or the rougher the surface of the paper). For this reason, the drop of water spreads out over the surface to a greater or lesser degree. In nature, this lotus effect is responsible for the surface of the leaf remaining for the most part clean (even if there are distinct differences in 'cleaning behaviour', which are the result of the different levels of water repellent properties of the surfaces). In this way, the plant preserves its full functional properties for photo synthesis and gas exchange.
2. The hydrophilic clay particles are absorbed by the drop because their adhesion to the leaf surface is lower than the adhesion to the water. Not until the drop is saturated with clay it is no longer able to exercise its cleaning effect. This means that the remaining clay particles will be left on the leaf surface (Figure 15.16).
3. The water repellent property has been neutralised on the damaged leaves. The drops stick to the leaf surface although their chemical characteristic has not changed. The leaf surfaces are now easily wetted. The micro or nanostructures which are otherwise responsible for the water-repellent effect of the surface in un-damaged leaves have been destroyed. By the way: In the case of living plants, after a while the lotus effect is re established, i.e., the plant regenerates and repairs the structure of the surface. This is known as self-structuring.

Figure 15.16 Cleaning effect of self cleaning surface (Photo: nanoX GmbH, Saoarbrucken)

Experiment 4.3 Development of artificial hydrophobic surfaces using nano particles

Description of Experiment

Development of artificial hydrophobic surfaces using nano-material coatings on substrates like such as wood or stone is the essence of this experiment.

Coating materials with functional properties: The coating covering the material is for the most part only a few nanometers thick and basically consists of one organic and one inorganic component Figure 15.17. An inorganic matrix is often composed of nano particles silicon dioxide, alluminium oxide, zirconium dioxide or titanium dioxide, which are integrated into an organic matrix (in most cases an organic solvent). These can be used to attain a multitude of functional properties. The key to success of these nano coatings depends on the correct combination of substances which, depending on the proportion of ingredients, determines which effect can be achieved. This knowledge is guarded in the same way as a treasure by research institutes and companies, who have it patented.

Thus nano coating basically composed of an organic-inorganic nanocomposite material in which the nano-particles are integrated into a bonding agent for which a general procedure is described below.

Materials: spray flask wood/stone coating, untreated wood or porous mineral substrate (e.g., pebbles, pumice, concrete or similar).

Procedure for preparation of organic-inorganic nano composite material: Silanes (silicic acid esters) with functional groups (epoxide functions, acrylate functions) are combined with nano-particles by hydrolysis and condensation reactions. In this process, the water reaction causes reactive Si-OH groups (hydrolysis) to emerge from the ethoxy residue of the Silane components, which can react with further OH groups by splitting off the water (condensation). An inorganic, glasslike network develops as a result of the hydrolysis and condensation reaction respectively. The silane functional group, which can be activated with appropriate initiators, takes the process a step further to generate an organic network: hence, the name 'organic-inorganic nano-composite (Figure 15.17).

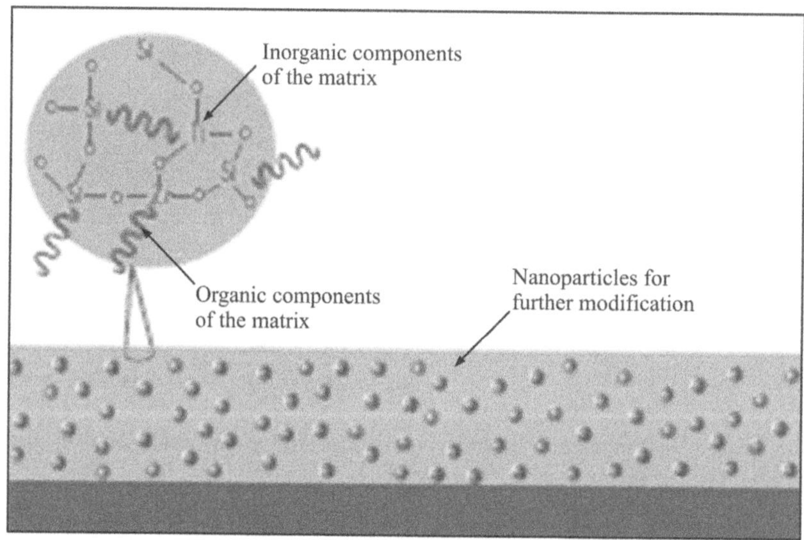

Figure 15.17 Composition of nano coating (Source: Sarastro GmbH Gottelborn)

By developing an organic and inorganic network in the coating, as well as integrating the nanoscale particles (15.18) which are also firmly attached to the matrix through chemical bonding, a layer is produced, which contains a combination of glasslike and polymer properties. Such coatings are known as hydrophobic, hard-elastic coatings. Their mechanical stability is so high that, even if the layer of the coating is extremely thin, steel wool will no longer scratch the surface, or at least will have difficulty in causing any damage.

In order to produce the coating the relevant silanes and nanoparticles are mixed with water to start the hydrolysis and condensation reaction. The reaction mixture, which was two phase when the water was first added, heats up and becomes single-phase. A reaction initiator is then added to cross-link the silane organic.

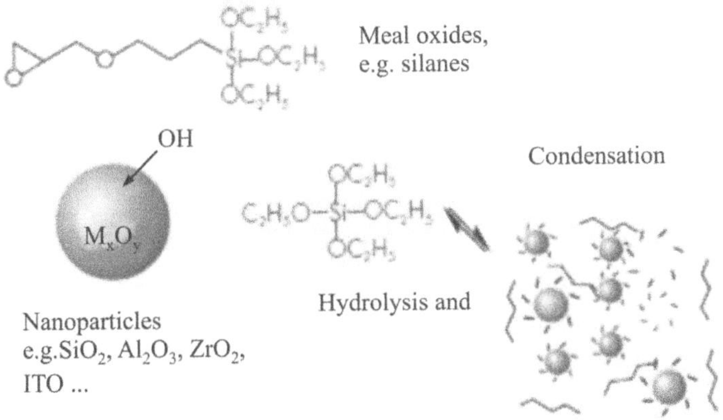

Meal oxides,
e.g. silanes

OH

Condensation

M_xO_y

Hydrolysis and

Nanoparticles
e.g.SiO_2, Al_2O_3, ZrO_2,
ITO ...

Figure 15.18 Integration of nano particles in the hydrolysis reaction

Procedure for development of synthetic hydrophobic surfaces by spraying organic-inorganic nanno coatings: Find an open-pored or absorbent substrate made of wood or stone, which has not been treated. The surface to be treated must be dry and clean; remove any deposits of loose dirt. Now spray the substrate with the coating solution until it is saturated. Allow it to dry completely. The solution can only really take effect when it has completely dried. The time needed for it to dry completely depends on the temperature (Please do not heat above 40 °C) 2 to 6 hours. The coated surface can now be tested with different liquids, for example waste water, coffee, lemonade, etc. Particular care is to be taken if façade paints are to be tested as some paints (*e.g.*, those with an acrylic base) may stick to the coating.

Observations: One can see a drip-off effect with a clear formation of drops. Due to its surface tension, the water tries to draw itself into a ball (with minimum surface and maximum volume). Once the coating has been applied and has hardened, the water-repellent molecular chains of the coating will give rise to non-abrasive layers with reduced surface energy. As already explained above, this is due to the van der Waals forces. If the surface has been made hydrophobic, the interactions are restricted to a minimum, which leads to a marked increase in the wetting angle of contact. When the coating has completely dried, it is abrasion-resistant; however, it can be destroyed again by chemical means.

Applications of Hydrophobic Surfaces

Many different effects can be achieved with nanolayers with such 'nanocoatings like easy to clean , corrosion protection, antifogging, self cleaning, anifungal, hydrophyllic/hydrphobic etc. A positive side effect of using lateral fluorinated chains in the coating material is a reduction in the infestation of algae and germs on such surfaces. In practice, hydrophobic surfaces are useful because any contamination which may occur can be washed away with very little detergent, the process being much less time consuming. A hydrophobic surface is not a self-cleaning surface as in the case of the lotus plant so you cannot avoid to clean these surfaces now and again.

Experiment 4.4 Development of Scratch-resistant coating of wood

Generally, MDF (middle-density fireboard) panels with a wood pattern are used for the production of interior doors, fronts and backs of furniture. The patterns are printed using various printing techniques, such as inkjet printing, onto the panels. A problem that is often linked to this process is that the surface is highly susceptible to mechanical strain and can, therefore, be easily scratched.

Materials: Pieces of wooden flooring, steel wool.

In order to protect the patterns on these MDF panels a transparent, 10 μm thin protective coating will be applied over one half, in which nanoparticles are integrated. This gives the panels greater stability; even if steel wool is scraped over the surface, the latter will not be scratched.

Procedure: Once the organic-inorganic nano coating material is produced as described earlier it is sprayed onto the wooden pieces of flooring. The pieces are subsequently dried in the air drier at a temperature of approx. 110 °C. This process is often called 'curing'. To be precise, the solvent is removed during the drying phase, i.e., really 'dried'. However, it is also 'cured' as the reaction initiator starts the polymerisation of the organic functional groups and, in addition, the glasslike network is completely condensed. The 'drying' is often referred as 'physical curing', the 'curing' as 'chemical curing'. In an air drier the entire process is completed in 20 minutes and the coating on the wooden pieces can be achieved with its highest level of scratch resistance. Naturally, this can also be radically accelerated by intelligent curing methods. This is particularly important for furniture production, in which large quantities have to be produced in a short period of time.

In order to test the mechanical stability the steel wool is scraped over the piece of wood several times - preferably over the border between the coated and untreated area. If you wish, you can draw a line with the marker over the piece of wood before doing the experiment.

A scratch can clearly be seen on the untreated part, while the coated side is un-damaged. If you drew a line over the piece of wood before the experiment, you will see that the colour on the coated side can be rubbed off with the steel wool, while it can no longer be removed from the untreated side.

15.6 Experiment 5: Application of Ferro-Fluids

Experiment 5.1: Preparation and Properties of Ferro Fluids

A ferrofluid is a specific type of liquid which responds to a magnetic field. Ferrofluids are composed of nanoscale magnetic particles suspended in a carrier fluid. The solid particles are generally stabilized with an attached surfactant layer. Ferrofluids are extremely stable meaning that they will not cluster together even in extremely strong magnetic fields

The following experiments introduce students to the synthesis of ferrofluids and their magnetic properties. Students will prepare: 1. A ferrofluid sample using ferrous and ferric chlorides, (2). The students will compare the spiking abilities of synthesized ferrofluids with a commercial grade one, varying the reactants to determine which produces best results, and developing possible explanations for the differences.

Theory

Particles less than 100 nm are considered nanoparticles. With the aid of a suitable carrier liquid, metal oxide nanoparticles can form a stable suspension of colloids. There must be a surfactant in order for this synthesized ferrofluid to remain suspended in aqueous or other medium. One of the ionic surfactants used is tetra methyl ammoniun hydroxide (TMAH) because it produces an electrostatic repulsion in an aqueous medium. The hydroxide anions bind to the iron atoms at the surface of the magnetic nanoparticles, and the tetramethyl ammonium cations form a sheath around the negatively charged particles through electrostatic attraction (Figure 15.19). These sheaths of positive charges, thus, prevent the particles from agglomerating, which makes it easier to investigate its magnetic ability. Other surfactants like *cis* oleic acid can be used for oil-based ferrofluids. The formation of ferrofluid involves various types of forces that hold the components together. For example, magnetite is held together by ionic interactions. Ionic attractions between hydroxide anions and tetramethyl ammonium cations allow colloidal suspension of the magnetite in the solution.

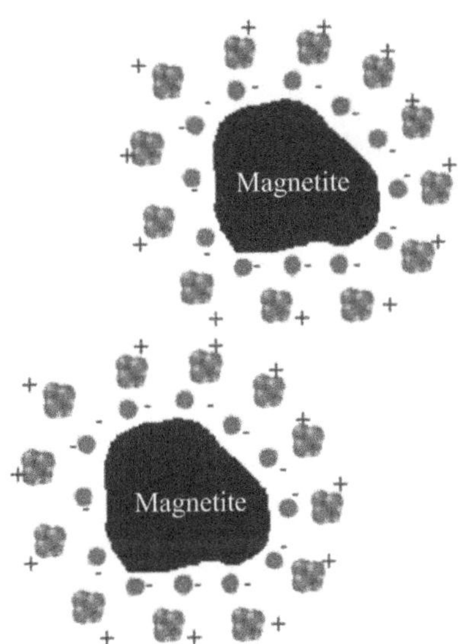

Figure 15. 19 How surfactants prevent agglomeration. Interaction of the hydroxide ions (red) with the magnetite and the interactions of the tetramethylammonium cations (blue and grey) with the water serving as the medium prevent the magnetite nanoparticles (black) from interacting with each other through electrostatic repulsion

Without the tetramethyl ammonium hydroxide as a surfactant, the magnetite nanoparticles tend to cluster together. Therefore it is necessary to have the appropriate surfactant to stabilize an aqueous ferrofluid.

There are two major steps in synthesizing a ferrofluid.

The first is to make the magnetic nanoparticles (~100 Å diameter) that will be dispersed in the colloidal suspension. These particles must be chemically stable in the liquid carrier. The magnetic particles in ferrofluid are generally magnetite, Fe_3O_4, although other magnetic particles can also can be used.

The second synthetic step is the dispersion of the magnetic particles into a carrier liquid by utilizing a surfactant to create a colloidal suspension. Surfactants are dispersion agents for particles in a liquid that work by adhering to the particles and creating a net repulsion between them (steric and/or coulombic) raising the energy required for the particles to agglomerate, and stabilizing the colloid.

Unique properties of Ferro fluids: Stick to magnets, Take 3 Dimensional shape (Spikes) in a magnetic field. Change density in proportion to magnetic field strength.

Materials Needed

Iron (II) chloride tetrahydrate, iron (III) chloride hexahydrate, can be purchased.

>25% tetramethyl ammonium hydroxide is commercially available.

Plastic weigh boats or Petri dishes, disposable gloves, magnetic stirrers and stir bars, strong craft magnets, cow magnets, 3% hydrogen peroxide can be purchased.

Iron oxide nanoparticles and Commercial ferrofluid can also be purchased.

Procedure for the Synthesis of Ferro fluids

Step 1: Dissolve 67.58 g of $FeCl_3.6H_2O$ in 250 mL of 2M HCl

Dissolve 39.6g of $FeCl_2.4H_2O$ in 100 mL of 2M HCl

Step 2: 1M.$FeCl_3$ should be used within one week of Preparation

2M.$FeCl_3$ should also be used within one week of preparation

Step 3: Combine 3mL of $FeCl_2$ solution and 12 mL of $FeCl_3$ solution and fill a burette with 150 ml of 0.7M ammonium hydroxide solution

Add ammonia very slowly whilst stirring. A black precipitate of magnetite will form

Step 4: After addition is complete stop stirring and use a strong magnet ($NdFe_{12}B$) to settle the black precipitate to the bottom of the flask

Decant off the water and add fresh water. Rinse the precipitate and again decant off the water. Repeat three times to remove excess ammonia.

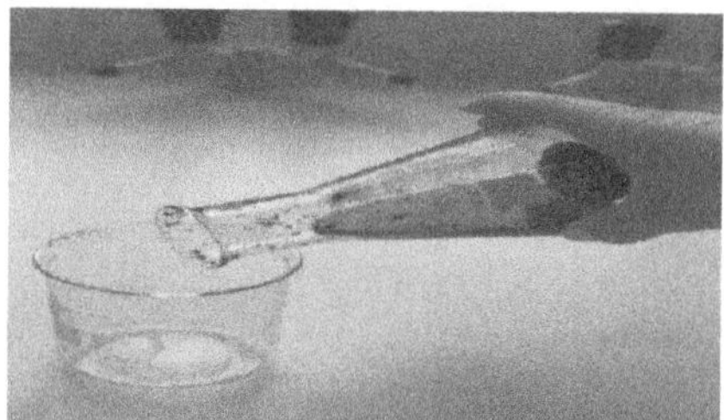

Step 5: Transfer the viscous liquid to weighing boat using a little extra water if necessary. Use a magnet on the base of the weighing boat to remove excess water.

Add 24 mL of tetramethyl ammonium hydroxide (25% solution) and stir with a glass rod

Step 6: Hold a magnet on the base of the weighing boat and let the solid settle to the bottom. Decant off any excess liquid to leave a very viscous black liquid.

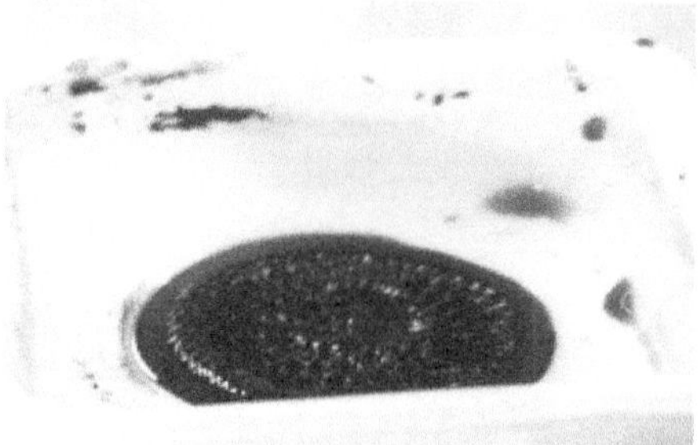

The viscous liquid will form spikes if a magnet is held underneath the weighing boat and one need to adjust the amount of water

Experiment 5.2 Examination of Spikes of Ferro-fluids under magnetic field

1. The ferrofluids show spiked structures in the presence of a magnetic field. Spikes are a pattern of uplifted particles that result from placing a magnet near the ferrofluid.
2. The extent of spikes of various ferrofluids prepared in the laboratory and commercially available can be compared as size of the magnetite particle decreases the extent of spike increases. Commercially available Ferrofluids show large spikes as they will have Particles of size (5-10 nm) while laboratory made ones will have low spikes as they will have magnetite particles of (30-50 nm). Further the surfactant attachments may also can be less efficient and the particles may have agglomerated.
3. The spikes are not permanent. They last only as long as there is a magnetic field in the proximity of the ferrofluid. They disappear when the magnet is removed.

The magnetic effect on the ferrofluid samples is a physical change because no new substance is produced (i.e., the particles are still made of magnetite).

Experiment 5.3 Variation of Density of ferro-fluid in a applied magnetic field

Description of Experiment: In this experiment a non magnetic coin which first sinks to the bottom of the liquid because of its density, swims on the top in the presence of a magnetic field

Materials Required

- ferrofluid
- petri dish
- coin (non-magnetic)

Procedure: First, place the coin in the Petri dish. Now use the pipette to carefully fill with ferrofluid.

The coin remains at the bottom and is completely immersed in the liquid. Now hold the magnet under the bowl. The ferrofluid is 'pulled to the bottom' and the coin rises from the bottom and can be 'rolled about' on the ferrofluid. The porcupine structure can also be seen here to some extent. Ferrofluid residue on the coin stands upright (Figure 15.20).

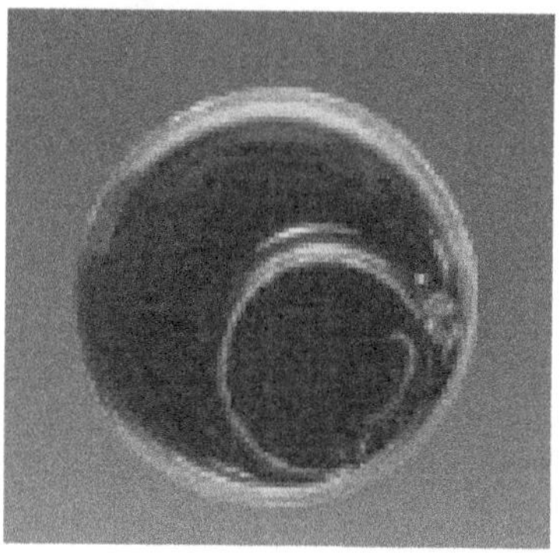 

Figure 15.20 Density separation process using ferrofluids under magnetic field

This experiment can also be varied with other non-magnetic materials (aluminium, stones, brass, etc.). You can pour the ferrofluid back into the bottle when the experiment is completed.

Observations: In this density separation, pressure is created by means of a magnetic field gradient with which you can make substances with a higher density 'swim' or separate substances with different densities. This process is used, for example, in the automobile industry for recycling metals or for extracting gold and precious stones. The magnet creates a force opposing gravity. The magnetite particles of the liquid are pulled to the magnet and, consequently, materials with a fairly low density begin to 'swim' on the ferrofluid.

You can vary this experiment with further non-magnetic materials. And always remember to be careful with magnets in the vicinity of ferrofluid solutions.

Experiment 5.4 Application of Ferro -fluids for Cleaning Oil Spills

Background

Oil spills at sea are a serious hazard, causing enormous ecological damage. Billions of dollars are spent on cleanup operations, which do not always completely remove the oil from the environment. Ferrofluids can be effectively used to clean oil spills.

Theory: Magnetic nanoparticles have great ability to stay suspended in liquids, creating a ferromagnetic fluid or ferrofluid. As discussed earlier the composition of Ferrofluid can be presented as shown in Figure 15.21 which includes a liquid (referred to as the carrier fluid), ferromagnetic nanoparticles, and a surfactant or soap-like substance to keep the nanoparticles from clumping together. Different types of oil or water can be used as carrier fluids.

When no strong *magnetic field* is present, the ferrofluid flows as we expect liquids to flow. However, when a strong magnet (or magnetic field) is introduced, the ferrofluid is pulled to the magnet and unexpected things happen. One can see the ferrofluid "jump" to the magnet, where it might create beautiful spike-like structures defying gravity.

Cleaning of Oil spill involve to make the spilled oil magnetic so that strong magnets could be used to separate the magnetic oil from the non-magnetic water.

Figure 15.22 shows the schematic of method of cleaning oil spills using ferro-fluids.

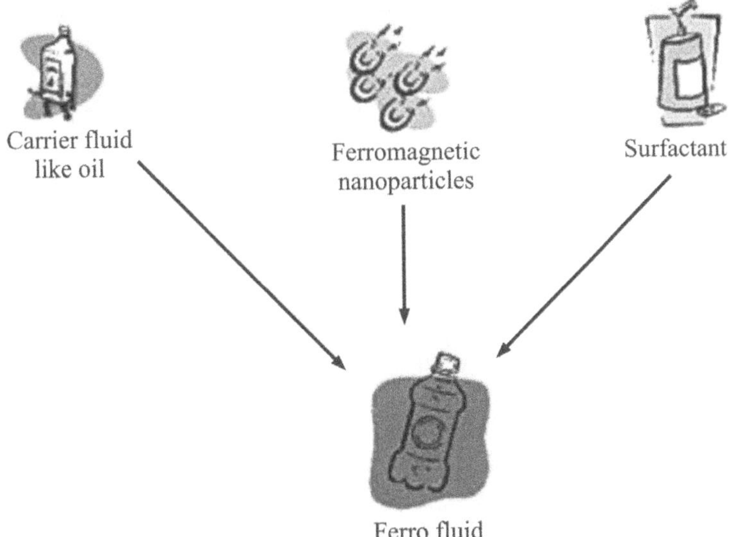

Figure 15.21 This diagram shows how ferrofluids are made from a carrier fluid in which ferromagnetic nanoparticles are suspended. A surfactant is added to prevent the nanoparticles from clumping together

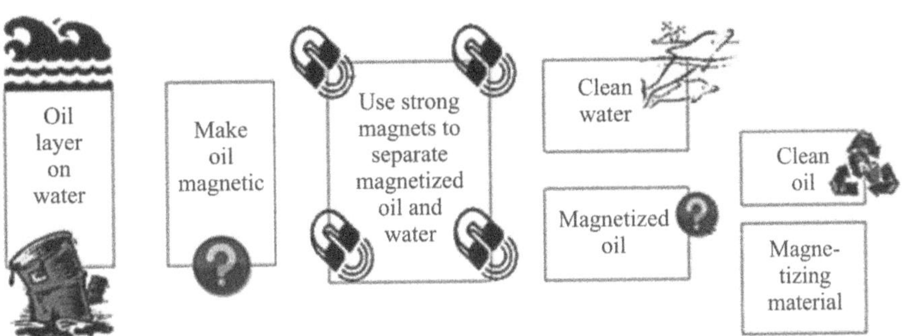

Figure 15.22 Flow chart visualizing the idea of separating oil (spills) from water using magnets by first making the oil magnetic

In this experiment the student will examine the possibility of using ferromagnetic fluids to literally magnetize the oil on a small scale and investigate how efficient the method is in removing oil from the surface of water, and what are its limitations and strengths.

In step one ferrofluid will be added to the spilled oil to make it magnetic which will allow the ferromagnetic particles from the ferrofluid spread and mix with the spilled oil, making it magnetic. If a carrier fluid is chosen such that it mixes well with the spilled oil, the ferrofluid will naturally spread and mix with the oil spill. In step two using a strong magnet remove the oil/ferrofluid mixture from the water.

The aim of the experiment is not only to investigate how the method works on a small scale, but it also seeks to quantify how efficient the method is in removing oil from water at this scale. The volume of oil removed using this clean-up procedure will be compared to the volume of the original oil spill and efficiency of separation can be quantified as shown in Equation 1

Equation 1:

$$\textit{Efficiency} = \text{volume of removed oil/volume of original spill}$$

An efficiency value close to 1 means that almost all of the oil has been removed; this indicates that the cleanup method works well in removing the spilled oil from the water. An efficiency value close to 0 means that little of the original oil has been removed indicating that the cleanup method does poorly in removing spilled oil from the surface of the water.

Given that it is easier to measure the volume of oil left after the cleanup procedure, one can also use alternate equations to estimate the process efficiency

Equation 2:

$$\textit{Efficiency} = (\text{volume of original spill} - \text{volume of leftover oil})/\text{volume of original spill}$$

or

Equation 3:

$$\textit{Efficiency} = 1 - \text{volume of left over oil/volume of original spill}$$

Materials

- Ferrofluid, 25 mL (ferrofluid must have mineral oil as its carrier fluid.)
- Mineral oil, 50 mL
- Neodymium magnet
- Petri dishes, 60 mm × 15 mm (10)
- Graduated cylinder, 25 mL volume
- Plastic transfer pipettes, graduated (9)
- Nitrile gloves, pairs (3)

Experimental Procedure

Prepare Oil spill in Petridishes: Take three petri -dishes and mark them as 5 drops Ferro-fluid, 1 drop Ferro-fluid and no- Ferro-fluid. Use a pipette to fill each petri dish with about 14 mL of the colored water. If you are using a different size of petri dish, fill the petri dishes until the water level is a couple of millimeters high. Using a graduated pipette add exactly 2.5 mL of mineral oil in the middle of the petri-dishes to each of the Petri dishes. which represents the oil spill

Make the oil magnetic by adding ferrofluid: Shake the bottle of ferrofluid before opening. For the petri dish marked "1 drop ferrofluid", place one drop of ferrofluid in the middle of the oil spill. For the petri dish in the "5 drops ferrofluid" column, place five drops of ferrofluid in the oil spill, preferably distributed over the oil surface.

Clean up the oil spill with a magnet: Open a clean plastic sandwich bag. Put the neodymium magnet in one of the corners of the plastic bag, as shown in Figure 15.23 Move the magnet enclosed in the sandwich bag through the oil in one movement. It works best to slightly submerge the magnet in the liquid and try to pass through the complete oil spill in one movement as best one can. Wipe the bag off on a paper towel or cloth. Repeat the process one more time.

Figure 15.23 Place the neodymium magnet in the corner of the sandwich bag It is now ready to pick up the magnetized oil spill in an oil cleanup procedure

Repeat the following cleanup procedure for all three petri dishes, starting with the one above the column titled "No ferrofluid added."

Measure how much oil is left on the water. Repeat the following procedure for all three petri dishes. Carefully transfer all of the leftover liquid (water, oil, and ferrofluid) from the petri dish to the graduated cylinder. Some oil will stick to the petri-dish. Try to get as much as possible in the cylinder. Do not use a funnel, as more oil would stick to the funnel and lower the readings even more. Wait till all of the oil settles on top of the water in the cylinder. Read the amount of oil left on top of the water. Make sure you have the oil layer level with your eye. The oil layer can have a curved shape. A close-up picture like the one shown in Figure15.24 can help you make readings more accurate.

Figure 15.24 This close-up picture of a graduated cylinder can help determine the volume of oil left on the water. Make sure your camera is level with the oil fluid and markings on the cylinder are visible before taking a picture

Data Analysis: Record your reading in a table for each type of petridish (1drop Ferrofluid, 5 drops ferrofluid and no drops of ferrofluid). Repeat steps two more times for a total of three tests for each cleaning procedure. Calculate the average volume of leftover mineral oil from the three tests and record the results in your data table. Calculate the efficiency of the cleanup procedures using equation given below and record the values in your data table.

$$\textit{Efficiency} = 1 - \text{volume of leftover oil (mL)}/2.5\ \text{mL}$$

Resources for Further Reading

1. Rudolf. J (2008, July) Oil Magnets,
   http://Sciencenews.org view/generic/id/34477/description/oil_magnets
2. Mahajan, Y.R (2011, February 12). Nanotechnology-based solutions for oil spills.
   http://www.nanowerk.com/spotlight/spotid=20215.php
3. Nicolades, G. (2012, September 17) Cleaning oil spills with magnets,
   CLEANMAG http://youtube/Uu1va7anw

15.7 Experiment 6: Synthesis and Characterization of CdSe Quantum Dots

Overview: In this Experiment students will receive an introduction to semiconductor quantum dots, namely CdSe. Students will study how surfactant -based chemistry can be used to synthesize CdSe quantum dots and study how the size of the quantum dots can be controlled by varying reaction time. The experiment will then demonstrate how the color of these quantum dots can be connected to the size of the nanoparticle by considering the electrons as freely moving particles in a box with the dimensions of the nanoparticle. Students will be able to estimate the size of quantum dots by using UV-VIS absorption spectroscopy

15.7.1 Theory

Quantum Dots: What are they and why are we interested in them?

Quantum Dots (QDs) are semi conducting nano crystals which are of the size 10–40 times the diameter of an atom. These dots are made of elements that, when combined, make semiconductors. For dots with dimensions in the range (1-100 nm) it is found that the optical properties vary with their size. This is illustrated in Figure 15.25, where we see fluorescence of five colloidal quantum dots solutions irradiated with UV light.

Generally speaking, the smaller the QD, the bluer the fluorescence spectrum. By controlling the size of QD made, it is possible to make dots with very specific colors, as well as conductivities. As such, QDs are finding applications in transistors, solar cells, LEDs and diode lasers.

The spectral properties of Quantutum dots are discussed in more detail in Chapter 11.

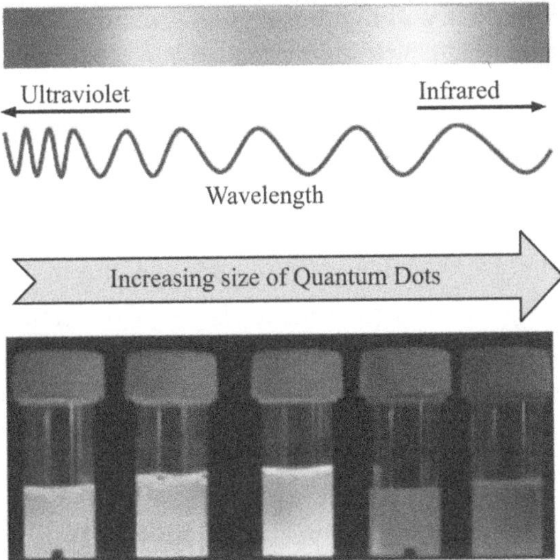

Figure 15.25 Quantum Dots irradiated with UV light. The different emission colors are generated by quantum dots of different sizes

15.7.2 Particle in Box Model for Explaining Quantum Mechanical behavior and Resultant Fluorescence Spectra of QDS

Let us take the example CdSe nanocrystals which have a crystal lattice of alternating cadmium and selenium atoms that exhibits the same structure as bulk CdSe. When a photon is absorbed by the bulk crystal, an electron-hole pair is created and the particles maintain a characteristic distance apart. However, when the size of the crystal is small enough (less than 112 Å in diameter for CdSe), the electron-hole pair can no longer achieve the desired distance and the particles become confined to a box. As a result, these nanocrystals demonstrate quantum mechanical behavior and their transition energies can be explained using the particle in a box model, where the electron is the particle and the nanocrystal is the box.

The energy gap (See Figure 11.2) for particle in a box model, modified for the case of a sphere, is given by equation.

$$E_n = \frac{hc}{\lambda_{max}} = \frac{h^2\left(n_f^2 - n_f^2\right)}{Sm_e \times R^2} \qquad .....(15.2)$$

where h = Planck's constant R = radius of the nanocrystal

n_f = the final energy level

n_i = the initial energy level

m_e* = effective mass of an electron.

The effective mass of an electron must be used, because conduction electrons have a smaller mass than regular electrons. Effective mass varies by material and is $0.13m_0$ (m_0 = mass of an electron) for CdSe.

In this experiment, the student will synthesize CdSe quantum dots of varying in diameter from approximately 1.8 nm to 4.0 nm and will then measure the absorbance and emission spectra of the quantum dots and use λ_{max} to calculate their size.

15.7.3 Synthesis of Quantum Dots

Safety Precautions

- One must wear a lab coat, chemical splash goggles, and gloves while in the lab.
- Wear proper heat-resistant gloves for transferring QD solution Chemicals Several of the chemicals used in this experiment can be hazardous if handled improperly. It is extremely important that all work be carried out in the hood and that the splash be kept as low as possible while working.

Chemicals

- **Octadecene** - Octadecene vapor should not be inhaled, and contact with skin and eyes should be avoided.
- **Trioctylphosphine** - Trioctylphosphine is corrosive and causes burns. Cadmium oxide - Cadmium oxide is highly toxic and cadmium compounds are known to be human carcinogens if inhaled or swallowed.
- **Oleic acid** - Oleic acid is air, light, and heat sensitive and irritating to eyes, the respiratory system, and the skin.
- **Elemental selenium** is toxic by inhalation, by absorption through the skin, and the effects of ingestion are cumulative.

Heat: In this experiment the reaction mixture will be heated to 225°C (more than double the boiling point of water!) and proper care should be taken while transferring hot solution from one vessel to another. One should wear nitrile gloves and heat resistant gloves during this step and focus on transferring the solution in a slow and controlled manner to avoid spilling any hot solution.

Materials and Equipment: Nitrile gloves, safety glasses, fume hood

- 25 mL or 50 mL round bottom flask (one neck or three necked),
- heating mantle works; you'll also need 2 Septa if using a 3 necked flask
- ring stand,
- 2 clamps,
- 400°C Hg thermometer
- hold stopper
- 0.13 g cadmium oxide
- 0.50 g myristic acid
- weigh paper or boat
- spatula
- 0.12 g selenium dioxide

- scales capable of measuring milligrams
- paper towel
- 7 mL octadecene (ODE)
- graduated cylinder
- 10-12 small test tubes or vials with stoppers
- test tube rack
- 9 in. Pasteur pipette with bulb,
- stopwatch (optional); students can use a smart phone or a clock in the laboratory that has a second hand
- 40 ml of hexane or decane as a solvent
- test tube brush and soapy water
- UV light source (e.g. LED UV Flashlight)
- UV/VIS spectrophotometer/Fluorescence meter

Procedure: In this experiment, one need to use a Pasteur Pipette to transferring reaction mixture (at very high temperature!) from a round bottom flask to a test tube. It is important to have the flask clamped at a height that makes it comfortable and safe to transfer liquid from the flask to the test tube.

Set up the apparatus as shown in Figure 15.26 using a clean 25 mL round bottom flask.

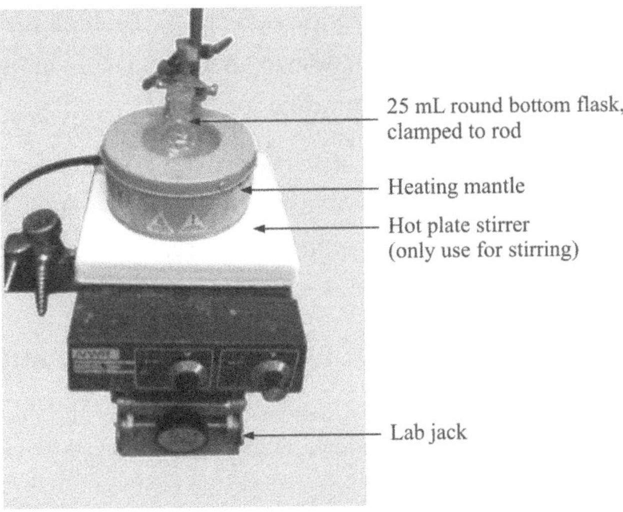

Figure 15.26 Reaction set-up for the synthesis of quantum dots

Fill the flask with ~ 10 mL of distilled water. Place a test tube holder and one test tube next to the set up. Using a Pasteur Pipette (Figure 15.27) practice pipetting from the flask to the test tube.

Adjust the height of the flask until it is at the most comfortable height for pipetting. Add a clamp above the round bottom flask (RBF) for holding a thermometer. When you are done practicing, remove the "test" flask by lowering the lab jack (Do not change the height of the clamp!). Dump out the water and put the flask on a rack to dry.

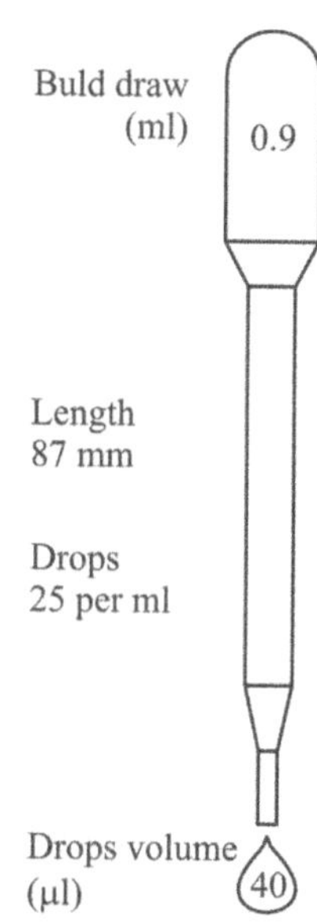

Figure 15.27 Pasteur Pipette

15.7.4 Preparation of CdSe Quantum Dots

Take a 25 mL round bottom flask containing approximately 13 mg of CdO. Clamp this flask above the heating mantle. Using a 1 mL syringe, add 0.6 mL oleic acid to the flask. Then add 10 mL octadecene using a graduated cylinder. Insert a thermometer capable of measuring 225 °C and turn the variac to 70 to begin heating the flask. While you are waiting for the reaction mixture to heat up, get a test tube rack and fill it with ten test tubes. Obtain a clean Pasteur Pipette.

Prepare a 0.038 M selenium stock solution by dissolving elemental selenium in octadecene and trioctylphosphine. As soon as the temperature of your reaction mixture reaches 225 °C, using a syringe, transfer 1 mL of the room temperature selenium stock solution to the 225 °C cadmium solution.

Because the size of the nanoparticles depends on reaction time, start timing when the selenium solution is added. Use a 9-inch Pasteur pipet to carefully remove approximately 1 mL samples at frequent intervals and quench by squirting into separate test tubes. You should obtain nine or ten samples within two to three minutes.

15.7.5 Spectral Characterization of QDs and Determination of Sizes

Observing Fluorescence: Place your test tubes in the UV box and record the color of the emitted light for each of your samples. Note the intensity of the light.

Measuring Absorbance and Emission

Obtain a cuvette holder and 5 cuvettes and caps. Choose your five most distinctly colored samples and, using clean pipettes, transfer your quantum dots to the cuvettes and cap them. Record a set of absorbance spectra and emission spectra for your samples using a Spectrometer and note the maximum wavelength. Print a copy of each set of spectra and attach it to your lab report.

Table 15.6 Spectral Characteristics of QDs

Sample	Absorption λmax	Emission Imax
1		
2		
3		
4		
5		

15.7.6 Data Analysis: Determination of the Size of the QDs

Since the frequency or wavelength for the onset of absorption is dependent on the size of the QDs, we should be able to determine the size of the dots for each absorption spectrum recorded. Using existing data, equation (2) relates the size of the dot to the threshold wavelength or frequency of absorption.

You will need to first extract an approximate threshold wavelength from the data. Do this in the manner that is demonstrated with the red and blue lines in Figure 15.28. and extend the straight line tangential to the curve a third of the way down on the red frequency side of the absorption line. Extrapolate the tangent down to where it intercepts the wavelength axis. The point of intercept is l threshold. Obtain a value for threshold for each sample (there will be 10).

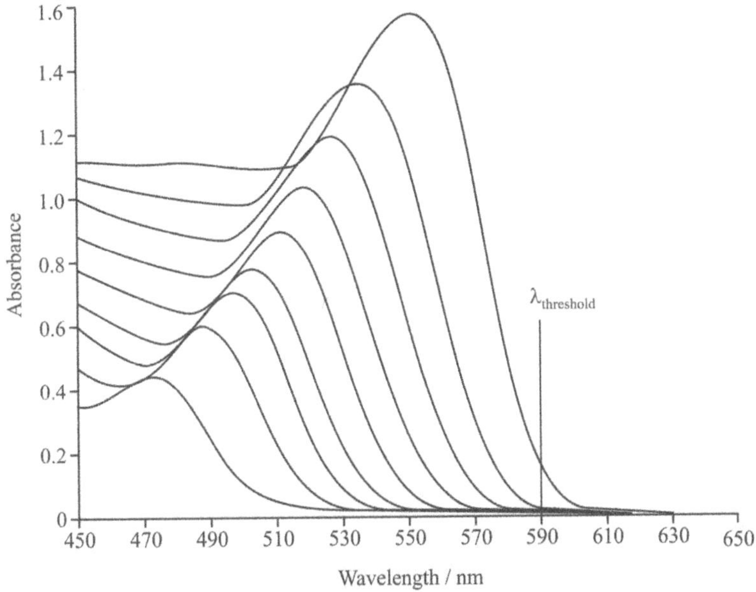

Figure 15.28 Visible absorption spectra of CdSe samples. Longer wave length, smaller energy emission corresponds with larger particle sizes. The red and blue lines illustrate how to estimate threshold wave length

Table 15.7 Threshold values and ΔE_{fi} values of QDs of different sizes

Sample number	Reaction (s)	Threshold (nm)	ΔE_{fi} (J)
1			
2			
3			
4			
5			
6			
7			
8			
9			
10			

Quantum Dots-Particles in a Box

The energy of the excited state is quantized and the energy required to excite an electron from the valence band to the lowest conduction state is given by equation (15.3).

$$\nu = \frac{c}{\lambda} = \frac{E_{gap}}{h} + \frac{h}{8m_0R^2}\left(\frac{1}{m_e} + \frac{1}{m_h}\right) - \frac{1.8e^2}{4\pi\varepsilon_0\varepsilon hR} \qquad .....(15.3)$$

where R is the radius of the QD, E_{gap} is the gap in energy between the top of the valence band and the bottom of the conduction band in bulk CdSe, m_0 is the mass of the electron, m_e and m_h are the reduced masses of the electron and the hole, e is the charge on a proton, h is Planck's constant, and ε and ε_0 are the dielectric constant and permittivity of free space. The energy needed to excite the QD is absorbed from a photon of light. The lowest frequency (the onset frequency) $\nu = c/\lambda$ (where c is the speed of light and λ the onset wavelength of the light) required to excite the QD is given by equation (15.4).

So if we measure either the onset wavelength or onset frequency of this light we can determine the Q-dot's radius *R*.

Assessing the accuracy of the Particle-in-a-Box Model and Creating a Calibration Curve

To ascertain the size of the quantum dots mono-dispersed in samples 1-10, it is necessary to be able to calculate the size of the QD that is needed for a given l threshold. We are going to estimate the size of the QDs by two methods one that constructs a calibration curve from previously measured size versus wavelength data, and one that uses Equation 15.2 with the parameters that are pertinent to CdSe (these are listed in Table 15.8).

$$\Delta E_{fi} = E_{gap} + \frac{h^2}{8m_0R^2}\left(\frac{1}{m_e} + \frac{1}{m_h}\right) - \frac{1.8e^2}{4\pi\varepsilon_0\varepsilon R} = h\nu \qquad .....(15.4)$$

Complete Table 15.7 (paying special attention to units) and make a plot of ΔE_{fi} vs. $1/R^2$, that compares the experimental ΔE_{fi} to that calculated in equation (3).

* Data for calculating ΔE_{fj} from equation (3) will need the data from Table 15.9

Use the ΔE_{fi} vs. $1/R^2$ plot as a calibration curve to ascertain the approximate size of the particles in samples 1-10, and complete Table 15.10

Table 15.8 Useful physical constants for CdSe along with some useful unit conversions

Constant	Value
m_0	$9.11e^{-31}$ kg
m_e	0.13
m_h	0.45
ε	10.72
ε_0	$8.85e^{-12}$ C^2/Nm
e	$1.601e^{-19}$ C
h	$6.626e^{-34}$ Js
c	$2.998e^{8}$ m/s
E_{gap}	1.8125eV
1 eV	$1.601e^{-19}$ J

Table 15.9 Experimental data for the radius of mono-dispersed Quantum dots versus the threshold transition energy. (Data taken from "Measurement and assignment of the size-dependent optical spectrum in CdSe quantum dots" by D.J. Norris and M.G. Bawendi, Physical Review B 33 (24) 16338-16346 (1996))

R (m)	ΔE_{fi} (eV)	$1/R^2$ (m^{-2})	ΔE_{fi} (J)	ΔE_{fi} (J) From eqn (15.2)
$4e^{-9}$	1.91			
$3e^{-9}$	2.06			
$2.5e^{-9}$	2.15			
$2.0e^{-9}$	2.30			
$1.5e^{-9}$	2.50			

Table 15.10 Estimated sizes and threshold values of quantum dots

Sample number	Reaction (s)	Threshold (nm)	R (nm)
1			
2			
3			
4			
5			
6			
7			
8			
9			
10			

Clean-up and Waste Disposal. Dump all waste into the designated waste container in the hood. Thoroughly clean all glassware with soapy water over a waste beaker. Rinse with distilled water. When you are done, dump the contents of waste beaker into the designated waste container. Place washed glassware into the blue bins. Thoroughly clean the surface of your bench using paper towels. Dispose of all solid waste (contaminated gloves and paper towels, empty cuvettes and caps) in the solid waste container.

Resources for Further Reading

1. Merkle R., Nanotechnology (hosted by Zyvex). There's Plenty of Room at the Bottom. (accessed August, 2012) http://www.zyvex.com/nanotech/feynman.html
2. Boatman E.M., Lisensky G.C. and Nordell K.J (2005). A Safer, Easier, Faster Synthesis W for CdSe Quantum Dot Nanocrystals, Journal of Chemical Education, 82: 1697-1699.
3. Li Y. A Visual Demonstration of "Particle in a Box" theory: Multicolor CdSe Quantum Dots. Department of Chemistry & Biochemistry University of California, Santa Cruz.
4. Lisensky G. University of Wisconsin. Synthesis of Cadmium Selenide Quantum Dot. Nanoparticles. (accessed August, 2012) http://mrsec.wisc.edu/Edetc/nanolab/CdSe/index.html
5. Norris D.J. and Bawendi M.G (1996). Measurement and Assignment of the Size-Dependent Optical Spectrum in CdSe Quantum Dots. Physical Review B, 33:16338-16346.
6. Chen O., Chen X., Yang Y., Lynch J., Wu H., Zhuang J., and Cao Y.C (2008). Synthesis of Metal–Selenide Nanocrystals Using Selenium Dioxide as the Selenium Precursor. Angew. Chem. Int. Ed., 47: 8638-8641.

Questions

Chapter 1: Nano Science and Technologies

1. Explain What is nano Science and what are nano Technologies. Compare the size ranges for several entities as compared to some nanotechnology devices.
2. Define Nano Science and nano Technologies. Nano Technologies are horizontal convergent multi disciplinary technologies. Discuss.
3. What is nano Technology? what is nano scale ? what is nano meter? Is nano Technology New? Where did it come from?
4. What are nano materials? Do they exist in nature?
5. What are nano particles, nano tubes, nano films, where nanotechnology used today? What are future uses of nano technology?
6. What is meant by Wave-corpuscle duality of matter?
7. What is meant by Quantum confinement in nano particles?
8. Explain the Size dependent quantization of Energy in nano particles.
9. Explain how random molecular motion become significant in nano materials.
10. What are three dimensional nano systems?
11. What are two dimensional Quantum well systems?

    What are one dimensional quantum wires?
12. The Quantum wires do not obey the general electrical resistivity law $R = \rho l/A$ (where ρ = resistivity, l is length and A is cross sectional area of wire). explain.
13. What are special characters of zero dimension quantum dots?
14. What is nano chemistry?
15. What are Engineered nano, materials?
16. The structure of the Lizard foot is one of the best example for the relationship between function and nanostructure. Explain.
17. Butterflies often display extraordinary colours and iridescence due to nano photonic crystals. Explain.
18. Discuss the Relation between Categories and Families of nano structured Materials.
19. "There is plenty of room at the bottom." This was stated by _____.

    [A] Eric Drexler [B] Richard Feynmann [C] Harold Croto [D] Richard Smalley

20. Who coined the word 'nanotechnology'?
    [A] Eric Drexler [B] Richard Feynmann [C] Sumio Tijima [D] Richard Smalley
21. When and where FEYNMAN delivered his lecture on nanotechnology and what is the name of his classical lecture?
22. Define top down and bottom up approach.
23. Give any two salient points addressed by Feynman.
24. Define nano structured material.
25. Classify nanomaterials and give examples for them.
26. Classify nanocomposites.
27. Discuss the unique (different) physical properties of nanomaterials in comparison with bulk based on size effects.
28. Explain how more %surface atoms exists in Nano systems when compared to bulk.
29. Explain the phenomenon of Coloumb Blackade observed in nano particles.
30. What is meant by giant magneto resistance effect (GMR) nano thin fims. How it is used in many applications?
31. Explain how nano materials can act as better catalysts.
32. Discuss various industrial process where nano technologies are being used for improving the product performance.
33. Discuss 'bottom-up', and 'top-down' approaches for production of Engineered nano materials.
34. What is molecular manufacturing?
35. What are Environmental beneficial effects of nano technologies?
36. What are environmental risks of nano materials?
37. What are expected adverse effects of nano systems on human health?
38. What are the mile stones in the evolution of nanotechnology?
39. "One nanometer is a magical point on the dimension scale." Why? Explain.
40. What are the salient points addressed by Feynman in his lecture "there is plenty of room at the bottom"?
41. What is the difference between nanotechnology & nanoscience?
42. What are the challenges faced by researchers in nanotechnology?
43. Why surface area/volume ratio is very large for nanoparticles?
44. Compared to bulk materials? Explain with a simple example. Highlight any two. problems associated with increase in surface area? Explain the role of bottom up & top down approaches in nanotechnology.
45. What is Tunneling effect? Explain the terms quantum confinement and quantization of energy levels in nano particles.

46. What is nano surface chemistry .How properties like catalytic reactivity, electrical resistivity, adhesion, gas storage and chemical reactivity depend on the nature of the interface?
47. What are special properties of nano materials which differ from bulk and make them technologically important?
48. Why melting points of nano particles decrease with decreasing size?
49. Carbon nano tubes exhibit exceptional mechanical properties based on their structure. Discuss.
50. How nano particles can be used to improve mechanical properties of existing materials?
51. Explain how nano particles exhibit size dependent optical properties. What is meant by Localised Surface Plasmon resonance, (LSPR)?
52. What is Surface Energy. How it changes with decreasing size in a nano particle?
53. Discuss how the special electrical and electronic properties of nano materials can be used in molecular electronics.
54. Explain how the surface to volume ratio changes with size in nano particles and facilitate their catalytic activity.
55. Give an overview of size effects on the various properties of nano materials and their various industrial applications.
56. What are natural and Engineered nano materials? Discuss how natural nano materials can provide a model for designing radically improved artificial materials for innovative technologies for a number of applications.
57. Explain Gleiter's classification of nanostructured materials. what are 3D, 2D, 1D nano structures?
58. Discuss salient features of Bottom-up and Top-down techniques in nano material production.
59. Explain Self Assembly and Position assembly techniques in bottom-up process of nano material synthesis.
60. What kind of precision engineering is expected in various nano production processes? What is molecular manufacturing?
61. Discuss Safety and heath risks of nano materials.
62. Discuss the Environmental benefits and risks of nano materials.

Multiple Answer Questions

1. The size of nanoparticles is between _____ nm.

   [A] 100 to 1000 [B] 0.1 to 10 [C] 1 to 100 [D] 0.01 to 1

2. "There is plenty of room at the bottom." This was stated by _____.

   [A] Eric Drexler [B] Richard Feynmann

   [C] Harold Croto [D] Richard Smalley

3. Who coined the word 'nanotechnology'?

[A] Eric Drexler [B] Richard Feynmann
[C] Sumio Tijima [D] Richard Smalley

4. Who was the first scientist to describe that substances having nanodimensions possess altogether different and unique properties?

[A] Richard Feynmann [B] Eric Drexler
[C] Archimedes [D] Michael Faraday

5. Which of the following does not apply to nanotechnology?

[A] It is a general-purpose technology.
[B] It can be called Green technology.
[C] Newtonian mechanics can describe it.
[D] It involves rearrangement of atoms.

6. The diameter of human hair is ____ nm.

[A] 50,000 [B] 75,000 [C] 90,000 [D] 1,00,000

7. The diameter of human hair is ____ m.

[A] 5×10^{-8} [B] 5×10^{-7} [C] 5×10^{-6} [D] 5×10^{-5}

8. The size of E.Coli bacteria is ____ nm.

[A] 2,000 [B] 5,000 [C] 50 [D] 90

9. The size of RBC is ____ nm.

[A] 50 [B] 90 [C] 2,000 [D] 5,000

10. The size of a virus is ____ nm.

[A] 2 [B] 20 [C] 50 [D] 2000

11. Which ratio decides the efficiency of nanosubstances?

[A] Weight/volume [B] Surface area/volume
[C] Volume/weight [D] Pressure/volume

12. The surface area to volume ratio of a sphere with radius 1 cm is R_1 and that of a sphere with radius 5 cm is R_2. Then R_1 = ____ R_2.

[A] 3 [B] 1/3 [C] 5 [D] 1/5

13. The surface area to volume ratio of a cube with side 1 unit is R_1 and that of a cube with side 10 units is R_2. Then R_2 = ____ R_1.

[A] 1/10 [B] 10 [C] 1/100 [D] 100

Answers

1. (c)	2. (b)	3. (b)	4. (c)	5. (a)	6. (a)
7. (d)	8. (c)	9. (c)	10.	11. (b)	12. (c)
13. (b)					

Chapter 2: Carbon Nanotube Technologies (CNT)

1. Distinguish the characteristics between Carbon nano tubes and Fullerenes.
2. Chiral vector (n) and unit vector (a)play important role in determining properties of Single Wall nano tubes (SWNT) Explain.
3. Give the equation for calculating SWNT diameter in terms of its nm indices. Discuss the characteristics of zig-zag, chiral, and arm chair structures of SWNTs. What are nano buds and nano fibers.
4. Compare and contrast between single wall nano tubes (SWNT) and multi walled nano tubes (MWNT).
5. Carbon nano tubes exhibit exceptional electrical, thermal and optical properties when compared to bulk materials. Explain.
6. sp^2 C-C bond of carbon nano tubes give them unusual mechanical strength. Explain.
7. Discuss the three critical steps in the growth mechanism of carbon nanotubes.
8. Discuss the salient features of arc discharge, laser ablation and chemical vapor deposition methods for the synthesis of SWNTs.
9. Explain the key steps in Chemical Vapor Deposition (CVD) method for production of SWNT and MWNTs. What are different types of CVD processes.
10. Discuss the role of Co and Mo catalysts in CoMo Cat process for production of SWNTs.
11. What are various methods available for purification of CNTs What methods are suitable for removal of carbon compounds and metal catlaysts from SWNTs?
12. Discuss various applications SWNT composites. What special properties of SWNT composites are interesting for industrial applications?
13. SWNT and MWNTs are low temperature Field Emission Materials. Discuss application of this property.
14. Discuss how SWNT and MWNTs are finding applications in high energy storage batteries.
15. How CNTs can improve efficiency of solar cells? Explain.
16. Covalent and no covalent bonding options can be used to functionalize SWNTs for improving bio compatibility and decreasing their toxicity for exploiting bio medical applications. Discuss.

17. What are various technological barriers to exploit the full potential of SWNT for industrial applications?
18. Discuss the salient features of special properties if CNTs like (i) chemical reactivity, (ii) electrical conductivity (iii) thermal conductivity (iv) optical properties (v) Mechanical strength. Explain with examples the unique and important application these properties.
19. What are various characterization techniques used to examine the morphology and purity of CNTs?
20. Discus various possible biomedical applications of CNTs. Explain how fictionalization of CNTs help in exploring various biomedical applications.

Short Answer Questions

21. What is the diameter of a bucky ball? How many pentagons and hexagons are there in a bucky ball?
22. The hardness of a standard SWNT is ____ Pa.
23. The bulk modulus of a standard SWNT is ____ that of diamond.
24. How much current can be passed through 1 cm^2 cross-section of a metal nanotube?
25. The electrical conductivity of a nanotube is ____ times that of silver.
26. An MWNT possesses electrical superconductivity up to temperature of____
27. At room temperature, the thermal conductivity of a copper wire is ____ watt/(m.K).
28. In radial direction, the thermal conductivity of a nanotube is ____ watt/(m.K).
29. The thermal stability of a nanotube is seen up to ____ K in vacuum.
30. The thermal conductivity of an SWNT along length is ____ watt/(m.K).
31. The size of a quantum dot is ____ nm.
32. The width of a carbon nanotube is ____ nm.
33. The thermal stability of a nanotube is seen up to ____ K in air.
34. Who prepared and explained nanotubes for the first time?[A] Sumio Tijima [B] Richard Smalley [C] Eric Drexler [D] Richard Feynmann
35. Which of the following statement/s is are true? (i)Volume to surface area ratio is very large for nanomaterials (ii) The cut-off limit of human eye is 10^{-5} m (iii).Hardness of a SWNT is about 63×10^9 Pa (iv) Carbon nanotubes are cylindrical fullerenes.
36. What are the types of carbon nanotubes?
37. List methods for producing carbon nanotubes?
38. Diameter of single nano tube is ____ nm) and inter tube spacing is ____ nm
39. What is a Fullerene or Bucky ball? on whose name they were named?
40. Why C-60 molecules are called as bucky balls? Give reasons
41. What is Chiral vector of a carbon nano tube?

42. What is the difference in the pattern of structural arrangements in (a) zigzag, (b) armchair and (c) chiral nanotubes ?
43. Explain how chiral vectors of nano tubes deterrmine the conducting or the wall.
44. What are single wall (SWNT)and multi walled nano tubes (MWNT)?
45. Nano Buds exhibit properties of both CNts and fullerenes. Explain.
46. What are carbon nano fibers?
47. Exlain how the electronic structure of carbon nano tubes help to exhibit better thermal and electrical conductivity than metals.
48. Raman spectroscopy will be useful to characterize CNTs. Explain.
49. Determine the diameter(nm) of a CNT with X value230 and with radial breathing mode (RBM) in Raman spectra found at ωRBM = 200 cm-1using the equation d (nm)= X/ω(RBM)cm-1.
50. The tensile strength of carbon nano tubes will be nearly 100 times greater than the same diameter steel. Explain the reasons.

Chapter 3: Nano/Micro Fabrication Technologies

1. What are the main challenges to be addressed by nano fabrication techniques and what are basic characters of nano materials to be achieved by any nano fabrication method?
2. Discuss the salient features of Top down and bottom-up approaches for the production of nano materials.
3. What is Lithography, and what are main categories of Lithography? Discuss the main steps of any Lithographic Process.
4. Discuss the salient features of Photo Lithography. What are main steps involved in Photo Lithogrphic process?
5. Explain the processes (a) e-beam lithography, (b) ion lithography, (c) Xray Lithography. Discuss the advantages of these processes over UV lithoghraphy.
6. Discuss various process components (i) micro-stamping, (ii) stencil patterning, and (iii) microfluidic patterning of soft Lithography.
7. Discuss the main principle of Electrically induced nano patterning. What is polymeric wave?
8. What are different etching processes? Explain the terms isotropic and anisotropic etching.
9. What is dry etching? Explain the processes (i) xenon difluoride etching (ii) reactive ion etching and plasma etching.
10. Explain the salient features of low pressure Chemical Vapor deposition and plasma enhanced chemical vapor deposition.
11. Explain the principles of Physical Vapor Deposition. What is sputtering?

12. What are various applications of Thin film doping? Explain the advantages of oxidation process for production of SiO_2 thin films.
13. Explain diffusion and ion implantation processes for doping.
14. Discuss the advantages o f Pulsed Laser Atomic Layer deposition over other Thinfilm deposition techniques.
15. Explain how colloidal particle self assembly will be usefull to produce 3D nano structures.
16. What is chemical self assembly? How binding and recognition sites for fabricating complex 3-D structures from colloidal nanoparticles can be produced by self assembly?
17. How scanning probe microscopes like STM &AFM will be useful for atomic scale manipulation for nano structure fabrication? Explain scanning probe induced oxidation for nano meter scale local oxidation. What is Dip-Pen Lithography?
18. Write a short notes on nanofabrication.
19. What is photolithography? Explain with a neat sketch Lithographic process.
20. What are the special attributes of Electron beam lithography? Explain with a neat sketch.
21. Explain the special features of ion beam lithography with a neat sketch.
22. What is X-ray Lithography? In which industry it is widely used?
23. What are the methods available for fabrication of MEMS? Explain.
24. Explain cantilever sensors (mechanical mems) with a neat sketch.
25. Explain the working of magnetic actuator (magnetic mems) with a neat sketch.
26. What are various lithographic techniques called soft lithography? What is meant by stencil patterning?
27. What are the advantages poly (dimethylsiloxane) (PDMS) as the mold, stamp, or mask to generate or transfer the pattern in Replica Molding?
28. What is Nano/Micro contact printing /Embossing Process? What are its advantages over Photo lithography?
29. What are thermoplasts? How they are useful in microfluidic patterning?
30. What are the advantages of micro/nano fabricated devices?
31. What are main challenges to be addressed by nan/micro fabrication techniques?
32. Discuss the salient features of Top down & Bottom-up methods in nano fabrication.
33. What are the differences between mask and scanning lithography?
34. What are the steps involved in photo lithographic process? What is a photo resist?
35. What are problems and advantages of lithographic process?
36. What is nano sphere lithography?
37. What is meant by colloidal lithography? Explain with a neat sketch various steps involved in colloidal lithography. What is the lowest resolution that can be achieved by this technique?

38. What is meant by electrically induced nano patterrning? Explain the essential components and working process of this technique. How etching process can be used as nano fabrication Technique?
39. Explain what is meant by selectivity and directionality in any etching process.
40. Explain the terms (a) wet Etching (b) Isotropic etching and anisotropic etching in nano fabrication.
41. Discuss various important dry etching process in nano fabrication.
42. Explain plasma etching and electrochemical etching process in nano fabrication.
43. Explain the general process involved in Chemical Vapor deposition with a neat sketch.
44. What is low pressure chemical vapor deposition? What are its advantages?
45. What is Plasma enhanced CVD? How it can be used at low temperatures at the substrate?
46. Explain how evaporation and sputtering process are used in physical vapor deposition in bottom-up methods in nano fabrication.
47. What is thin film deposition? What are specific areas they find application?
48. Thermal oxidation is a special type of thin film deposition method-Explain.
49. Explain the diffusion and implantation process used in doping. How doping finds application in nano fabrication?
50. Discuss the unique advantages of Pulsed laser and atomic layer deposition techniques over other thin film deposition methods.
51. What are the advantages of Atomic Layer Deposition (ALD) over regular CVD?
52. Discuss the four man steps involved in the growth of monolayers of ALD. What are the various types ALD reactors used ?
53. What are the advantages and disadvantages of ALD process?
54. What is electro deposition process? Discuss with a neat sketch the functioning of direct current electro deposition and reverse pulse electro deposition process.
55. What is meant by strain induced self assembly? Explain Stranski–Krastanow growth mode process.
56. Explain Physical self assembly process. What are its advantages? Where this process is specifically useful?
57. What is chemical self assembly? What are the most commonly used organic mono layers in this process?
58. What is meant by molecular self assembly? What are the forces involved in providing thermodynamically feasible interactions between large molecules like peptides, proteins, DNA etc. in this process?
59. What are the advantages of Molecular self assembly?
60. What is scanning probe lithography? How it is useful for atomic scale manipulation and nano machining for production of nano devices?
61. What is meant by dip pen nano lithography?

Chapter 4: Micro Electro Mechanical Systems (MEMS) and Nano Electro Mechanical Systems (NEMS)

1. What are Micro Electro Mechanical Systems (MEMS) and nano electro mechanical systems (NEMS)? Discuss their functional elements.
2. Discuss some generic and Defense applications of MEMS and NEMS.
3. What are various commercially available MEMS available in present markets and their potential for future development?
4. What are current limitations/hurdles in NEMS/MEMS markets?
5. Discuss General Process Flow for Planar processing of MEMS.
6. Explain the processes Bulk micromachining and surface micromachining in MEMS Fabrication.
7. What are various Wafer bonding methods for fabrication of 3D structures? Explain LIGA Process.
8. Discuss various Challenges/issues in Industrial scale Production of MEMS.
9. What are various Wafer bonding methods for fabrication of 3D structures? Explain LIGA Process.
10. Discuss various Challenges/issues in Industrial scale Production of MEMS.
11. Discuss various generic applications of NEMS. What are current Technical hurdles in commercialization of NEMS.
12. What are main approaches practiced for fabrication of NEMS? Explain how carbon nano tubes and Graphene will be better materials for NEMS.
13. What are expected Market Trends in MEMS & NEMS?
14. (a) Write a short note on microfabrication.

    (b) Define MEMS. What are the types and applications of MEMS?
15. (a) Write a short notes on nanofabrication.

    (b) What are the applications of MEMS?
16. Explain photolithography with a neat sketch.
17. Explain Electron beam lithography with a neat sketch.
18. Explain atomic beam lithography with a neat sketch.
19. What are the methods available for fabrication of MEMS? Explain.
20. Explain cantilever sensors (mechanical mems) with a neat sketch.
21. Explain the working of magnetic actuator (magnetic mems) with a neat sketch.
22. Write a short note on thermal mems and thermal actuators.
23. Explain working of MEMS gyroscope with a neat sketch.
24. Explain Mechanical actuator used in mechanical mems.
25. Explain (a) acoustic mems (b) mems gyroscopes.

Chapter 5: Quality Control, Standards, and Metrology for Nano-Technology

1. Discuss the Ojkowski and Fecht classification of nano materials useful for quality control process in nano technology.
2. What are important characteristics/parameters of nano materials which need quality control protocols for their measurement?
3. What are main concepts on which nano standards are being developed? What are current challenges for developing nano standards?
4. Explain the terms (a) calibration, (b) Traceability (c) tolerance used in developing nano standards.
5. What are leading standards setting organizations and their relevant nanotechnology committees in different countries in the world.
6. What is nano metrology. Discuss the role of AFM and SEM in nano metrology.
7. Discuss the functioning of AFM and SEM and their relative merits and demerits.
8. Compare the AFM and SEM technologies with respect to 3 factors: 1. Surface Structure, 2. Composition, and 3. Environment.
9. Discuss the principle of working of Scanning Tunneling Microscopy (STM) and discuss how it will be useful for 3-D Tropology of nano materials.

Chapter 6: Applications of Polymers/Organics, Liquid Crystals, Nano-Composites and Ferro-Fluids in Nano Technology

1. Discuss the application of Polymers in in top down and bottom-up approaches in nano technologies
2. What is a supramolecular Polymer? Supramolecular polymers and block copolymers are promising candidates for bottom-up nanotechnology, since their morphology can be tuned by changes in molecular design or assembly conditions .Explain
3. Discuss the various applications of Polymers in nano medicine
4. What are the challenges in the application of polymers in nano technology
5. Discuss how nano composites play significant role in the application of medicine and reinforcements of mechanical properties
6. What are different types of Liquid Crystals and their fascinating characteristics which find wide application in displays
7. What are different display technologies using Liquid Crystals
8. Explain how Organic light emitting diodes (OLED) display works. Discuss its advantages and disadvantages over Liquid Crystal (LC) technologies.
9. What are various barriers in the application of OLED technologies
10. Explain how OLED active matrix work

11. What is a ferrofuid? What are the forces that operate for the sustainability of a Ferro-fluid?
12. What are surfacted ferro fluids and ionic ferrofluids? What is their technological importance?
13. Explain the behaviour of Ferrofluids in the absence and presence of External magneticfield? What is meant by magnetostriction?
14. Explain the terms (a) spiking effect (b) Mechanic-Elastic Qualities (Viscosity) of ferrofluids in applied magnetic field.
15. What are various Ferromagnetic materials? Explain the essential features of synthesis Ferrofluids.
16. What are the basic properties of Ferrofluids exploited for various applications? Discuss how the Magnetostriction properties of Ferrofuids explored for various technological applications.
17. Explain the following terms related to Magentstriction effects and their applications (i) JouleEffect (ii) Villari effect (iii) E-Effect-Change of Youngs modulus with magnetic field and (iv) Wiedeman Effect and Matteuci Effects.
18. Explain the principles on which Ferrofluids find application in (a) Ferrohydrodynamic sealing, (b) Ferrohydrodynamic damper (c) electrodynamic speakers (d) Heat transfer in minature micro scale devices (d) treatment of tumors and medical applications.

Chapter 7: Role of Surface and Colloid Chemistry in Nano-Technology

1. What are various types of colloids and interfaces? Explain their importance in nano technology.
2. What are the different characteristics of colloidal dispersions make colloids industrially important?
3. As the particle size reduces the surface area increase which affect the surface properties like catalytic reactivity, electrical resistivity, adhesion, gas storage and chemical reactivity of a colloid -Explain.
4. What are different mechanisms the dispersed particles acquire surface electrical charge when brought into contact with a polar medium?
5. What is Surface Free Energy? Why it is high for nano particle than bulk? What are various methods to moderate surface energy for bringing stability to nano systems?
6. What are important types of particle–particle interaction forces that can exist in a colloid?
7. What are various types of van der Waals forces? van der Waals interaction (attraction) between colloidal bodies decays more rapidly as the separation between them increases. Explain.
8. Hamaker constant A is a well-established parameter that can provide the relative picture and magnitude of the van der Waals interaction between colloidal objects. Explain.

9. What are different mechanism through the interfaces in colloid can acquire electrical charge? Explain the structure of electrical double layer and variation of electric potential with changing inter surface distance?
10. Steric force is mainly induced by polymers, polyelectrolytes, or biomacromolecules that are adsorbed or grafted onto the surface of colloidal objects. Explain.
11. What are solvation and hydration forces in colloids?
12. Explain how surfactants can moderate hydrophobic forces in a colloid.
13. Hydrogen bond provides the stability and the directionality to the molecular self-assembly. Explain.
14. What are various Liquid phase methods commonly used for nano fabrication? Discuss the main steps and mechanisms involved in the processes.

Chapter 8: Molecular Manufacturing, Future Nano Technologies and Dexler-Smalley Debate

1. What is molecular manufacturing? What are possible Physical Techniques for controlling factors for engineered nano/molecular production Systems?
2. How the precise control the sequence and/or position of the reactions at atomic or molecular lever can be achieved in molecular manufacturing? What is relative productivity?
3. What are benefits and risks with molecular manufacturing?
4. What are different possible approaches for realizing molecular manufacturing in large scale?
5. What are different Molecular Manufacture Enabling Tools/Technologies presently available?
6. Discuss various approaches for Self assembly production of Molecular /nano structure using -Bottom-Up methods. What are challenges to be addressed in Self Assembly?
7. Discuss Current Practices and applications of nano bio systems/bio nano technology.
8. Discuss the application of molecular/nano bio systems in Tissue Engineering, cancer therapy, optical coding for bio assays.
9. Discuss the salient points of Drexler-Smalley debate - on future projections on Molecular nano Technology and molecular manufacturing.
10. Give an overview of future molecular manufacturing based on the points of views expressed by Drexler and Smalley. What are future areas need to be pursued for realizing fruits of molecular manufacturing in large scale for the benefit of jumanity?

Chapter 9: Semiconductors, Nano Size Effects and Moore's Law

1. What is energy band? On the basis of energy band diagrams, distinguish between metals, insulators and semiconductors.

2. What are intrinsic semiconductors? Explain the formation of a hole in an intrinsic semiconductor. Draw the energy level diagram.
3. What is extrinsic semiconductor? Distinguish between n-type and p-type semiconductors. Draw relevant energy level diagrams.
4. What is a p-n junction? Explain the formation of the depletion region in a p-n junction. How does the width of this region change when the junction is (i) forward biased, (ii) reverse biased. Explain.
5. Explain the terms band structure and effective mass of a semiconductor.
6. What is a Fermi level energy in a Intrinsic and extrinsic Semiconductor? Give the corresponding Energy equations.
7. Discuss the Energy band gap differences in bulk semi conductors and nano sized crystals of semiconductors.What are the effects of Quantization on the electron and hole spatial motion in a semiconductor?
8. Discuss the Optical and electronic properties semiconductor nano crystal and their applications.
9. What is Moore's Law? How it helped electronic industry? What is its present status under the scenario of nano technologies deeply penetrating into this industry?

Chapter 10: Quantum Computing

1. What is a quantum computer? How it is different from classical computer? What is a quantum de coherence?
2. What is a Qubit? What are different methods to realize it? How Quantum Dots will be useful as Qubits in quantum computing?
3. Explain the following phenomenon (a) superposition (b) entanglement (c) Quantum decoherence.
4. Explain how Quantum computing using Shors algorithms will be advantageous over classical computation in factoring large numbers.
5. Discuss some of the expected problems in quantum computing and possible solutions.

Chapter 11: Quantum Dots, Nano-Photonics and their Applications

1. Discuss the structure of quantum dots. What is quantum confinement and Exiton Bohrs Radius (EDR)?
2. How you can tune band gap in QDs to achieve size dependent Fluorescence?
3. Discuss nucleation and contoured growth techniques for the synthesis of QDs.
4. What are different types of Core-Shell Quantum Dots? How this classification will be usefill for varying band gap and fluorescent intensity of QDs?
5. Discuss various potential Defense applications of QDs.
6. How QDs can improve efficiency of Photovoltaic Cells?

7. Discuss various methods that can be adopted for achieving bio compatibility of QDs.
8. Discus various biological and medical applications of QDs.
9. What is nano photonics? How the photonic band gap can be manipulated? Discuss various applications of nano photonics.
10. Explain how Photonic crystals can b designed to confine, manipulate and control the propagation of photons in three dimensions and can find applictions as wave guides.
11. Discuss various approaches for Fabrication of Photonic nano Crystals with propagation of photons in all three dimensions. Discus various major areas the application of nano photonic is being explored.
12. A gap (defect) in the crystal traps light in the photonic crystal while a line defect acts as a wave guide which has great potential for applications in optical integrated circuits Explain.

Chapter 12: Thin Films

1. What are various categories of vacuum deposition processes? Discuss some major areas of thin film applications.
2. What are desired characteristics of thin films to be achieved by any deposition process. What are various Physical Vacuum Deposition (PVD) and Chemical Deposition (CVD) methods are in use for production of Thin Films.
3. What are various advantages of PVD methods? Discuss key steps involved in PVD methods.
4. Discuss the mechanism of film deposition and growth in any PVD method. What are various control parameters for maintaining thin film characteristics in a PVD method?
5. What are various types of evaporation techniques used in PVD methods?
6. What is thermally induced evaporation? Discuss various methods by which heat can be delivered to the charge to cause vaporization.
7. Discuss the following process in any PVD method (a) Molecular Beam Epitaxy (MBE) (b) Induction Heated Thermal Evaporation (c) Electron Beam Evaporation (d) Laser Ablation.
8. What is Sputter Deposition of Thin Films? Explain the terms diodic Sputtering and sputtering yield.
9. What are essential characteristics of chemical vapor deposition (CVD) method for thin film deposition? Explain the salient features of the following CVD methods (i) Chemical Bath deposition (ii) Arrested Precipitation Technique (iii) Anodic oxidation (iv) Cathodic deposition.
10. What is electro chemical deposition? What are steps involved in it? What are its advantages?
11. Discuss the mechanism of electro chemical deposition. Explain how Faraday's laws help in estimating the amount of thin film deposited in a give time. Discuss various pathways in electro deposition growth.

12. What is electrical double layer? Discuss its role in electro deposition of thin films.
13. Explain Sterns model of electrode-electrolyte interface reactions. What are various governing factors in electro deposition?
14. Discuss various application of electro deposition of thin films in nano technology.
15. What is Plasma deposition/plasma enhanced surface modification? Discuss the Advantages of plasma surface modification or polymerization.
16. Discuss the following Plasma enhanced Surface modification processes (i) Contamination Removal (ii) Surface Activation (iii) Etch (iv) Cross Linking.
17. Discuss various Colloidal Phase Synthesis of Nano Structured Thin films.
18. What is the difference between in-situ and ex-situ methods for characterization of nano thin films? Discuss the following techniques (a) Sheet Resistivity, (b) Optical Transmission.

Chapter 13: Nano-Bio-Info-Cogno (NBIC) Convergent Technologies

1. Explain the term convergence in Science and Technologies. What are convergent technologies NBIC? Explain the scope of various convergent technologies as defined by various international organizations.
2. What is meant by enabling technologies? Explain how nanotechnology, biotechnology, and information technology act as enable technologies to enable each other.
3. Discuss various the future potential of various Converging Technologies in various thrust areas identified by US NSF.
4. Discuss Future Visions for the application of Convergent Technologies for improvement of Human Performance and Quality of Life.
5. Discuss the limitations in the application of Convergent Technologies with examples.

Chapter 14: Opportunities and Challenges in Newly Emerging Nano-Technologies and Strategies for their Utilization for Sustainable Development

1. Discuss various new opportunities and concerns in the application of nano technologies with reference to new developments like NBIC.
2. What are grand challenges identified by US-NSF to be addressed for rapid growth of nanoscale science and engineering at commercial level?
3. Discuss various Emerging nano-technologies, Specific challenges/barriers/solutions in their application to various fields.
4. Discuss the impact of emerging nanotechnologies on their applications for (a) Sustainable Processes-Materials and Resources (b) Energy Technologies and (c) Environmental study and monitoring. What are challenges and barriers in these applications?

5. Discuss the role of emerging nano technologies on global sustainable agriculture, Health Care, Medicine, Bio-diagnostics and in global climatic processes and challenges in their applications?
6. Discuss the possible risks of nano materials. What are possible risks of Carbon nanotubes? How the risks of carbon nano tubes (CNT) are managed?
7. How the global challenges in nano technologies application can be addressed with reference to (a) international collaboration (b) international competition and patent rights and (c) Regulatory issues?
8. What are special Challenges for developing countries in promotion of nanotechnologies for development? How they can be addressed?
9. What are Important Strategies to be adopted for realizing the benefits of newly emerging nano technologies? How the challenges in their implementation for sustainable growth can be addressed?
10. Discuss the relative role various Governments in developing nano technologies for the welfare of people. To what extent Pivate/Industry participation an help nano technology growth globally?
11. What are the challenges in (a) Education (b) Research Infrastructure and (c) Funding faced by developing countries, in promoting nano technology growth for sustainable developed? How they can be addressed?
12. How the Safety and regulatory and policy issues can be addressed on the application of nano technologies by developing countries?
13. Discuss India's Status in nano technologies development. What are the challenges and how they can be addressed?
14. Discuss the Scope of generation of career opportunities due to emerging applications of nano technologies in various fields.
15. What is the impact new Convergent Technologies on career opportunities?
16. Discuss the potential for Future Career opportunities in Material Processing Informatics and quantum computation.

Chapter 15: Laboratory Exercises

1. Prepare a graph showing the relationship between the independent and dependent variables.
2. Describe the relationship between independent and dependent variables.
3. What is the control in the experiment? What is the purpose of this control?
4. Why is the aseptic technique critical for accurate results? Give examples of aseptic practices used in this experiment.
5. Suggest possible sources of error in your experiment. These sources of error may NOT be mistakes in reading the measurements or in calculations.
6. Why are silver nanoparticles more effective than bulk silver in inhibiting bacterial growth?

Index

D

E

F

G

H

I

R

S

T

U

V

W

X

Y

Z

Printed at Repro India Ltd.